Proceedings of the 5th Asia-Pacific

BIOINFORMATICS

Conference

SERIES ON ADVANCES IN BIOINFORMATICS AND COMPUTATIONAL BIOLOGY

ISSN: 1751-6404

Vol. 1: Proceedings of the 3rd Asia-Pacific Bioinformatics Conference
Eds: Yi-Ping Phoebe Chen and Limsoon Wong

Vol. 2: Information Processing and Living Systems
Eds: Vladimir B. Bajic and Tan Tin Wee

Vol. 3: Proceedings of the 4th Asia-Pacific Bioinformatics Conference
Eds: Tao Jiang, Ueng-Cheng Yang, Yi-Ping Phoebe Chen and Limsoon Wong

Vol. 4: Computational Systems Bioinformatics
Eds: Peter Markstein and Ying Xu

Vol. 5: Proceedings of the 5th Asia-Pacific Bioinformatics Conference
Eds: David Sankoff, Lusheng Wang and Francis Chin

Series on Advances in Bioinformatics and Computational Biology – Volume 5

Proceedings of the 5th Asia-Pacific

BIOINFORMATICS

Conference

Hong Kong 15 – 17 January 2007

editors

David Sankoff
University of Ottawa, Canada

Lusheng Wang
City University of Hong Kong, Hong Kong

Francis Chin
The University of Hong Kong, Hong Kong

ICP

Imperial College Press

Published by

Imperial College Press
57 Shelton Street
Covent Garden
London WC2H 9HE

Distributed by

World Scientific Publishing Co. Pte. Ltd.
5 Toh Tuck Link, Singapore 596224
USA office: 27 Warren Street, Suite 401-402, Hackensack, NJ 07601
UK office: 57 Shelton Street, Covent Garden, London WC2H 9HE

British Library Cataloguing-in-Publication Data
A catalogue record for this book is available from the British Library.

PROCEEDINGS OF THE 5TH ASIA-PACIFIC BIOINFORMATICS CONFERENCE

ISBN-13 978-1-86094-783-4
ISBN-10 1-86094-783-2

Printed in Singapore by B & JO Enterprise

PREFACE

High-throughput sequencing and functional genomics technologies have given us the human genome sequence as well as those of other experimentally, medically and agriculturally important species, and have enabled large-scale genotyping and gene expression profiling of human populations. Databases containing large numbers of sequences, polymorphisms, structures and gene expression profiles of normal and diseased tissues are rapidly being generated for human and model organisms. Bioinformatics is thus rapidly growing in importance in the annotation of genomic sequences, in the understanding of the interplay among and between genes and proteins, in the analysis the genetic variability of species, in the identification of pharmacological targets and in the inference of evolutionary origins, mechanisms and relationships.

The Asia-Pacific Bioinformatics Conference series is an annual forum for exploring research, development, and novel applications of bioinformatics. It brings together researchers, professionals, and industrial practitioners for interaction and exchange of knowledge and ideas. The Fifth Asia-Pacific Bioinformatics Conference, APBC2007, was held in Hong Kong 15–17 January, 2007.

A total of 104 papers were submitted to APBC 2007. These submissions came from Bangladesh, China, Hong Kong, India, Japan, Korea, Malaysia, Singapore, Taiwan, Thailand, Australia, New Zealand, Denmark, France, Germany, Hungary, Italy, Israel, Portugal, UK, Canada, Mexico and USA. We assigned each paper to at least three members of the programme committee. Although not all members of the programme committee managed to review all the papers assigned to them, a total of 317 reviews were received, so that there were about three reviews per paper on average.

A total of 35 papers (33%) were accepted for presentation and publication in the proceedings of APBC 2007. Based on the affiliation of the authors, 1.25 of the accepted papers were from China, 1.46 were from Hong Kong, 3 were from Japan, 0.83 were from Korea, 1 were from Singapore, 2.30 were from Australia, 2 were from Denmark, 0.15 were from France, 4.5 were from Germany, 0.5 were from Italy, 1 were from Portugal, 1.66 were from UK, 4.57 were from Canada, 10.78 were from USA.

In addition to the accepted papers, the scientific programme of APBC 2006 also included three keynote talks, by Jennifer A. Marshall Graves, Joseph H. Nadeau and Pavel A. Pevzner, as well as tutorial and poster sessions. The presentations were of very high quality. Almost a third focused on evolution and phylogeny, largely at the genome level, a similar number dealt with protein structure and proteomics more generally, and a good proportion studied various aspects of pathways, networks, transcriptomics and microarray technology. A range of other topics in bioinformatics and computational biology were also covered, ranging from motif and gene recognition, through haplotypes and population genetics, to databases and text mining. Much of this work featured techniques of sequence analysis, while many of the papers included applications to biology and medicine.

We had a great time in Hong Kong, enhancing the interactions between many researchers and practitioners, and reuniting the Asia-Pacific bioinformatics community in the context of an international conference with worldwide participation.

Finally, we wish to express our gratitude to the authors of the submitted papers, the members of the programme commitee and their subreferees, the members of the organizing committee, Phoebe Chen and Limsoon Wong (our liaisons in the APBC steering committee), the keynote speakers, our generous sponsors, and supporting organizations for making APBC 2007 a great success.

David Sankoff
Lusheng Wang
Francis Chin

17 January 2007

APBC2007 ORGANIZATION

Conference Chair

Francis YL Chin, The University of Hong Kong, Hong Kong

Organizing Committee

Francis YL Chin (Chair), The University of Hong Kong
David Smith (Poster Session), The University of Hong Kong
H.F. Ting (Finance and Registration), The University of Hong Kong
Lusheng Wang (Publication), The City University of Hong Kong
Xiaowen Liu (Publication), The City University of Hong Kong
S.M. Yiu (Local Arrangement and Tutorial), The University of Hong Kong
Daniel Hung (Webmaster), The University of Hong Kong
Samson Sin (Webmaster), The University of Hong Kong

Steering Committee

Phoebe Chen (Chair), Deakin University, Australia
Sang Yup Lee, KAIST, Korea
Satoru Miyano, University of Tokyo, Japan
Mark Ragan, University of Queensland, Australia
Limsoon Wong, National University of Singapore

Program Committee

David Sankoff (Chair), The University of Ottawa
Lusheng Wang (Chair), The City University of Hong Kong
Tatsuya Akutsu, Kyoto University
Miguel Andrade, Ottawa Health Research Institute
Stephane Aris-Brosou, University of Ottawa
Joel Bader, Johns Hopkins University
Serafim Batzoglou, Stanford University
David Bryant, University of Auckland
Jeremy Buhler, Washington University in St. Louis
Peter Donnelly, University of Oxford
Dannie Durand, Carnegie Mellon University
Nadia El-Mabrouk, University of Montreal
Robert Giegerich, Bielefeld University
Carole Goble, University of Manchester
Concettina Guerra, Universita' di Padova
Dan Gusfield, University of California, Davis
Michael Hallett, McGill University
Sridhar Hannenhalli, University of Pennsylvania
Daniel Huson, Tbingen University
Gavin Huttley, Australian National Univeristy
Jenn-Kang Hwang, National Chiao Tung University
Tao Jiang, University of California - Riverside
Uri Keich, Cornell University
Anand Kumar, University of Leipzig
Tak Wah Lam, The University of Hong Kong
Doheon Lee, KAIST, Korea
Jinyan Li, Institute for Infocomm Research
Wentian Li, Feinstein Institute for Medical Research
Guohui Lin, University of Alberta
Michal Linial, The Hebrew University of Jerusalem
Zhijie Liu, University of Georgia
Bin Ma, University of Western Ontario
Satoru Miyano, The University of Tokyo
Laxmi Parida, IBM Thomas J. Watson Research Center
Mark Ragan, University of Queensland
Marie-France Sagot, University Claude Bernard, Lyon I
Akinori Sarai, Kyushu Institute of Technology
Vincent Schachter, Genoscope
Steven Skiena, State University of New York at Stony Brook
Edward Susko, Dalhousie University
Yun Song, University of California, Davis
Robert Stevens, University of Manchester
Alfonso Valencia, Centro Nacional de Biotechnologia
Michael Waterman, University of Southern California

Ken Wolfe, University of Dublin
Stacia Wyman, Williams College
Hong Yan, City University of Hong Kong
Qiang Yang, Hong Kong University of Science and Technology
Kaizhong Zhang, University of Western Ontario
Liqing Zhang, Virginia Tech
Louxin Zhang, National University of Singapore

Additional Reviewers

Shandar Ahmad
Marcos Jesus Arauzo Bravo
Alexander Auch
Iris Bahir
Richard Bean
Pierre-Yves Bourguignon
Shihyen Chen
Matteo Comin
Mike Cornell
Tobias Dezulian
Zhihong Ding
Maxime Durot
James Eales
Logan Everett
Paul Fisher
Morihiro Hayashida
Cornelia Hedeler
Duncan Hull
Helen Hulme
Shane Jensen
Noam Kaplan
Gunnar W. Klau
Kiyoung Lee
Yaoyong Li
Jingping Liu
Yaniv Loewenstein
Suryani Lukman
Julia Mixtacki
Jose Carlos Nacher
Niranjan Nagarajan
Hiroshi Nakashima
Kay Nieselt
Elon Portugaly
Magnus Rattray
Jonathan Schug
Charles Semple
Baozhen Shan
Yun S. Song
Kristian Stevens
Ashish V. Tendulkar
Victor Tong
Roy Varshavsky
Balaji Venkatachalam
Li-San Wang
Michael Wilson
Katy Wolstencroft
Yufeng Wu
Lei Xin
Zheng Yuan
Guanglan Zhang

CONTENTS

EXPLORING GENOMES OF DISTANTLY RELATED MAMMALS

JENNIFER A. MARSHALL GRAVES

ARC Centre for Kangaroo Genomics, Research School of Biological Sciences
Australian National University, Canberra, ACT 2601, Australia

There are three groups of extant mammals, two of which abound in Australia. Marsupials (kangaroos and their relatives) and monotremes (echidna and the fabulous platypus) have been evolving independently for most of mammalian history. The genomes of marsupial and monotreme mammals are particularly valuable because these alternative mammals fill a phylogenetic gap in vertebrate species lined up for exhaustive genomic study. Human and mice ($\sim$70MY) are too close to distinguish signal, whereas mammal/bird comparisons ($\sim$310MY) are too distant to allow alignment. Kangaroos (180 MY) and platypus (210 MY) are just right. Sequence has diverged sufficiently for stringent detection of homologies that can reveal coding regions and regulatory signals. Importantly, marsupials and monotremes share with humans many mammal-specific developmental pathways and regulatory systems such as sex determination, lactation and X chromosome inactivation.

The ARC Centre for Kangaroo Genomics is characterizing the genome of the model Australian kangaroo Macropus eugenii (the tammar wallaby), which is being sequenced by AGRF in Australia, and Baylor (funded by NIH) in the US. We are developing detailed physical and linkage maps of the genome to complement sequencing, and will prepare and array cDNAs for functional studies, especially of reproduction and development. Complete sequencing of the distantly related Brazilian short-tailed opossum Monodelphis domestica by the NIH allows us to compare distantly related marsupials. Sequencing of the genome of the platypus, Ornithorhynchus anatinus by Washington University (funded by the NIH) is complete, and our lab is anchoring contigs to the physical map. We have isolated and completely characterized many BACs and cDNAs containing kangaroo and platypus genes of interest, and demonstrate the value of comparisons to reveal conserved genome organization and function, and new insights in the evolution of the mammalian genome, particularly sex chromosomes.

BUGS, GUTS AND FAT - A SYSTEMS APPROACH TO THE METABOLIC 'AXIS OF EVIL'

JOE NADEAU

Department of Genetics and Center for Computational Genomics and Systems Biology
Case Western Reserve University, Cleveland, Ohio, USA

Rapidly growing evidence suggests that complex and variable interactions between host genetic and systems factors, diet, activity and lifestyle choices, and intestinal microbes control the incidence, severity and complexity of metabolic diseases. The dramatic increase in the world-wide incidence of these diseases, including obesity, diabetes, hypertension, heart disease, and fatty liver disease, raises the need for new ways to maintain health despite inherited and environmental risks. We are pursuing a comprehensive approach based on diet-induced models of metabolic disease. During the course of these studies, new and challenging statistical, analytical and computational problems were discovered. We pioneered a new paradigm for genetic studies based on chromosome substitution strains of laboratory mice. These strains involve systematically substituting each chromosome in a host strain with the corresponding chromosome from a donor strain. A genome survey with these strains therefore involves testing a panel of individual, distinct and non-overlapping genotypes, in contrast to conventional studies of heterogeneous populations. Studies of diet-induced metabolic disease with these strains have already led to striking observations. We discovered that most traits are controlled by a many genetic variants each of which has unexpectedly large phenotypic effects and that act in a highly non-additive manner. The non-additive nature of these variants challenges conventional models of the architecture of complex traits. At every level of resolution from the entire genome to very small genetic intervals, we discovered comparable levels of genetic complexity, suggesting a fractal property of complex traits. Another remarkable property of these large-effect variants is their ability to switch complex systems between alternative phenotypic states such as obese to lean and high to low cholesterol, suggesting that biological traits might be organized in a small number of stable states rather than continuous variability. Moreover, by studying correlations between non-genetic variation in pairs of traits (the genetic control of non-genetic variation), we discovered a new way to dissect the functional architecture of biological systems. Finally, a neglected aspect of these studies of metabolic disease involves the intestingal microbes. Early studies suggest that diet and host physiology affect the numbers and kinds of microbes, and that these microbes in turn affect host metabolism. These interactions between 'bugs, guts and fat' extend systems studies from conventional aspects of genetics and biology to population considerations of the functional interactions between hosts, diet and our microbial passengers. With these models of diet-induced metabolic disease in chromosome substitution strains, we are now positioned find ways to tip complex systems from disease to health.

PROTEIN IDENTIFICATION VIA SPECTRAL NETWORKS ANALYSIS

PAVEL PEVZNER

Department of Computer Science and Engineering
University of California, San Diego

Advances in tandem mass-spectrometry (MS/MS) steadily increase the rate of generation of MS/MS spectra. As a result, the existing approaches that compare spectra against databases are already facing a bottleneck, particularly when interpreting spectra of modified peptides. We introduce a new idea that allows one to perform MS/MS database search without ever comparing a spectrum against a database. We propose to slightly change the experimental protocol to intentionally generate spectral pairs - pairs of spectra obtained from overlapping (often non-tryptic) peptides or from unmodified and modified versions of the same peptide. While seemingly redundant, spectral pairs open up computational avenues that were never explored before. Having a spectrum of a modified peptide paired with a spectrum of an unmodified peptide, allows one to separate the prefix and suffix ladders, to greatly reduce the number of noise peaks, and to generate a small number of peptide reconstructions that are likely to contain the correct one. The MS/MS database search is thus reduced to extremely fast pattern matching (rather than time-consuming matching of spectra against databases). In addition to speed, our approach provides a new paradigm for identifying post-translational modifications and shotgun protein sequencing via spectral networks analysis. This is a joint work with Nuno Bandeira, Karl Clauser, Ari Frank and Dekel Tsur.

METAGENOME ANALYSIS USING MEGAN

DANIEL H. HUSON and ALEXANDER F. AUCH

Center for Bioinformatics, Tübingen University, Sand 14, 72076 Tübingen, Germany

JI QI and STEPHAN C. SCHUSTER

310 Wartik Laboratories, PennState University, Center for Comparative Genomics, Center for Infectious Disease Dynamics, University Park, PA 1803, USA

In metagenomics, the goal is to analyze the genomic content of a sample of organisms collected from a common habitat. One approach is to apply large-scale random shotgun sequencing techniques to obtain a collection of DNA reads from the sample. This data is then compared against databases of known sequences such as NCBI-nr or NCBI-nt, in an attempt to identify the taxonomical content of the sample. We introduce a new software called MEGAN (Meta Genome ANalyzer) that generates species profiles from such sequencing data by assigning reads to taxa of the NCBI taxonomy using a straight-forward assignment algorithm. The approach is illustrated by application to a number of datasets obtained using both sequencing-by-synthesis and Sanger sequencing technology, including metagenomic data from a mammoth bone, a portion of the Sargasso sea data set, and several complete microbial test genomes used for validation proposes.

1. Introduction

Genomics is the study of the genome sequence of individual organisms. Most genome sequences available in databases today were obtained by "Sanger sequencing", using a shotgun approach that involves cloning small inserts of DNA and then determining their sequence using fluorescent dideoxynucleotides for termination and electrophoresis for measurement[7]. The NCBI website (`www.ncbi.nlm.nih.gov`) lists hundreds of bacterial, tens of archaeal and about one hundred eukaryotic genomes as being completely sequenced, or in the process of being sequenced.

Metagenomics has been defined as "the genomic analysis of microorganisms by direct extraction and cloning of DNA from an assemblage of microorganisms"[5], and its importance stems from the fact that 99% or more of all microbes are deemed unculturable. If we take a genome to be the entire genetic information of a single organism, then a *metagenome* can be defined as the entire genetic information of an ensemble of organisms, living in a common habitat. The aim of metagenomics is to understand the genetic diversity of a metagenome, ideally, by identifying the (relative abundances of) species present. Metagenomics promises to lead to the discovery of new genes that have useful applications in biotechnology and medicine[10].

One main technique in metagenomics is to apply large-scale random shotgun sequencing. A number of recent projects use Sanger sequencing to create datasets in this way, for example, from an acid mine biofilm[12], sea-water samples [13], deep sea sediment[4], or soil and whale falls[11]. Recently, a new sequencing approach "sequencing-by-synthesis" was published that uses emulsion-based PCR application of a large number of DNA fragments and high-throughput parallel pyro-sequencing[6]. A single instrument is able to sequence 25 million bases within four hours, at a lower price, per base, than Sanger-based methods. Current drawbacks of the method are short read lengths of $\approx$ 100bp, in contrast to $\approx$ 800bp using Sanger sequencing, and a higher error rate. Moreover, sequencing of pair-ended reads is not yet possible.

Given present-day technology, obtaining the complete sequences of all genomes present in a metagenome is not feasible, even using Sanger sequencing and paired-end reads, as the amount of data required is too large and the assembly problem too difficult. More realistic goals are to determine the presence or absence of specific species of interest, or to obtain a rough overview of the taxa represented in a given metagenome.

In this paper we present a straight-forward approach to the latter problem. We describe a strategy for processing DNA reads collected within the frame-work of a metagenomics project and provide a new program called *MEGAN* (MEtaGenome ANalyser) that can be used to explore a metagenomics data set in a taxonomical context. The program employs a combinatorial algorithm, which we call "LCA-assignment", to estimate the taxonomical content of a metagenome, based on sequence comparisons.

We first illustrate this approach by application to a set of $302,692$ reads obtained from a sample of mammoth bone[8], using the sequencing-by-synthesis approach. We then address the question whether species can be identified with confidence from short reads of length 100. Finally, to demonstrate the applicability of the approach to data sets obtained using other sequencing approaches, we apply it to a subset of the Sargasso sea data[13].

Ease-of-use is a main design criterion of MEGAN. An analysis is initiated by simply opening a BlastX, BlastN or BlastZ file and is then performed interactively. For maximum portability, the program is written in Java and installers for Linux/Unix, MacOS and Windows are freely available for academic use from:
`http://www-ab.informatik.uni-tuebingen.de/software/megan.`

2. Processing metagenomic data

The following simple approach to metagenome analysis is a typical starting point for more sophisticated strategies (see Figure 1): First, randomly sequence a collection of DNA reads from the given sample. Second, perform Blast[1] comparisons of the reads against one or more reference databases, such as NCBI-nr, NCBI-nt, NCBI-env-nr, NCBI-env-nt[2], and additional genome specific databases, when appropriate. (Sequence comparison is the main computational bottle neck, which will grow more serve as the sizes of datasets and databases continue to grow.) Third, analyze the output of these comparisons and then assign individual reads to taxa, including higher-order taxa. Finally, for each taxon implicated, evaluate the provided evidence for its presence in the sample.

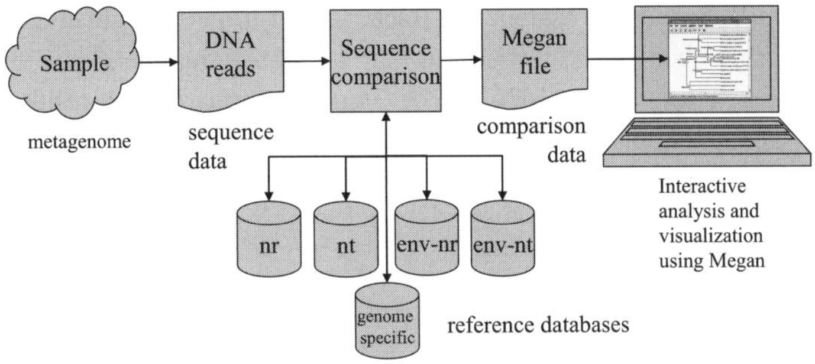

Figure 1. For a given sample of organisms, a randomly selected collection of DNA fragments is sequenced. The resulting reads are then compared with one or more reference databases using an appropriate sequence comparison program such as Blast. The resulting data is processed by MEGAN to produce an interactive analysis of the taxonomical content of the sample.

2.1. *Analysis of the Mammoth dataset*

As an example, we recently[8] used a metagenomics approach to analyze the DNA present in a sample of one gram of bone taken from a mammoth that was preserved in permafrost for 27,000 years. The project proceeded in the following steps. First, we used the Roche GS20 sequencing technology to randomly collect DNA from the sample, obtaining $302,692$ reads of mean length 95 base pairs (bp). We will refer to this as the *Mammoth dataset*.

To identify those reads that come from the mammoth genome, we performed BlastZ[9] comparisons of genome sequences for elephant, human and dog, downloaded from `www.genome.ucsc.edu`. As a result of this computation, in the mentioned paper[8] we estimate that at least 45.4% of the reads represent mammoth DNA.

We were interested in determining the possible sources of the remaining reads, as they probably represent micro-organisms that were present at, or immediately after, the time of the mammoth's death. To this end, we first used BlastX to compare all reads against the NCBI-nr ("non-redundant") protein database[2]. This resulted in a file of size 1.4GB containing $2,911,587$ local alignments of reads to sequences in the database. Of the $302,692$ reads, only $52,179$ have one or more alignments. We then loaded the results of the BlastX search into a preliminary version of MEGAN (then called GenomeTaxonomyBrowser[8]) and applied the LCA-assignment algorithm to compute an assignment of reads to taxa, thus obtaining an estimation of the taxonomical content of the sample.

Here we repeat this analysis, but are slightly more conservative and now employ a threshold of 30 for the bit score of matches, and will discard any isolated assignments, that is, any taxon that has only one read assigned to it. (We remove isolated assignments to avoid false positive identification of taxa due to sequencing errors or chance matches.) The LCA-assignment algorithm assigns $50,093$ reads to taxa and $2,086$ remain unassigned either because the bit score of their matches fall below the threshold or because they give rise to an isolated match.

A total of $19,841$ reads are assigned to Eukaryota, of which $7,969$ are assigned to Gnathostomata (jawed vertebrates) and thus presumably come from mammoth genes. Further, a total of 16, 972 reads are assigned to Bacteria, 761 to Archea and 152 to viruses, respectively. These numbers are slightly lower than previously reported[8] due to our slightly more conservative settings. MEGAN can be used to summarize the results at different levels of the NCBI taxonomy, see Figures 2 and 3.

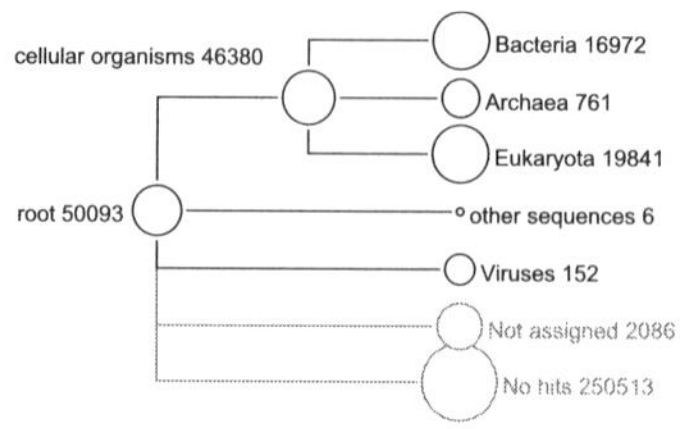

Figure 2. High-level summary of a MEGAN analysis of the mammoth dataset, based on a BlastX comparison of the 302, 692 reads against the NCBI-nr database. In all figures, each circle represents a taxon in the NCBI taxonomy and is labeled by its name and the number of reads that are assigned either directly to the taxon, or indirectly via one of its sub-taxa. The size of the circle is scaled logarithmically to represent the number of reads assigned directly to the taxon.

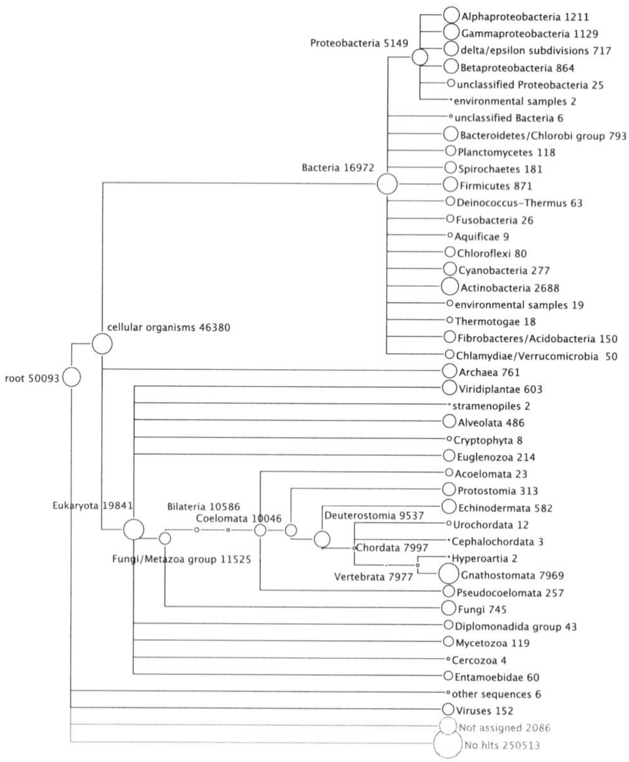

Figure 3. A more detailed MEGAN analysis of the mammoth dataset.

2.2. *Identifiability of species from short reads*

The average read length currently obtainable using Roche GS20 sequencing technology is $\approx$ 100bp. The question arises whether this sequence length is long enough to provide meaningful information on the taxonomical content of a metagenome. This can be addressed by collecting a set of reads from a known genome and then processing the data as a metagenome dataset.

In a first experiment, we considered 2000 reads sequenced from *E. coli K12*, using Roche GS20 sequencing technology. As *E. coli* is widely used in laboratories, this dataset may potentially give rise to many false positive identifications, as parts of its sequence occur by error in a number of different genome sequences.

In Figure 4 we show the resulting MEGAN analysis, based on a BlastX comparison of the reads against the NCBI-nr database, using a bit score threshold of 35 and discarding any isolated assignments. Of the 2000 reads, approximately 25% (448) have no hits and 116 reads are not assigned. Of the remaining 1436 reads, approximately 50% (699) are assigned to *Enterobacteriaceae*, thus making a correct assignment up to the family level. All other reads, except two, are assigned to super taxa, thus producing correct, if increasingly weak, predictions.

The two false positive assignments to *Haemophilus somnus* appear to be due to false entries in the NCBI-nr database: one of the assigned reads has a perfect BlastN match to 16S rRNA sequence in *E. coli* and the other has a perfect BlastN match to 23S rRNA sequence in *E. coli*. On the other hand, the matched sequences representing *Haemophilus somnus* in NCBI-nr are both labeled "hypothetical" proteins.

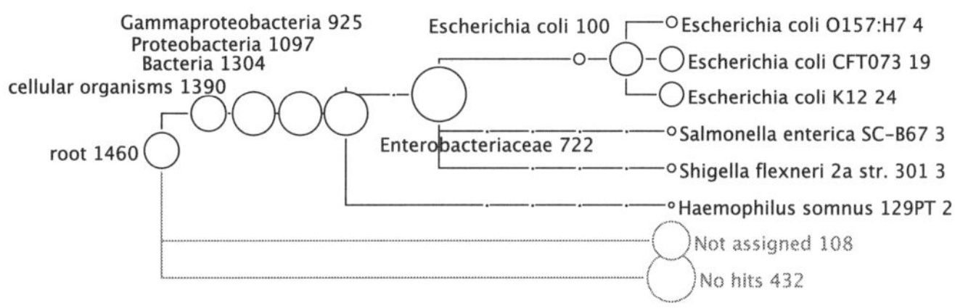

Figure 4. MEGAN analysis of 2000 reads collected from *E. coli K12* using Roche GS20 sequencing, based on a BlastX comparison with the NCBI-nr database.

In a second experiment, we considered 2000 reads sequenced from *Bdellovibrio bacteriovorus HD100* using Roche GS20 sequencing technology. In Figure 5(a) we show the resulting MEGAN analysis, based on a BlastX comparison of the reads against the NCBI-nr database, using a bit score threshold of 35 and discarding any isolated assignments. Of the 2000 reads, approximately 20% (397) have no hits and 5% (105) are not assigned. Of the remaining 1498 reads, approximately 70% (1062) are assigned to *Bdellovibrio bacteriovorus HD100*. All other reads are assigned to super taxa, thus producing correct, if increasingly weak, predictions. There are no false positive predictions.

In Figure 5(b) we show the MEGAN analysis obtained when using a copy of the NCBI-nr database from which all sequences representing *Bdellovibrio bacteriovorus HD100* have been removed. This mimics the case in which reads are obtained from a genome that is not represented in the database. Of the 2000 reads, approximately 65% (1361) have no hits and approximately 15% (272) are not assigned. A small number of false positives occur up to the level of bacteria.

(a)

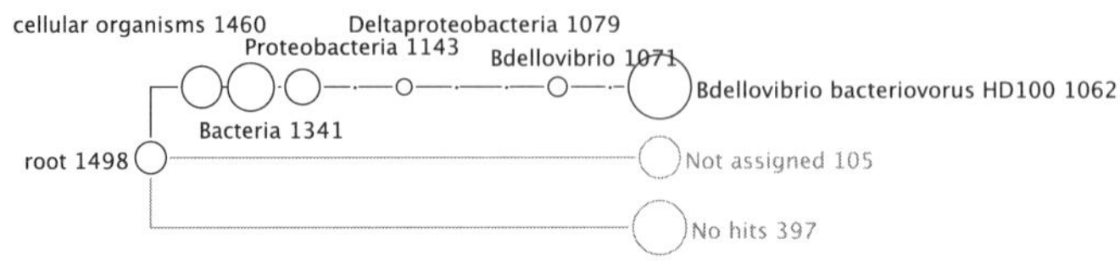

(b)

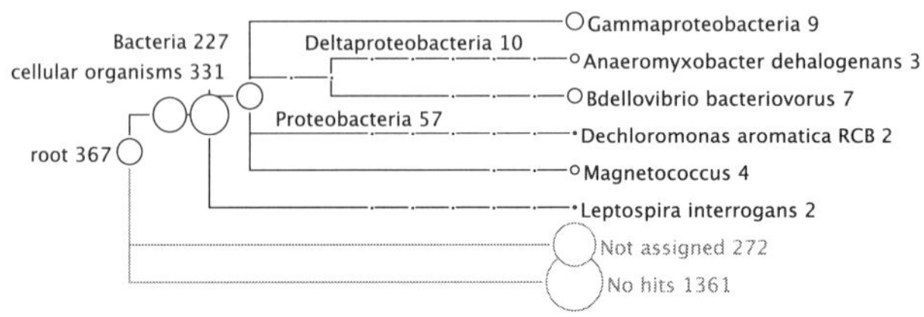

Figure 5. MEGAN analysis of 2000 reads collected from *Bdellovibrio bacteriovorus HD100* using Roche GS20 sequencing technology. (a) Analysis based on a BlastX comparison with NCBI-nr. (b) Similar analysis, but with all hits to database sequences representing *Bdellovibrio bacteriovorus HD100* removed, mimicking the situation in which the reads originate from a genome that is not represented in NCBI-nr.

These two experiments show that the LCA-assignment algorithm is quite conservative, avoiding false positive assignments at the price of producing quite large numbers of inspecific assignments. Further, they also indicate that the performance of this type of approach depends heavily on the set of sequences represented in the reference database. In particular, if close relatives are missing in the database, then reads from an unknown organism will give rise to many unassigned reads and possibly some false positive assignments, as well.

2.3. *Analysis of the Sargasso Sea data set*

In the Sargasso sea project[13], samples of sea water were collected and organisms of size $0.1 - 3.0\ \mu$m were extracted to produce a metagenome dataset. In 4 separate experiments, approximately 1.9 million reads of average length 818 bp were collected using Sanger sequencing.

To explore the application of MEGAN to such data, we downloaded the first 10,000 reads from `https://research.venterinstitute.org/sargasso/` and ran BlastX to compare the data against the NCBI-NR database. Only 1% (13) of the reads had no hits. A MEGAN analysis of the data using a bit score threshold of 100 and discarding all isolated assignments, assigned approximately 90% (8,977) to taxa, a majority of which (6811) were assigned to bacteria. The results are summarized in Figure 6. Interestingly, this analysis of a small portion of the Sargasso sea dataset is compatible with the analysis reported by Venter *et al.*[13], (although *Firmicutes* are missing, probably due to the small size of the sub-sample), and also confirms findings that parts of the data set is contaminated with *Shewanella* and *Burkholderia*[3].

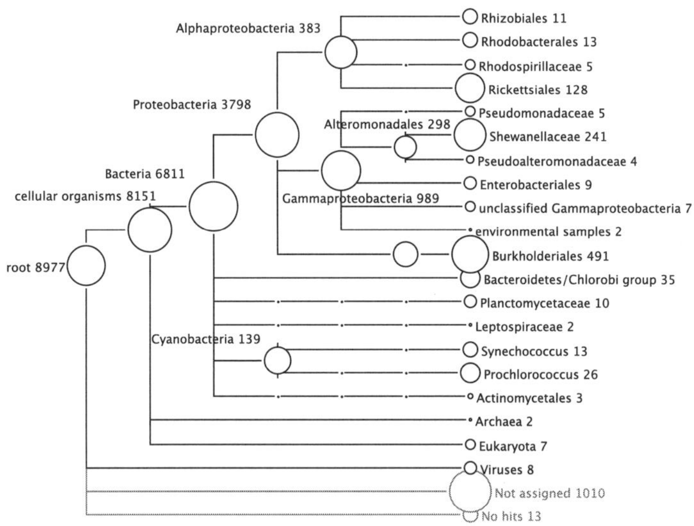

Figure 6. MEGAN analysis of 10, 000 reads of Sargasso sea data.

3. Analysis using MEGAN

At startup, MEGAN loads the complete NCBI taxonomy, currently containing over 280, 000 taxa, which can then be interactively explored, using customized tree-navigation features. However, the main application of MEGAN is to process results files generated by a comparison of sequencing reads with a database of annotated sequences. The program parse files generated by BlastX, BlastN or BlastZ, and saves the results as a series of read-taxon matches in a program-specific format. (Additional parsers may be added to process the results generated by other sequence comparison methods.)

The program assigns reads to taxa using the LCA-assignment algorithm (described in detail below) and then displays the induced taxonomy. Nodes in the taxonomy can be collapsed or expanded to produce summaries at different levels of the taxonomy. Additionally, the program provides a Find tool to search for specific taxa and an Inspector tool to view individual Blast matches (see Figure 7).

The approach uses a number of thresholds. First, a *min-score* threshold defines the minimum bit score that must be attained by a Blast alignment so that a read r is considered to *match* a given taxon t. Second, the *min-support* threshold specifies how many reads must be assigned to a specific taxon, or any taxon below it in the taxonomy, so that the taxon is identified as present.

The result of the LCA-assignment algorithm is presented to the user as the partial taxonomy T that is induced by the set of taxa that have been identified (see Figure 2). The

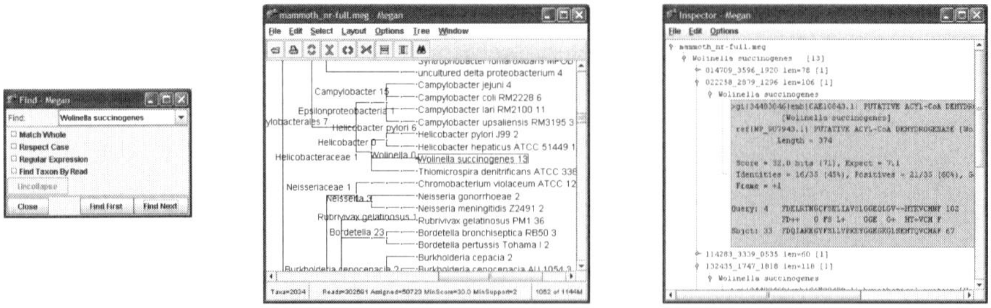

(a) (b) (c)

Figure 7. (a) MEGAN provides a Find tool to search for specific taxa of interest. (b) The result of a search is highlighted in a detailed summary of the analysis. (c) MEGAN provides an Inspector tool to view the individual sequence comparisons upon which the assignment of a particular read to a particular taxon is based.

program allows the user to explore the results both at a high level, and also a very detailed level, by providing methods for collapsing and expanding different parts of T. Each node in T represents a taxon t and can be queried to determine which reads have been assigned directly to t, and how many have been assigned to taxa below t. Additionally, the program allows the user to view the Blast alignments upon which a specific assignments is based (see Figure 7(3)).

4. Assignment of reads to taxa

MEGAN currently uses a simple combinatorial algorithm, which we call "LCA-assignment", in association with a number of different thresholds, to assign each read to a taxon at some level of the NCBI taxonomy.

The *LCA-assignment* algorithm operates as follows. Consider a read r and assume that the Blast computation has established matches to sequences representing a set of taxa $t(r) = \{t_1, t_2, \ldots, t_k\}$. We assign the read r to the *lowest common ancestor (LCA)* of $t(r)$ in the NCBI taxonomy. For example, if r matches *Campylobacter lari*, *Helicobacter hepaticus* and *Wolinella*, then r is assigned to the taxon *Campylobacterales*. If r does not match any sequence in the given reference database, that is, if $t(r) = \emptyset$, then r is assigned to the special taxon *no hits*. If r cannot be assigned to a taxon for other reasons, e.g. the read only matches sequences for which the taxon is unknown, then r is assigned to another special taxon *Not assigned*.

In this way, each read r in the dataset is assigned to one or more NCBI taxa, or to one of either special taxa. If the Blast matches computed for r involve only one or a few closely related species, then r will be assigned to a taxon near the tips of the taxonomy. If, on the other hand, r matches a wider range of taxa, then r will be assigned to a higher-level taxon. The read may even be assigned to the root of the taxonomy, if the sequence is completely unspecific.

To implement the LCA-assignment algorithm, we assign a binary *address* $a(t)$ to each every t in such away that if taxon s is an ancestor of taxon t, then the address $a(s)$ is a prefix of the address $a(t)$. Using this scheme, we can easily determine the lowest common ancestor of a set of n taxa $\{t_1, \ldots, t_n\}$ by determining the longest common prefix of the corresponding set of addresses $\{a(t_1), \ldots, a(t_n)\}$, in $O(n \times \log K)$ steps, where K is the maximum depth of the taxonomy.

4.1. *Under- and over prediction*

We say that a read gives rise to an *under prediction*, if it is assigned to a taxon that lies above the true taxon in the taxonomy. Under prediction happens when a read comes from a gene that is widely conserved. We say that a read gives rise to a *false prediction*, if it is assigned to a taxon that is not the true taxon, nor one of its ancestors in the taxonomy. W say that a false prediction is an *over prediction*, if it is caused by the fact that the true sequence is not represented in the employed databases.

For example, all reads analyzed in Figure 5(a)-(b) come from the genome of *Bdellovibrio bacteriovorus HD100*. However, there is a substantial amount of under prediction both in (a) and (b), in particular of the taxon *Bacteria*, and a number of cases of over prediction in (b), ranging from *Anaeromyxobacter dehalogenans* and *Gammaproteobacteria* to *Leptospira interrogans*.

As a simple combinatorial method, the LCA-assignment algorithm is susceptible to both types of errors. We hope to develop a more sophisticated approach that will not only take the presence or absence of matches into account, but also will make use of the quality of the matches and the levels of similarly that are typical for given genes in given clades of sequences.

5. Summary

A metagenomics project aims at understanding the taxonomical content of an ensemble of organisms. The approach described in this paper is to use sequencing techniques to produce DNA reads, to perform similarity searches in databases of known sequences and then to analyze and explore the resulting comparison data using software such as MEGAN.

MEGAN is based on a robust algorithm for assigning reads to taxa and is designed as an easy-to-use exploration tool that quickly produces summaries of the data at different taxonomical levels. It offers tools to search for specific taxa in the data and to inspect the evidence supporting the presence of any given taxon. This software provides a solid starting point for producing a first analysis of a metagenomic dataset.

References

1. S. F. Altschul, W. Gish, W. Miller, E. W. Myers, and D. J. Lipman. Basic local alignment search tool. *Journal of Molecular Biology*, 215:403–410, 1990.
2. D.A. Benson, I. Karsch-Mizrachi, D.J. Lipman, J. Ostell, and D.L. Wheeler. Genbank. *Nucleic Acids Res*, 1(33 (Database issue)):D34–38, 2005.
3. E.F. DeLong. Microbial community genomics in the ocean. *Nat Rev Microbiol.*, 3(6):459–69, 2005.
4. S. J. Hallam, N. Putnam, C.M. Preston, J.C. Detter, and D. Rokhsar. Reverse methanogenesis: Testing the hypothesis with environmental genomics. *Science*, 305:1457–62, 2004.

5. J. Handelsman. Metagenomics: Application of genomics to uncultured microorganisms. *Microbiology and Molecular Biology Reviews*, 68(4):669–685, 2004.
6. M. Margulies and *et al.* Genome sequencing in microfabricated high-density picolitre reactors. *Nature*, 437(7057):376–380, 2005.
7. D. Meldrum. Automation for genomics, part two: Sequencers, microarrays, and future trends. *Genome Research*, 10(9):1288–1303, 2000.
8. H. N. Poinar, C. Schwarz, Ji Qi, B. Shapiro, R. D. E. MacPhee, B. Buigues, A. Tikhonov, L. P. Tomsho D.H. Huson, A. Auch, M. Rampp, W. Miller, and S. C. Schuster. Metagenomics to Paleogenomics: Large-Scale Sequencing of Mammoth DNA. *Science*, 331:392–394, 2006.
9. S. Schwartz, W.J. Kent, A. Smit, Z. Zhang, R. Baertsch, R. C. Hardison, D. Haussler, and W. Miller. Human-mouse alignments with BLASTZ. *Genome Res.*, 13:103 – 107, 2003.
10. H.L. Steele and W.R. Streit. Metagenomics: Advances in ecology and biotechnology. *FEMS Microbiology Letters*, 247(2):105–111, 2005.
11. S. G. Tringe, C. von Mering, A. Kobayashi, A. A. Salamov, K. Chen, H. W. Chang, M. Podar, J. M. Short, E. J. Mathur, J. C. Detter, P. Bork, P. Hugenholtz, and E. M. Rubin. Comparative metagenomics of microbial communities. *Science*, 308:554–557, 2005.
12. G. W. Tyson, J. Chapman, P. Hugenholtz, E. E. Allen, R. J. Ram1, P. M. Richardson, V. V. Solovyev, E. M. Rubin, D. S. Rokhsar, and J. F. Banfield. Community structure and metabolism through reconstruction of microbial genomes from the environment. *Nature*, 428:37–43, 2004.
13. J. C. Venter, K. Remington, J. F. Heidelberg, A. L. Halpern, D. Rusch, J. A. Eisen, D. Wu, I. Paulsen, K. E. Nelson, W. Nelson, D. E. Fouts, S. Levy, A. H. Knap, M. W. Lomas, K. Nealson, O. White, J. Peterson, J. Hoffman, R. Parsons, H. Baden-Tillson, C. Pfannkoch, Y. Rogers, and H. O. Smith1. Environmental genome shotgun sequencing of the sargasso sea. *Science*, 304(5667):66–74, 2004.

ALGORITHMIC APPROACHES TO SELECTING CONTROL CLONES IN DNA ARRAY HYBRIDIZATION EXPERIMENTS*

Q. FU†, E. BENT‡, J. BORNEMAN ‡, M. CHROBAK† and N. YOUNG†

We study the problem of selecting control clones in DNA array hybridization experiments. The problem arises in the OFRG method for analyzing microbial communities. The OFRG method performs classification of rRNA gene clones using binary fingerprints created from a series of hybridization experiments, where each experiment consists of hybridizing a collection of arrayed clones with a single oligonucleotide probe. This experiment produces analog signals, one for each clone, which then need to be classified, that is, converted into binary values 1 and 0 that represent hybridization and non-hybridization events. Besides the sample clones, the array contains a number of control clones needed to calibrate the classification procedure of the hybridization signals. These control clones must be selected with care to optimize the classification process. We formulate this as a combinatorial optimization problem called *Balanced Covering*. We prove that the problem is $\mathbb{NP}$-hard and we show some results on hardness of approximation. We propose an approximation algorithm based on randomized rounding and we show that, with high probability, it approximates well the optimum. The experimental results confirm that the algorithm finds high quality control clones. The algorithm has been implemented and is publicly available as part of the software package called CloneTools.

1. Introduction

Background. We study the problem of selecting control clones for DNA array hybridization experiments. The specific version of the problem that we address arises in the context of the OFRG (Oligonucleotide Fingerprinting of Ribosomal RNA Genes) method. OFRG[8,9,10] is a technique for analyzing microbial communities that classifies rRNA gene clones into taxonomic clusters based on binary fingerprints created from hybridizations with a collection of oligonucleotide probes. More specifically, in OFRG, clone libraries from a sample under study (e.g., fungi or bacteria from an environmental sample) are constructed using PCR primers. These cloned rRNA gene fragments are immobilized on nylon membranes and then subjected to a series of hybridization experiments, with each experiment using a single radiolabeled DNA oligonucleotide probe. This experiment produces analog signals, one for each clone, which then need to be classified, that is, converted into binary values 1 and 0 that represent hybridization and non-hybridization events. Overall, this process creates a hybridization fingerprint for each clone, which is a vector of binary values indicating which probes bind with this clone and which do not. The clones are then identified by clustering their hybridization fingerprints with those of known sequences and by nucleotide sequence analysis of representative clones within a cluster.

*Research supported by NSF grants CCR-0208856 and BD&I-0133265.

†Department of Computer Science, University of California, Riverside, CA 92521.
‡Department of Plant Pathology, University of California, Riverside, CA 92521.

Besides the sample clones, the array contains a number of *control clones*, with known nucleotide sequences, used to calibrate the classification of hybridization signals. For each probe, two distributions of signal intensities are obtained: one from control clones that match the probe (e.g., they contain its complement and thus should hybridize with it) and the other from clones that do not. This information is then used to determine the signal intensity threshold for clone-probe hybridization: signals above this threshold are interpreted as clone-probe hybridization events, while those below correspond to non-hybridizations.

The quality of information obtained from hybridizations depends critically on the accuracy of the signal classification process. Ideally, each probe should match (bind with) roughly half of the control clones. In prior OFRG work, control clones were selected arbitrarily, often producing control clones with skewed distribution of binding/non-binding with some probes. As an example, from a set of 100 control clones, only two might bind with a specific probe. The signal classification for this probe would be very unreliable, as it would be based on signal intensities from hybridization with only two control clones.

Problem formulation. Our control-clone selection problem can be then formulated as a combinatorial optimization problem called *Balanced Covering* [a]. The instance is given as a pair $\langle G, s\rangle$, where $G = (C, P, E)$ is a bipartite graph and $s \leq |C|$ is an integer. C represents the candidate control clone set, P is the probe set, and the edges in E represent potential hybridizations between clones and probes, that is, $(c, p) \in E$ iff p should hybridize with c. For $p \in P$ and $D \subseteq C$, let $\deg_D(p)$ be the number of neighbors of p in D (that is, the number of clones in D that hybridize with p). Throughout the paper, unless stated otherwise, by m and n we will denote the cardinality of C and P, respectively.

Generally, as indicated earlier, our goal is to find a set $D \subseteq C$ of cardinality s such that, for each $p \in P$, $\deg_D(p)$ is close to $s/2$. Several objective functions can be studied. To measure the deviation from the perfectly balanced cover, for a given probe p, we can compute either $\min\{\deg_D(p), s - \deg_D(p)\}$ or $|\deg_D(p) - s/2|$. The objective function can be obtained by considering the sum of these values or the worst case over all probes. This gives rise to four objective functions: (1) maximize $\mathcal{C}_{min}(D) = \min_{p\in P} \min\{\deg_D(p), s - \deg_D(p)\}$; (2) maximize $\mathcal{C}_{sum}(D) = \sum_{p\in P} \min\{\deg_D(p), s - \deg_D(p)\}$; (3) minimize $\mathcal{D}_{max}(D) = \max_{p\in P} |\deg_D(p) - s/2|$; (4) minimize $\mathcal{D}_{sum}(D) = \sum_{p\in P} |\deg_D(p) - s/2|$, where each function needs to be optimized over all choices of D. By BCP_$\mathcal{C}_{min}$, etc., we denote the four optimization problems corresponding to these functions. We focus on BCP_$\mathcal{C}_{min}$ and BCP_$\mathcal{D}_{sum}$.

Results. In this paper we show several analytical and experimental results on Balanced Covering. In Section 2 we prove that all versions of Balanced Covering are $\mathbb{NP}$-hard. In particular, it is $\mathbb{NP}$-complete to decide whether there is a perfectly balanced cover with s clones. This immediately implies that (unless $\mathbb{P}$=$\mathbb{NP}$), there is no polynomial-time approximation algorithm for BCP_$\mathcal{D}_{max}$ and BCP_$\mathcal{D}_{sum}$. For BCP_$\mathcal{D}_{sum}$, we further show that even if we allow an additional additive term $O((nm)^\epsilon)$ and randomization, approximating the

[a] There have been some discussions on the *Balanced Set Cover* (see [1], [6]) problem, however, they are not directly related to *Balanced Covering* problem discussed in this paper.

optimum is still hard. For BCP_$\mathcal{C}_{min}$, we prove that it cannot be approximated efficiently within the bound $\text{OPT}_{\min}(G, s) - \frac{1}{2}(1-\epsilon)\ln n$, unless $\mathbb{NP}$ has slightly superpolynomial time algorithms.

Then, in Section 3, we propose a polynomial-time randomized rounding algorithm RRBC for BCP_$\mathcal{C}_{min}$. The algorithm solves the linear relaxation of the integer program for BCP_$\mathcal{C}_{min}$, and then uses the solution to randomly pick an approximately balanced cover. We show that, with probability at least $\frac{1}{2}$, RRBC's solution deviates by no more than $O(\sqrt{z^* \ln n} + \sqrt{s})$ from the optimal solution z^* of the linear program.

In Section 5, we present an experimental study of RRBC's performance which shows that its solutions are close to the optimum. More specifically, in 92.8% cases of real data sets, RRBC found solutions at least as large as 97% of the linear relaxation's optimum. The implementation of RRBC is publicly available at the OFRG website as part of the CloneTools software package, see `http://algorithms.cs.ucr.edu/OFRG/`.

Algorithm RRBC performs well for input instances where the optimum is relatively large, but the performance bound given in Section 4 is not quite satisfactory when z^* is small compared to s, say for $z^* = o(\sqrt{s})$. In Section 6, we present another algorithm called RRBC2 which, with probability at least $\frac{1}{2}$, computes a solution that deviates by no more than $O(\sqrt{z^* \ln n})$ from the optimum.

Relation to other work. Note that OFRG differs from other array-based analysis approaches that, typically, involve a single microarray experiment where one clone of interest is hybridized against a collection of arrayed probes, each targeting a specific sequence. These experiments include control clones as well, but these control clones are used to test whether they bind as predicted to particular microarray probes (see [7], for example). In contrast, OFRG uses a small set of probes (roughly 30-50) to coordinately distinguish a much larger set of sequences (e.g., all bacterial rRNA genes). Each probe is used in one hybridization experiment, and the unknown DNA clones are immobilized on the array.

2. Hardness Results

In this section we prove several hardness results for Balanced Covering. We will first show that all versions of Balanced Covering are $\mathbb{NP}$-hard.

$\mathbb{NP}$-hardness. Given a bipartite graph $G = (C, P, E)$ and an even integer s, define a *perfectly balanced cover* in G to be a subset $D \subseteq C$ with $|D| = s$ such that $\deg_D(p) = s/2$ for each $p \in P$.

Theorem 2.1. *The following decision problem is $\mathbb{NP}$-complete: given a bipartite graph $G = (C, P, E)$ and an even integer s, is there a perfectly balanced cover in G? Consequently, Balanced Covering is $\mathbb{NP}$-hard for all objective functions $\mathcal{C}_{min}$, $\mathcal{C}_{sum}$, $\mathcal{D}_{max}$ and $\mathcal{D}_{sum}$.*

Proof. The proof is by a reduction from X3C (Exact Cover by 3-Sets) [5], which is known to be $\mathbb{NP}$-complete. The instance of X3C consists of a finite set X of $3m$ items, and a collection T of n 3-element subsets of X that we refer to as *triples*. We assume $n \geq m \geq 2$. The objective is to determine whether T contains an *exact cover of* X, i.e., a sub-collection $T' \subseteq T$ such that every element of X occurs in exactly one triple in T'.

The reduction is defined as follows. Given an instance $\langle X, T\rangle$ of X3C above, we construct an instance $\langle G = (T \cup W, X, E), s\rangle$ of Balanced Covering, where W is a set that contains $m-2$ new vertices. For $t \in T$ and $x \in X$, we create an edge $(t, x) \in E$ if $x \in t$. Further, we create all edges $(w, x) \in E$ for $x \in X$ and $w \in W$. We let $s = 2m - 2$.

It remains to show that this construction is correct, namely that $\langle X, T\rangle$ has an exact cover iff $\langle G, s\rangle$ has a perfectly balanced cover.

($\Rightarrow$) If $\langle X, T\rangle$ has an exact cover T', we claim that $D = T' \cup W$ is a perfectly balanced cover for $\langle G, s\rangle$. To justify this, note first that $|T'| = m$ and $|W| = m - 2$, and thus $|D| = 2m - 2 = s$. Further, each vertex $x \in X$ has exactly one neighbor in T' and $m-2$ neighbors in W, so x has $m - 1 = s/2$ neighbors in D, as required.

($\Leftarrow$) Suppose that $\langle G, s\rangle$ has a perfectly balanced cover $D \subseteq T \cup W$. Denoting $W' = D \cap W$, $k = |W'|$, and $T' = D \cap T$. We first show that D must contain all vertices in W.

We count the edges between D and X. There are $3km$ edges between W' and X, since each vertex in W' is connected to all $3m$ vertices in X. There are $3(s-k)$ edges between T' and X, since each vertex in T has degree 3. On the other hand, there must be $3m(m-1)$ edges between X and D, since each vertex in X must be connected to exactly $s/2 = m-1$ vertices in D. Together, this yields $3(2m-2-k) + 3km = 3m(m-1)$. Solving this equation, we get $k = m-2$, which means that $W' = W$. So T' contains exactly $s-k = m$ vertices, as claimed. Each vertex $x \in X$ is adjacent to all vertices in W, so it has exactly $s/2 - (m-2) = 1$ neighbor in T'. This means that T' is an exact cover of X. □

Approximation of BCP_$\mathcal{D}_{sum}$. Theorem 2.1 immediately implies that BCP_$\mathcal{D}_{sum}$ (as well as BCP_$\mathcal{D}_{max}$) cannot be approximated with any finite ratio. We show that even if we allow an additive term in the approximation bound, achieving such bound is still $\mathbb{NP}$-hard.

Given an algorithm $\mathcal{A}$ for BCP_$\mathcal{D}_{sum}$, denote by $\mathcal{D}^{\mathcal{A}}_{sum}(G, s)$ the value of the objective function computed by $\mathcal{A}$ on instance $\langle G, s\rangle$, that is, $\mathcal{D}^{\mathcal{A}}_{sum}(G, s) = \sum_{p \in P} |\deg_D(p) - s/2|$, where D is the set computed by $\mathcal{A}$. By $\mathcal{D}^{*}_{sum}(G, s)$ we denote the optimal value. Recall that $m = |C|$ and $n = |P|$.

Theorem 2.2. *Assuming $\mathbb{P} \neq \mathbb{NP}$, there is no polynomial-time algorithm $\mathcal{A}$ for BCP_$\mathcal{D}_{sum}$ that for any instance $\langle G, s\rangle$ satisfies $\mathcal{D}^{\mathcal{A}}_{sum}(G, s) \le \alpha \cdot \mathcal{D}^{*}_{sum}(G, s) + \beta(mn)^{\epsilon}$, for any $\alpha, \beta > 0$ and $0 < \epsilon < 1$.*

Proof. Suppose, towards contradiction, that for some α, β and ϵ there is a polynomial-time algorithm $\mathcal{A}$ satisfying the theorem. We show that this would imply the existence of a polynomial-time algorithm that decides if there is a perfectly balanced covering, contradicting Theorem 2.1.

Given an instance $\langle G, s\rangle$ of BCP_$\mathcal{D}_{sum}$, where $G = (C, P, E)$, convert it into another instance $\langle G^r, s\rangle$ of BCP_$\mathcal{D}_{sum}$, where $G^r = (C, P', E')$ is obtained by creating r identical copies of each probe $p \in P$ (that is, with the same neighbors in C). Thus $|P'| = rn$. Without loss of generality, assume $\beta > 1$ and $mn \ge 2$. We choose $r = \lceil (mn)^{\delta} \rceil + 1$, for $\delta = (\epsilon + \log \beta)/(1 - \epsilon)$. For this r, we have $r > \beta(mnr)^{\epsilon}$. Therefore $\langle G^r, s\rangle$ satisfies:

• If $\langle G, s\rangle$ has a perfectly balanced cover (that is, $\mathcal{D}^*_{sum}(G, s) = 0$) then $\mathcal{D}^*_{sum}(G^r, s) = 0$ as well, and thus $\mathcal{D}^{\mathcal{A}}_{sum}(G^r, s) \leq \beta(mnr)^\epsilon < r$.

• If $\langle G, s\rangle$ does not have a perfectly balanced cover (that is, $\mathcal{D}^*_{sum}(G, s) \geq 1$) then $\mathcal{D}^{\mathcal{A}}_{sum}(G^r, s) \geq \mathcal{D}^*_{sum}(G^r, s) = r \cdot \mathcal{D}^*_{sum}(G, s) \geq r$.

So we get $\mathcal{D}^{\mathcal{A}}_{sum}(G^r, s) < r$ if and only if $\mathcal{D}^*_{sum}(G, s) = 0$. Since G^r can be obtained from G in polynomial time, this would yield a polynomial-time algorithm that determines the existence of a perfectly balanced cover, contradicting Theorem 2.1. □

Next, we show that approximating BCP_$\mathcal{D}_{sum}$ is hard even with randomization. Recall that class $\mathbb{RP}$ (randomized polynomial time) is the complexity class of decision problems $\mathcal{P}$ which have polynomial-time probabilistic Turing machines M such that, for each input I, (i) if $I \in \mathcal{P}$ then M accepts I with probability at least $\frac{1}{2}$, and (ii) if $I \notin \mathcal{P}$ then M rejects I with probability 1. It is still open whether $\mathbb{RP} = \mathbb{NP}$.

Theorem 2.3. *Assuming $\mathbb{RP} \neq \mathbb{NP}$, there is no randomized polynomial-time algorithm $\mathcal{A}$ for BCP_$\mathcal{D}_{sum}$ that for any instance $\langle G, s\rangle$ satisfies Exp$[\mathcal{D}^{\mathcal{A}}_{sum}(G, s)] \leq \alpha \cdot \mathcal{D}^*_{sum}(G, s) + \beta(mn)^\epsilon$, for any $\alpha, \beta > 0$ and $0 < \epsilon < 1$.*

Proof. Suppose, towards contradiction, that there exists a randomized polynomial-time algorithm $\mathcal{A}$ that satisfies the theorem. Given an instance $\langle G, s\rangle$ of BCP_$\mathcal{D}_{sum}$, we again convert it into an instance $\langle G^r, s\rangle$ of BCP_$\mathcal{D}_{sum}$, as in the proof of Theorem 2.2. Without loss of generality, assume $\beta > 1$ and $mn \geq 2$. We choose $r = \lceil (mn)^\delta \rceil + 1$, for $\delta = (\epsilon + \log\beta + 1)/(1-\epsilon)$. Note that $r > 2\beta(mnr)^\epsilon$. We now make the following observations.

• If $\langle G, s\rangle$ has a perfectly balanced cover (that is, $\mathcal{D}^*_{sum}(G, s) = 0$) then $\mathcal{D}^*_{sum}(G^r, s) = 0$, and therefore 2Exp$[\mathcal{D}^{\mathcal{A}}_{sum}(G^r, s)] \leq 2\beta(mnr)^\epsilon < r$. Using Markov's inequality, this implies that Pr$[\mathcal{D}^{\mathcal{A}}_{sum}(G^r, s) < r] \geq \frac{1}{2}$.

• If $\langle G, s\rangle$ does not have a perfectly balanced cover (that is, $\mathcal{D}^*_{sum}(G, s) \geq 1$) then $\mathcal{D}^{\mathcal{A}}_{sum}(G^r, s) \geq \mathcal{D}^*_{sum}(G^r, s) \geq r \cdot \mathcal{D}^*_{sum}(G, s) \geq r$, with probability 1.

Thus from $\mathcal{A}$ we could obtain a randomized polynomial-time algorithm that determines the existence of a perfectly balanced cover, contradicting Theorem 2.1. □

Approximation of BCP_$\mathcal{C}_{min}$. Next we show that BCP_$\mathcal{C}_{min}$ cannot be approximated efficiently with the bound $\mathcal{C}^*_{min}(G, s) - \frac{1-\epsilon}{2} \ln n$, unless $\mathbb{NP}$ has slightly superpolynomial time algorithms. Recall that $\mathcal{C}^*_{min}(G, s)$ is the optimal value of $\mathcal{C}_{min}$ on an instance $\langle G, s\rangle$ of BCP_$\mathcal{C}_{min}$. $\mathcal{C}^{\mathcal{A}}_{min}(G, s)$ is the value of the objective function computed by an algorithm $\mathcal{A}$.

Theorem 2.4. *Unless problems in $\mathbb{NP}$ have $n^{O(\log\log n)}$-time deterministic algorithms, there is no polynomial-time algorithm $\mathcal{A}$ for BCP_$\mathcal{C}_{min}$ that, for some $0 < \epsilon < 1$, for any instance $\langle G, s\rangle$, satisfies $\mathcal{C}^{\mathcal{A}}_{min}(G, s) \geq \mathcal{C}^*_{min}(G, s) - \frac{1-\epsilon}{2} \ln n$.*

Proof. Suppose, towards contradiction, that there exits a polynomial-time algorithm $\mathcal{A}$ that satisfies the theorem. We show that this would imply the existence of a polynomial-time $((1 - O(1)) \ln n)$-approximation algorithm $\mathcal{B}$ for the Set Cover problem, which would imply in turn that problems in $\mathbb{NP}$ have $n^{O(\log\log n)}$-time deterministic algorithms [4].

Given an instance $\langle Q, X\rangle$ of set cover, where Q is a collection of sets over universe $X = \{1, 2, \ldots, n\}$, the algorithm $\mathcal{B}$ first reduces it to an instance $\langle G, s\rangle$ of BCP_$\mathcal{C}_{min}$, then calls algorithm $\mathcal{A}$ on input $\langle G, s\rangle$, and finally converts the solution of $\mathcal{A}$ to a set cover of X. We claim that this cover will be of size at most $(1 - \Omega(\epsilon)) \ln(n)b$, where b is the size of the minimum set cover of X. Below we assume that, without loss of generality, algorithm $\mathcal{B}$ knows the optimal value b of $\langle Q, X\rangle$. Otherwise, $\mathcal{B}$ can simply try each $b \in \{1, 2, ..., n\}$, and choose the smallest set cover.

$\mathcal{B}$ reduces $\langle Q, X\rangle$ to an instance $\langle G = (T \cup W, P, E), s\rangle$ of BCP_$\mathcal{C}_{min}$, where $P = X \cup \{0\}$, T contains k vertices $q_1, q_2, ..., q_k$ for each set $q \in Q$ and $k = \lceil \frac{\ln n}{2} \rceil$, and W is a set containing k new vertices. For each $q \in Q$ and $i = 1, 2, ..., k$, we create an edge $(q_i, 0) \in E$ and edges $(q_i, x) \in E$ if $x \in q$. We let $s = kb + k$.

We claim that $\mathcal{C}^*_{min}(G, s) \geq k$ if $\langle Q, X\rangle$ has a set cover Q' of size b. To justify this, from Q', we build the following balanced cover D: add k vertices $q_1, q_2, ..., q_k$ to D if $q \in Q'$, and all vertices in W to D; thus $|D| = kb + k = s$. For each $x \in X$, the k copies of Q' ensure that $\deg_D(x) \geq k$, while the k vertices in W ensure that $\deg_D(x) \leq s - k$.

If $\langle Q, X\rangle$ has a set cover of size b, then algorithm $\mathcal{A}$ on input $\langle G, s\rangle$ will return a balanced cover D with objective function value at least $\mathcal{C}^*_{min}(G, s) - \frac{1-\epsilon}{2} \ln n \geq k - (1 - \epsilon)k = \epsilon k$. We have $|D \cap W| \geq \epsilon k$, because $0 \in X$ is adjacent to every vertex in T and D has at least ϵk vertices not adjacent to 0. Therefore $|D \cap T| \leq s - \epsilon k = kb + (1 - \epsilon)k$. Thus, since each $x \in P$ is adjacent to at least one vertex in D (in fact, at least ϵk), the collection of sets $\{q : (\exists i) q_i \in D\}$ forms a set cover of X of size at most $kb + (1 - \epsilon)k \leq (2 - \epsilon)kb \leq (1 - \epsilon/2)(\ln(n) + O(1))b \leq (1 - \Omega(\epsilon)) \ln(n)b$, as claimed.

The algorithm $\mathcal{B}$ clearly runs in polynomial time, and is a $((1 - O(1)) \ln n)$-approximation algorithm for the Set Cover problem. Thus the theorem follows. □

3. A Randomized Rounding Algorithm

In this section we present our randomized algorithm RRBC for BCP_$\mathcal{C}_{min}$. Given $G = (C, P, E)$, let A be the $m \times n$ adjacency matrix of G, that is $a_{ij} = 1$ iff $(c_i, p_j) \in E$; otherwise $a_{ij} = 0$. Then BCP_$\mathcal{C}_{min}$ is equivalent to the integer linear program MinIP:

$$\begin{aligned} \text{maximize:} \quad & z \\ \text{subject to:} \quad z &\leq \textstyle\sum_{i=1}^m a_{ij} x_i \quad \forall j = 1, ..., n \\ z &\leq \textstyle\sum_{i=1}^m (1 - a_{ij}) x_i \quad \forall j = 1, ..., n \\ \textstyle\sum_{i=1}^m x_i &\leq s \\ x_i &\in \{0, 1\} \quad \forall i = 1, ..., m \end{aligned}$$

RRBC first relaxes the last constraint to $0 \leq x_i \leq 1$ to obtain the linear program MinLP, and then computes an optimal solution x_i^*, $i = 1, 2, ..., m$, of MinLP. Next, applying randomized rounding, RRBC computes an integral solution $X_1, ..., X_m$ by choosing $X_i = 1$ with probability x_i^* and 0 otherwise. Note that this solution may not be feasible since $\sum_{i=1}^m X_i$ may exceed s. Let $L = \sum_{i=1}^m X_i - s$. If $L > 0$, RRBC changes L arbitrary variables $X_i = 1$ to 0, obtaining a feasible solution $\tilde{X}_1, ... \tilde{X}_m$.

4. Analysis of RRBC for BCP_$\mathcal{C}_{min}$

We denote by $\mathcal{C}^{\text{RRBC}}_{min}(G,s)$ the value of the objective function computed by RRBC, that is $\mathcal{C}^{\text{RRBC}}_{min}(G,s) = \tilde{Z} = \min_{j=1}^{n}\{\sum_{i=1}^{m} a_{ij}\tilde{X}_i, \sum_{i=1}^{m}(1-a_{ij})\tilde{X}_i\}$.

Lemma 4.1. *For any instance $\langle G,s\rangle$ of BCP_$\mathcal{C}_{min}$, with probability at least $\frac{1}{2}$, $\mathcal{C}^{\text{RRBC}}_{min}(G,s) \geq \mathcal{C}^{*}_{min}(G,s) - O\left(\sqrt{\mathcal{C}^{*}_{min}(G,s)\ln n} + \sqrt{s}\right)$.*

Proof. Let $z^* = \min_{j=1}^{n}\{\sum_{i=1}^{m} a_{ij}x_i^*, \sum_{i=1}^{m}(1-a_{ij})x_i^*\}$ be the optimum solution of MinLP. Let also $Z = \min_{j=1}^{n}\{\sum_{i=1}^{m} a_{ij}X_i, \sum_{i=1}^{m}(1-a_{ij})X_i\}$.

The $\{X_i\}$ are independent random variables with $\text{Exp}[X_i] = x_i^*$ so, for each j, $\text{Exp}[\sum_{i=1}^{m} a_{ij}X_i] = \sum_{i=1}^{m} a_{ij}x_i^* \geq z^*$. By a standard Chernoff bound, we get $\Pr[\sum_{i=1}^{m} a_{ij}X_i \leq (1-\lambda)z^*] \leq e^{-\lambda^2 z^*/2}$, where $0 < \lambda \leq 1$. Similarly, for all j, $\Pr[\sum_{i=1}^{m}(1-a_{ij})X_i \leq (1-\lambda)z^*] \leq e^{-\lambda^2 z^*/2}$. By the naive union bound, $\Pr[Z \leq (1-\lambda)z^*] \leq 2ne^{-\lambda^2 z^*/2}$.

Likewise, $\text{Exp}[\sum_{i=1}^{m} X_i] = \sum_{i=1}^{m} x_i^* \leq s$. Thus, by the Chernoff bound, $\Pr[\sum_{i=1}^{m} X_i \geq (1+\epsilon)s] \leq e^{-\epsilon^2 s/4}$, where $0 < \epsilon \leq 2e-1$. Letting $L = \sum_{i=1}^{m} X_i - s$, we have $\Pr[L \geq \delta\sqrt{s}] \leq e^{-\delta^2/4}$, where $0 < \delta \leq (2e-1)\sqrt{s}$.

Since $\tilde{Z} \geq Z - L$, and using the above estimates, we get $\Pr[\tilde{Z} \leq (1-\lambda)z^* - \delta\sqrt{s}] \leq \Pr[Z \leq (1-\lambda)z^*] + \Pr[L \geq \delta\sqrt{s}] \leq 2ne^{-\lambda^2 z^*/2} + e^{-\delta^2/4}$.

Choosing $\lambda = \sqrt{2\ln(8n)/z^*}$ and $\delta = \sqrt{4\ln 4}$, we get that $\tilde{Z} \geq z^* - \sqrt{2\ln(8n)z^*} - \sqrt{4\ln(4)s}$ with probability at least $\frac{1}{2}$, as long as when $z^* \geq 2\ln(8n)$. This inequality is also trivially true for $z^* < 2\ln(8n)$. This, together with the bound $\mathcal{C}^{*}_{min}(G,s) \leq z^*$, imply the lemma. □

5. Experimental Analysis

We implemented Algorithm RRBC using *LP_SOLVE* solver [2], and tested its performance on both synthetic and real data sets.

Synthetic data. We used random data sets composed of four adjacency matrices of sizes $(m,n) = (100,30), (100,100), (200,60), (200,200)$, respectively, where each element of the matrix is chosen to be 1 or 0 with probability $\frac{1}{2}$. We ran RRBC for $s = 20, 21, ..., 90$, and compared its solution to the optimal solution of the linear program MinLP.

Table 1. Performance of RRBC on synthetic data with $m = 100$ and $n = 30$.

s	20	25	30	35	40	45	50	55	60	65	70	75	80	85	90
MinLP	10	12.5	15	17.5	20	22.5	25	27.5	29.82	31.89	33.88	35.76	37.54	39.11	40
RRBC	7	10	13	15	18	19	23	25	28	30	32	34	35	38	40

Table 1 shows some results on the comparison of RRBC's solution and the MinLP solution from the experiment in which $m = 100$ and $n = 30$. This table presents only the performance of a single run of RRBC, so the results are likely to be even better if we run RRBC several times and choose the best solution.

We also repeated our simulation test 10 times for each of the above settings and took the average of them. Figure 1 illustrates results of these experiments.

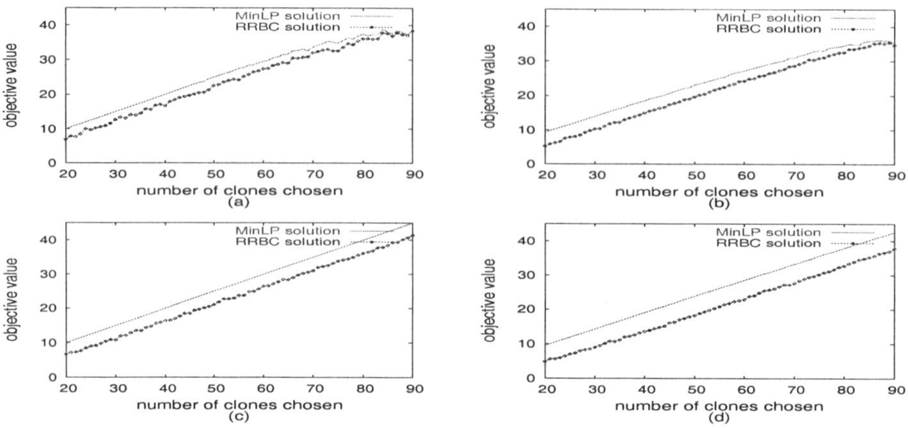

Figure 1. RRBC's performance on synthetic data for four matrices: (a) $(m, n) = (100, 30)$; (b) $(m, n) = (100, 100)$; (c) $(m, n) = (200, 60)$; (d) $(m, n) = (200, 200)$.

Note that (on average) RRBC was always able to find the solution close to that of MinLP. Furthermore, the true optimum (integral solution) could be smaller than the MinLP solution, so our approximation could be even closer to the true optimum than it appears.

Real data. To test the performance of RRBC with real data, we used four clone-probe adjacency matrices. The first two matrices were obtained from 500 bacterial clones extracted from rRNA genes analyzed in [10], along with two sets of 30 and 40 probes designed using the algorithm in [3], respectively. The other two matrices have similar settings, but used fungal clones of rRNA genes studied in [9]. For each of these four data sets we tested RRBC for $s = 200, 210, ..., 400$, running it 10 times and taking the average. We observe that in 92.8% cases, RRBC found a solution at least as large as 97% of MinLP's optimum. The results are shown in Figure 2.

Figure 2 shows that solutions found by RRBC are even closer to MinLP solution than those for synthetic data, for both bacterial and fungal data sets, sometimes even coinciding with MinLP solutions, i.e., RRBC achieved the optimum in some cases.

Our experiments were performed on a machine with Intel Pentium 4 2.4GHz CPU and 1GB RAM. The total running time for each single run of RRBC on these synthetic and real data sets was in the range of $20 - 80$ seconds, which is practically acceptable.

6. An Alternative Algorithm

The performance bound given for RRBC is not quite satisfactory for instances where the optimum is small compared to s. We now provide an alternative algorithm RRBC2, which

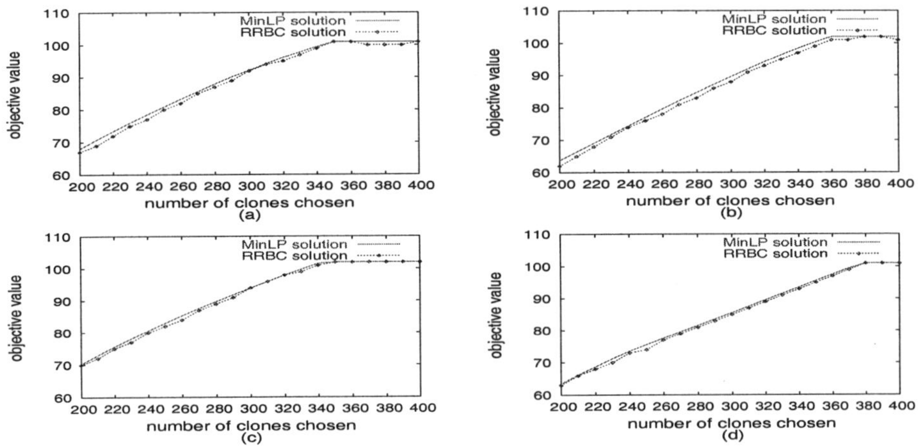

Figure 2. RRBC's performance for real data: (a) 500 bacterial clones and 30 probes; (b) 500 bacterial clones and 40 probes; (c) 500 fungal clones and 30 probes; (d) 500 fungal clones and 40 probes.

is identical to RRBC in all steps except for the rounding scheme: choose $X_i = 1$ with probability $(1-\epsilon)x_i^*$, and 0 otherwise, where $\epsilon = \min\left\{2\sqrt{\ln(4n+2)/z^*}, 1\right\}$. All notations are defined similarly as these in Section 4.

Lemma 6.1. *For any instance $\langle G, s\rangle$ of BCP_$\mathcal{C}_{min}$, with probability at least $\frac{1}{2}$,* $\mathcal{C}^{RRBC2}_{min}(G,s) \geq \mathcal{C}^*_{min}(G,s) - O\left(\sqrt{\mathcal{C}^*_{min}(G,s)\ln n}\right)$.

Proof. The $\{X_i\}$ are independent random variables with Exp$[X_i] = (1-\epsilon)x_i^*$, and $\sum_{i=1}^m x_i^* \leq s$. By linearity of expectation, Exp$[\sum_{i=1}^m X_i] \leq (1-\epsilon)s$. Thus, by the Chernoff bound, Pr$[\sum_{i=1}^m X_i \geq s] \leq$ Pr$[\sum_{i=1}^m X_i \geq (1+\epsilon)(1-\epsilon)s] \leq e^{-\epsilon^2(1-\epsilon)s/4}$.

Likewise, for each j, Exp$[\sum_{i=1}^m a_{ij}X_i] = \sum_{i=1}^m a_{ij}(1-\epsilon)x_i^* \geq (1-\epsilon)z^*$. Thus by the Chernoff bound, Pr$[\sum_{i=1}^m a_{ij}X_i \leq (1-\epsilon)^2 z^*] \leq e^{-\epsilon^2(1-\epsilon)z^*/2}$. Similarly, for all j, Pr$[\sum_{i=1}^m (1-a_{ij})X_i \leq (1-\epsilon)^2 z^*] \leq e^{-\epsilon^2(1-\epsilon)z^*/2}$.

As $z^* \leq s/2$, we have $s/4 \geq z^*/2$. Hence, by the union bound, the probability that any of these $2n+1$ events happens is at most $(2n+1)e^{-\epsilon^2(1-\epsilon)z^*/2}$. Since $(1-\epsilon)^2 \geq 1-2\epsilon$, for $\epsilon < \frac{1}{2}$, we have $\tilde{Z} \geq z^* - 4\sqrt{\ln(4n+2)z^*}$ with probability at lest $\frac{1}{2}$. This bound is also trivially true for $\epsilon \geq \frac{1}{2}$. Finally, since $\mathcal{C}^*_{min}(G,s) \leq z^*$, the lemma follows. □

7. Concluding Remarks

Our work demonstrated that randomized rounding is an effective method for solving the BCP_$\mathcal{C}_{min}$ version of Balanced Covering, especially on real data sets. In the actual implementation available at `http://algorithms.cs.ucr.edu/OFRG/`, the solution of RRBC is fed as an initial solution into a simulated annealing algorithm. We found out that the simulated annealing rarely produces any improvement of this initial solution, which provides further evidence for the effectiveness of randomized rounding in this case.

We have several additional results related to this work that are not described in the paper due to lack of space. For BCP_$\mathcal{C}_{min}$ we improved the hardness approximation bound to $\mathcal{C}^*_{min}(G,s) - O(\log n \log\log n)$. We can show that it is hard to approximate BCP_$\mathcal{C}_{sum}$ with the bound $\mathcal{C}^*_{sum}(G,s) - \beta(mn)^\epsilon$ for any constants $\beta > 0$ and $0 < \epsilon < 1$ even with randomization. Finally, we have a randomized algorithm for BCP_$\mathcal{D}_{max}$ whose solution deviates from the optimum by no more than $O(\sqrt{s}\ln n)$ with probability at least $\frac{1}{2}$. These results will appear in the full version of this paper.

References

1. B. Berger, J. Rompel, and P. Shor. Efficient NC algorithms for set cover with applications to learning and geometry. *30th Annual Symposium on the Foundations of Computer Science*, pages 54–59, 1989.
2. M. Berkelaar, K. Eikland, and P. Notebaert. *Lp_solve mixed integer linear programming solver 5.5*, 2004. Available at `http://lpsolve.sourceforge.net/5.5`.
3. J. Borneman, M. Chrobak, G.D. Vedova, A. Figueroa, and T. Jiang. Probe selection algorithms with applications in the analysis of microbial communities. *Bioinformatics*, 17(1):S39–S48, 2001.
4. U. Feige. A threshold of ln n for approximating Set Cover. *Journal of the ACM (JACM)*, 45(4):634–652, 1998.
5. M.R. Garey and D.S. Johnson. *Computers and Intractability. A Guide to the Theory of NP-Completeness*. W.H.Freeman, New York, 1979.
6. L. Gargano, AA. Rescigno, and U. Vaccaro. Multicasting to groups in optical networks and related combinatorial optimization problems. *International Parallel and Distributed Processing Symposium*, 2003.
7. J. Schuchhardt, D. Beule, A. Malik, E. Wolski, H. Eickhoff, H. Lehrach, and H. Herzel. Normalization strategies for cDNA microarrays. *Nucleic Acids Res.*, 28(10):e47, 2000.
8. L. Valinsky, A. Scupham, G.D. Vedova, Z. Liu, A. Figueroa, K. Jampachaisri, B. Yin, E. Bent, R. Mancini-Jones, J. Press, T. Jiang, and J. Borneman. Oligonucleotide fingerprinting of ribosomal RNA genes (OFRG). In G.A. Kowalchuk, F.J. de Bruijn, I.M. Head, A.D. Akkermans, and J.D. Van Elsas, editors, *Molecular Microbial Ecology Manual*, pages 569–585. Kluwer Academic Publishers, Dordrecht, The Netherlands, 2nd edition, 2004.
9. L. Valinsky, G. Della Vedova, T. Jiang, and J. Borneman. Oligonucleotide fingerprinting of ribosomal RNA genes for analysis of fungal community composition. *Applied and Environmental Microbiology*, 68(12):5999–6004, 2002.
10. L. Valinsky, G. Della Vedova, A. Scupham, S. Alvey, A. Figueroa, B. Yin, R. Hartin, M. Chrobak, D. Crowley, T. Jiang, and J. Borneman. Analysis of bacterial community composition by oligonucleotide fingerprinting of rRNA genes. *Applied and Environmental Microbiology*, 68(7):3243–3250, 2002.

SUBTLE MOTIF DISCOVERY FOR DETECTION OF DNA REGULATORY SITES

MATTEO COMIN*
Dept. Information Engineering, University of Padova, Italy
E-mail: ciompin@dei.unipd.it

LAXMI PARIDA
IBM T. J. Watson Research Center, Yorktown Heights, NY 10598, USA.
E-mail: parida@us.ibm.com

We address the problem of detecting consensus motifs, that occur with subtle variations, across multiple sequences. These are usually functional domains in DNA sequences such as transcriptional binding factors or other regulatory sites. The problem in its generality has been considered difficult and various benchmark data serve as the litmus test for different computational methods. We present a method centered around unsupervised combinatorial pattern discovery. The parameters are chosen using a careful statistical analysis of consensus motifs. This method works well on the benchmark data and is general enough to be extended to a scenario where the variation in the consensus motif includes indels (along with mutations). We also present some results on detection of transcription binding factors in human DNA sequences.
Availability: The system will be made available at www.research.ibm.com/computationalgenomics.
Keywords: pattern discovery, subtle motifs, consensus motifs, transcription factors, binding sites.

1. Introduction

The problem of detecting common motifs across DNA sequences for locating regulatory sites, transcription binding factors or even drug target binding sites is of prime importance. The main difficulty is that these motifs have subtle variations at each occurrence. This problem has been of interest to both biologists and computer scientists. A satisfactory practical solution has been elusive although the problem is defined very precisely:

Problem 1. ***(The Consensus Motif Problem):*** Given t sequence s_i on an alphabet Σ, a length $l > 0$ and a distance $d \geq 0$, the task is to find all patterns p, of length l that occur in each s_i such that each occurrence p_i' on s_i has at most d mismatches with p.

The problem in this form made its first appearance in 1984 [16]. In this discussion, the alphabet Σ is $\{A, C, G, T\}$ and the problem is made difficult by the fact that each occurrence of the pattern p may differ in some d positions and the occurrence of the consensus pattern p may not have $d = 0$ in any of the sequences. In the seminal paper [16], Waterman and coauthors provide exact solutions to this problem by enumerating *neighborhood* patterns, i.e., patterns that are at most d Hamming distance from a candidate pattern. Sagot gives a good summary of the (computational) efforts in [14] and offers a solution that improves the time complexity of the earlier algorithms by the use of generalized

*Work done during internship at IBM Thomas J. Watson Research Center, Yorktown Heights, NY 10598, USA.

suffix trees. These clever enumeration schemes, though exact, have a drawback that they run in time exponential in the pattern length.

This problem of detecting common subtle patterns across sequences is nevertheless of great interest and various statistical and machine learning approaches, which are inexact but more efficient, have been proposed [9,10,3,6]. One of the questions that can be asked to compare and test the efficacy of such methods of consensus motif detection systems is: *Given a set of sequences that harbor (with mutations) k motifs , what percentage of the k motifs does the system recover?* When k is large, all of the above approaches give good average-case performance under this criterion.

Yet another question to ask is : *Given a set of sequences that harbor (with mutations) ONE motif p , does the system recover p?* This is a rather difficult criterion to meet since these algorithms use some form of local search based on Gibbs sampling or expectation maximization or even clever heuristics. Hence it is not surprising that they may miss p. However, a question of this form is a biological reality. Consider the following, somewhat contrived, variation of Problem 1 which is an attempt at simplifying the computational problem.

Problem 2. ***(The Planted** (l,d)**-motif problem):*** Given t sequence s'_i on Σ, a pattern p of length l is embedded in s'_i, with exactly d errors (mutations), to obtain the sequence s_i of length n, for each $1 \leq i \leq t$. The task is to recover p, given s_i, $1 \leq i \leq t$ and the two numbers l and d.

Pevzner and Sze tantalized the community with the *challenge problem*, which was Problem 2 with parameters $n = 600$, $t = 20$, $l = 15$ and $d = 4$ [11]. A thrust of this paper also was the need for the deployment of combinatorial approaches to tackle this thorny problem. One of the algorithms they presented was an exact algorithm where the challenge problem was reduced to finding a t-sized clique in a t-partite graph with at most $n - l + 1$ vertices in each partition. Even the best known heuristics for clique finding problem failed to detect the clique corresponding to the signal. The second algorithm was based on enumerating possible patterns and checking their candidacy for being the subtle pattern using clever heuristics and an exhaustive search in a reduced space. A similar algorithm, with different heuristics was presented in [12,7].

One of the most effective algorithms, we found, was the one discussed by Buhler and Tompa [4]. The probabilistic algorithm uses a random projection h and hashes each input l-mer x into bucket $h(x)$. Any hash bucket with sufficiently many entries is explored as a potential embedded motif. This approach solved the challenge problem and some more. There has been a flurry of activity around this problem of subtle motifs [5,7,8]. See also [13] for some practical implementations of exact approaches.

Overview of our approach. We first clarify the different "motifs" used in this paper: Our central goal is to detect the *consensus* or the *embedded* or the *planted* motif in the given data sets which is also sometimes referred to as the *signal* in the data or the *subtle signal.* When a motif is not qualified with these terms, it refers to a substring that appears in multiple sequences, with possible wild cards.

We propose an approach that uses unsupervised motif discovery to solve Problem 2. We show that this method works well for the more general Problem 1 as well. Recall that the signal ("subtle motifs") is embedded in t random sequences. The problem is compounded by the fact that although the consensus motif is solid (i.e., an l-mer without wild cards or dont-care characters), it is not necessarily contained in any of the t sequences. However, if we can obtain a correct alignment of the m sequences, then it is relatively easy to extract the consensus motif satisfying the (l, d) constraint. In other words, one of the difficulties of the problem is that the sequences are unaligned. The extent of similarity across the sequences is so little that any global alignment scheme cannot be employed. So we tackle this problem in two steps: First, we identify *potential signal (PS)* segments of interest in the input sequences. This is done by using the imprints of the discovered motifs on the input. Second, amongst these segments, we carry out an exhaustive comparison and alignment to extract the consensus motif. This delineation into two steps helps us also address a more realistic version of the problem that includes insertion and deletion in the consensus motif:

Problem 3. ***(The Indel Consensus Motif Problem):*** Given t sequence s_i on an alphabet Σ, a length $l > 0$ and a distance $d \geq 0$, the task is to find all patterns p, of length l that occur in each s_i such that each occurrence p'_i on s_i is at an edit distance (mutation, insertion, deletion) at most d from p.

The main focus of our method is in obtaining good quality PS segments and restricting the number of such segments to keep the problem tractable. The Type I error or false negative errors, in detecting PS segments, are reduced by using appropriate parameters for the discovery process based on a careful statistical analysis of consensus motifs which is discussed in Section 2. The Type II error or false positive errors are reduced by using irredundant motifs [2] and their statistical significance measures [1] discussed in Section 3.1. Loosely speaking, irredundancy helps to control the extent of over-counting of patterns and the pattern-statistics helps filter the true signal from the signal-like-background. In the scenario where indels (insertions and/or deletes) are permitted along with mutations, the unsupervised discovery process detects *extensible* motifs (instead of *rigid* motifs that have a fixed imprint length in all the occurrences). Also, the second step uses *gapped* alignments.

2. Statistics of consensus motifs

Here we make some calculations, under simplifying assumptions, to justify our unsupervised motif discovery approach to the problem. We consider the most general version of the problem which is formally stated as Problem 3 in the last section. Recall that this setting permits insertion and deletion as well as mutation in the embedded motif.

Given t sequences of length l each, a pattern satisfies *quorum* K if it occurs in $K' \geq K$ of the given t sequences. Further it is of *maximal* size h, if in each of the K' occurrences, the size cannot be increased without decreasing the number of occurrences K' (see [1] for a more rigorous definition).

For simplicity, the sequences are the same length l as the consensus motif and all the t sequences are aligned and we will further assume that a pattern occurs at most once in each sequence. Let q be the probability of any position in the input data to be contained in a pattern and let $P_{maximal}(K, H, q)$ be the probability that a pattern with maximal H solid characters and quorum K occurs in the input data. Then [a]

$$P_{maximal}(K, H, q) = \sum_{k=K}^{t} \binom{t}{k} \left(1 - q^H\right)^{t-k} q^{H\,k}(1 - q^k)^{l-H} \tag{1}$$

Let $Z_{K,q}$ be a random variable denoting the number of maximal motifs with quorum K and q as defined above, and, $E(Z_{K,q})$ denotes the expectation of $Z_{K,q}$. Note that for maximal motifs, it is the case that the occurrences of two distinct motifs are independent events. Further, using linearity of expectations, we obtain (for a fixed t and l),

$$E(Z_{K,q}) = \sum_{h=1}^{l} \binom{l}{h} P_{maximal}(K, h, q) \tag{2}$$

Computing q. Consider the case where the embedded motif is constructed with some d edit operations. Let the edit operations be (1) mutation, (2) deletion and (3) insertion. Let q_M be the probability of mutation, q_X the probability of deletion and q_I the probability of insertion with $q_M + q_X + q_I = 1$.

Note that for simplicity we have assumed that the t sequences are aligned. For example, the table on the left below shows exactly one edit applied to the signal motif and the table on the right shows

[a] Note that if the motifs are not maximal then $P(K, H, q) = \sum_{k=K}^{t} \binom{t}{k} \left(1 - q^H\right)^{t-k} q^{H\,k}$. Also if motif m_1 is a maximal version of motif m_2 then the occurrences of m_1 and m_2 are not independent.

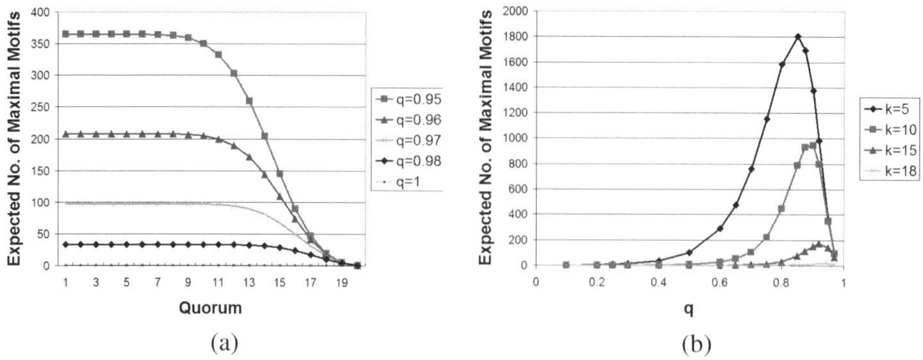

Figure 1. For $t = 20$, $l = 20$, the expected number of maximal motifs $E(Z_{K,q})$, is plotted against (a) quorum K shown along the X-axis (for different values of q), and, (b) against q shown along the X-axis (for different values of quorum K).

the alignment of the embedded motifs.

Edits	signal = ACGTAC						
M	A	C	G	T	C	C	
X	A	G	T	A	C		
I	A	C	G	A	T	A	C
M	A	C	C	T	A	C	
M	G	C	G	T	A	C	

Alignment						
A	C	G	-	T	c	C
A	-	G	-	T	A	C
A	C	G	a	T	A	C
A	C	c	-	T	A	C
g	C	G	-	T	A	C

Assume that d out of the l positions are picked at random on the embedded motif for exactly one of the edit operations, insertion, deletion or mutation. Then it is easy to see that probability q of a position to be contained in a motif is:

$$q = 1 - \frac{d}{l}\left(q_M + q_X\right) \tag{3}$$

Now, it is straightforward to compute the value of q, to estimate $E(Z_{K,q})$ of Equation (2), given different scenarios. For example, consider the following cases.

(1)	Exactly d mutations	$q_M = 1$	$q = 1 - d/l$
(2)	Exactly d edits	$q_M = q_X = q_I = 1/3$	$q = 1 - 2d/3l$

Also note that when *no more than d' edit operations* are carried out on the embedded motif, it is usually interpreted as each collection of $0, 1, 2, \ldots, d'$ positions being picked with equal probability, and thus $d = d'/2$ for Equation (3).

2.1. *Rationale for using unsupervised motif discovery*

A motif of length l that occurs across $t' \leq t$ sequences provides a local alignment of length l for the t' sequences which, in a sense, justifies the simplified scenario of the last section. The best case scenario, for our problem, is when the embedded motif m is identical in all t sequences and the discovery process detects this *single* maximal motif with quorum t. So the scenarios closer to the best case should have fewer (but important) maximal motifs. Figure 1(a) shows the expected number of motifs with different values of q and quorum K. Notice that the expected number of motifs saturates for small values of K and falls dramatically as K increases. The saturation at lower values occurs since we are seeking *maximal* motifs. Thus as q increases the saturation occurs at a higher value of K. Figure 1(b) shows the variation of the expected number of maximal motifs with q which is unimodal, for different values of K. The value of q is determined by the given problem scenario and thus a large value of K is a good handle on controlling the number and "quality" of maximal motifs.

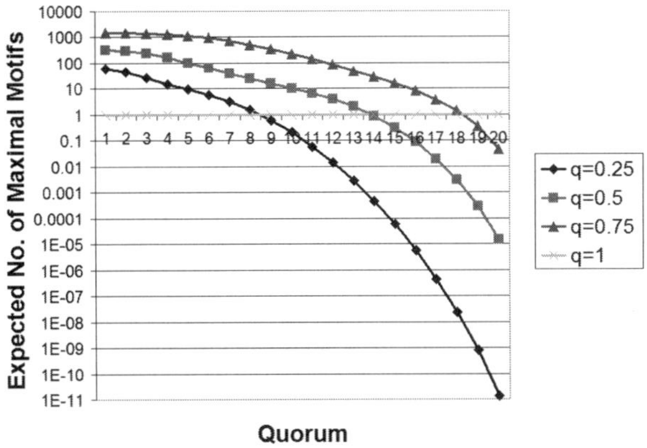

Figure 2. For $t = 20$, $l = 20$, the expected number of maximal motifs $E(Z_{K,q})$, is plotted against quorum K shown along the X-axis, for different values of q, in a logarithmic scale. Notice that when $q = 1$, the curve is a horizontal line at $y = 1$. Note that for DNA sequences, $q = 0.25$ corresponds to the random input case.

The *signal* is embedded in the *background* and it is important to exploit the characteristics that distinguishes one from the other. In our case, we assume that the background is random, in other words it is assumed to be randomly generated using an i.i.d process. Under this condition, it is easy to see that $q = 1/4$. Thus we need to compare $E(Z_{K,q})$ with $E(Z_{K,1/4})$, the expectation for the random case. To compare these expectation curves, particulary around small values (close to 1 in the Y-axis), we study the plots of $\log(E(Z_{K,q}))$ against quorum K in Figure 2.

For example, consider the case when $q = 0.75$; this is the value of q for the *challenge problem* of Section 1. In Figure 2, this is shown by the red curve and for large K, say $K \geq 16$, the expected number of motifs is small. Also, the corresponding expected numbers for the random case is extremely low, thus providing a strong contrast in the number of expected motifs. Hence the reasonable choice for the quorum parameter K is 16 or more, in the unsupervised discovery process.

Before we conclude this section, we must point out that in the case where the embedded motif is changed with insertions and/or deletions (indels), the q value is computed appropriately using Equation (3) and the corresponding expectation curve in Figure 2 is studied. However, the burden is heavier for the unsupervised discovery process and we use the extensible (or, variable-sized gaps) motif discovery capability in Varun [1].

3. Subtle Motif Finder: Our Approach

Here we present our approach, *SubtleMotif*, which detects the consensus motifs in two steps. We first locate regions in the sequence called *potential signal* (PS) segments. The statistical analysis of the previous section suggests that the detection of PS segments via unsupervised motif discovery is indeed possible. The two important parameters in the combinatorial discovery process are K and D: K is the quorum or the minimum number of sequences where the pattern must occur and D is the size of the gap between any two solid characters in the pattern. In the second step we carry out a local alignment of these short segments and extract the consensus motif.

3.1. *(Step 1) Detecting PS segments*

As seen in the last section, we expect to see more maximal patterns in the signal region than the background in an appropriate range of quorum K. We extract all common motifs across sequences using an unsupervised combinatorial motif discovery process. We use the system Varun [1] for this purpose. This allows us to discover motifs with "dont-cares" or wild-cards. The number of such characters is controlled by the parameter D in Varun, which is a bound on the number of "dont-cares" between any two solid characters in a pattern. Next, we simply count the number of motifs that cover a position i on the input. The first prediction of the PS segments are the positions (i's) with high counts.

This elementary rule works well for simple cases like Problem 2 with $n = 600$, $t = 20$, $l \leq 10$ and $d = 2$. Here the PS segments are predicted accurately. However, for $d > 2$ we found it is difficult to distinguish the true from the false PS segments using this simple approach. To weed out these wrong PS segments, we explored other means of pruning the motifs using some combinatorial and statistical approaches. Firstly, we use the idea of *irredundant* or *basis* motifs [2], to avoid overcounting of patterns that cover the same region multiple times on the sequence. Secondly, we consider only those motifs that have a significant z-score and also, biased the motif count at a position i on the input with the probability of the occurrence of that motif. Due to space constraints, we omit the discussion on irredundant motifs and their statistical significance evaluations and instead direct the reader to [2,1]. We use Varun to discover irredundant motifs in the input data. In the right, the motif discovery parameters are K and D and $l = 15$, $d = 4$, $t = 20$, $n = 600$ and the value of q is $11/15 \approx 0.73$, using Equation (3). Column I shows the number of correct PS segments predicted using *all* motifs and column II shows the same using only *irredundant motifs*. In all the cases, there is an increase in the number of correctly detected positions for the latter.

K	D	I	II
20	2	2	3
19	2	0	1
20	3	3	4
19	3	1	2
20	4	2	5

We compute the z-score of each irredundant motif using our previous result (Equation (5) in [1]) and filter these motifs based on a cut-off threshold z-score. We further use a weighted count for each input position in the imprint of the motif m, where the weight is $(1/p_m)$ and p_m is computed as in Equation (4) in [1]. Figure 3 shows the results for a variety of settings comparing the use of statistical methods (both z-score and weighted counting), called Method II, with the one that does not use them, called Method I.

Notice that using Method II, we can restore all 10 positions of the $n = 200$, $t = 20$, $l = 10$ and $d = 2$ of Problem 2. In the experiments for $l = 15$ and $d = 4$, we can recover 4 positions correctly out of 20. We find that only in two cases Method I recovers more PS segment positions than Method II. However, in all the remaining 22 cases, Method II outperforms Method I.

Since it is very difficult to detect 100% of the PS segments correctly in this step alone, we use these partial PS segments in the next step to reconstruct the true signal.

3.2. *(Step 2) Processing PS Segments*

In the previous step we identified the potential signal (PS) segments in the input. Next, we merge the information form each sequence by combining different PS segments. Assuming that the PS segment is predicted correctly, the planted motif is embedded in this segment. If the length of the consensus motif is known, say l, then the PS segment is constrained to be substring of length $2 \times l$. Thus given a candidate position i in sequence s, the signal is contained in the interval $s[i - l, i + l]$.

We next pick one PS segment from each sequence to "locally align" the segments across some C sequences. We enumerate all the $\binom{t}{C}$ configurations here. Let the C PS segments, each from a distinct sequence, be given as $(s_{i_1}[b_{i_1}, e_{i_1}], s_{i_2}[b_{i_2}, e_{i_2}], ..., s_{i_C}[b_{i_C}, e_{i_C}])$. We make the assumption that the starting position x_{i_j} of the consensus motif in sequence s_{i_j} lies in the substring $s_{i_j}[b_{i_j}, e_{i_j}]$, i.e., $b_{i_j} \leq x_{i_j} \leq e_{i_j}$. We are seeking all possible alignments of length l using these PS segments. We use the following measure to evaluate an alignment. The *majority* string s_m, of length l, is simply the string obtained by using the majority base at each aligned position (column). The score f is the sum total of the aligned positions in all the C segments that agree with s_m. For example, consider the aligned segments here below where $l = 8$, $d = 3$, $C = 5$, and $f = 27$.

$$\begin{array}{ll}
(1) & \text{---}A\ C\ T\ G\ C\ T\ C\ C\text{---} \\
(2) & \text{---}A\ G\ G\ G\ T\ T\ G\ A\text{---} \\
(3) & \text{---}C\ C\ G\ G\ T\ T\ G\ A\text{---} \\
(4) & \text{---}C\ C\ T\ C\ T\ A\ C\ A\text{---} \\
(5) & \text{---}A\ C\ G\ G\ T\ -\ C\ A\text{---} \\
s_m = & \mathbf{A\ C\ G\ G\ T\ T\ C\ A}
\end{array}$$

Since our first step is very tightly controlled, we found in practice that there are only a few candidate PS segments. Also, in the model that uses insertion and deletion (i.e., the length of the imprint of the occurrence of the consensus motif in each sequence is not necessarily l), we use the same score by keeping track of the alignment columns: deletions and insertions result in gaps in some sequences in the alignment (see sequence 5 in the above example). We consider all those alignments, whose score f exceeds a fixed threshold T_C. In all our experiments we have used $C = 3$ and the values of T_C are reported in the experiments.

Extracting the consensus motif across t sequences. At the previous step, we have multiple alignments, where each alignment is across some $C(\leq t)$ sequences. From these we need to extract the consensus motif across all the t sequences. For each alignment, we designate the majority substring s_m (see last section) as the putative consensus motif. Then we scan all the t input strings for the occurrence of s_m with at most d errors which can be done in linear time. For each sequence, we pick the best occurrence, i.e., the one with the minimum edit distance from s_m. In practice, this step very quickly discards the erroneous consensus motifs and quickly converges to the one(s) satisfying the distance constraint of d.

$l = 10, d = 2$

Motif params K	D	M	Methods I	II
10	2	95	8	7
9	2	236	8	10
8	2	434	7	8

(a) $n = 100, t = 10$

Motif params K	D	M	Methods I	II
20	2	281	10	10
19	2	459	12	13
18	2	588	18	18

(b) $n = 200, t = 20$

$t = 20, l = 15, d = 4$

Motif params K	D	M	Methods I	II
20	2	539	2	4
19	2	647	6	7
18	2	837	12	12
20	3	952	5	6
19	3	1164	11	10
18	3	1582	13	13
20	4	1454	8	9
19	4	1832	9	10
18	4	2577	11	11

(c) $n = 300$

Motif params K	D	M	Methods I	II
20	2	1588	2	2
19	2	3526	1	1
18	2	5456	1	1
16	2	7316	1	2
20	3	3348	4	4
19	3	7885	2	2
18	3	12444	1	1
17	3	15318	2	3
16	3	17017	0	1

(d) $n = 400$

Figure 3. Number of PS segment positions predicted correctly using Methods I and II for different parameters. The motif discovery parameters are K and D and M is the total number of irredundant motifs discovered in the input. The values of q, obtained using Equation 3, are as follows: (a) & (b) $q = 0.8$, (c) & (d) $q = 0.73$.

4. Results

Let P be the set of all positions covered by the prediction and S be the same set for the embedded motif. The score of the prediction P, with respect to the embedded motif, can be given as (see [15]): $score = \frac{|P \cap S|}{|P \cup S|}$. The score is 1 if the prediction is 100% correct. However, even for values much smaller than 1, the embedded motif may be computed correctly. This measure is rather stringent and so we use yet another measure, the *solution coverage* (SC) score. This is defined as the number of sequences that contains at least one occurrence of the predicted motif whose distance from the prediction is within the problem constraint i.e., bounded by d. Again if the coverage is equal to the total number of sequences t, then the prediction can be considered 100% correct.

Results on benchmark synthetic data. We report our results in terms of these two measures in Figure 4 averaged over eight random experiments. Each experiment is defined by the four parameters n, t, l and d. In the unsupervised motif discovery process of the first step we use parameters $K = t = 20$ and $0 < D < 4$. The high K value was suggested by the statistical analysis in Section 2 and confirmed by our experiments in Section 3.1. In the second step we use $C = 3$ based on our experiments reported in Figure 3. In Figure 4(a),(b) and (c), we show the performance measures for various instances of Problem 2. We compare our results with what we found as the best performing algorithm, PROJECTION [4]. In all cases our best results are similar, or slightly better, than PROJECTION as shown in Figure 4. We observe that as we increase the number of gaps D, the $score$ improves. In particular if $D = 0$ (i.e., solid motifs), the chances of success drops dramatically. We observe a similar tendency in Problem 3 as shown in Figure 4(d) and (e). Although this version of the problem, with indels, should be harder, we find that the method gives surprisingly good results.

K	D	N	Score	SC
20	1	2	0.066	10
20	2	2	0.415	12
20	3	4	0.95	20
20	4	3	0.94	20

(a) $l = 15, d = 4$
$Score_{PRJ} = 0.93$

K	D	N	Score	SC
20	0	0	0.02	10
20	1	1	0.49	11
20	2	1	0.8	20
20	3	1	0.93	20
20	4	2	0.91	20

(b) $l = 17, d = 5$
$Score_{PRJ} = 0.93$

K	D	N	Score	SC
20	1	2	0.75	11
20	2	2	0.95	20
20	3	4	0.95	20

(c) $l = 19, d = 6$
$Score_{PRJ} = 0.96$

K	D	N	Score	SC
20	0	1	0.05	5
20	1	3	0.75	20
20	2	3	0.81	20

(d) $l = 15$, 3 mutations & 1 indel

K	D	N	Score	SC
20	0	1	0.09	8
20	1	3	0.68	20
20	2	4	0.78	20

(e) $l = 15$, 2 mutations & 2 indels

Figure 4. In all cases, $t = 20$, $n = 600$. The motif discovery parameters are K and D and we use $C = 3$ and the values of T_C are as follows: (a) 32 (b) 36 (c) 40 (d) & (e) 30. The results are are averaged over 8 random problem instances. N is the total number of PS segments predicted correctly. See text for definitions of $Score$ and SC. $Score_{PRJ}$ is the score for the PROJECTION algorithm by Tompa et al.

Results on Human hm01r data. We have tested the system on various real data sets and we give details of one such case- that of detecting transcription binding factors on human DNA sequences on the data set suggested by Tompa [15]. The details are as follows:

No	pos	$Predictions$	M	I
0	-101	$T\ G\ A\ C\ G\ T\ C\ A$	$-$	1
1	-299	$T\ G\ C\ -\ G\ T\ C\ A$	1	$-$
2	-71	$T\ G\ A\ C\ A\ T\ C\ A$	1	1
3	-69	$A\ T\ G\ A\ -\ G\ T\ C\ A\ G$	$-$	2
4	-527	$T\ G\ C\ G\ A\ T\ G\ A$	2	1
6	-173	$T\ G\ A\ -\ C\ T\ A\ A$	2	$-$
7	-1595	$T\ G\ A\ -\ A\ T\ G\ A$	2	$-$
8	-221	$T\ G\ G\ -\ G\ T\ C\ T$	2	$-$
9	-69	$T\ G\ A\ -\ C\ T\ G\ C$	3	$-$
10	-105	$T\ G\ A\ -\ A\ T\ C\ A$	1	$-$
12	-780	$T\ G\ C\ -\ G\ T\ C\ A$	1	$-$
14	-1654	$A\ T\ G\ A\ -\ A\ T\ C\ A$	1	1
15	-69	$A\ T\ G\ A\ -\ G\ T\ C\ A\ A$	$-$	2
16	-97	$T\ G\ A\ -\ G\ T\ A\ A$	1	$-$
17	-1936	$A\ T\ G\ A\ -\ A\ T\ C\ A$	1	1
$signal$		**T G A G T C A**		

The parameters for this data set are $n = 2000$, $t = 18$. Note that we had to estimate l and d through a series of trials. l was estimated to be 7 and d to be 3. We use parameters $K = 18$ and $D = 1$ in the motif discovery process in Step 1 and use $C = 3$ and $T_C = 12$ in Step 2. We identify the signal in 15 of the 18 sequences at positions given in the pos column. We miss the signal in only one sequence (sequence no 5) and the signal is absent in two other sequences (no 11 and 13). We reconstruct the consensus sequence as $TGAGTCA$ which is at most 3 edit distance away from the "embedded" signals. In the table M denotes number of mutations and I the number of insertions; no deletions were found.

5. Concluding Remarks

The problem of detecting subtle consensus motifs is tricky and a purely combinatorial or a purely statistical approach has been unsatisfactory (see Section 1). It appears it requires a delicate combination of the two methods. We have presented a method that uses unsupervised combinatorial pattern discovery, followed by a careful statistical refinement and processing. Since we use tried-and-tested tools such as pattern discovery, in the first step, and local alignment, in the second step, we have focussed more on choosing and combining appropriate parameters. Also, extension of the method to handling a more general scenario such as inclusion of indels (insertion and/or deletion) in the embedded motif has been relatively straightforward. We achieved this by using extensible motifs in the pattern discovery process of the first step and gapped alignment in the second step. The results on benchmark data and some real DNA sequences have been very encouraging. We are looking at the yet harder instance of the problem which is the task of finding subtle motifs within the same sequence.

References

1. A. Apostolico, M. Comin, and L. Parida. Conservative extraction of over-represented extensible motifs. *ISMB (Supplement of Bioinformatics)*, 21:9–18, 2005.
2. A. Apostolico and L. Parida. Incremental paradigms for motif discovery. *Journal of Computational Biology*, 11(4):15–25, 2004.
3. T. L. Bailey and C. Elkan. Fitting a mixture model by expectation maximization to discover motifs in biopolymers. pages 28–36, 1994.
4. Buhler and Tompa. Finding motifs using random projections. In *Proceedings of the Annual Conference on Computational Molecular Biology (RECOMB01)*, pages 69–75. ACM Press, 2001.

5. Eleazar Eskin and Pavel Pevzner. Finding composite regulatory patterns in DNA sequences. In *Bioinformatics*, volume 18, pages 354–363, 2002.
6. G. Z. Hertz and G. D. Stormo. Identifying DNA and protein patterns with statistically significant alignments of multiple sequences. *Bioinformatics*, 15:563–577, 1999.
7. Keich and Pevzner. Finding motifs in the twilight zone. In *Annual International Conference on Computational Molecular Biology*, pages 195–204, Apr, 2002.
8. Uri Keich and Pavel Pevzner. Subtle motifs: defining the limits of motif finding algorithms. In *Bioinformatics*, volume 18, pages 1382–1390, 2002.
9. C. E. Lawrence and Reilly A.A. An expectaion maximization (EM) algorithm for the identification and characterization of common sites in unaligned biopolymer sequences. *Proteins: Structure, Function and Genetics*, 7:41–51, 1990.
10. C. E. Lawrence, S. F. Altschul, M. S. Boguski, J. S. Liu, A. F. Neuwald, and J. C. Wootton. Detecting subtle sequence signals: A Gibbs sampling strategy for multiple alignment. *Science*, 262:208–214, Oct, 1993.
11. P. A. Pevzner and S.-H. Sze. Combinatorial approaches to finding subtle signals in DNA sequences. In *Proceedings of the Eighth International Conference on Intelligent Systems for Molecular Biology*, pages 269–278. AAAI Press, 2000.
12. Alkes Price, Sriram Ramabhadran, and Pavel Pevzner. Finding subtle motifs by branching from sample strings. In *Bioinformatics*, number 1, pages 149–155, 2003.
13. S. Rajasekaran, S. Balla, and C.-H Huang. Exact algorithms for planted motif problems. *Journal of Computational Biology*, 12(8):1117–1128, 2005.
14. M. F. Sagot. Spelling approximate repeated or common motifs using a suffix tree. *Latin 98: Theoretical Informatics, Lecture Notes in Computer Science*, 1380:111–127, 1998.
15. Martin Tompa, Nan Li, Timothy L Bailey, George M Church, Bart De Moor, Eleazar Eskin, Alexander V Favorov, Martin C Frith, Yutao Fu, W James Kent, Vsevolod J Makeev, Andrei A Mironov, William Stafford Noble1, Giulio Pavesi, Graziano Pesole, Mireille Rgnier, Nicolas Simonis, Saurabh Sinha, Gert Thijs, Jacques van Helden, Mathias Vandenbogaert, Zhiping Weng, Christopher Workman, Chun Ye, and Zhou Zhu. Assessing computational tools for the discovery of transcription factor binding sites. *Nature Biotechnology*, 23:137–144, 2005.
16. M.S. Waterman, R. Aratia, and D.J. Galas. Pattern recognition in several sequences: Consensus and alignment. *Bulletin of Mathematical Biology*, 46(4):515–527, 1984.

AN EFFECTIVE PROMOTER DETECTION METHOD USING THE ADABOOST ALGORITHM *

XUDONG XIE

Department of Electronic Engineering, City University of Hong Kong, Hong Kong
Department of Electronic Engineering, Tsinghua University, Beijing, China

SHUANHU WU

Department of Electronic Engineering, City University of Hong Kong, Hong Kong
School of Computer Science and Technology, Yantai University, China

KIN-MAN LAM

Department of Electronic and Information Engineering,
The Hong Kong Polytechnic University, Hong Kong

HONG YAN

Department of Electronic Engineering, City University of Hong Kong, Hong Kong
School of Electronic and Information Engineering, University of Sydney,
NSW 2006, Australia

In this paper, an effective promoter detection algorithm, which is called PromoterExplorer, is proposed. In our approach, various features, i.e. local distribution of pentamers, positional CpG island features and digitized DNA sequence, are combined to build a high-dimensional input vector. A cascade AdaBoost based learning procedure is adopted to select the most "informative" or "discriminating" features to build a sequence of weak classifiers. A number of weak classifiers construct a strong classifier, which can achieve a better performance. In order to reduce the false positive, a cascade structure is used for detection. PromoterExplorer is tested based on large-scale DNA sequences from different databases, including EPD, Genbank and human chromosome 22. The proposed method consistently outperforms PromoterInspector and Dragon Promoter Finder.

1 Introduction

In the past decade, many reliable methods have been developed for protein-coding regions prediction [1] in human genome annotation. However, for regulatory regions of genes, exact promoter detection still remains a challenge, and relatively few algorithms have been proposed to tackle the problem. A promoter is the region of a genomic sequence which is close to a gene's transcription start site (TSS), and it largely controls the biological activation of the gene [2]. Therefore, promoter detection can be considered a fundamental and important step in gene annotation.

* This work is supported by research grants from City University of Hong Kong (Projects 9010003 and 9610034).

In order to discriminate a promoter region from non-promoters regions, many different features are considered, such as CpG islands [3, 4], TATA boxes [5, 6], CAAT boxes [5, 6], some specific *transcription factor binding sites* (TFBSs) [5, 6, 7], pentamer matrix [8] and oligonucleotides [9]. And also, various pattern recognition technologies are adopted for classifying, e.g. neural networks [3, 5, 6, 8], linear and quadratic discriminant analyses [4, 7], interpolated Markov model [6], *independent component analysis* (ICA) [10, 11] and *non-negative matrix factorization* (NMF) [11]. The experimental results and analyses in [12] show that selection of the right biological signals to be implemented in promoter prediction programs still remains an open issue. In fact, none of these signals can cover all promoter representations, and each feature abstracted from promoter sequences has its own limitation.

2 Feature Extraction from DNA Sequence

In our method, we consider three different kinds of features, i.e. local distribution of pentamers, positional CpG island features and digitized DNA sequence, which are described in the following sections respectively.

2.1. *Local Distribution of Pentamers*

In our method, we select pentamers as input features. For an input DNA sequence, a set of pentamers a_i, $i = 1, 2, \ldots, W$, can be obtained, where the maximal value of W is $4^5 = 1024$. In order to select the most infor.mative pentamers for discriminating promoters and non-promoters, we consider the posterior probability of I given a_i, $P(I \mid a_i)$, where I is an indicator which equals 1 when the input sequence is a promoter, otherwise $I = 0$. If $P(I=1 \mid a_i) > P(I=0 \mid a_i)$, the input sequence should be a promoter with a higher probability, and *vice versa*. Define

$$\eta = \frac{P(I=1 \mid a_i)}{P(I=0 \mid a_i)}, \qquad i = 1,2,\cdots W, \tag{1}$$

and compute the value of η for each pentamer. According to the Bayes' Theorem, we have

$$P(I=1 \mid a_i) = \frac{P(a_i \mid I=1)P(I=1)}{P(a_i)}, \qquad i = 1,2,\cdots W, \tag{2}$$

and

$$P(I=0 \mid a_i) = \frac{P(a_i \mid I=0)P(I=0)}{P(a_i)}, \qquad i = 1,2,\cdots W. \tag{3}$$

From Eq. (1) ~ Eq. (3), we can obtain that

$$\eta = \frac{P(a_i \mid I=1)P(I=1)}{P(a_i \mid I=0)P(I=0)}, \qquad i = 1,2,\cdots W. \tag{4}$$

Considering $P(I=1)$ and $P(I=0)$ are constant, we define η as following equation, i.e.

$$\eta = \frac{P(a_i \mid I=1)}{P(a_i \mid I=0)}, \qquad i = 1,2,\cdots W, \tag{5}$$

and then the pentamers are ranked according to their η values. The pentamers which have the highest 250 η values are selected to combine a pentamer set *Pset*.

In order to solve the small sample problem, all pentamers in *Pset* are considered as one class, and the others as another class. In other words, 1024 pentamer patterns are converted into two kinds of patterns: pentamers in *Pset* and pentamers out of *Pset*. For each position of a DNA sequence, not only the pentamer at the position concerned, but also the pentamers within its neighborhoods are considered. A window of 51 bp moves across the sequence at 1 bp intervals and the number of pentamers in *Pset* within this window is taken as a feature at the center of the window. Therefore, for a DNA sequence with a length l, the number of features, which represent local distributions of pentamers in *Pset*, is $l-4$.

2.2. *Positional CpG Island Features*

CpG islands are regions of DNA near and in the promoter of a mammalian gene where a large concentration of phosphodiester-linked cytosine (C) and guanine (G) pairs exist. The usual formal definition of a CpG island is a region with at least 200 bp and with a GC percentage greater than 50% and with an observed/expected CpG ratio greater than 0.6 [13]. CpG islands can be used to locate promoters across genomes [2, 3, 4]. The most widely used CpG island features are GC percentage (*GCp*) and observed/expected CpG ratio (*o/e*), which are defined as follows:

$$GCp = P(C) + P(G), \tag{6}$$

and

$$o/e = \frac{P(CG)}{P(C) \times P(G)}, \tag{7}$$

where $P(CG)$, $P(C)$ and $P(G)$ are percentages of CG, C and G in a DNA sequence, respectively.

GCp and *o/e* are two global features for G+C rich or G+C related promoters. However, for the promoters that are G+C poor, CpG island features cannot be used to predict the position of a promoter. It is a reasonable assumption that there are some short regions, which are G+C rich, in a G+C poor promoter sequence, and then these regions can be used for promoter detection. In other words, if we consider *GCp* and *o/e* a sequence of local features instead of global features, more promoters can be described based on CpG islands. Similar to pentamer feature extraction, a sliding window 51 bp in length is used, and *GCp* and *o/e* are calculated for each window.

Then, for an l-length DNA sequence, the number of extracted positional CpG island features is $2l - 8$.

2.3. *Digitized DNA Sequence*

Beside the local distribution of pentamers and the positional CpG island features, we also adopt the digitized DNA sequence as input features. In our method, each nucleotide is represented using a single integer as given by: A = 0; T = 1; G = 2; and C = 3.

From the discussion above, we can see that for an l-length input DNA sequence, the number of extracted features, including local distribution of pentamers, positional CpG island features and digitized DNA sequence, is $l - 4 + 2\,l - 8 + l = 4l - 12$. These features are concatenated to form a high-dimensional vector, and then a cascade AdaBoost learning algorithm is used for feature selection and classifier training for promoter detection.

3 Feature Selection and Classifier Training with AdaBoost

AdaBoost (Adaptive Boosting) is a boosting algorithm [14], which runs a given weak learner several times on slightly altered training data, and combines the hypotheses to one final hypothesis, in order to achieve higher accuracy than the weak learner's hypothesis would have [15]. The main idea of AdaBoost is that each example of the training set should act a different role for discrimination at different training stage. The examples, which can be easily recognized, should be considered less in the following training; while the examples, which are incorrectly classified in previous rounds, should be paid more attention to. In this way, the weak learner is forced to focus on the informative or the "difficult" examples of the training set. The importance of each example is represented by a weight.

As discussion in Section 2, for an input DNA sequence with a length l, the number of features extracted is $N = 4l - 12$, e.g. if $l = 250$, then $N = 988$. We assume that of these features, only a small number are necessary to form an effective strong classifier. We therefore define our weak classifier as follows:

$$h_j(\mathbf{X}) = \begin{cases} 1 & \text{if } x_j > \theta_j \\ -1 & \text{otherwise} \end{cases} \qquad j = 1,2,\cdots,N\,, \tag{8}$$

where $\mathbf{X}$ is an input feature vector, x_j is the jth feature of $\mathbf{X}$, and θ_j is a threshold. Suppose we have a set of training samples: $(\mathbf{X}_1, y_1), \ldots, (\mathbf{X}_m, y_m)$, where $\mathbf{X}_i \in \mathbf{X}$ and $y_i \in \mathbf{Y} = \{1,-1\}$ ('1'denotes positive examples and '–1' is used for negative examples). In order to create a strong classifier, the following procedure is used:

1. Initialize the weights for each training example:

$$w_{1,i} = \begin{cases} \frac{1}{2N^+} & y_i = 1 \\ \frac{1}{2N^-} & y_i = -1 \end{cases} \qquad i = 1,2,\cdots,m\,, \tag{9}$$

where N^+ and N^- are the number of positives and negatives respectively.

1. For $t = 1, \ldots, T$:
   1. For each feature x_j, train a classifier $h_j(\mathbf{X})$, which implies selecting the optimal θ_j to produce the lowest error. The error for the classifier h_j considers all input samples with the condition of θ_j, which is defined as $\varepsilon_j = \sum_{i=1}^{m} w_t [y_i \neq h_j(\mathbf{X}_i)]$.
   2. Find the classifier $h_t : \mathbf{X} \to \{1,-1\}$ that minimizes the error with respect to the distribution w_t: $h_t = \arg\min_{h_j \in \mathrm{H}} \varepsilon_j = \sum_{i=1}^{m} w_t [y_i \neq h_j(\mathbf{X}_i)]$. Here $\varepsilon_t = \min_{h_j \in \mathrm{H}} \varepsilon_j$ should be larger than 0.5.
   3. Update the weights of the examples for the next round $w_{t+1,i} = w_{t,i}\beta_t^{1-e_i}$, where $e_i = 0$ if example $\mathbf{X}_i$ is correctly classified, otherwise $e_i = 1$, and $\beta_t = \frac{\varepsilon_t}{1-\varepsilon_t}$.
   4. Normalize the weights to make w_{t+1} a probability distribution: $w_{t+1,i} = \frac{w_{t+1}}{\sum_{j=1}^{m} w_{t+1,j}}$.

After T iterations, the resulting strong classifier is:

$$h(\mathbf{X}) = \begin{cases} 1 & \sum_{t=1}^{T} \alpha_t h_t(\mathbf{X}) \geq \frac{1}{2}\sum_{t=1}^{T} \alpha_t\,, \\ -1 & \text{otherwise} \end{cases} \tag{10}$$

where $\alpha_t = \log\frac{1}{\beta_t}$. The procedure described above not only selects features, which produce the lowest error ε_t when computing weak classifiers, but also trains the weak classifiers and the combined strong classifiers, i.e. the optimal values of θ_j, w_t and α_t are determined based on the training set.

For an input DNA sequence, the number of non-promoter segments is much larger than the number of promoters. Therefore, it is best to remove as many non-promoter segments from consideration as possible early on. Then we can cascade our classifiers to filter out most of the non-promoters, where a number of strong classifiers are used. In the early stage, few features, or weak classifiers, are considered, which can rapidly filter out most of the non-promoters while maintain most of the promoters. In the later stages, increasingly more complex features are adopted. For each stage, total positive samples (promoters) and only the negative samples (non-promoters), which are incorrectly classified in the previous stage, are used for training. In our method, a five-layer cascade is used, and for each strong classifier, the number of weak classifiers is 10, 20, 50, 100 and 200, respectively.

4 Experimental Results

In this section, we will evaluate the performance of the proposed algorithm, namely PromoterExplorer, for promoter detection based on different databases. The training set is from the Eukaryotic Promoter Database (EPD), Release 86 [16], and the testing databases include EPD, six Genbank genomic sequences and human chromosome 22 (http://www.sanger.ac.uk/HGP/Chr22/).

For training, the positive samples are 2,426 promoter sequences in EPD, which are from 200 bp upstream to 50 bp downstream of the TSS. The negative samples are randomly extracted 11,515 sequences of 250 bp, which are out of the range [–1000, 1000] relative to the TSS locations. In our experiments, all training sequences are constructed only by A, T, G and C; in other words, the sequence which includes the letter 'N' is excluded.

When perform testing, an input DNA sequence is divided into a set of segments of 250 bp, which are overlapped each other with a 10 bp shift. As described in Section 2, features including the local distribution of pentamers, the positional CpG island features and the digitized DNA sequence are obtained, following a cascade AdaBoost for classification. From Eq. (10), if the final output is larger than zero, a TSS candidate is marked. Those TSS candidates, which have no more than 1,000 nucleotides apart from their closest neighboring prediction, should be merged into a cluster. Then a new TSS prediction is used to represent this cluster, which is obtained by averaging all TSS candidates within the cluster. In order to fairly compare the performance of PromoterExplorer with other methods, a similar merging mechanism is also adopted for PromoterInspector and DPF. As the criteria proposed in [12], when one or more predictions fall in the region [-2000, +2000] relative to the reference TSS location, a true positive is counted; otherwise the predictions are denoted as false positives. When the known gene is missed by this count, it represents a false negative. Sensitivity (S_n) and specificity (S_p) are two criteria widely used to evaluate the performance of promoter prediction program, which are defined as following:

$$S_n = \frac{TP}{TP + FN}, \tag{11}$$

$$S_p = \frac{TP}{TP + FP}, \tag{12}$$

where *TP*, *FP* and *FN* denote number of true positives, false negative and false negative, respectively. For DPF, the values of S_n can be preset which is used to control the predictions. In our algorithm, the sensitivity can be modulated by the number of TSS candidates within a cluster. For each cluster to be merged, if the number of TSS candidates within this cluster is larger than a threshold, the merged TSS prediction is considered a true prediction; otherwise, the cluster is removed from the output. Various thresholds will result in different outputs.

4.1. *Experimental Results Based on EPD*

In this section, we will test the PromoterExplorer based on the whole 2,541 vertebrate promoters in EPD, which include a total of 40,656,000 base pairs in the genome sequences. The sensitivity- specificity curve is shown in Figure 1. We can see that when the sensitivity is 68.5%, the specificity is about 68.6%. In this case, the average of the distance between the predicted TSS and the real TSS location is about 320. This result is better than the performances evaluated in [12], where no program simultaneously achieves sensitivity and specificity >65%.

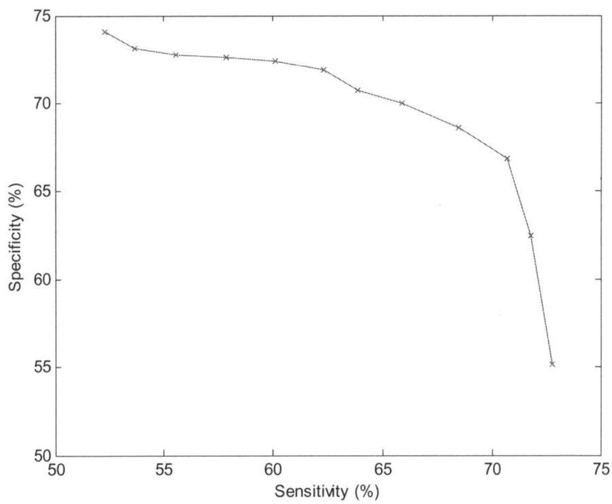

Figure 1. The sensitivity-specificity curve based on EPD using PromoterExplorer.

4.2. *Experimental Results Based on Genbank*

We also evaluate PromoterExplorer on another test set, which contains six Genbank genomic sequences with a total length of 1.38 Mb and 35 known TSSs in these sequences. In Figure 2, the sensitivity-specificity curves of PromoterExplorer, PromoterInspector and DPF are shown.

From Figure 2, we can see that PromoterExplorer outperforms others. When the sensitivity of the prediction is about 35%, the specificity of PromoterExplorer, PromoterInspector and DPF is about 52.0%, 46.4% and 41.4%, respectively; also, the corresponding average distance between the predicted TSS and real TSS is 467, 472, and 486, respectively.

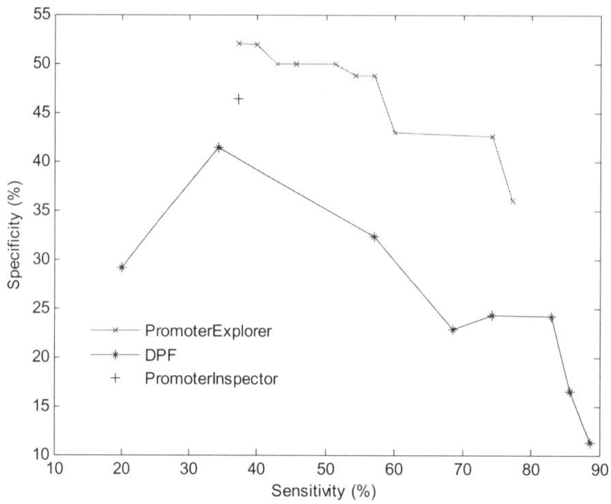

Figure 2. The sensitivity-specificity curves based on Genbank.

4.3. *Experimental Results Based on Human Chromosome 22*

Finally, we evaluate PromoterExplorer on Release 3 of the human chromosome 22, which includes 34,748,585 base pairs and 393 known genes. The comparative experimental results are shown in Figure 3. Similar to the observation in Section 4.2, PromoterExplorer performs better than PromoterInspector and DPF. The average distance between the predicted TSS and real TSS for PromoterExplorer, PromoterInspector and DPF is 306 (S_n=63.9), 351 (S_n=63.6), and 315 (S_n=67.7), respectively.

5 Conclusions

In this paper, we have proposed an effective promoter detection algorithm, which is called PromoterExplorer. In our approach, different kinds of features, i.e. local distribution of pentamers, positional CpG island features and digitized DNA sequence, are extracted and combined. Then a cascade AdaBoost algorithm is

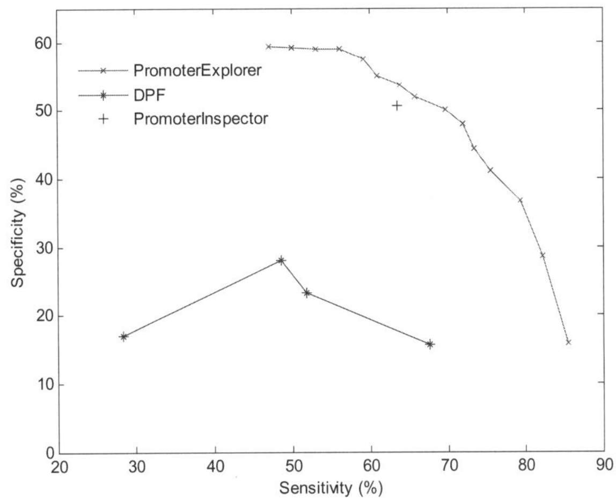

Figure 3. The sensitivity-specificity curves based on Human Chromosome 22.

adopted to perform feature selection and classifier training. An advantage of our algorithm is the most "informative" features in different classifying stages can be selected for classification. PromoterExplorer is tested based on large-scale DNA sequences, which are from different databases. It has superior performance to existing techniques. Our method can achieve a balance between the sensitivity and specificity of the predictions; therefore it can be used to detect unknown prompter locations in a new DNA sequence.

References

1. Claverie J.M., Computational methods for the identification of genes in vertebrate genomic sequences. *Human Molecular Genetics*., 6:1735-1744, 1997.
2. Pedersen A.G., Baldi P., Chauvin Y., Brunak S., The biology of Eukaryotic promoter prediction: a review. *Computers & Chemistry*, 23:191-207, 1999.
3. Bajic V.B. and Seah S.H., Dragon gene start finder: an advanced system for finding approximate locations of the start of gene transcriptional units. *Genome Research*, 13:1923-1929, 2003.
4. Davuluri R.V., Grosse I., Zhang M.Q., Computational identification of promoters and first exons in the human genome. *Nature Genetics*, 29:412-417, 2001.
5. Knudsen S., Promoter2.0: for the recognition of PolII promoter sequences. *Bioinformatics*, 15:356-361, 1999.

6. Ohler U., Liao G.C., Niemann H., Rubin G.M., Computational analysis of core promoters in the *Drosophila* genome. *Genome Biology*, 3(12):RESEARCH0087, 2002.
7. Solovyev V.V. and Shahmuradov I.A., PromH: promoters identification using orthologous genomic sequences. *Nucleic Acids Research*, 31:3540-3545, 2003.
8. Bajic V.B., Chong A., Seah S.H., Brusic V., An intelligent system for vertebrate promoter recognition. *IEEE Intelligent Systems Magazine*, 17(4):64-70, 2002.
9. Scherf M., Klingenhoff A., Werner T., Highly specific localization of promoter regions in large genomic sequences by PromoterInspector: a novel context analysis approach. *Journal of Molecular Biology*, 297:599-606, 2000.
10. Matsuyama Y. and Kawamura R., Promoter recognition for E. coli DNA segments by independent component analysis. *Proceedings of the Computational Systems Bioinformatics Conference*, 2004:686-691, 2004.
11. Hiisila H. and Bingham E., Dependencies between transcription factor binding sites: comparison between ICA, NMF, PLSA and frequent sets. Proceedings. *IEEE International Conference on Data Mining*, 4:114-121, 2004.
12. Bajic V.B., Tan S.L., Suzuki Y., Sugano S., Promoter prediction analysis on the whole human genome. *Nature Biotechnology*, 22:1467-1473, 2004.
13. Gardiner-Garden M. and Frommer M., CpG islands in vertebrate genomes. *Journal of molecular biology*, 196(2):261-282, 1987.
14. Duda, R.O., Hart, P.E., and Stork, D.G., Pattern Classification, Second Edition, John Wiley & Sons Inc., 2001.
15. Freund, Y. and Schapire, R.E., A decision-theoretic generalization of on-line learning and an application to boosting. *Journal of Computer and System Sciences*, 55(1):119-139, 1997.
16. Schmid, C.D., Périer R., Praz, V., Bucher, P., EPD in its twentieth year: towards complete promoter coverage of selected model organisms. *Nucleic Acids Research*, 34:82-85, 2006.

A NEW STRATEGY OF GEOMETRICAL BICLUSTERING FOR MICROARRAY DATA ANALYSIS*

HONGYA ZHAO

Department of Electronic and Engineering, City University of Hong Kong, Hong Kong

ALAN W. C. LIEW

Department of Computer Science and Engineering, Chinese University of Hong Kong, Hong Kong

HONG YAN

Department of Electronic and Engineering, City University of Hong Kong, Hong Kong
School of Electronic and Information Engineering, University of Sydney, NSW2006, Sydney, Australia

In this paper, we present a new biclustering algorithm to provide the geometrical interpretation of similar microarray gene expression profiles. Different from standard clustering analyses, biclustering methodology can perform simultaneous classification on the row and column dimensions of a data matrix. The main object of the strategy is to reveal the submatrix, in which a subset of genes exhibits a consistent pattern over a subset of conditions. However, the search for such subsets is a computationally complex task. We propose a new algorithm, based on the Hough transform in the column-pair space to perform pattern identification. The algorithm is especially suitable for the biclustering analysis of large-scale microarray data. Our simulation studies show that the method is robust to noise and computationally efficient. Furthermore, we have applied it to a large database of gene expression profiles of multiple human organs and the resulting biclusters show clear biological meanings.

1. Introduction

DNA microarray technology is a high-throughput and parallel platform that can provide expression profiling of thousands of genes in different biological conditions, thereby enabling the rapid and quantitative analysis of gene expression patterns on a global scale. It aids the examination of the integration of gene expression and function at the cellular level, revealing how multiple gene products work together to produce physical and chemical responses to both static and changing cellular needs [14]. As an increasing number of large-scale microarray experiments are carried out, analysis of the expression data produced by these experiments remains a major challenge. A key step of the analysis is the identification of groups of genes that exhibit similar expression patterns. Therefore cluster analysis has emerged as one of the most valuable tools to elicit complex structures and gather information about how genes work in combination with microarray data. A large number of clustering methods have been proposed for the analysis of gene function

* This work is supported by grant 9010003 of City University of Hong Kong interdisciplinary research and grant CityU122005 of Hong Kong Research Grant Council.

on a global scale [17]. Usually, gene expression data are arranged in a matrix, where each gene corresponds to one row and each condition to one column. Each element of the matrix represents the expression level of a gene under an experimental condition. Thus, clustering methods can be applied to group genes by comparing rows or conditions by comparing columns. However, conventional clustering methods have their limitations: they require that the related genes (conditions) behave similarly across all measured conditions (genes) in one cluster. In fact, many activation patterns are common to a group of genes only under specific experimental conditions. As such, an interesting cellular process may be involved in a subset of genes co-regulated or co-expressed only under a subset of conditions, but to behave almost independently under other conditions. Discovering such local expression patterns may be the key to uncovering many genetic pathways that are not apparent otherwise. Thus it is highly desirable to move beyond the clustering paradigm, and to develop approaches capable of discovering local patterns in microarray data.

Beyond the traditional clustering method, the term 'biclustering', also called coclustering, bidimentional clustering, and subspace clustering, was first formulated by Hartigan [6]. It was first applied to expression matrices for simultaneous clustering of both genes and conditions by Cheng and Church [3]. Since then different kinds of algorithm are proposed [5, 9, 11, 12]. And biculster was recently summarized in two papers [10, 15]. The general strategy in these algorithms can be described as adding or deleting rows and/or columns in the data matrix in some optimal ways such that a merit function is improved by the action. In contrast, a different viewpoint of the biclustering is in terms of the spatial geometrical distribution of points in data space [4]. The biclustering problem is tackled as the identification and division of coherent sub-matrices of data matrices into geometrical structures (lines or planes) in data space. This novel perspective opened a door to the performance of biclustering using the methodology of detecting the geometric lines or planes within a unified framework. In the framework, a series of the well-known Hough transforms are conducted to detect lines and planes. No explicit cost function is required to define the procedure. As such, the Hough transform is noted for its ability to detect lines and planes in noisy data [7]. Thus, it is especially suitable for biclustering analysis since noise is one of the major issues in microarray data. However, if the number of conditions is small, the speed of the geometric biclustering algorithm is acceptable. With the augmentation of the dimension, vote accumulators in the Hough transform use so much memory that the computation time is significantly increased and become ineffective. In order to overcome the difficulty, a novel strategy is proposed in this paper based on geometric biclustering. In our algorithm, the Hough transform is only performed in the column-pair space. Instead of computing all genes, only useful genes (features) are extracted for the combination of the following iterations.

The paper is organized as follows. First, we demonstrate that all biclustering patterns of interest in data matrices can be formulated with the linear relation in column-pair space. Based on this premise, a visualization tool, the AMPP plot, is proposed to separate the genes into different groups of biclusters. Then, the complete algorithm is given on the basis of the Hough transform and AMPP in Sec. 3. The characteristics of the algorithm are discussed in simulation study. Lastly, we apply the algorithm to bicluster the microarray expression matrix of multiple human organs. The genes in the different biclusters are further analyzed with the gene ontology (GO) tool to infer their biological process, molecular function and cellular component.

2. Linear Pattern of Biclusters in Column-pair Space

An interesting criterion to evaluate in a biclustering algorithm is the identification of type of biclusters. In this paper we focus on five major classes corresponding to significant gene expression. Table 1 shows five different types of biclusters that are of interest in microarray analysis. These biclusters are: (a) constant bicluster, (b) constant rows, (c) constant columns, (d) additive coherent values, where each row or column can be obtained by adding the constant to another row or column, (e) multiplicative coherent values, where each row or column can be obtained by multiplying another row or column by a constant value. In the case of gene expression data, constant biclusters reveal subsets of genes with similar expression values within a subset of conditions. A bicluster with constant values in the rows identifies a subset of genes with similar expression values across a subset of conditions, allowing the expression levels to differ from gene to gene. Similarly, a bicluster with constant columns identifies a subset of conditions within which a subset of genes present similar expression values assuming that the expression values may differ from condition to condition. However, one may be interested in identifying more complex relations between the genes and the conditions, such as coherent values on both rows and columns. In these cases, we can consider additive and multiplicative relations between rows or columns.

Obviously, it is unnecessary to show the relation of all columns together within a bicluster. It is enough to describe a bicluster pattern using an equation of two variables, as shown in the bottom rows of Table 1. Furthermore, it is advantageous to bicluster microarray data matrices in a column-pair space. Firstly, it is obvious that the first three classes of biclusters are special cases of the additive and multiplicative models when b_{ij} = 0 or a_{ij} = 1 in the column-pair space. Secondly, all five patterns in Table 1 can be generalized into the linear relation $x_j = a_{ij} x_j + b_{ij}$ although they appear to be substantially different from each other. Little attention is paid to the equation with two parameters because there is no corresponding biological meaning in gene expression. As such, we are more interested in the additive and multiplicative patterns, which are described by $x_j = x_j + b_{ij}$ and $x_j = a_{ij} x_j$, respectively. Thirdly, instead of the computation for all genes and conditions, the computation complexity and time are significantly decreased and become operable in the column-pair space. Of course, it is absolutely necessary to compare all pairs of conditions and combine the similar subblocks in order to identify biclusters. Compared to other methods, the geometric perspective we present here allows us to better detect linear relations that define various different bicluster patterns using a generic linear finding algorithm. The algorithm is provided in Sec. 3.

Table 1. Classes of different biclusters: (a) Constant bicluster (b) Constant rows (c) Constant columns (d) Additive coherent values (e) Multiplicative coherent values.

Constant			
x_1	x_2	x_3	x_4
50	50	50	50
50	50	50	50
50	50	50	50
50	50	50	50
$x_i=x_j=a$			

Constant Rows			
x_1	x_2	x_3	x_4
10	10	10	10
25	25	25	25
35	35	35	35
50	50	50	50
$x_i=x_j$			

Constant Columns			
x_1	x_2	x_3	x_4
10	25	35	50
10	25	35	50
10	25	35	50
10	25	35	50
$x_i=a_i$, $x_j=a_j$ $(a_i \neq a_j)$			

Additive			
x_1	x_2	x_3	x_4
15	12	20	35
20	17	25	40
25	22	30	45
30	27	35	50
$x_j=x_i+b_{ij}$ $(b_{ij} \neq 0)$			

Multiplicative			
x_1	x_2	x_3	x_4
2	6	4	12
5	15	10	30
6	18	12	36
8	24	16	48
$x_j= a_{ij}x_i$ $(a_{ij} \neq 1)$			

3. Geometric Algorithm of Biclustering

Based on the linear structures discussed above, we propose a new biclustering algorithm. First we identify genes of interest with linear structures discussed above and divide them into different patterns using the additive and multiplicative pattern plot (AMPP) described below in the column-pair space. Then these genes in the same patterns are combined step by step to form new biclusters. A robust method of line detection in the column-pair space is a key step in the proposed framework. The Hough transform (HT) is an effective, powerful, and robust technique widely used for line detection in 2-D images [1]. In this section, we first introduce the HT and then propose the AMPP as a visualization tool to separate the genes into corresponding additive and multiplicative patterns. The biclustering algorithm will then be developed based on the HT and AMPP.

3.1. Hough Transform and Line Detection

The Hough transform is a methodology that detects analytic lines and curves in images through a voting process in parameter space [1]. A line in x-y data space is defined by

$$y = kx + b \tag{1}$$

Note that a line in x-y space as defined by Eq. (1) corresponds to point (k,b) in k-b parameter space. Conversely, the line in Eq. (1) in k-b space corresponds to point (x,y) in x-y space. If n points $\{(x_i,y_i):i=1,\ldots,n\}$ on a line in the x-y space are known, the line obtained from each such point should pass through the same point in k-b space, which is the point defining the line in x-y space. Therefore, to determine lines from points, we can initialize all entries of k-b space to 0 and increment an entry by 1 when the line representing a point in x-y space passes through it, and then find the entry in k-b space that has the highest count. If more than one line is to be detected, entries with local peak counts in k-b space are located and their coordinates are used as the slopes and y-intercepts of the line. The accumulator array in parameter space may be very large because the range of the slope is large, especially for vertical lines. Alternatively, the polar form can be used to describe a line:

$$y = x\cos\theta + y\sin\theta \tag{2}$$

where ρ is the distance of a line to the original point and θ is the angle of the normal to the line with the x axis. Since ρ is limited from $-\sqrt{x^2+y^2}$ to $\sqrt{x^2+y^2}$ and θ is limited from $-\pi/2$ to $\pi/2$, the dynamic ranges of the parameters are compressed and a small accumulator array is sufficient to find all lines. Note that if the polar equation of a line is used, for each point in x-y space, a sinusoidal curve rather than a line can be drawn in the accumulator array. Again, array entries with local peak counts should be identified and used to detect the lines [7, 18].

3.2. Additive and Multiplicative Pattern Plot (AMPP)

Given points on a line in column-pair data space, we need to classify their corresponding genes into the additive or multiplicative patterns. We develop a visualization tool, named the AMPP for this task. As discussed in Sec. 2, only the additive and multiplicative patterns are of our interest in microarray analysis, so the difference in the patterns of gene expression is of concern. For example, given $\{(x_i,y_i):i=1,\ldots,k\}$ which are the expressions of k genes under two conditions, we assume that the k points are on a line detected using the HT. Now we try to cluster them into two types of expression patterns. We employ $d_i=x_i-y_i$ and $r_i=\arctan(x_i/y_i)$ to show the difference in the additive and multiplicative models. Again, we use $r_i=\arctan(x_i/y_i)$ instead of the direct ratio x_i/y_i to reduce the dynamics range of the ratio. We plot d_i against r_i $(i=1,\ldots,k)$ in the AMPP. In the plot, the horizontal axis represents the change of additive patterns, and the vertical axis the multiplicative patterns. Based on the AMPP, we employ the boxplots to obtain the points in the additive and multiplicative models. The Boxplot, also called box-and-winker plot, was first proposed by John Tukey, as simple graphical summaries of the distribution of variables [2]. In a boxplot, the middle line represents the median, and the lower and upper sides of the rectangle show the medians of the lower and upper halves of the data. Along the horizontal boxplot, the points in the box are considered to be shifted with their median in the additive model and the points in box of the vertical boxplot are considered to be multiplied by their median in multiplicative model. The points in their intersect set are considered as the overlapped genes in the two patterns. The method is used in the following algorithm to recognize patterns after line detection with HF in column-pair space.

3.3. Biclustering Algorithm

Given expression data matrix $D_{N\times n}$ with N genes and n experimental conditions, we denote the index of rows (genes) with $\mathbf{G}=\{g_1,\cdots,g_N\}$ and the index of columns (conditions) with $\mathbf{C}=\{c_1,\cdots,c_n\}$. We denote the expression matrix with the row and column index as $\mathbf{D}=(\mathbf{G},\mathbf{C})$, then the bicluster is defined as $\mathbf{B}=(\mathbf{I},\mathbf{J})$, where $\mathbf{I}=\{g_{i_1},\cdots,g_{i_s}\}$ is a subset of $\mathbf{G}$ and $\mathbf{J}=\{c_{j_1},\cdots,c_{j_t}\}$ is a subset of $\mathbf{C}$. Based on the line detection with the HF and AMPP in the column-pair space, we propose the following algorithm to identify a set of biclusters $\{\mathbf{B}_k:\mathbf{B}_k=(\mathbf{I}_k,\mathbf{J}_k)\}$.

Parameters to be predetermined:

- Resolutions: quantization step size for voting in parameter space;
- Minimum number of rows δ (genes to form one bicluster);
- Minimum number of columns ζ (conditions to form one bicluster) ;

1. Select any two columns from $\mathbf{C}$ as $\mathbf{J}_s=\{c_{s1},c_{s2}\}$, where $s=1,\cdots,M=\binom{\|\mathbf{C}\|}{2}$, where ||C|| denotes the number of elements in the set C, and transform each of $\{\mathbf{B}_s:\mathbf{B}_s=(\mathbf{G},\mathbf{J}_s)\}$ in the column-pair space to the polar parameter space.
2. Given $\mathbf{J}_s$ in $\mathbf{B}_s=(\mathbf{G},\mathbf{J}_s)$, there are $\|\mathbf{G}\|$ sinusoidal curves corresponding to $\mathbf{B}_s$ in the parameter space. Then perform a voting count in the quantized parameter space to find the accumulator array p_s. Similar procedures are applied M times to $\mathbf{J}_s$. Denote the series of the curves passing p_s as $\mathbf{G}_s$, their corresponding accumulated count as $\|\mathbf{G}_s\|$ and $\Re=\arg_s\{\|\mathbf{G}_s\|\geq\delta:s=1,\cdots,M\}$, we set the corresponding bicluster as $\mathbf{B}_r=(\mathbf{G}_r,\mathbf{J}_r)(r\in\Re)$ in the column-pair space.
3. With the help of AMPP in the column-pair space, we separate the genes in $\mathbf{G}_r$ into three parts of the additive set $\mathbf{I}_r^A$, multiplicative $\mathbf{I}_r^M$ and their overlap $\mathbf{I}_r^{AM}$. Therefore, three patterns are obtained, denoted as $\mathbf{B}_r^A=(\mathbf{I}_r^A,\mathbf{J}_r^A)$, $\mathbf{B}_r^M=(\mathbf{I}_r^M,\mathbf{J}_r^M)$, and $\mathbf{B}_r^{AM}=(\mathbf{I}_r^{AM},\mathbf{J}_r^{AM})$ where $\mathbf{J}_r=\mathbf{J}_r^A=\mathbf{J}_r^M=\mathbf{J}_r^{AM}$ $(r\in\Re)$.
4. First we consider the combination steps for the additive pattern. Set $i=1$ and begin with the set of $\mathbf{B}_i^A=\{(\mathbf{I}_i^A,\mathbf{J}_i^A):\mathbf{I}_i^A=\mathbf{I}_r^A,\mathbf{J}_i^A=\mathbf{J}_r^A\}$ including $\|\Re\|$ subclusters.
5. We unite any two elements $\mathbf{B}_{iu}^A=(\mathbf{I}_{iu}^A,\mathbf{J}_{iu}^A)$ and $\mathbf{B}_{iv}^A=(\mathbf{I}_{iv}^A,\mathbf{J}_{iv}^A)$ of $\mathbf{B}_i^A$ every time. We consider their intersection of rows and union of columns as a new subcluster. Denote the biculsters $\mathbf{B}_{i+1}^A=\{(\mathbf{I}_{i+1}^A,\mathbf{J}_{i+1}^A):\mathbf{I}_{i+1}^A=\mathbf{I}_{iu}^A\bigcap\mathbf{I}_{iv}^A,\mathbf{J}_{i+1}^A=\mathbf{J}_{iu}^A\bigcup\mathbf{J}_{iv}^A\}$.
6. Repeat Step (5) until there is no new combined subcluster and $\|\mathbf{I}_i^A\|\geq\delta,\|\mathbf{J}_i^A\|\geq\zeta$. From $\mathbf{B}^A=\{(\mathbf{I}^A,\mathbf{J}^A):\|\mathbf{I}^A\|\geq\delta,\|\mathbf{J}^A\|\geq\zeta,\mathbf{B}^A\in\mathbf{B}_k^A,k=1,\cdots,i-1\}$ we consider the biclusters obtained from the last step as the largest one.

7. As to the other two cases, $\{\mathbf{B}_r^M : r \in \Re\}$ and $\{\mathbf{B}_r^{AM} : r \in \Re\}$, the combination steps are similar.

4. Simulation Study

Gene expression data from microarray experiments are often degraded by noise. Furthermore, it is important to find the multiple biclusters with overlapping patterns. Therefore, in the simulation study, the following two questions were investigated: robustness to noise and ability to identify the multiple overlapping biclusters.

4.1. Synthetic Data with Noisy Biclusters

Here we investigate the performance of our algorithm on noisy data. We use an additive pattern as an example. We embed an additive bicluster pattern of 20 rows by 8 columns into a dataset of size 100 by 30. One column of the additive pattern is generated from U(-5, 5) and the additive factors of other columns are randomly obtained from U(-5, 5). The background is also generated from the uniform distribution U(-5, 5). Gaussian noise with variance from 0.3 to 0.9 is generated to degrade additive patterns in the bicluster. We apply the new algorithm to the simulated data with parameters $\delta = 15$ and $\zeta = 5$. It is anticipated that our algorithm is robust to noise. In fact, in HT the accumulator arrays are used instead of a point, which accommodates the noise in the data, and thus all genes of interest are already included in the biclusters in the column-pair space. With the combination of clusters step by step, the exact 20x8 additive biclusters are identified after six iterations. Besides the additive bicluster one multiplicative bicluster is also discovered with our methodology, however, it is completely overlapped by the additive bicluster. Thus, more patterns may be discovered with our algorithm.

4.2. Synthetic Data with Multiple Overlapping Biclusters

To show that our algorithm can successfully detect biclusters of different types, we have generated biclusters with additive, multiplicative and their overlapping patterns. Furthermore, we have also examined the ability of our algorithm to simultaneously identify multiple biclusters, especially when the overlap is present. We embed two overlapped biclusters into noisy background generated by the uniform distribution U(-5, 5). Gaussian noise with variance from 0.3 to 0.9 is generated to degrade the bicluster data. The dataset has 100 rows by 30 columns, and two embedded biclusters have the following sizes: 30x6 Bicluster 1 of additive pattern, 25x8 Bicluster 2 of multiplicative and their overlap is a 10x3 submatrix. The random row and column permutations are then performed to obtain the final testing dataset. In this experiment, Bicluster 1 is found with all six conditions and 27 rows. Bicluster 2 is found with eight conditions and 24 rows. The overlap part is identified with all three columns and 10 rows.

5. Applications

We apply our algorithm to the gene expression database of multiple human organs [13]. The database captures 18,927 unique genes for 19 different organs from 158 normal human tissues from 30 donors. In fact, there are several replicated tissues for every organ and they are considered as the replicated samples of the same organ. So in every organ we take the median of the replicated measurements of every gene for further analysis. We also filter the genes with low variance and entropy methods and have obtained one 5298x19 expression matrix of 5298 genes under 19 experimental conditions for the following bicluster analysis. In the first step, 19x18/2=171 subclusters can be obtained in the column-pair space. In all these subclusters, the number of columns is two and that of rows is the largest number of lines passing one accumulator array after HT in the corresponding parameter space. We demonstrate all subclusters in Fig. 1. The indexes of rows and columns in the square show 19 different organs. The values of the cross points are the number of genes in every bicluster in the column-pair space. We set the diagonal value to zero. Obviously, the square matrix is symmetric. We use different gray scales to represent different count values. The darker the color is, the larger the value is. For example, the largest value in the square matrix is 468 between the comparison of colon and ileum, that is, their gene expression patterns are very similar, which are in logical agreement with known organ functions.

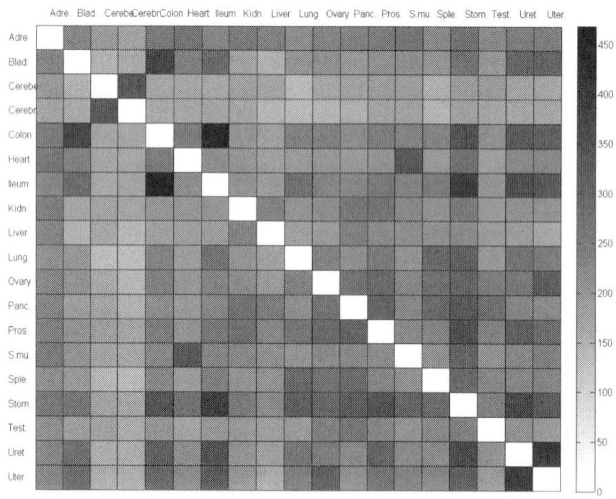

Figure 1. Heat map of the symmetric square matrix of highest count in the column-pair space. The rows and columns represent 19 organs.

In the following steps, we combine the small subclusters in the column-pair space into larger biclusters. We discover that the results of the procedures coincide with the corresponding organ function. For example, we combine colon, ileum, bladder and stomach into one significant bicluster with the largest number of common genes in some iteration. In the combination steps, the number of conditions is increased and the number of genes is decreased until the stop criteria. At last, one additive and six multiplicative biclusters are recorded with the parameters $\delta = 25$ and $\zeta = 5$. The largest additive and one multiplicative biclusters are shown in Fig. 2

Given the biclusters, we try to test the results and infer their organ functions. Gene Ontology (GO) has become a well accepted standard in organizing gene function categories [16]. Thus GO provides us with a systematic tool to determine the functional and biological significance of genes and gene products in our results. GO describes the attributes of genes in three key domains, molecular function, biological process and cellular component. We calculate p-values of each gene in the biclusters with the hypergeometric probability distribution. In one additive bicluster of six organs, bladder, colon, ileum, stomach, ureter, and uterus, the smallest p-value is 0.0004 corresponding to the GO term 0004479, which is related to methionyl-tRNA formyltransferase activity in molecular function. In one multiplicative bicluster of 11 organs, bladder, colon, heart, ileum, heart, lung, ovary, prostate, stomach, ureter and uterus, the smallest p-value is 0.0003 corresponding to GO term 0030145, which is related to cell differentiation in biological process. We discovered that significant genes in the bicluster patterns are mostly related to molecular function rather than biological process analyzed in [13].

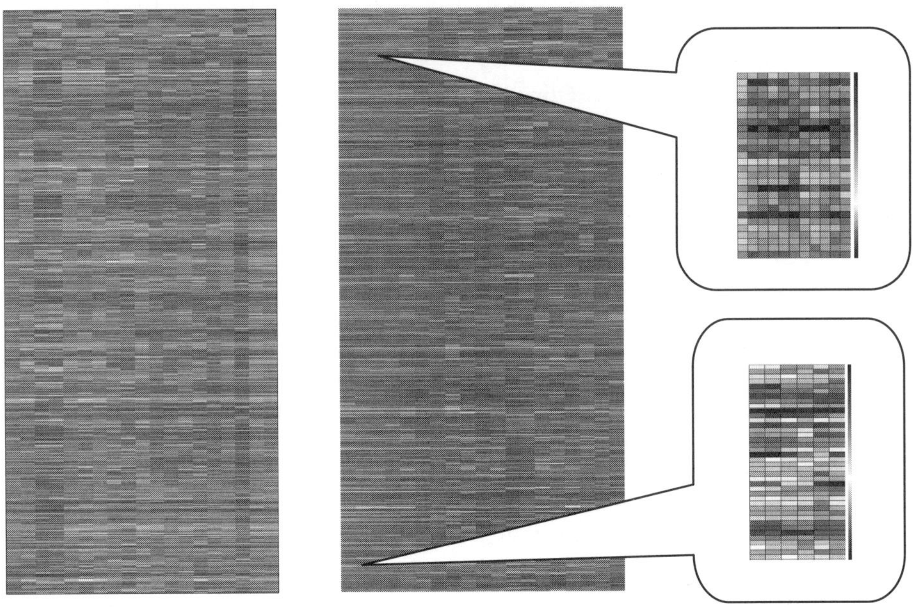

Figure 2. Bicluster analysis of 5298x19 gene expression data matrix of multiple human organs. The original data matrix is left one; Two bicluster patterns after permutation of genes and conditions in right: one multiplicative bicluster 28x11 in the upper-left corner and the other additive 40x6 one in the left bottom part.

References

1. D. H. Ballard and C. M. Brown. Computer Vision. Prentice Hall, Englewood Cliffs, NJ, 1982.
2. W. S. Celveland. Visualizing data. Murray Hill, N.J.: At & T Bell Labloratories, 1993.
3. Y. Cheng and G. M. Church. Biclustering of expression data. *Proceedings of the Eighth International Conference on Intelligent Systems for Molecular Biology (ISMB)*, 93-103, 2000.
4. X. Gan, A. W. C. Liew and H. Yan. Biclustering gene expression data based on high dimensional geometric method. *Proc. Int'l Conf. Machine Learning and Cybernetics*, IEEE SMC Society, **6**: 3388-3393, 2005.
5. G. Getz, E. Levine and E. Domany Coupled two-way clustering analysis of gene microarray data. *Proc. Natural Academy of Sciences USA*, **97**: 12079-12084, 2000.
6. J. A. Hartigan. Direct clustering of a data matrix. *J. American Statistical Association*, **67**(337), 123-129, 1972.
7. J. Illingworth and J. Kittler. A survey of the Hough transform. *Computer Vision, Graphics, and Image Processing*, **44**(1):87-116, 1988.
8. Y. Kluger, R. Basri, J. T. Chang and M. Gerstein. Spectral biclustering of microarray data: coclustering genes and conditions. *Genome Research*, **13:**703-716, 2003.
9. Lazzeroni, L and Owen, A. Plaid models for gene expression data. Statistica Sinica, 12:61-86, 2002.
10. S. C. Madeira and A. L. Oliveira. Biclustering algorithms for biological data analysis: a survey. *IEEE/ACM Trans. Computational Biology Bioinformatics*, **1:**24-45, 2004.
11. E. Segal, B. Taskar, A. Gasch, N. Friedman and D. Koller. Rich probabilistic models for gene expression. *Bioinformatics*, **17**: S243-S252, 2001.
12. Q. Sheng, Y. Moreau and B. De Moor. Biclustering microarray data by Gibbs sampling. *Bioinformatics,* **19**:Sii196-Sii205, 2003.
13. C. Son, S. Bilke, S. Davis, B.Greer, J. Wei, C. Whiteford, Q. Chen, N. Cenacchi and J. Khan. Database of mRNA gene expression profiles of multiple human organs. *Genome Research*, **15**: 443-450, 2005.
14. R. B. Stoughton. Applications of DNA Microarrays in Biology. *Annu Rev Biochem*, **74**:53-82, 2004.
15. A. Tanay, R. Sharan and R. Shamir. Biclustering Algorithms: A Survey**.** In *Handbook of Computational Molecular Biology*. Edited by: Aluru S. Chapman & Hall/CRC Computer and Information Science Series, 2005.
16. The Gene Ontology Consortium. Gene ontology tool for the unification of biology. *Nat. Genet.*, **25:**25-29, 2000.
17. S. Wu, A. W. C. Liew, H. Yan, H. Yang. Cluster analysis of gene expression data based on self-splitting and merging. *IEEE Trans. Information Technology Biomedicine*, **8**:5-15, 2004.
18. L. Xu and E. Oja. Randomized Hough Transform (RHT): Basic Mechanisms, Algorithms, and Computational Complexities. *CVGIP: Image Understanding*, **57**:131-154, 1993.

USING FORMAL CONCEPT ANALYSIS FOR MICROARRAY DATA COMPARISON

V. CHOI *

Department of Computer Science,
Virginia Tech,
660 McBryde Hall,
Blacksburg, VA 24061, USA
E-mail: vchoi@cs.vt.edu

Y. HUANG

Department of Computer Science,
Rutgers University,
110 Frelinghuysen Road,
Piscataway, NJ 08854, USA

V. LAM, D. POTTER, R. LAUBENBACHER, K. DUCA

Virginia Bioinformatics Institute,
Washington Street, MC 0477
Virginia Tech, Blacksburg, VA 24061
Blacksburg, VA 24060, USA

Microarray technologies, which can measure tens of thousands of gene expression values simultaneously in a single experiment, have become a common research method for biomedical researchers. Computational tools to analyze microarray data for biological discovery are needed. In this paper, we investigate the feasibility of using Formal Concept Analysis (FCA) as a tool for microarray data analysis. The method of FCA builds a (concept) lattice from the experimental data together with additional biological information. For microarray data, each vertex of the lattice corresponds to a subset of genes that are grouped together according to their expression values and some biological information related to gene function. The lattice structure of these gene sets might reflect biological relationships in the dataset. Similarities and differences between experiments can then be investigated by comparing their corresponding lattices according to various graph measures. We apply our method to microarray data derived from influenza infected mouse lung tissue and healthy controls. Our preliminary results show the promise of our method as a tool for microarray data analysis.

1. Introduction

Microarray technologies, which can measure tens of thousands of gene expression values simultaneously in a single experiment, across different conditions and over time, have been

*Corresponding author

widely used in biomedical research. They have found many applications, such as classification of tumors, assigning functions to previously unannotated genes, grouping genes into functional pathways etc (see [16] for a review). A large collection of database is available in the public domain (e.g. see [8, 15]). A wealth of methods [13, 5] has been proposed for analyzing these datasets to gain biological insights. A main method for analyzing these microarray data is based on clustering, which groups set of genes, and/or groups of experimental conditions, that exhibit similar expression patterns. These include single clustering algorithms, such as hierarchical clustering, k-means, self-organizing map (SOM) algorithms (see [10] for a review and references therein); and biclustering algorithms (see [12] and references therein). However, the challenge to derive useful knowledge from microarray data still remains. For example, see [3] for a recent biclustering algorithm that is based on non-smooth non-negative matrix factorization. In this paper, we propose another method that is based on Formal Concept Analysis (FCA) [18, 6] as an alternative to the clustering approach.

Our Approach. The method of FCA builds a (concept) lattice from the experimental data together with additional biological information. Each vertex of the lattice corresponds to a subset of genes that are grouped together according to their expression values and some biological information related to gene function. See Section 2 for the background of FCA. The lattice structure of these gene sets might reflect biological relationships in the dataset. Similarities and differences between experiments can then be investigated by comparing their corresponding lattices according to various graph measures. In the high level description, our method consists of the following three main steps:

(1) Build a binary relation (cross-table) for each experiment. The objects of the binary relation are genes; and there are two types of attributes: gene expression attributes and biological attributes. The gene expression attributes are obtained by a *discretization* procedure on gene expression values. The biological attributes can be any biological properties related to gene function.
(2) Construct a Galois/concept lattice for each experiment's binary relation using the efficient Galois/concept lattice algorithm described in [4].
(3) Define a distance measure and compare the lattices.

Note that the biological attributes of genes are invariant/constant for all experiments and they can be preprocessed. The ability to integrate these constant biological attributes is one of the advantages of our method over clustering methods. This is because the constant information will be canceled out in clustering methods and thus do not add any contributions.

Related work of using FCA for microarray data mining. Using FCA for microarray data comparison was proposed in D. Potter's thesis [14]. Based on the framework proposed there, we more rigorously develop each step. In particular, we more carefully discretize the gene expression values to suit our purposes, namely, close gene expression values should share the same attribute; and distance measure is also more rigorously defined and better results are obtained. Also, our concept lattice construction algorithm is very efficient (within 1 second) while our data sets were too large for the program in[14] to handle. We should also mention that using FCA or Concept lattice approach to mine microarray data were also studied in [1, 2]. The goal there was to extract local patterns in the microarray data and no biological attributes were employed.

Outline. The paper is organized as follows. In Section 2, we review some background and notation on FCA. In Section 3, we describe our method in details. In Section 4, we describe our data and present our preliminary results applying our method to the data. We conclude with future work in Section 5.

2. Background on FCA

Formal Concept Analysis (FCA) [6] is a method that is based on lattice theory for the analysis of binary relational data. It was introduced by Rudolf Wille in 1980s. Since its introduction, FCA has found many applications in data mining, knowledge discovery and machine learning etc [18].

The input of FCA consists of a triple $(\mathcal{O}, \mathcal{M}, \mathcal{I})$, called *context*, where $\mathcal{O} = \{g_1, g_2, \ldots, g_n\}$ is a set of n elements, called *objects*; $\mathcal{M} = \{1, 2, \ldots, m\}$ is a set of m elements, called *attributes*; and $\mathcal{I} \subseteq \mathcal{O} \times \mathcal{M}$ is a binary relation. The context is often represented by a *cross-table* as shown in Figure 1. A set $X \subseteq \mathcal{O}$ is called an *object set*, and a set $J \subseteq \mathcal{M}$ is called an *attribute set*. Following the convention, we write an object set $\{a, c, e\}$ as ace, and an attribute set $\{1, 3, 4\}$ as 134.

For $i \in \mathcal{M}$, denote the adjacency list of i by $\mathsf{nbr}(i) = \{g \in \mathcal{O} : (g, i) \in \mathcal{I}\}$. Similarly, for $g \in \mathcal{O}$, denote the adjacency list of g by $\mathsf{nbr}(g) = \{i \in \mathcal{M} : (g, i) \in \mathcal{I}\}$.

Definition 2.1. The function $\mathsf{attr} : 2^{\mathcal{O}} \longrightarrow 2^{\mathcal{M}}$ maps a set of objects to their common attributes: $\mathsf{attr}(X) = \cap_{g \in X} \mathsf{nbr}(g)$, for $X \subseteq \mathcal{O}$. The function $\mathsf{obj} : 2^{\mathcal{M}} \longrightarrow 2^{\mathcal{O}}$ maps a set of attributes to their common objects: $\mathsf{obj}(J) = \cap_{j \in J} \mathsf{nbr}(j)$, for $J \subseteq \mathcal{M}$.

It is easy to check that for $X \subseteq \mathcal{O}$, $X \subseteq \mathsf{obj}(\mathsf{attr}(X))$, and for $J \subseteq \mathcal{M}$, $J \subseteq \mathsf{attr}(\mathsf{obj}(J))$.

Definition 2.2. An object set $X \subseteq \mathcal{O}$ is *closed* if $X = \mathsf{obj}(\mathsf{attr}(X))$. An attribute set $J \subseteq \mathcal{M}$ is closed if $J = \mathsf{attr}(\mathsf{obj}(J))$.

The composition of obj and attr induces a *Galois connection* between $2^{\mathcal{O}}$ and $2^{\mathcal{M}}$. Readers are referred to [6] for properties of the Galois connection.

Definition 2.3. A pair $C = (A, B)$, with $A \subseteq \mathcal{O}$ and $B \subseteq \mathcal{M}$, is called a *concept* if $A = \mathsf{obj}(B)$ and $B = \mathsf{attr}(A)$.

For a concept $C = (A, B)$, by definition, both A and B are closed. The set of all concepts of the context $(\mathcal{O}, \mathcal{M}, \mathcal{I})$ is denoted by $\mathcal{B}(\mathcal{O}, \mathcal{M}, \mathcal{I})$ or simply $\mathcal{B}$ when the context is understood.

Let (A_1, B_1) and (A_2, B_2) be two concepts in $\mathcal{B}$. Observe that if $A_1 \subseteq A_2$, then $B_2 \subseteq B_1$. We order the concepts in $\mathcal{B}$ by the following relation $\prec$:

$$(A_1, B_1) \prec (A_2, B_2) \Longleftrightarrow A_1 \subseteq A_2 \text{ (and } B_2 \subseteq B_1).$$

It is not difficult to see that the relation $\prec$ is a partial order on $\mathcal{B}$. In fact, $\mathcal{L} = <\mathcal{B}, \prec>$ is a complete lattice and it is known as the *concept* or *Galois* lattice of the context $(\mathcal{O}, \mathcal{M}, \mathcal{I})$. For $C, D \in \mathcal{B}$ with $C \prec D$, if for all $E \in \mathcal{B}$ such that $C \prec E \prec D$ implies that $E = C$ or $E = D$, then C is called the *successor* [a](or *lower neighbor*) of D, and D is called the *predecessor* (or *upper neighbor*) of C . The diagram representing an ordered set (where only successors/predecessors are connected by edges) is called a *Hasse diagram* (or a line diagram). See Figure 1 for an example of the line diagram of a Galois lattice.

When the binary relation is represented as a bipartite graph (see Figure 1), each concept corresponds to a maximal bipartite clique (or maximal biclique). There is also a one-one correspondence between a closed itemset [17] studied in data mining and a concept in FCA. The one-one correspondence of all these terminologies – concepts in FCA, maximal bipartite cliques in theoretical computer science, and closed itemsets in data mining – was known, e.g. [17]. There is extensive work of the related problems in these three communities, see [4] for related literature. The current fastest algorithm given in [4] takes $O(\sum_{a \in \mathsf{ext}(C)} |\mathsf{cnbr}(a)|)$ polynomial delay for each concept C, where $\mathsf{cnbr}(a)$ is the *reduced* adjacency list of a. Readers are referred to [4] for the details.

	1	2	3	4
a	X		X	
b	X	X		X
c	X		X	
d		X		X

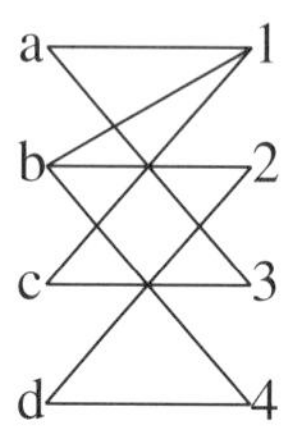

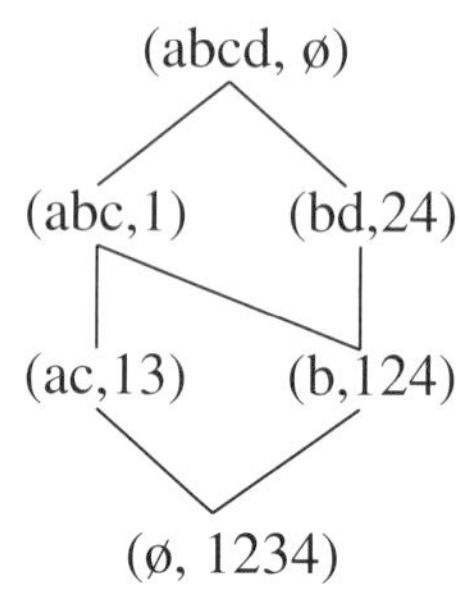

Figure 1. Left, a context $(\mathcal{O}, \mathcal{M}, \mathcal{I})$ with $\mathcal{O} = \{a, b, c, d\}$ and $\mathcal{M} = \{1, 2, 3, 4\}$. The cross $\times$ indicates a pair in the relation $\mathcal{I}$. Middle, the bipartite graph corresponding to the context. Right, the corresponding Galois/concept lattice.

3. Methods

3.1. *Building Binary Relations*

In this section, we describe how to the construct the *context* $(\mathcal{O}, \mathcal{M}, \mathcal{I})$ for each experiment. Here the object set $\mathcal{O}$ consists of a set of genes. There are two types of attributes in the

[a]Some authors called this as immediate successor.

attribute set $\mathcal{M}$: biological attributes and gene expression attributes. For a gene $g \in \mathcal{O}$ and an attribute $a \in \mathcal{M}$, $(g, a) \in \mathcal{I}$ if g has the attribute a.

3.1.1. *Biological Attributes*

Any gene function related properties can be used as biological attributes. For instance, one can use protein motif families as such attributes: a gene has the attribute if its corresponding protein belongs to the motif family. As an example, the protein motif family *oxidoreductases* can be such an attribute, and a gene has this attribute if it is one of the oxidoreductases. There are many other possible biological attributes, such as functional characteristics of a gene, chromosomal location of a gene, known association with disease states etc.

3.1.2. *Discretization of Gene Expression Values*

The data obtained from a microarray experiment consists of a set of genes and each gene has a *gene expression value*. The gene expression values are continuous real numbers. In order to represent microarray data in a binary relation, it requires to discretize the continuous gene expression values into a finite set of values that correspond to attributes. Intuitively, we would like to discretize the gene expression values such that two close values share the same attribute. The straightforward method will be dividing the gene expression values according to large "gaps". First, we sort all the expression values in the increasing order. Let the ordered values be $y_1 < y_2 < \ldots < y_m$. Then we compute the gaps $\delta_i = y_{i+1} - y_i$, for $i = 1, \ldots, m-1$. The idea then is to divide the gene expression values into t subintervals according to the largest $t-1$ gaps. However, the empirical results showed that majorities of these gene expression values were very close. For example, if we partitioned the values according to the largest 4 gaps, more than 75% of these values would belong to one subinterval. If we were to recursively partition this large subinterval, then again, a large portion of the values concentrated on one big subinterval. Instead, after partition the gene expression values into t subintervals, we partition the largest subinterval into s even subintervals.

Recall that our idea is to discretize the gene expression values such that close gene expression values share the same attribute (or a subinterval). However, our even partition might not achieve this goal, for example, two close genes might belong to the same subinterval in one experiment but belong to two consecutive subintervals in another experiment. To overcome this problem, instead of assigning the gene expression value to only one subinterval, if the gene expression value from one of the subintervals is within 50% of the neighbor subinterval, we assign both subintervals to this gene expression value.

We illustrate the discretization procedure by an example in Figure 2. In this figure, $t = 5$ and $s = 4$. That is, we first partition the gene expression values into five subintervals $I_1, I_2, \ldots, I_5$ according to the four largest gaps. And then partition the largest subinterval (I_2 in this example) into four even subintervals. There are total of eight subintervals: $I_1, I_{2_1}, I_{2_2}, I_{2_3}, I_{2_4}, I_3, I_4, I_5$, and each subinterval in the order corresponds to one gene expression attribute a_i, for $i = 1$ to 8. The gene expression value that falls in the subintervals I_1 (I_3, I_4, I_5 respectively) is assigned to its corresponding attribute a_1 (a_6, a_7, a_8 respectively). The gene expression value that falls in I_2 is assigned to one or two consecutive attributes depending the region it falls. If it falls in the subintervals J_{ii+1} (the region overlapping with 50% of two neighboring subintervals) as shown in Figure 2, then it is assigned to two attributes a_i and a_{i+1}. For example, if a gene expression value falls in the subinterval J_{23} then it gets both attributes a_2 and a_3.

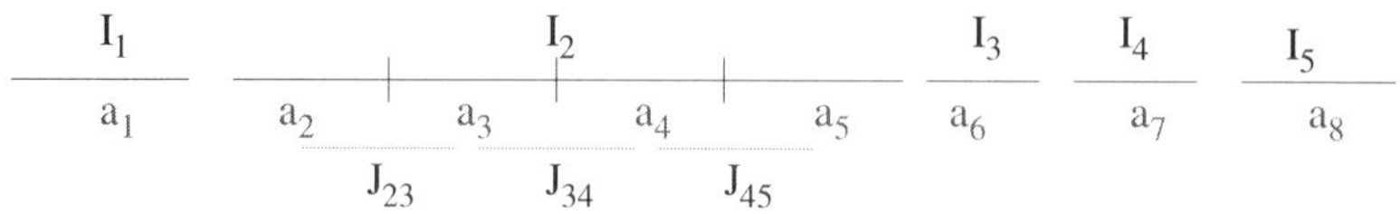

Figure 2. Discretization example. We partition the sorted gene expression values into five subintervals, $I_1, I_2, \ldots, I_5$ according to the four largest gaps. We further partition the largest subinterval (I_2 in this example) into four even subintervals. There are total of eight disjoint subintervals and each subinterval corresponds to one gene expression attribute a_i. The gene expression value that falls in the subintervals I_1 (I_3, I_4, I_5 respectively) is assigned to its corresponding attribute a_1 (a_6, a_7, a_8 respectively). The gene expression value that falls in I_2 is assigned to one or two consecutive attributes depending on the region it falls. If it falls in the subintervals J_{ii+1}, then it is assigned to two attributes a_i and a_{i+1}, for $i = 2, 3, 4$.

3.2. *Concept Lattices Construction*

Once we have the binary relations, we then build a concept lattice for each binary relation, using the efficient algorithm described in [4].

3.3. *Distance Measures for Lattices Comparison*

Given two lattices $L_1 = (V_1, E_1)$ and $L_2 = (V_2, E_2)$, there are many possible distance measures that one can define to measure the similarities or differences of these two lattices. The simplest distance maybe is the one based on common subgraphs. Recall that each vertex in our lattice is a concept and it is labeled by a subset of genes and a subset of attributes. For our purpose, we will ignore all attributes, that is, each vertex is labeled by its object set of genes only. See Figure 3 for an example. A vertex v is called a *common vertex* if v appears in both V_1 and V_2. Let $V_C = V_1 \cap V_2$ be the set of common vertices. For $u, v \in V_C$, if $e = \{u, v\}$ is in both E_1 and E_2. Then e is called a *common edge*. Let E_C be the set of common edges. The distance distance(L_1, L_2) is then defined by

$$\frac{|L_1 \setminus L_2| + |L_2 \setminus L_1|}{|L_1 \cup L_2|}$$

where $|L_1 \setminus L_2| = |V_1 \setminus V_C| + |E_1 \setminus E_C|$, $|L_2 \setminus L_1| = |V_2 \setminus V_C| + |E_2 \setminus E_C|$ and $|L_1 \cup L_2| = |L_1| + |L_2 \setminus L_1|$ with $|L_1| = |V_1| + |E_1|$.

Since the data is not perfect, we relax our requirement of the definition of a common vertex. Instead of requiring the exact matching of the gene sets of the vertices, we consider

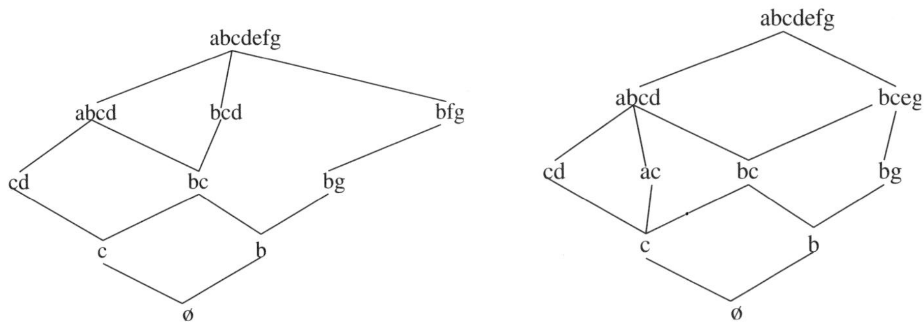

Figure 3. Lattices Comparison. How similar/different are these two lattices?

v_1 and v_2 the same if their gene sets share more than ξ of the maximum size of the two gene sets, i.e., $|\mathsf{obj}(v_1) \cap \mathsf{obj}(v_2)| \geq \xi max(|\mathsf{obj}(v_1)|, |\mathsf{obj}(v_2)|)$.

Many other possible distance measures will be investigated in the future. For example, spectral distance or maximal common sublattice distance that were also mentioned in [14].

4. Our Experiments

4.1. *Our Data*

Our microarray data [11] were derived from the lung tissue of mouse under four different conditions : (1) Control: the mouse was normal and healthy; (2) Flu: the mouse was infected by influenza (H3N1); (3) Smoking: the mouse was forced to smoke for four consecutive days, with nine packs cigarette per day; (4)SmokeFlu: the mouse was both infected with flu and smoking. For each condition, the gene expression values on 6 different time points — 6 hours, 20 hours, 30 hours, 48 hours, 72 hours and 96 hours — were measured. At each time point, there were three replicates, which were used to clean up the noise in the data. Also, the expression values were measured on *probes* and several probes can correspond to a same gene. After cleaning up (through a clean-up procedure written in perl scripts), one gene expression value was obtained for each gene of each sample. There are total of $11,051$ genes for all 24 samples (=6 time points $\times$ 4 conditions).

4.2. *Applying Our Method to the Data*

In this section, we describe the parameters used in our method when it is applied to the above data.

Biological Attributes. We used the protein motif families obtained from PROSITE [9] as our biological attributes. In particular, for our gene sets, we used the stand-alone tools from PROSITE [7] and identified 21 PROSITE families for our gene set. That is, we have 21 biological attributes. Note these biological attributes are experiment independent and thus they only need to be computed once for all experiments.

Discretization of Gene Expression Values. We performed the discretization procedure for the gene expression values (after taking log transformation) of each sample. The parameter t was set to 5, and s was set to 4. That is, there were total of 8 gene expression attributes. We have tested different parameter values and found that increasing the values (and thus number of attributes) did not significantly change the results.

After discretization, we had total of 24 binary relations over the 11051 genes and the 29 attributes. We then applied our lattice construction algorithm on these binary relations to construct the corresponding lattices. There is a small variation of these lattices in terms of its size: each of them has around 530 vertices and 1500 edges. Using the program in [4], it took less than one second to construct each lattice on a Pentium IV 3.0GHz computer with 2.0G memory running under Fedora Core 2 Linux OS.

Distances for Comparing Lattices. The parameter in defining common vertices ξ was set to 70%. That is, $v_1 \in V_1$ and $v_2 \in V_2$, v_1 and v_2 were considered the same if $|\mathsf{obj}(v_1) \cap \mathsf{obj}(v_2)| \geq \xi max(|\mathsf{obj}(v_1)|, |\mathsf{obj}(v_2)|)$, where $\xi = 70\%$. We have tested various relaxation of ξ. This value seems to be best in our current application, that is, it clearly separates different conditions (see Section 4.3 for details).

4.3. *Our Results*

First, using Control sample at 6 hours as a reference, we computed the distances between all other Control samples and all Flu samples to this reference sample. The results are shown in Figure 4. Since the distance measure is not transitive, we also computed all

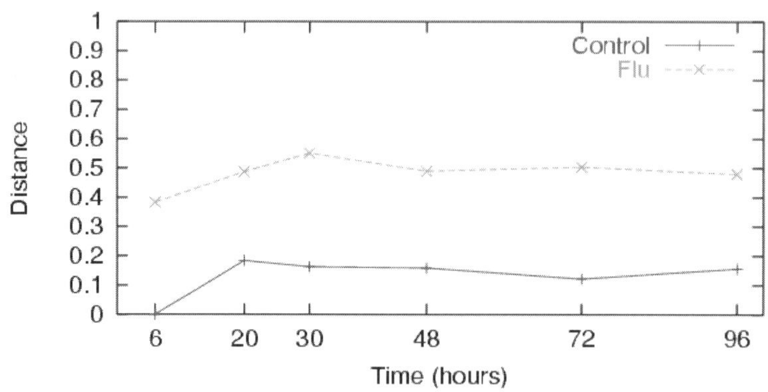

Figure 4. Distance to Control sample at 6 hours (the reference sample). Each data point represents the distance from a sample (Control shown in red, Flu shown in green) to the reference sample. There are small distances from other Control samples to the reference sample. And there is a clear separation between Flu samples from the Control sample at 6 hours.

distances when one of the Control samples was taken as a reference. See Figure 5. The results showed that there was a clear separation between Flu samples from Control samples (regardless which time point of Control was taken as the reference). This is a encouraging result and we are currently investigating what substructures in the lattices that contribute to the differences which might shed some new biological insights. The results of comparing

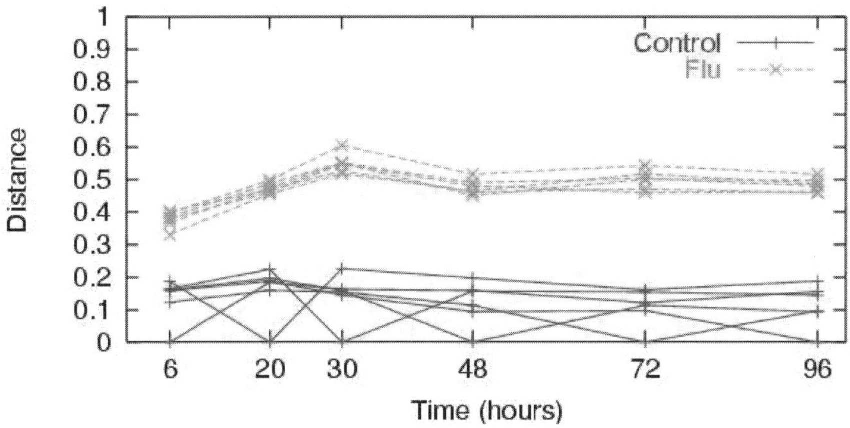

Figure 5. Distance to Control at each time point. There is a clear separation between Flu samples from Control samples regardless which time point of Control is taken as the reference..

other 2 conditions (Smoking/SmokeFlu) is shown in Figure 6.

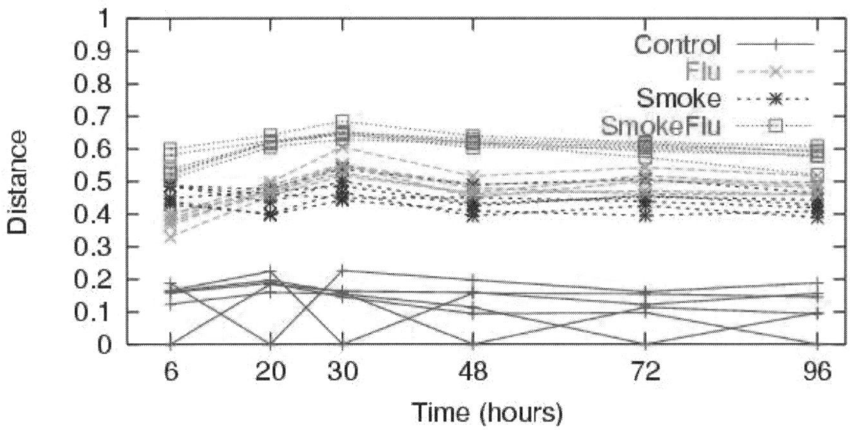

Figure 6. Distance to Control from Flu/Smoking/SmokeFlu.

5. Conclusion and Future Work

Our current preliminary results showed the promise of using FCA approach for microarray data analysis. The distance measure we employed is quite basic and it has not utilized the properties of the lattice structure. One can investigate other possible distance measures, such as spectral distance, and distance based on maximal common sublattice. These distances will take advantage of the lattice structures and might provide better distinction to aid analyzing the differences between experiments. Beside the global lattice comparison, one can also investigate the local structures of the lattices. These local structures, or sublattices, can be obtained by context decomposition or lattice decomposition. A study of sublattices may assist identification of particular biological pathways or substructures of functional importance.

References

1. J. Besson, C. Robardet, J-F. Boulicaut. Constraint-Based Mining of Formal Concepts in Transactional Data. *PAKDD*, 615–624, 2004.
2. J. Besson, C. Robardet, J-F. Boulicaut, S. Rome. Constraint-based concept mining and its application to microarray data analysis. *Intell. Data Anal.*, 9(1): 59–82, 2005.
3. P. Carmona-Saez, R.D. Pascual-Marqui, F. Tirado, J.M Carazo and A. Pascual-Montano. Biclustering of gene expression data by non-smooth non-negative matrix factorization. *BMC Bioinformatics*, 7:78, 2006.
4. V. Choi. Faster Algorithms for Constructing a Galois/Concept Lattice. Available at arXiv:cs.DM/0602069.
5. R.C. Deonier, S. Tavare, M.S. Waterman. Computational Genome Analysis: An Introduction. Springer Verlag, 2005.
6. B. Ganter, R. Wille. Formal Concept Analysis: Mathematical Foundations. Springer Verlag, 1996 (Germany version), 1999 (English version).
7. A. Gattiker, E. Gasteiger and A. Bairoch. ScanProsite: a reference implementation of a PROSITE scanning tool *Applied Bioinformatics*, 1:107–108, 2002.
8. L.L. Hsiao, F. Dangond, T. Yoshida, R. Hong, R.V. Jensen, J. Misra, et al. A compendium of gene expression in normal human tissues. *Physiol Genomics*, 7: 97–104, 2001.
9. N. Hulo, A. Bairoch, V. Bulliard, L. Cerutti, et al. The PROSITE database. *Nucleic Acids Res.* 34:227–230, 2006.
10. A. Kjersti. Microarray data mining: a survey. SAMBA/02/01. Availalble at http://nr.no/files/samba/smbi/microarraysurvey.pdf
11. V. Lam, K. Duca. Mouse gene expression data (Unpublish data).
12. S.C. Madeira and A.L. Oliveira. Biclustering Algorithms for Biological Data Analysis: A Survey. *IEEE/ACM Transactions on Computational Biology and Bioinformatics*, 1(1), 24–45, 2004.
13. G. Piatetsky-Shapiro and P. Tamayo. Microarray Data Mining: Facing the Challenges. *SIGKDD Explorations*, 2003.
14. D.P. Potter. A combinatorial approach to scientific exploration of gene expression data: An integrative method using Formal Concept Analysis for the comparative analysis of microarray data. Thesis dissertation, Department of Mathematics, Virginia Tech, August 2005.
15. R. Shyamsundar, Y.H. Kim, J.P. Higgins, K. Montgomery, M. Jorden, et al. A DNA microarray survey of gene expression in normal human tissues. *Genome Biol*, 6:22, 2005.
16. R.B. Stoughton. Applications of DNA Microarrays in Biology. *Annu Rev Biochem.*, 2004.
17. M.J. Zaki, M. Ogihara. Theoretical foundations of association rules. *Proc. 3rd ACM SIGMOD Workshop on Research Issues in Data Mining and Knowledge Discovery*, 1–7, 1998.
18. A Formal Concept Analysis Homepage. http://www.upriss.org.uk/fca/fca.html

AN EFFICIENT BICLUSTERING ALGORITHM FOR FINDING GENES WITH SIMILAR PATTERNS IN TIME-SERIES EXPRESSION DATA*

SARA C. MADEIRA
INESC-ID / IST
University of Beira Interior
Rua Marquês D'Ávila e Bolama, 6200-001 Covilhã, Portugal
E-mail: smadeira@di.ubi.pt

ARLINDO L. OLIVEIRA
INESC-ID / IST
Rua Alves Redol, 9, 1000-039 Lisbon, Portugal
E-mail: aml@inesc-id.pt

Biclustering algorithms have emerged as an important tool for the discovery of local patterns in gene expression data. For the case where the expression data corresponds to time-series, efficient algorithms that work with a discretized version of the expression matrix are known. However, these algorithms assume that the biclusters to be found are perfect, in the sense that each gene in the bicluster exhibits exactly the same expression pattern along the conditions that belong to it. In this work, we propose an algorithm that identifies genes with similar, but not necessarily equal, expression patterns, over a subset of the conditions. The results demonstrate that this approach identifies biclusters biologically more significant than those discovered by other algorithms in the literature.

1. Introduction

Several non-supervised machine learning methods have been used in the analysis of gene expression data. Recently, biclustering[4], a non-supervised approach that performs simultaneous clustering on the row and column dimensions of the data matrix, has been shown to be remarkably effective in a variety of applications. The advantages of biclustering (when compared to clustering) in the discovery of local expression patterns have been extensively studied and documented[4,14,1,6,10]. These expression patterns can be used to identify relevant biological processes involved in regulatory mechanisms. Although, in its general form, biclustering is NP-complete, in the case of time-series expression data the interesting biclusters can be restricted to those with contiguous columns leading to a tractable problem.

In this context, CCC-Biclustering[11] is a recent proposal of an algorithm that finds and reports all maximal contiguous column coherent biclusters (CCC-Biclusters) in time linear on the size of the expression matrix by manipulating a discretized matrix using string

*This work was partially supported by projects POSI/SRI/47778/2002, BioGrid and POSI/EIA/57398/2004, DBYeast, financed by FCT, Fundação para a Ciência e Tecnologia, and the POSI program.

processing techniques based on suffix trees. Each expression pattern shared by a group of genes in a contiguous subset of time-points is a potentially relevant biological process.

However, discretization may limit the ability of the algorithm to discover biologically relevant patterns due to the noise inherent to most microarray experiments. To overcome this problem we present a new algorithm, e-CCC-Biclustering, that finds CCC-Biclusters with up to a given number of errors per gene in their expression pattern (e-CCC-Biclusters). These errors can, in general, be substitutions of a symbol in the expression pattern by other symbols in the alphabet (measurement errors), or restricted to the lexicographically closer discretization symbols (discretization errors).

We present results using a well known gene expression dataset that support the view that allowing errors in CCC-Biclusters improves the ability of the algorithm to discover more relevant biological processes, either by adding genes to the CCC-Bicluster that had been excluded due to errors, or by adding columns (up to the number of errors allowed) either at the left or at the right of the expression pattern of CCC-Bicluster.

The paper is organized as follows: Section 2 presents definitions needed to state the problem and construct the algorithm, as well as related work on biclustering in time-series expression data. Section 3 presents the algorithm and Section 4 describes the experimental results. Finally, Section 5 states some conclusions and outlines future work.

2. Definitions and Related Work

2.1. *Strings and Suffix Trees*

This section revises basic concepts about strings and suffix trees that will be needed throughout the paper.

Definition 2.1. A **string** S is an ordered list of symbols (over an alphabet Σ) written contiguously from left to right[5]. For any string S, $S[i..j]$ is its (contiguous) **substring** starting at position i and ending at position j. The **suffix** of S that starts at position i is $S[i..|S|]$.

Definition 2.2. The ***e*-Neighborhood** of a string S of length $|S|$ ($e \geq 0$) defined over the alphabet Σ, $N(e, S)$, is the set of strings S_i, such that: $|S| = |S_i|$ and $Hamming(S, S_i) \leq e$. This means that the Hamming distance between S and S_i is no more than e, that is, we need at most e substitutions to obtain S_i from S. The e-Neighborhood of a string S contains the following number of elements: $\nu(e, |S|) = \sum_{j=0}^{e} C_j^{|S|}(|\Sigma| - 1)^j \leq |S|^e|\Sigma|^e$.

Definition 2.3. A **suffix tree** of a string S is a rooted directed tree with exactly $|S|$ leaves, numbered 1 to $|S|$. Each internal node, other than the root, has at least two children. Each edge is labeled with a nonempty substring of S (**edge-label**), and no two edges out of a node have edge-labels beginning with the same character. Its key feature is that for any leaf i, the label of the path from the root to the leaf (**path-label**) spells out the suffix of S starting at position i. Each leaf is identified by the starting position of the suffix it corresponds to.

In order to enable the construction of a suffix tree obeying this definition when one suffix of S matches a prefix of another suffix of S, a character terminator, that does not appear anywhere else in the string, is added to its end.

Definition 2.4. A **generalized suffix tree** is a suffix tree built for a set of strings S_i. Each leaf is now identified by two numbers, one identifying the string and the other the suffix.

Suffix trees (generalized suffix trees) can be built in time linear on the size of the string (sum of the sizes of the strings), using several algorithms[5]. Ukkonen's algorithm[16], used in this work, uses *suffix links* to achieve a linear time construction. An example of a generalized suffix tree built for the set of strings that correspond to the rows of the right matrix in Figure 1 is presented in Figure 2.

Definition 2.5. There is a **suffix link** from node v to node u, (v, u), if the path-label of node u represents a suffix of the path-label of node v and the length of the path-label of u is equal to the length of the path-label of v minus 1.

2.2. *Gene Expression Data and Matrix Discretization*

Let A' be a gene expression matrix defined by its set of rows (genes), R, and its set of columns (conditions), C. In this context, A'_{ij} represents the expression level of gene i under condition j, which is usually a real value corresponding to the logarithm of the relative abundance of mRNA in gene i under condition j. Let A'_{iC} and A'_{Rj} denote row i and column j of matrix A', respectively. Moreover, consider that $|R|$ is the number of rows and $|C|$ is the number of columns in A'.

In this work, we are interested in the case where the gene expression levels in A' can be discretized to a set of symbols of interest, Σ, that represent distinct activation levels. In the simpler case, Σ may contain only two symbols, one used for *no-regulation* and other for *regulation*. Another widely used possibility, is to consider a set of three symbols, $\{D, N, U\}$, meaning *DownRegulation*, *NoChange* and *UpRegulation* In other applications, the values in matrix A' may be discretized to a larger set of symbols. After discretization, A' is transformed into matrix A and $A_{ij} \in \Sigma$ represents the discretized value of A'_{ij}.

	C1	C2	C3	C4	C5
G1	0.07	0.73	−0.54	0.45	0.25
G2	−0.34	0.46	−0.38	0.76	−0.44
G3	0.22	0.17	−0.11	0.44	−0.11
G4	0.70	0.71	−0.41	0.33	0.35

	C1	C2	C3	C4	C5
G1	N	U	D	U	N
G2	D	U	D	U	D
G3	N	N	N	U	N
G4	U	U	D	U	U

	C1	C2	C3	C4	C5
G1	N1	U2	D3	U4	N5
G2	D1	U2	D3	U4	D5
G3	N1	N2	N3	U4	N5
G4	U1	U2	D3	U4	U5

Figure 1. Toy example: (left) original expression matrix, (middle) discretized matrix and (right) discretized matrix after alphabet transformation.

Figure 1 (middle) represents a possible discretization of the expression values in the left matrix in the same figure. In this example, the alphabet $\Sigma = \{D, N, U\}$ was used and an expression level was considered as *NoChange* if it falls in the range $[-0.3, 0.3]$.

Consider now the **alphabet transformation** that consists, essentially, in appending the column number to each symbol in the matrix. This corresponds to considering a new alphabet $\Sigma' = \Sigma \times \{1, \ldots, |C|\}$, where each element Σ' is obtained by concatenating one symbol in Σ and one number in the range $\{1 \ldots |C|\}$. In order to do this we use a function $f : \Sigma \times \{1, \ldots, |C|\}$ defined by $f(a, k) = a|k$, where $a|k$ represents the character in Σ' obtained by concatenating the symbol a with the number k. For example, if $\Sigma = \{D, N, U\}$ and $|C| = 3$, then $\Sigma' = \{D1, D2, D3, N1, N2, N3, U1, U2, U3\}$. As examples, $f(D, 2) = D2$ and $f(U, 1) = U1$.

Consider also that Σ is always given in lexicographic order. The function $f_j : \Sigma \times \{j\}$ defined by $f_j(a, j) = a|j$, where $a|j$ represents the character in Σ'_j obtained by concatenating the symbol a with the number j, is used to define the possible alphabet for a specific column j. Moreover, $\Sigma'_j[p]$ is defined as the p element of Σ'_j. For instance, $\Sigma'_1 = \{D1, N1, U1\}$ is the possible set of symbols in column 1 and $\Sigma'_1[2] = N1$.

In this setting, consider also the **set of strings** $S_i = \{S_1, \ldots, S_{|R|}\}$ **obtained by mapping each row** A_{iC} **in matrix** A **to string** S_i such that $S_i[j] = f(A_{ij}, j)$. Each string S_i has exactly $|C|$ symbols which correspond to the symbols in row A_{iC}. After this transformation, the middle matrix in Figure 1 becomes the right matrix in Figure 1.

2.3. *Biclusters in Gene Expression Data*

Consider now the matrix A, corresponding to the discretized version of matrix A'. This matrix is defined by the discretized versions of the set of rows and the set of columns in A': $\{A_{iC}, 1 \leq i \leq |R|\}$ and $\{A_{Rj}, 1 \leq j \leq |C|\}$. Let $I \subseteq R$ and $J \subseteq C$ be subsets of the rows and columns, respectively. Then, $A_{IJ} = (I, J)$ is a submatrix of A that contains only the elements A_{ij} belonging to the submatrix with set of rows I and set of columns J.

Definition 2.6. A **bicluster** is a subset of rows that exhibit similar behavior across a subset of columns, and vice-versa. The bicluster A_{IJ} is thus a subset of rows and a subset of columns where $I = \{i_1, ..., i_k\}$ is a subset of the rows in R ($I \subseteq R$ and $k \leq |R|$), and $J = \{j_1, ..., j_s\}$ is a subset of the columns in C ($J \subseteq C$ and $s \leq |C|$). As such, the bicluster A_{IJ} can be defined as a k by s submatrix of matrix A.

Given this definition and a data matrix, A', or its discretized version, A, the goal of biclustering algorithms is to identify a set of biclusters $B_k = (I_k, J_k)$ such that each bicluster satisfies specific characteristics of homogeneity. These characteristics vary from approach to approach enabling the discovery of many types of biclusters by analyzing directly the values in matrix A or using its discretized version[10]. In this paper we will deal with biclusters that exhibit coherent evolutions, characterized by a specific property of the symbols in the discretized matrix. We are interested in column coherent biclusters:

Definition 2.7. A **CC-Bicluster**, column coherent bicluster, A_{IJ}, is a subset of rows $I = \{i_1, \ldots, i_k\}$ and a subset of columns $J = \{j_1, \ldots, j_s\}$ from the matrix A such that $A_{ij} = A_{lj}$, for all $i, l \in I$ and $j \in J$.

2.4. *Biclusters in Time-Series Gene Expression Data*

When analyzing time-series gene expression data we can restrict the attention to biclusters with contiguous columns[11,17,8]. This leads us to the definition of CCC-Bicluster and other relevant definitions related to it (already defined in previous work [11]), such as, trivial CCC-Biclusters and row-maximal, left-maximal, right-maximal, and maximal CCC-Biclusters:

Definition 2.8. A **CCC-Bicluster**, contiguous column coherent bicluster, A_{IJ}, is a subset of rows $I = \{i_1, \ldots, i_k\}$ and a **contiguous** subset of columns $J = \{r, r+1, \ldots, s-1, s\}$ from matrix A such that $A_{ij} = A_{lj}, \forall i, l \in I$ and $j \in J$.

In this settings, each CCC-Bicluster A_{IJ} defines a string S corresponding to a contiguous **expression pattern** that is common to every row in the CCC-Bicluster, between columns r and s of matrix A. This means there exists a string $S = S_i[r...s], \forall i \in I$.

Definition 2.9. A CCC-Bicluster A_{IJ} is **trivial** if it has only one row or only one column.

Definition 2.10. A CCC-Bicluster A_{IJ} is **row-maximal** if no more rows can be added to its set of rows I while maintaining the coherence property in Def. 2.8.

Definition 2.11. A CCC-Bicluster A_{IJ} is **right-maximal** if its expression pattern S cannot be extended to the right by adding one more symbol at its end (the column contiguous to the last column of A_{IJ} cannot be added to J without removing genes from I).

Definition 2.12. A CCC-Bicluster A_{IJ} is **left-maximal** if its expression pattern S cannot be extended to the left by adding one more symbol at its beginning (the column contiguous to the first column of A_{IJ} cannot be added to J without removing genes from I).

Given the three definitions above we can intuitively say that a maximal CCC-Bicluster is a CCC-Bicluster that is row-maximal, left-maximal and right-maximal. This means that no more rows or contiguous columns (either at right or at left) can be added to it while maintaining the coherence property in Def. 2.8.

Definition 2.13. A CCC-Bicluster A_{IJ} is **maximal** if no other CCC-Bicluster exists that properly contains it, that is, if for all other CCC-Biclusters A_{LM}, $I \subseteq L$ and $J \subseteq M \Rightarrow I = L \wedge J = M$.

Given these definitions we can now define the type of biclusters we are interested in this work, *e*-CCC-Biclusters and maximal *e*-CCC-Biclusters:

Definition 2.14. An ***e*-CCC-Bicluster**, contiguous column coherent bicluster with e errors, A_{IJ}, is a CCC-Bicluster where all the strings S_i that define the expression patterns of each of the genes in I are in the e-Neighborhood of an expression pattern S that defines the e-CCC-Bicluster, that is, $S_i \in N(e, |S|), \forall i \in I$. The definition of *0*-CCC-Bicluster is equivalent to the definition of a CCC-Bicluster (Def. 2.8).

Definition 2.15. An *e*-CCC-Bicluster, A_{IJ}, is **maximal** if it is row-maximal, left-maximal and right-maximal. This means that no more rows or contiguous columns can be added to it while maintaining the coherence property in Def. 2.14.

The goal of the *e*-CCC-Biclustering algorithm we propose in this work can now be defined: find and report all maximal *e*-CCC-Biclusters given a discretized version A of the original gene expression matrix A'.

2.5. *Related Work on Biclustering Algorithms for Time-Series Expression Data*

Although several algorithms have been proposed to address the general problem of biclustering[10], to our knowledge, only three recent proposals have addressed this problem in the specific case of time-series expression data[17,8,11].

Zhang et. al[17] proposed to modify the heuristic algorithm of Cheng and Church[4], by restricting it to add and/or remove only columns that are contiguous to the partially constructed bicluster thus forcing the resulting bicluster to have only contiguous columns. Multiple biclusters are identified (as in the approach of Cheng and Church) by masking the biclusters found so far with random values. This method has one strong limitation, however. The greedy row and column addition and removal, that is already likely to find sub-optimal biclusters in general expression data, does not work well in time-series gene expression data. In fact, the restriction imposed on the columns that can be removed makes the algorithm converge, in many cases, to a local minimum, from which it does not escape.

A different approach, from Ji and Tan[8], also works with a discretized data matrix. As in CCC-Biclustering[11], an $O(|R||C|)$ algorithm that has been recently proposed and that will be described in the end of this section) they are also interested in identifying biclusters formed by consecutive columns. Therefore, their idea generates exactly the same biclusters as the ones generated by CCC-Biclustering. With an appropriate implementation (not described by the authors) their sliding window approach can have its complexity reduced to $O(|R||C|^2)$, a complexity that is still of the order of $|C|$ higher than that of CCC-Biclustering. However, they propose to use a naive algorithm that, as made available by the authors[7], requires time and space exponential on the number of columns, when applied to the generation of all CCC-Biclusters. In practice, it cannot be applied to generate biclusters with more than 10 or 11 time-points.

CCC-Biclustering[11], finds and reports all CCC-Biclusters in time linear on the size of the expression matrix by manipulating a discretized version A of the original matrix A' and using string processing techniques based on suffix trees. Let T be the generalized suffix tree obtained from the set of strings S obtained after the matrix transformation explained in Sec. 2.2. Let v be a node of T and let $P(v)$ be the **path-length** of v, that is, the number of symbols in the string that labels the path from the root to node v (**path-label**). Additionally, let $E(v)$ be the **edge-length** of v, that is the number of symbols in the edge that leads to v (**edge-label**), and $L(v)$ the number of leaves in the sub-tree rooted at v, in case v is an internal node. The CCC-Biclustering algorithm is based on the following theorem:

Theorem 2.1. *Let v be a node in the generalized suffix tree T. If v is an internal node, then v corresponds to a maximal CCC-Bicluster iff $L(v) > L(u)$ for every node u such that there is a suffix link from u to v. If v is a leaf node, then v corresponds to a maximal CCC-Bicluster iff the path-length of v, $P(v)$, is equal to $|S_i|$ and the edge-label of v has symbols other than the string terminator, that is, $E(v) > 1$. Furthermore, every maximal CCC-Bicluster in the matrix corresponds to a node v satisfying one of these conditions.*

Figure 2 illustrates that every node in the generalized suffix tree corresponds to one CCC-Bicluster. However, the rules in Theorem 2.1 have to be applied to extract only the maximal. In this case (no errors allowed) each CCC-Bicluster is *perfect*, in the sense of having no errors, and is identified by exactly one node in the suffix-tree. We will see that this is no longer true when our goal is to extract *e*-CCC-Biclusters with $e > 0$.

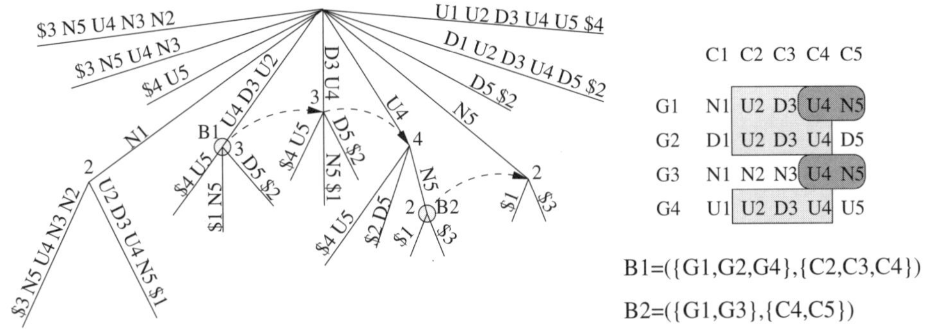

Figure 2. (Left) Generalized suffix tree for the right matrix in Figure 1 used by CCC-Biclustering. The circles identify the Maximal Non-Trivial CCC-Biclusters (B1 and B2). (Right) The CCC-Biclusters B1 and B2 are showned in the matrix as subsets of rows and columns (I, J). The strings m =[U2 D3 U4] and m =[U4 N5] correspond to the expression patterns of B1 and B2, respectively (called valid models/motifs in Section 3.1).

3. Finding and Reporting all Maximal *e*-CCC-Biclusters

3.1. *Finding e-CCC-Biclusters and the Common Motifs Problem*

SPELLER[13] is an algorithm that extracts common motifs from a set of N sequences using a generalized suffix tree. The motifs searched correspond to *words* which occur with at most e mismatches in $1 \leq q \leq N$ distinct sequences. The words representing the motifs may not be present exactly in the sequences (see SPELLER for details). As such a **motif** is seen as an "external" object and denoted by the term **model**. In order to be considered a **valid model**, a given model m of length $|m|$ has to verify the **quorum** constraint, that is, it must belong to the e-neighborhood of a word w, $N(e, w)$, in at least q distinct sequences.

The **common motifs problem** is as follows[13]: given a set of N sequences S_i $(1 \leq i \leq N)$, and two integers $e \geq 0$ and $2 \leq q \leq N$, where e is the number of errors allowed and q is the required quorum, find all models m that appear in at least q distinct sequences of S_i.

SPELLER solves the problem above starting by building a generalized suffix tree T of the sequences S_i and then, after some further preprocessing, using this tree to "spell" the valid models. When $e = 0$ spelling this models leads to a node v in T such that $L(v)$ is at least q (similarly to CCC-Biclustering). When errors are allowed, spelling all the occurrences of a model m leads to a set of nodes v_1,...,v_k in T for which $\sum_{j=1}^{k} L(v_j)$ is at least q[13]. Since in SPELLER the occurrences of a model m are in fact nodes of the generalized suffix tree T, they are called node-occurrences:

Definition 3.1. A **node-occurrence** of a model m is a triple (v, v_{err}, p), where v is a node in the generalized suffix tree, v_{err} is the number of mismatches between m and the path-label from the $root$ to node v and p identifies a position in the generalized suffix tree: if $p = 0$, we are exactly at node v, if $p > 0$ we are at a point p between two symbols in $label_b$ $(1 \leq p < |label_b|)$, where b is the edge between nodes $father_v$ and v.

The goal of SPELLER is to identify all valid models by extending them in the generalized suffix tree and to report the results using their set of node-occurrences. Note that in SPELLER, a node-occurrence is defined by a pair (v, v_{err}) and not by a triple (v, v_{err}, p) (for simplicity, the algorithm was exemplified in an uncompacted version of the generalized suffix tree, that is, a trie). However, as pointed out by the author, when using a generalized suffix tree, as we do, we need to know whether we are at a node v or at edge b between two nodes. Moreover, when we traverse T with a symbol α we also need to know if we get to a node v or if we stay inside an edge b. We use p to deal with these questions.

Consider that m is a model, α is a symbol in Σ', v is a node in T, $father_v$ is its father, b is the edge between $father_v$ and v and $label_b$ is the edge-label of b with length $|label_b|$. The algorithm we propose is based on the following Lemmas (adapted from SPELLER):

Lemma 3.1. *$(v, v_{err}, 0)$ is a node-occurrence of a model $m' = m\alpha$, if, and only if: **(1)** $(father_v, v_{err}, 0)$ is a node-occurrence of m and $label_b$ is α **or** $(v, v_{err}, |label_b| - 1)$ is a node-occurrence of m and the last symbol in $label_b$ is α **(match)**; **(2)** $(father_v, v_{err} - 1, 0)$ is a node-occurrence of m and $label_b$ is $\beta \neq \alpha$ **or** $(v, v_{err} - 1, |label_b| - 1)$ is a node-occurrence of m and the last symbol in $label_b$ is $\beta \neq \alpha$ **(substitution)**.*

Lemma 3.2. *$(v, v_{err}, 1)$ is a node-occurrence of a model $m' = m\alpha$, if, and only if: **(1)** $(father_v, v_{err}, 0)$ is a node-occurrence of m and the first symbol in $label_b$ is α **(match)**; **(2)** $(father_v, v_{err} - 1, 0)$ is a node-occurrence of m and $label_b[1] = \beta \neq \alpha$ **(substitution)**.*

Lemma 3.3. *(v, v_{err}, p), $2 \leq p < |label_b|$ is a node-occurrence of a model $m' = m\alpha$, if, and only if: **(1)** $(v, v_{err}, p - 1)$ is a node-occurrence of m and $label_b[p] = \alpha$ **(match)**; **(2)** $(v, v_{err} - 1, p - 1)$ is a node-occurrence of m and $label_b[p] = \beta \neq \alpha$ **(substitution)**.*

SPELLER can be adapted to extract all right-maximal *e*-CCC-Biclusters from the transformed matrix A. In fact, given the set of $|R|$ strings S_i of Section 2.2, $e \geq 0$ and $1 \leq q \leq |R|$, what we want to find is the set of all models m (identifying expression patterns) that are present in at least q distinct rows S_i **starting and ending at the same columns**. The set of node-occurrences of each model m and the model itself identifies one *e*-CCC-Bicluster with a maximum length $|C|$. Furthermore, it is possible to find all maximal *e*-CCC-Biclusters (without restricting the number of genes) by setting q to 1.

Figure 3 shows the generalized suffix tree used by SPELLER when $q = 1$ and $e = 1$ and two maximal *1*-CCC-Biclusters (B1 and B2) identified by two valid models. It is also possible to observe this fact in the matrix in Figure 4 where it is also clear that a model can be valid without being right/left maximal. Additionally, several valid models may identify the same e-CCC-Bicluster, when $e \geq 1$. For example, m=[N1 U2 D3] is valid but it is not right-maximal, m=[N2 D3 U4 N5] is also valid but it is not left-maximal, and finally the models m = [D1 U2 D3 U4 D5] and m=[D1 U2 D3 U5 N5] are both valid but identify the same *1*-CCC-Bicluster (B1). Similarly, m = [U2 D3 U4 D5], m = [U2 D3 U4 N5] and m = [U2 D3 U4 U5] are all valid models that represent B2.

The next section explains how SPELLER was adapted to extract exactly one valid model for each maximal *e*-CCC-Bicluster.

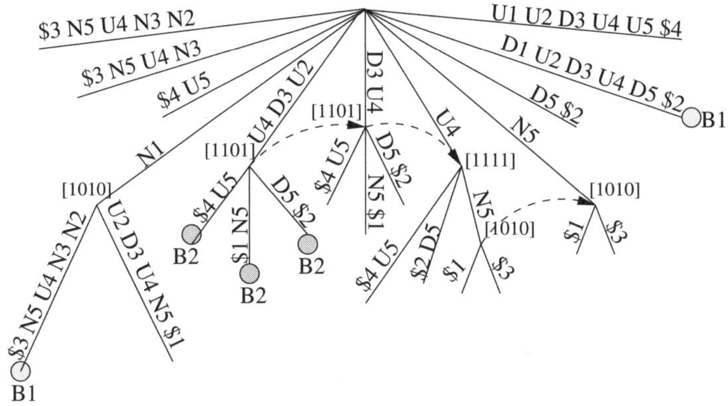

Figure 3. Generalized suffix tree for the right matrix in Figure 1 used in e-CCC-Biclustering when $e > 0$ (when $e = 0$ the suffix tree used is in Figure 2). The circles B1 and B2 identify two *1*-CCC-Biclusters (see Figure 4)

	C1	C2	C3	C4	C5
G1	N1	U2	D3	U4	N5
G2	D1	U2	D3	U4	D5
G3	N1	N2	N3	U4	N5
G4	U1	U2	D3	U4	U5

B1=({G1,G2},{C1,C2,C3,C4,C5})

	C1	C2	C3	C4	C5
G1	N1	U2	D3	U4	N5
G2	D1	U2	D3	U4	D5
G3	N1	N2	N3	U4	N5
G4	U1	U2	D3	U4	U5

B2=({G1,G2,G4},{C2,C3,C4,C5})

Figure 4. (Left) Maximal *1*-CCC-Bicluster corresponding to the valid model/motif m = [D1 U2 D3 U4 D5]. (Right) Maximal *1*-CCC-Bicluster corresponding to the valid model/motif m = [U2 D3 U4 D5]. Note that these two 1-CCC-Biclusters correspond, respectively, to two and three node-occurrences (B1 and B2) in Figure 3.

3.2. *Algorithm Description*

This section presents e-CCC-Biclustering (see Algorithm 1) and describes its main steps.

The first step stores all valid models, m, and its node-occurrences, Occ_m, that correspond to right-maximal e-CCC-Biclusters in the list $modelsOcc$. In order to do this it uses an adaptation of SPELLER with two basic changes:

(1) Check if a model m corresponds to a right-maximal CCC-Bicluster using the procedure CHECKRIGHTMAXIMALITY which works as follows: if one of the models that result from the extension of a model m with a symbol α, $m\alpha$, is also a valid model and has as many genes in its node-occurrences as its father m, then m does not correspond to a right maximal CCC-Bicluster. As such, it is removed from the stored models. In order to compute the genes in the node-occurrences of a model m (returned as a bit vector, $genesOcc_m$), the function COMPUTEGENES(Occ_m) performs a bitwise or between the bit vectors $colors_v$ of all node-occurrences (v, v_{err}, p) of m in case $e > 0$, and uses the function NUMBEROFLEAVES(v) in case $e = 0$ (in this case Occ_m has only one node-

occurrence). The function NUMBEROFGENES($genesOcc_m$) is then used to compute the number of genes.

(2) The extensions, Ext_m, of a given model m are restricted according to the level of the model in the suffix tree. For example, if $\Sigma = \{D, N, U\}$ and model m is being extended descending from the root, m can only be $m = D1$, $m = N1$ or $m = U1$, and the possible symbols α with which it can be extended to $m\alpha$ are in $\Sigma'_2 = \{D2, N2, U2\}$.

Algorithm 1: Algorithm to Find and Report all Maximal e-CCC-Biclusters

input: A, Σ, e, q

$S \longleftarrow \{S_1, \ldots, S_{|R|}\}$, $S_i[j] = f(A_{ij}, j)$, $1 \leq i \leq |R|$ **and** $1 \leq j \leq |C|$
$T_{right} \longleftarrow$ CONSTRUCTGENERALIZEDSUFFIXTREE (S)
ADDNUMBEROFLEAVES(T_{right}) //Adds $L(v)$ to each node v in T_{right}.
$m \longleftarrow$ "" //model m is a string $[m_1 \ldots m_{|m|}]$
$length_m \longleftarrow 0$
$father_m \longleftarrow$ "" //$father_m$ is a string $[m_1 \ldots m_{|m|-1}]$
$numberOfGenesOcc_{father_m} \longleftarrow 0$
$Occ_m \longleftarrow \{\text{ROOT}(T_{right}), 0, 0)\}$ //Set of node-occurrences of model m.
$modelsOcc \longleftarrow \{\}$ //List of (m, Occ_m, $genesOcc_m$, $numberOfGenesOcc_m$).
if $e = 0$ **then**
  $Ext_m \longleftarrow \{\}$
  forall *edges* b *leaving node* ROOT(T_{right}) **do**
    if $label_b[1]$ *is not a string terminator* **then**
      $Ext_m \longleftarrow Ext_m \cup label_b[1]$ //Ext_m is the set of possible symbols α to extend the model m.
else
  ADDCOLORARRAY(T_{right}) //Adds $colors_v$ to each node v in T_{right}
  //$colors_v[i] = 1$, if there is a leaf in the subtree rooted at v that is a suffix of S_i.
  //$colors_v[i] = 0$, otherwise.
  $Ext_m \longleftarrow \Sigma'$
SPELLMODELS (Σ, e, q, $length_m$, m, Occ_m, Ext_m, $father_m$, $numberOfGenesOcc_{father_m}$, $modelsOcc$)
DELETEMODELSNOTLEFTMAXIMALBICLUSTERS($modelsOcc$)
if $e > 0$ **then**
  DELETEMODELSREPRESENTINGSAMEBICLUSTERS($modelsOcc$)
REPORTMAXIMALBICLUSTERS($modelsOcc$)

The second step removes from the models stored in $modelsOcc$ (right-maximal e-CCC-Biclusters) those not corresponding to left maximal e-CCC-Biclusters. Non left-maximal biclusters are removed by first building a trie with the reversed patterns of all models m and storing the number of genes in Occ_m in its corresponding node in the trie. After this, it is sufficient to mark as non-maximal any node in the trie that has at least one child with as many genes as itself. This is easily achieved by performing a DFS of the trie and computing, for each node, the maximum value of all its children.

In the case where errors are allowed, different models may identify the same e-CCC-Bicluster. The third step uses an hash tree to remove from $modelsOcc$ (maximal *e*-CCC-Biclusters) repeated *e*-CCC-biclusters. The idea is that all models m with equal first and last columns and set of genes represent the same maximal CCC-Bicluster.

Finally, all maximal *e*-CCC-Biclusters in $modelsOcc$ are reported.

3.2.1. *e-CCC-Biclustering with Restricted Errors*

The *e*-CCC-Biclustering algorithm presented above allows general errors, that is substitutions of the symbols A_{ij} of the CCC-Bicluster $B = (I, J)$ by any of the symbols in the alphabet Σ'_j except A_{ij}. This kind of errors are specially relevant to identify measurement errors that occurred during the microarray experiments. However, if we are specially interested in identifying discretization errors we can consider restricted errors, that is, substitutions of the symbols A_{ij} by the lexicographically closer symbols in Σ'_j.

For example, when general errors are allowed, $\Sigma = \{D, N, U\}$, and m = [U2 D3 U4 D5], symbol D5 can be substituted by N5 and U5 in Σ'_5 leading to the *1*-CCC-Bicluster B2=({G1,G2,G4},{C2,C3,C4,C5}) in Figure 3 and Figure 4(b). However, if only restricted errors were allowed, symbol D5 could only be substituted by N5 leading to the *1*-CCC-Bicluster B=({G1,G2},{C2,C3,C4,C5}).

In general, when restricted errors are considered, the allowed substitutions for any symbol A_{ij} are in $\Sigma_j'^{Rest} = \{\Sigma'_j[p-1], \ldots, \Sigma'_j[p-z], \Sigma'_j[p+1], \ldots, \Sigma'_j[p+z]\}$, where p is the position of A_{ij} in Σ'_j and $z \in \{1, ..., |C|\}$. If $z = |C|$ then the errors are not restricted.

It is easy to modify Algorithm 1 to restrict the allowed errors.

3.2.2. *Complexity Analysis*

The construction of T_{right} and the computation of $L(v)$ for all its nodes take $O(|R||C|)$ time each, using the Ukkonen's algorithm with appropriate data structures, and a DFS, respectively. Adding the color array to all nodes in T_{right} (needed only when $e > 0$) takes $O(|R|^2|C|)$ time, and the remaining procedures in the algorithm take $O(|R|^2|C|^{1+e}|\Sigma|^e)$ time each. Therefore, the complexity of *e*-CCC-Biclustering is $O(|R|^2|C|^{1+e}|\Sigma|^e)$, when general errors are allowed, and $O(|R|^2|C|^{1+e}|\Sigma^{Rest}|^e)$, in the case of restricted errors.

When $e = 0$, Theorem 2.1 and CCC-Biclustering[11] can be used to obtain $O(|R||C|)$.

4. Experimental Results

In order to validate the quality of the results of the e-CCC-Biclustering algorithm we used the yeast cell-cycle dataset publicly available[3], described by Tavazoie et al.[15] and processed by Cheng and Church[4]. This dataset contains the expression profiles of more than 6000 yeast genes measured at 17 time-points over two complete cell cycles. We used 2884 genes selected by Cheng and Church[4] (as in Tavazoie et al.[15]) and removed the genes with missing values. The matrix with the remaining 2268 genes was discretized using 3 *equal frequency intervals*. We have also used *smoothing* as a preprocessing step to discretization with a window of size 5 and the set of weights $\beta_k = \{0.05, 0.2, 0.5, 0.2, 0.05\}$.

$$A_{ij} = \sum_{k=-w}^{+w} \beta_{j+k} A'_{i(j+k)} \tag{1}$$

Smoothing has been used by a number of authors in order to reduce the effect of experimental errors in the gene expression levels[2,9,11]. In order to minimize these errors each expression value A'_{ij} in matrix A' is smoothed using Equation (1), where $2w + 1$ is the window size. The values β_k are the weights given to the expressions values around A'_{ij} in a window of size $2w + 1$. These values control how much smoothing is applied to the data.

After the discretization process, we applied *e*-CCC-Biclustering with $e = 1$ with restricted errors and $z = 1$ (see Section 3.2.1) to the discretized matrix described above. We have also computed the results when $e = 0$. We argue that allowing errors in the pattern of the *0*-CCC-biclusters found by *0*-CCC-Biclustering should improve the biological significance of the biclusters by minimizing the effect of possible discretization errors. In fact, in the specific case of allowing 1 error in the pattern of a *0*-CCC-Bicluster one of the following three situations can happen: (1) the *1*-CCC-Bicluster remains equal to the *0*-CCC-Bicluster; (2) one or more genes excluded from the *0*-CCC-Bicluster (due to a single error) could be added to the *1*-CCC-Bicluster; or (3) the pattern of the *0*-CCC-Bicluster could be extended by adding one column either at its beginning or at its end (leading to a *1*-CCC-Bicluster with as many genes as the *0*-CCC-Bicluster but with one more column).

In order to validate the biological relevance of the *e*-CCC-Biclusters discovered we used the Gene Ontology (GO) annotations and associations files together with Ontologizer 2.0[12]. The goal was to evaluate the biological significance by computing p-values obtained when the hyper-geometric distribution is used to model the probability of observing at least k genes, from a CCC-Bicluster with $|I|$ genes, by chance, in a GO category containing c genes from the total number of genes ($|R|$) in each dataset. We used the functions from the three GO categories, biological process, molecular function and cell component.

We used the following procedure to validate the thesis that allowing errors in the CCC-Biclusters can improve the quality of the results: for each *0*-CCC-Bicluster with at least 4 genes and 2 conditions, we computed the number of GO functions enriched, in a statistically significant way, after Bonferroni correction, and stored this value together with the p-value of the most enriched function, that is, the lower GO p-value. The *0*-CCC-Biclusters were sorted according to their statistical significance. This significance was computed by evaluating the probability that a bicluster of that size appeared by chance in a matrix where the symbols are generated by a Markov chain, whose transition probabilities are obtained from the values in the matrix. Biclusters that were redundant, since their similarity[a] to biclusters already reported was above 50%, were removed from the analysis. We have then computed the *1*-neighborhood of the pattern of the top 10 *0*-CCC-Biclusters.[b]

[a] The similarity was measured by the number of common genes and conditions.

[b] This computation, although simple, is not described in this paper, since the purpose of this analysis is to evaluate the relative quality of *0*-CCC-Biclusters and *1*-CCC-Biclusters regarding their biological significance.

Table 1. Comparison between the top 10 *0*-CCC-Bicluster (sorted according to the statistical p-value) and the best *1*-CCC-Bicluster found by the 1-CCC-Biclustering algorithm whose pattern is a *1*-neighbor of the CCC-Bicluster without errors. Column 6 shows the best p-value computed using the hypergeometric distribution and the Gene Ontology, column 7 shows the Bonferroni correction of the previous value and finally column 8 shows the number of GO functions that are significantly enriched after Bonferroni correction (corrected p-value smaller or equal to 0.01).

ID	e	PATTERN	CONDITIONS	#GENES	P-VALUE	CORRECTION	#FUNCTIONS
2237	0	DDNNNDD	11-17	164	2.8E-12	2.6E-9	11
58537	1	**DDNNDDD**	11-17	**531**	1.1E-16	1.2E-17	17
767	0	UNDDDNNNDD	8-17	81	2.2E-11	1.5E-8	9
18544	1	UUDDDNNNDD	8-17	**108**	1.7E-14	1.3E-11	15
3041	0	NNNDD	13-17	340	1.2E-11	1.2E-8	9
63752	1	**DNNNDD**	**12-17**	**343**	3.5E-16	4.8E-13	17
260	0	UUUNDDDNNNDD	6-17	57	1.4E-8	1.7E.7	7
11361	1	UUUNDDDNNNDD	6-17	**144**	6.1E-13	5.2E-10	12
3220	0	NDD	15-17	772	3.5E-8	7.1E-5	3
68222	1	**DDD**	15-17	**1321**	2.2E-10	5.4E-7	7
3071	0	UNNDD	13-17	164	1.0E-5	1.1E-1	**0**
63773	1	**DUNNDD**	**12-17**	**249**	1.1E-9	1.4E-6	12
3035	0	NDDDD	13-17	161	3.6E-4	3.7E-1	**0**
66652	1	**NNDDD**	13-17	**608**	2.2E-9	4.0E-6	11
3217	0	DDD	15-17	492	2.2E-4	3.9E-1	**0**
68222	1	DDD	15-17	**1321**	2.2E-10	5.4E-7	7
627	0	NNDDDNNNDD	8-17	65	8.1E-4	4.1E-1	**0**
24875	1	NNDDDNNNDD	8-17	**165**	6.5E-14	6.0E-11	14
3191	0	UUND	14-17	183	1.0E-3	9.3E-1	**0**
66883	1	**UUUND**	**13-17**	**311**	4.6E-7	5.9E-4	6

Table 1 reports these results. Each row contains one *0*-CCC-Bicluster followed by the best *1*-CCC-Bicluster, as measured by the number of GO functions enriched. These results show that the *1*-CCC-Biclusters tend to have higher biological significance, since their sets of genes are more significantly enriched than those of the best *0*-CCC-Biclusters. Even in cases where the *0*-CCC-Bicluster does not pass the statistical test for biological significance, there exists a *1*-CCC-Bicluster in its *1*-neighborhood that has biological significance.

For lack of space, we do not report here additional results that support the view that *e*-CCC-Biclusters are biologically more relevant than perfect CCC-Biclusters.

5. Conclusions and Future Work

In this work, we presented and analyzed a new algorithm, *e*-CCC-Biclustering, for the identification of groups of genes that exhibit similar activities in a subset of conditions, in time-series expression data. The algorithm finds and reports, in time polynomial on the size of the matrix, all e-CCC-Biclusters that correspond to these groups of genes. By selecting the e-CCC-Biclusters that are statistically more significant, it is possible to identify potentially relevant biological processes. The algorithm avoids the limitations that previous methods exhibit since they cannot consider genes that have small deviations from the central pattern of expression. Moreover, the results demonstrate that this approach identifies biclusters that are biologically more significant than those identified by existing algorithms.

In future work, we plan to build a graphical user interface to e-CCC-Biclustering, make it available to the scientific community, and use the algorithm to identify regulatory modules (sets of co-regulated genes that share a common function) in gene regulatory networks.

References

1. A. Ben-Dor, B. Chor, R. Karp, and Z. Yakhini. Discovering local structure in gene expression data: The order–preserving submatrix problem. In *Proc. of the 6th International Conference on Computacional Biology*, pages 49–57, 2002.
2. T. Chen, V. Filkov, and S. Skiena. Identifying gene regulatory networks from experimental data. In *Proceedings of the 3rd International Conference on Research in Computational Molecular Biology*, pages 99–103, 1999.
3. Y. Cheng and G. M. Church. Biclustering of expression data - supplementary information. http://arep.med.harvard.edu/biclustering/, [July 15, 2006].
4. Y. Cheng and G. M. Church. Biclustering of expression data. In *Proc. of the 8th International Conference on Intelligent Systems for Molecular Biology*, pages 93–103, 2000.
5. D. Gusfield. *Algorithms on strings, trees, and sequences*. Computer Science and Computational Biology Series. Cambridge University Press, 1997.
6. P. De Boeck I. Van Mechelen, H. H. Bock. Two-mode clustering methods: a structured overview. *Statistical Methods in Medical Research*, 13(5):979–981, 2004.
7. L. Ji and K. Tan. Identifying time-lagged gene clusters using gene expression data - supplementary information. http://www.comp.nus.edu.sg/∼jiliping/p2.htm, [July 15, 2006].
8. L. Ji and K. Tan. Identifying time-lagged gene clusters using gene expression data. *Bioinformatics*, 21(4):509–516, 2005.
9. A. Kwon, H. Hoos, and R. Ng. Inference of transcriptional regulation relationships from gene expression data. *Bioinformatics*, 19(8):905–912, 2003.
10. S. C. Madeira and A. L. Oliveira. Biclustering algorithms for biological data analysis: a survey. *IEEE/ACM Transactions on Computational Biology and Bioinformatics*, 1(1):24–45, 2004.
11. S. C. Madeira and A. L. Oliveira. A linear time biclustering algorithm for time series expression data. In *Proc. of 5th Workshop on Algorithms in Bioinformatics*, pages 39–52. Springer Verlag, LNCS/LNBI 3692, 2005.
12. U. Bohme B. Beattie P. N. Robinson, A. Wollstein. Ontologizing gene-expression microarray data: characterizing clusters with gene ontology. *Bioinformatics*, 20(6):979–981, 2004.
13. M.-F. Sagot. Spelling approximate repeated or common motifs using a suffix tree. In *Proc. of Latin'98*, pages 111–127. Springer Verlag, LNCS 1380, 1998.
14. A. Tanay, R. Sharan, and R. Shamir. Discovering statistically significant biclusters in gene expression data. *Bioinformatics*, 18(1):136–144, 2002.
15. S. Tavazoie, J. D. Hughes, M. J. Campbell, R. J. Cho, and G. M. Church. Systematic determination of genetic network architecture. *Nature Genetics*, 22:281–285, 1999.
16. E. Ukkonen. On-line construction of suffix trees. *Algorithmica*, 14:249–260, 1995.
17. Y. Zhang, H. Zha, , and C. H. Chu. A time-series biclustering algorithm for revealing co-regulated genes. In *Proc. of the 5th IEEE International Conference on Information Technology: Coding and Computing*, pages 32–37, 2005.

SELECTING GENES WITH DISSIMILAR DISCRIMINATION STRENGTH FOR SAMPLE CLASS PREDICTION

ZHIPENG CAI, RANDY GOEBEL, MOHAMMAD R. SALAVATIPOUR, YI SHI, LIZHE XU, AND GUOHUI LIN*

Department of Computing Science, University of Alberta
Edmonton, Alberta T6G 2E8, Canada
Email: zhipeng, goebel, mreza, ys3, ghlin@cs.ualberta.ca, lizhe.xu@dhs.gov

One of the main applications of microarray technology is to determine the gene expression profiles of diseases and disease treatments. This is typically done by selecting a small number of genes from amongst thousands to tens of thousands, whose expression values are collectively used as classification profiles. This gene selection process is notoriously challenging because microarray data normally contains only a very small number of samples, but range over thousands to tens of thousands of genes. Most existing gene selection methods carefully define a function to score the differential levels of gene expression under a variety of conditions, in order to identify top-ranked genes. Such single gene scoring methods suffer because some selected genes have very similar expression patterns so using them all in classification is largely redundant. Furthermore, these selected genes can prevent the consideration of other individually-less but collectively-more differentially expressed genes. We propose to cluster genes in terms of their class discrimination strength and to limit the number of selected genes per cluster. By combining this idea with several existing single gene scoring methods, we show by experiments on two cancer microarray datasets that our methods identify gene subsets which collectively have significantly higher classification accuracies.

1. Introduction

DNA microarrays provide the opportunity to measure the expression levels of thousands of genes simultaneously. This novel technology supplies us with a large volume of data to systematically understand various gene regulations under different conditions. As one of the main applications, it is very important to determine the gene expression profiles of diseases and disease treatments. Among the thousands of genes in the arrays, many of them do not have their expression values distinguishably changed across different condition, e.g., so-called "house keeping" genes. These genes certainly would not be very useful in profiling since they do not contribute much to disease or treatment class recognition. In practice, a small number, typically in the tens, of genes that are highly differentially expressed across different conditions are to be selected to compose profiles for the purpose of class prediction. This process is known as *gene selection*; there are many existing methods, which typically define a function to score the level of how differentially expressed a gene is, under different conditions, and identify those top ranked genes.[1,2,3,4,5] Such single gene scoring

*To whom correspondence should be addressed.

methods typically suffer the problem that some selected genes have very similar expression patterns, therefore using them all in classification is largely redundant, and those selected genes prevent other individually-less but collectively-more differentially expressed genes from being selected.

Several other gene selection methods have recognized the problem with the redundancy of some highly expressed genes, and look for a subset of genes that collectively maximize the classification accuracy. For example, Xiong *et al.* define a function to measure the classification accuracy for individual genes and select a subset of genes through *Sequential Forward* [*Floating*] *Selection* (SF[F]S),[6] which was developed decades ago for general feature selection. Guyon *et al.* propose another method that uses *Support Vector Machines* (SVMs) and *Recursive Feature Elimination* (RFE).[7] In terms of effectiveness, these gene selection methods perform much better than those single gene scoring methods, since they measure the classification strength of the whole set of selected genes. Computationally, they are essentially heuristics which replace exhaustive enumeration of an optimal subset of genes which typically takes much longer to return a solution. The inefficiency of these methods actually prevent them from being used in practice. Nevertheless, there are alternative implementations of the key idea, which is to exclude a gene when there is already a similar gene selected.

We propose another implementation which first clusters genes according to their class discrimination strength, namely, two genes that have very close class discrimination strength are placed in a common cluster; we then limit the number of genes per cluster to be selected. This provides a more efficient clustering process which, when combined with a single gene scoring method, leads to an efficient and effective gene selection algorithm. We call our method an *EEGS-based gene selection method.* In the next section, we present the details of a novel measure of class discrimination strength difference between two genes, using their expression values. With this distance measure, we briefly explain how to adopt the *k-means* algorithm[8] to cluster genes. We also briefly introduce three single gene scoring methods, namely F-test[3], Cho[4] and GS,[5] and two classifiers, namely, a linear kernel SVM classifier[7] and a k Nearest Neighbor (KNN) classifier.[3] Finally, we outline a complete high level description of the EEGS-based gene selection methods. In Section 3, we briefly introduce our performance measurements, followed by the dataset descriptions, and our experimental results. Section 4 discusses parameter selection, the effects of variety within data sets, classifiers and the performance measurements, and finally, the overall results compared to single gene scoring methods. Section 5 summarizes our main contributions, our conclusions on the suitable datasets for the EEGS-based methods, and some plans for future work.

2. The EEGS-Based Gene Selection Methods

There are two challenges in microarray data classification. One is class discovery to define previously unrecognized classes. The other is to assign individual samples to already-defined classes, which is the focus here.

2.1. *The Performance Measurements*

The genes selected by a method are evaluated by their class discrimination strength, measured by the classification accuracy, defined as follows. For gene selection purposes, a number of microarray samples with known class labels are provided, which form a *training dataset*. The selected genes are then used for building a classifier, which can take a new microarray sample and assign it a class label. The set of such samples for testing purpose is referred to as the *testing dataset*, and the percentage of the correctly labeled samples is defined as the *classification accuracy* of the method (on this particular testing dataset). Note that we have to have the class labels for the samples in the testing dataset in order to calculate the classification accuracy. For computational convenience, given a microarray dataset whose samples all have known class labels, only a portion of it is used to form the training dataset; the rest of the samples have their class labels removed and are used to form the testing dataset. There are two popular cross validation schemes adopted in the literature to evaluate a method, which are *ℓ-Fold* and *Leave One Out* (LOO). We adopt the ℓ-Fold cross validation in this work, in which the whole dataset is (randomly) partitioned into ℓ equal parts and, at one time, one part is used as testing dataset and the other $\ell - 1$ parts are used as training dataset. The process is repeated for each part and the average classification accuracy over these ℓ ones is taken as the final classification accuracy. Here set $\ell = 5$ and repeat the process for 20 iterations. Therefore, the final classification accuracy is the average over 100 values. We report the 5-Fold classification accuracies for all the six tested gene selection methods in Section 3.

2.2. *The Classifiers*

We adopt two classifiers in our study. One is a linear kernel SVM classifier that has been used in Guyon *et al.*[7] and the other is a KNN classifier that has been used in Dudoit *et al.*[3] Essentially, with a given set of selected genes determined by some gene selection method, the SVM classifier, which contains multiple SVMs, finds decision planes to best separate the labeled samples based on the expression values of these selected genes. Subsequently, it uses this set of decision planes to predict the class label of a test sample. For a more detailed explanation of how the decision planes are constructed, the readers are referred to Guyon *et al.*[7] The KNN classifier predicts the label of a testing sample in a different way. Using the expression values of (only) the selected genes, the classifier identifies the k most similar samples in the training dataset. It then uses the class labels of these k similar samples through a majority vote. In our experiments, we set the value of k to be 5 as default, after testing for several values in the range 4 to 10.

2.3. *The Single Gene Scoring Methods*

Many of the existing gene selection methods are single gene scoring methods that define a function to approximate the class discrimination strength of a gene.[1,2,3,4,5] Typically, an F-test gene selection method[2,9] is presented, which basically captures the variance of the class variances of the gene expression values in the dataset. A bigger variance indicates that a gene is more differentially expressed and thus ranked higher. Because class sizes might differ a lot, Cho *et al.*[4] proposed a weighted variant, which was further refined by Yang *et al.*[5] We denote these three single gene scoring methods as F-test, Cho and GS, respectively, and combine the EEGS idea with them to have the EEGS-based methods, denoted as EEGS-F-test, EEGS-Cho and EEGS-GS, respectively.

2.4. *Gene Clustering*

Gene clustering in microarray data analysis is an independent research subject, in which genes having a similar expression pattern are clustered for certain applications. In our work here, we are particularly interested in the class discrimination strength of the genes, since we do not want to select too many genes that have similar class discrimination strength. Note that genes having a similar expression pattern would certainly have similar class discrimination strength, but the other way around is not necessarily true. Therefore, we define a new measure trying to better capture the difference in the class discrimination strength between two genes.

Assume there are p genes and n samples in the microarray training dataset, and these n samples belong to L distinct classes. Let a_{ij} denote the expression value of gene i in sample j. This way, the training dataset can be represented as a matrix $A_{p\times n} = (a_{ij})_{p\times n}$. Let $C_1, C_2, \ldots, C_L$ denote the L classes, and $n_q = |C_q|$, for $q = 1, 2, \ldots, L$. Let $\overline{a}_{iq}$ be the mean expression value of gene i in class C_q: $\overline{a}_{iq} = \frac{1}{n_q}\sum_{j\in C_q} a_{ij}$, for $q = 1, 2, \ldots, L$. The centroid matrix is thus $\overline{A}_{p\times L} = (\overline{a}_{iq})_{p\times L}$. The *discrimination strength vector* of gene i is defined as $v_i = \langle |\overline{a}_{iq_1} - \overline{a}_{iq_2}| \mid 1 \leq q_1 < q_2 \leq L\rangle$, where the order of $\frac{1}{2}L(L-1)$ vector entries is fixed the same for all genes, for example the lexicographical order. After all the discrimination strength vectors have been calculated, the k-means algorithm[8] is applied to cluster these p genes into k clusters using their discrimination strength vectors. Essentially, k-means is a centroid-based clustering algorithm that partitions the genes based on their pairwise distances. We adopt both the Euclidean distance and the Pearson correlation coefficient in our experiments. Again, we have tested several values of k in the k-means algorithm (cf. Section 4.1) and we have set it to 100 as default.

2.5. *The Complete EEGS-Based Gene Selection Methods*

Given a microarray training dataset containing p genes and n samples in L classes, an EEGS-based gene selection method first calls the k-means algorithm (with $k = 100$) to cluster genes. Next, depending on the detailed single gene scoring method integrated in the method, which is one of F-test, Cho and GS, it calls the single gene scoring method to score all the genes and sort them into non-increasing order. Using this gene order and the gene cluster information, the EEGS-based method selects a pre-specified number, x, of top ranked genes with the constraint that there are at most T genes per cluster can be selected. In more details, it scans through the gene order and picks up a gene only if there are less than T genes from the same cluster selected. These x selected genes are then fed to classifier construction, either the SVM classifier or the KNN classifier. In our experiments, we have tested x ranging from 1 to 80 and several values for T (cf. Section 3.2). We have set $T = 1$ as default (cf. Section 4.1).

Depending on the single gene scoring method integrated into the EEGS-based gene selection method, which is one of F-test, Cho and GS, the method is referred to as EEGS-F-test, EEGS-Cho and EEGS-GS, respectively.

3. Experimental Results

We compare the three EEGS-based gene selection methods with the three ordinary gene selection methods, measured by the 5-Fold cross validation classification accuracy. Note that we have adopted two distance measures in the k-means clustering algorithm. We have a broader collection of experimental results, but here report only those based on the Euclidean distance, as there is essentially no difference between the results based on the Pearson correlation coefficient (cf. Section 4.1). Note also that we have adopted two classifiers, a linear kernel SVM classifier and a KNN classifier. We choose to plot their classification accuracies together labeled by different notations, for instance, EEGS-Cho-KNN labels the accuracies of the KNN classifier. The experiments are done on two real cancer microarray datasets, CAR[10] and LUNG,[9] whose details are described in the following subsection.

3.1. *Dataset Descriptions*

The CAR dataset contains 174 samples in eleven classes: *prostate*, *bladder/ureter*, *breast*, *colorectal*, *gastroesophagus*, *kidney*, *liver*, *ovary*, *pancreas*, *lung adenocarcinomas*, and *lung squamous cell carcinoma*, which have 26, 8, 26, 23, 12, 11, 7, 27, 6, 14, and 14 samples, respectively.[10] Each sample originally contained 12,533 genes. We preprocessed the dataset as described in Su *et al.*[10] to include only those probe sets whose maximum hybridization intensity in at least one sample is ≥ 200; Subsequently, all hybridization intensity values ≤ 20 were raised to 20, and the values were log transformed. After preprocessing, we obtained a dataset of 9,182 genes.

The LUNG dataset[9] contains in total 203 samples in five classes: *adenocarcinomas*, *squamous cell lung carcinomas*, *pulmonary carcinoids*, *small-cell lung carcinomas* and *normal lung*, which have $139, 21, 20, 6, 17$ samples, respectively. Each sample originally had 12,600 genes. A preprocessing step which removed genes with standard deviations smaller than 50 expression units, produced a dataset with 3,312 genes.[9]

3.2. *Cross Validation Classification Accuracies*

The classification accuracies reported here were obtained under the default setting which uses Euclidean distance, $k = 100$ in the k-means clustering algorithm, and at most $T = 1$ gene per cluster could be selected. On each of the two datasets, all six gene selection methods, F-test, Cho, GS, EEGS-F-test, EEGS-Cho, and EEGS-GS, were run and the 5-Fold cross validation classification accuracies were collected and plotted in Figure 1.

Obviously, these plots show that regardless of which cross validation scheme and which classifier were used, the classification accuracies of the EEGS-based gene selection methods were significantly higher than that of their non-EEGS-based counterparts. Typically, on the CAR dataset, the classification accuracies of the EEGS-based methods were significantly higher — though the difference between the classification accuracies became smaller with the increasing number of selected genes, it remained to be more than 10%. From Figure 1, among the single gene scoring methods, another observation is that the GS method performed better than the Cho method and the F-test method. The EEGS-based methods had the same performance tendency on the CAR dataset. On the LUNG dataset, similar results were obtained, although the performance differences between the EEGS-based methods and the non-EEGS-based methods were smaller than those on the CAR dataset.

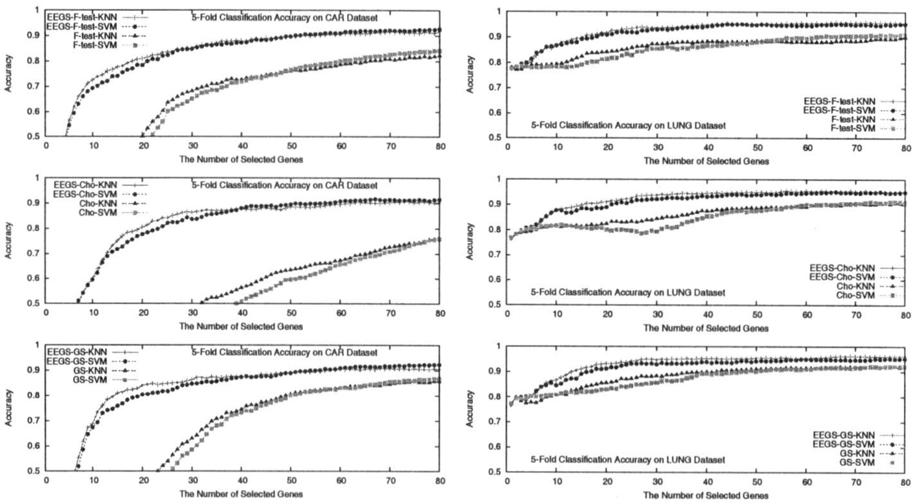

Figure 1. The 5-Fold classification accuracies of the six gene selection methods on the CAR and LUNG datasets.

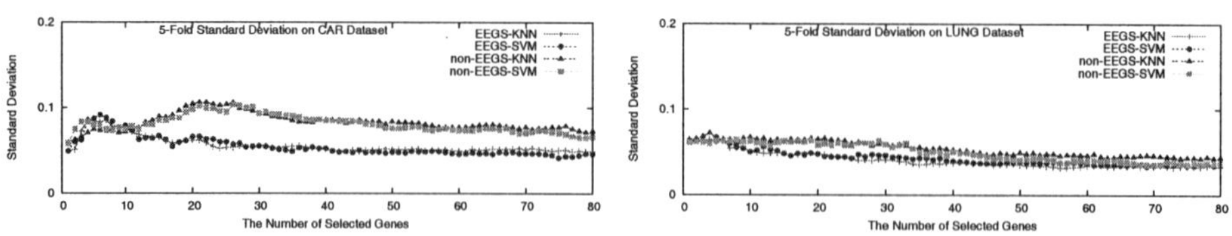

Figure 2. Plots of average standard deviations of the 5-Fold classification accuracies of the EEGS-based and the non-EEGS-based gene selection methods, combined with the KNN and the SVM classifiers, on the CAR dataset (left) and LUNG dataset (right), respectively.

We have also calculated the standard deviations of the 5-Fold cross validation classification accuracies. Note that the accuracies plotted in Figure 1 were averages over 100. Figure 2 plots the average standard deviations of the EEGS-based methods and the non-EEGS-based methods, on the CAR and the LUNG datasets, respectively. Namely, the EEGS-KNN plot records the average standard deviations of the three EEGS-based methods (EEGS-F-test, EEGS-Cho and EEGS-GS) combined with the KNN classifier, and the non-EEGS-SVM plot records the average standard deviations of the three non-EEGS-based methods (F-test, Cho and GS) combined with the SVM classifier, and so on. These results show that the standard deviations of the classification accuracies of the EEGS-based methods were even smaller than those of the non-EEGS-based methods, indicating that the EEGS-based methods performed more consistently. The statistical significance of the outperformance of the EEGS-based methods over the non-EEGS-based methods was done and the p values in the Analysis of variance (ANOVA) was always less than 0.001. (For the complete results, the readers might refer to supplementary materials at http://www.cs.ualberta.ca/~ghlin/src/WebTools/cgs.php.)

4. Discussion

4.1. *Gene Clustering*

We adopted the k-means algorithm for gene clustering, in which k, the number of expected clusters, has to be set beforehand. Obviously, the value of k will affect the sizes of resultant clusters, and therefore will affect T ultimately, which is the maximum number of genes per cluster to be selected. We chose to empirically determine these two values. To this end, we experimented with 15 values for k: from 10 to 150 (in the tens), and five values for T: 1, 2, 3, 4 and 5. All three EEGS-based methods combined with two classifiers were tested on the CAR dataset, under the 5-Fold cross validation, for each combination of k and T. Associated with each value of T, a classification accuracy is defined as the mean value of $100 \times 3 \times 2 \times 15 = 9,000$ values, where there are 100 runs in the 5-Fold cross validation, 3 EEGS-based methods, 2 classifiers, and 15 values of k in the test. These classification accuracies, with respect to the number of selected genes, are plotted in Figure 3 (left), where $T = 1$ clearly performed the best. Similarly, associated with each value of k, a

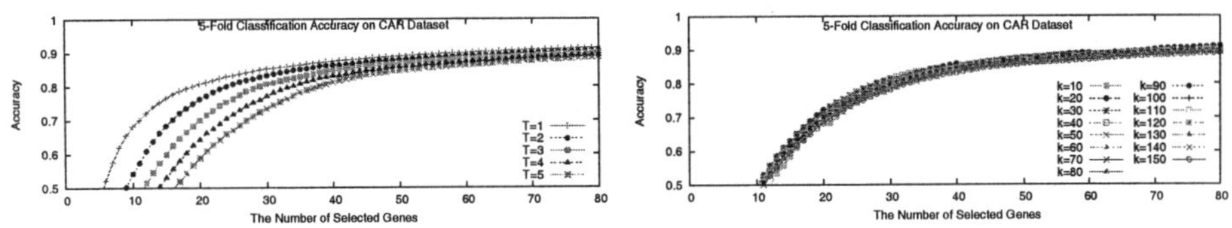

Figure 3. The effects of the number of clusters in gene clustering and the maximum number of genes per cluster can be selected.

classification accuracy is defined as the mean value of $100 \times 3 \times 2 \times 5 = 3,000$ values, where there are 5 values of T in the test. Again, these classification accuracies, with respect to the number of selected genes, are plotted in Figure 3 (right), where we can see that the value of k didn't really affect the performance. Since we decided to set the maximum number of selected genes to be 80, we determined to set $k = 100$ and $T = 1$ as default.

Another important factor in gene clustering that might affect its performance is the

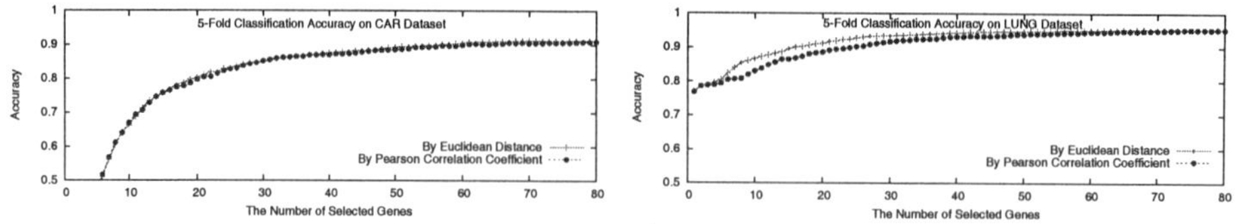

Figure 4. The effects of the Euclidean distance and the Pearson correlation coefficient in gene clustering.

distance measure, for which the Euclidean distance and the Pearson correlation coefficient are two most commonly adopted ones. We have experimented with both of them in the k-means clustering algorithm on the CAR and the LUNG datasets. With the default setting for k and T, we collected the 5-Fold classification accuracies which are the mean values of $100 \times 3 \times 2 = 600$ values and plotted them in Figure 4. It can be clearly seen that, the detailed distance measure did not seem to affect the overall performance of the EEGS-based methods, in terms of their classification accuracy. Therefore, we chose the Euclidean distance as our default setting.

4.2. *Datasets*

Note that in the EEGS-based gene selection methods, a discrimination strength vector is computed for every gene, and genes are clustered using the Euclidean distance defined on their discrimination strength vectors. The main intention for such clustering is to limit the number of genes that have very similar class discrimination strength to be selected, and thus to provide space for other individually-less but collectively-more differentially expressed genes to participate in the class prediction. This goal would not be achieved when there are only two classes in the dataset (binary classification), which would mean that the discrimination strength vectors have only one entry and the EEGS-based method reduces to its component basic gene selection method. For similar reasons, we suspect that the EEGS-based gene selection methods would work well when the number of classes in the dataset is three. The CAR and the LUNG datasets contain eleven and five classes, respectively, and therefore the discrimination strength vectors have 55 and 10 entries, respectively. The EEGS-based gene selection methods all performed excellent on them.

For various reasons, microarray datasets are often imbalanced, that is, the sizes of the classes are highly variable. For example, in the LUNG dataset, the maximum class size is 139 while the minimum class size is only 6. Since it is possible that during the 5-Fold cross validation the random partition produces a training dataset containing only a few samples, or maybe even none, from a small class, the testing would make mistakes on the samples from the same class. To verify how much the dataset unbalance would affect the performance of a gene selection method, we removed the classes of sizes smaller than 10 from the CAR and the LUNG datasets to produce two reduced but more balanced datasets, denoted as CAR^r and LUNG^r, respectively. Consequently, the CAR^r and the LUNG^r datasets contain 153 samples in 8 classes and 197 samples in 4 classes, respectively. We then ran all six methods combined with both the KNN classifier and the SVM classifier on the full and the reduced datasets, and plotted the average classification accuracies (each over three methods with two classifiers, i.e., six values) in Figure 5. In the figure, one can see that the performance of the EEGS-based methods did not change a lot on the reduced CAR^r and the LUNG^r datasets, compared with their performance on the full datasets. Interestingly, for the non-EEGS-based methods, their performance increased significantly on the CAR^r dataset, but not on the LUNG^r dataset. Nevertheless, these results show that the EEGS-based methods performed more stable (and better) than the non-EEGS-based methods on imbalanced datasets. One of the possible reasons is that the EEGS-based methods might be able to select some genes that are signatures of the samples in the small classes, for which further studies are needed to understand better.

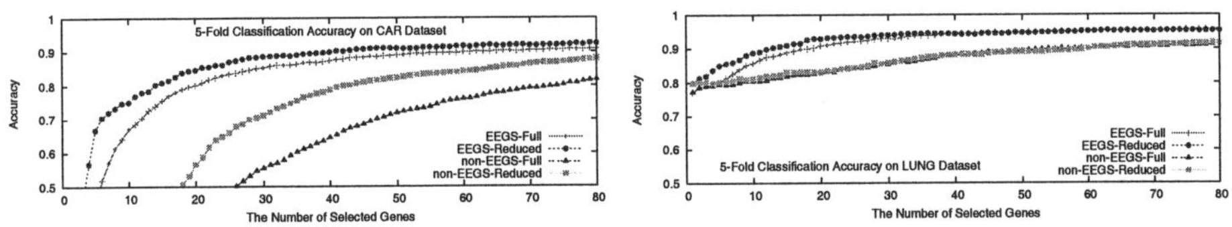

Figure 5. Classification accuracies of the EEGS-based and the non-EEGS-based methods on the full and reduced datasets, where EEGS-Full plots the average classification accuracies of the EEGS-based methods on the full dataset.

Our further statistics on the classification *precision* and *recall* for each class in the 5-Fold cross validation shows that they seemed independent of the class size (data not shown but can be accessed through http://www.cs.ualberta.ca/~ghlin/src/WebTools/cgs.php).

4.3. *A Case Study on the CAR Dataset*

The EEGS-based gene selection methods are designed to not select too many genes having similar class discrimination strength, so as to consider individually-less but collectively-more differentially expressed genes. In this sense, some selected genes might be lower in the gene order, but have strength to discriminate some classes on which its preceding genes might not do well. To examine if this indeed happened, we collected more detailed results on the first 10 genes selected by F-test and EEGS-F-test, respectively. The 5-Fold cross validation classification accuracies of the SVM classifier built on the first x genes were also collected, for $x = 1, 2, \ldots, 10$. We summarized the results in Table 1, in which column 'Probe Set' records the probe set (gene) id in the CAR dataset, column 'R' records the rank of the gene in the gene order by F-test, and column 'Accuracy' records the classification accuracy of the gene subset up to the gene at the row.

Note that the third gene (probe set) selected by EEGS-F-test, *765_s_at*, has a rank 17, which was thus not selected by F-test. The classification accuracy of the top 10 genes selected by F-test was only 30.63%, while adding the third gene *765_s_at* in EEGS-F-test lifted the classification accuracy to 42.18%, already significantly higher than 30.63%. On average, the contribution of each gene, except the first, selected by EEGS-F-test was 6.10% in terms of classification accuracy; the contribution of each gene, except the first, selected by F-test was only 1.23%. These figures suggested that when the number of selected genes was fixed, the genes selected by EEGS-F-test had much higher class discrimination strength compared to the genes selected by F-test.

Table 1. The first 10 genes selected by the EEGS-F-test and the F-test methods on the CAR dataset, respectively, and the respective 5-Fold cross validation classification accuracies of the SVM classifiers built on the genes. Column 'R' records the rank of the gene in the gene order by F-test, and column 'Accuracy' records the classification accuracy of the gene subset up to the gene at the row.

EEGS-F-test-SVM			F-test-SVM		
Probe Set	R	Accuracy	Probe Set	R	Accuracy
40794_at	1	19.54%	40794_at	1	19.54%
41238_s_at	4	28.91%	660_at	2	25.46%
765_s_at	17	**42.18%**	32200_at	3	25.40%
1500_at	21	60.52%	41238_s_at	4	30.00%
35220_at	24	64.54%	34941_at	5	30.69%
32771_at	27	68.62%	41468_at	6	30.52%
34797_at	38	70.75%	36141_at	7	30.12%
35194_at	50	73.56%	617_at	8	30.35%
36806_at	52	73.16%	37812_at	9	30.29%
40511_at	63	74.48%	217_at	10	30.63%

Acknowledgments. This research is supported in part by AHFMR (to LX), AICML (to RG), CFI (to GL), iCore (to YS and RG), NSERC (to RG, MS and GL) and the University of Alberta (to MS).

References

1. T. R. Golub *et al.* Molecular classification of cancer: Class discovery and class prediction by gene expression monitoring. *Science*, 286:531–537, 1999.
2. P. Baldi and A. D. Long. A Bayesian framework for the analysis of microarray expression data: Regularized t-test and statistical inferences of gene changes. *Bioinformatics*, 17:509–519, 2001.
3. S. Dudoit, J. Fridlyand, and T. P. Speed. Comparison of discrimination methods for the classification of tumors using gene expression data. *Journal of the American Statistical Association*, 97:77–87, 2002.
4. J. Cho, D. Lee, J. H. Park, and I. B. Lee. New gene selection for classification of cancer subtype considering within-class variation. *FEBS Letters*, 551:3–7, 2003.
5. K. Yang, Z. Cai, J. Li, and G.-H. Lin. A stable gene selection in microarray data analysis. *BMC Bioinformatics*, 7:228, 2006.
6. M. Xiong, X. Fang, and J. Zhao. Biomarker identification by feature wrappers. *Genome Research*, 11:1878–1887, 2001.
7. I. Guyon, J. Weston, S. Barnhill, and V. Vapnik. Gene selection for cancer classification using support vector machines. *Machine Learning*, 46:389–422, 2002.
8. F. Blanchot-Jossic *et al.* Up-regulated expression of ADAM17 in human colon carcinoma: co-expression with EGFR in neoplastic and endothelial cells. *Oncogene*, 207:156–163, 2005.
9. A. Bhattacharjee *et al.* Classification of human lung carcinomas by mRNA expression profiling reveals distinct adenocarcinoma subclasses. *Proceedings of National Academy of Sciences of the United States of America*, 98:13790–13795, 2001.
10. A. I. Su *et al.* Molecular classification of human carcinomas by use of gene expression signatures. *Cancer Research*, 61:7388–7393, 2001.

COMPUTING THE ALL-PAIRS QUARTET DISTANCE ON A SET OF EVOLUTIONARY TREES

M. STISSING, T. MAILUND,* C. N. S. PEDERSEN AND G. S. BRODAL

Bioinformatics Research Center and Dept. of Computer Science, University of Aarhus, Denmark

R. FAGERBERG

Dept. of Mathematics and Computer Science, University of Southern Denmark, Denmark

We present two algorithms for calculating the quartet distance between all pairs of trees in a set of binary evolutionary trees on a common set of species. The algorithms exploit common substructure among the trees to speed up the pairwise distance calculations thus performing significantly better on large sets of trees compared to performing distinct pairwise distance calculations, as we illustrate experimentally, where we see a speedup factor of around 130 in the best case.

1. Introduction

In biology, trees are widely used for describing the evolutionary history of a set of taxa, e.g. a set of species or a set of genes. Inferring an estimate of the (true) evolutionary tree from available information about the taxa is an active field of research and many reconstruction methods have been developed, see e.g. Felsenstein[1] for an overview. Different methods often yield different inferred trees for the same set of species, as does using different information about the species, e.g. different genes, when using the same method. To reason about the differences between estimates in a systematic manner several distance measures between trees have been proposed.[2–6] Given a set of estimates of evolutionary trees on the same set of species, the differences between the trees can then be quantified by all the pairwise distances under an appropriate distance measure.

This paper concentrates on computing the distance between all pairs of trees in a set of unrooted fully resolved (i.e. binary) evolutionary trees on a common set of species using a distance measure called the quartet distance. For an evolutionary tree, the *quartet topology* of four species is determined by the minimal topological subtree containing the four species. The four possible quartet topologies of four species are shown in Fig. 1. In a fully resolved tree only the three fully quartet topologies can of course occur. Given two evolutionary trees on the same set of n species, the *quartet distance* between them is the number of sets of four species for which the quartet topologies differ in the two trees. For binary trees, the quartet distance can be calculated in time $O(n \log n)$, where

*Current affiliation: Dept. of Statistics, University of Oxford, UK.

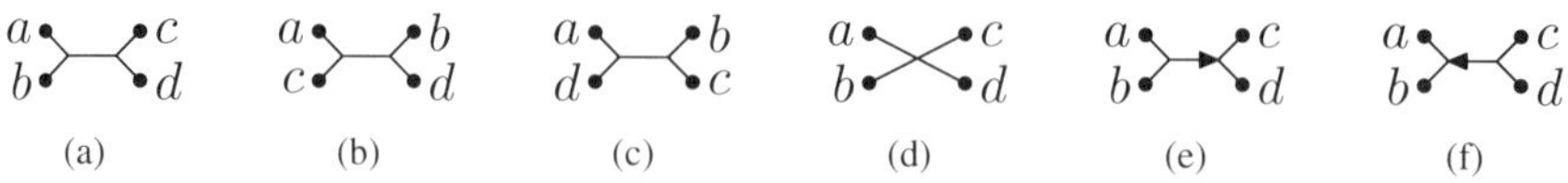

Figure 1. Figures (a)–(d) show the four possible quartet topologies of species a, b, c, and d. Topologies $ab|cd$, $ac|bd$, and $ad|bc$ are *butterfly* quartets, while topology ${}^{a}_{b} \times {}^{c}_{d}$, is a *star* quartet. For binary trees, only the butterfly quartets are possible. Figures (e) and (f) show the two ordered butterfly quartet topologies induced by the butterfly quartet topology in (a).

n is the number of species.[7] A simpler algorithm with running time $O(n \log^2 n)$ has been implemented in the tool *QDist*.[8] For arbitrary degree trees, the quartet distance can be calculated in time $O(n^3)$ or $O(n^2 d^2)$, where d is the maximum degree of any node in any of the two trees.[9]

As fully resolved trees are the focus of most reconstruction algorithms for evolutionary trees, we, in this paper, develop an algorithm for computing the pairwise distances between k fully resolved trees on a common set of species that exploits similarity between the input trees to improve on the straightforward approach of performing the k^2 pairwise comparisons independently which has $O(k^2 t)$ running time, where t is the time for comparing two trees, $O(n \log n)$ if the algorithm from Brodal et al.[7] is used. The worst case running time of our approach remains $O(k^2 t)$, if the input trees share very little structure, but experiments show that the our algorithm achieves significant speedups of more than a factor 100 on more related trees. Our method has been implemented as an extension to the existing tool QDist Mailund and Pedersen[8] as is available for download at the QDist www-page.

2. The Pair-wise Quartet Distance

The $O(n \log^2 n)$ algorithm developed by Brodal et al.[10] serves as a basis for our all-pairs algorithm. The $O(n \log n)$ works just as well, but as there is no known working implementation of it, we settled for the simpler $O(n \log^2 n)$. In this section we describe the algorithm for computing the quartet distance between a pair of trees. Two basic ideas are used in the algorithm: instead of counting the number of quartet topologies that differs between the two trees, the algorithm counts the number of quartet topologies that are shared and subtract this number from the total number possible of quartets, $\binom{n}{4}$, and instead of counting shared quartet topologies, it counts the shared *ordered* topologies. An (unordered) butterfly quartet topology, $ab|cd$ induces two ordered quartet topologies $ab \to cd$ and $ab \leftarrow cd$, by the orientation of the middle edge of the topology, as shown in Fig. 1, (e) and (f). Clearly there are twice as many ordered quartet topologies, and the quartet distance can thus be obtained by subtracting from $2\binom{n}{4}$ the number of shared ordered quartet topologies and then dividing by 2.

Given an ordered edge, e, let A denote the tree behind e, and B and C denote the two trees in front of e. Each such edge induces a number of oriented quartets: $aa' \to bc$ where $a, a' \in A$, $b \in B$, and $c \in C$, and each oriented quartet is induced by exactly one oriented edge in this way. The number of oriented quartets induced by e is $\binom{|A|}{2} \cdot |B| \cdot |C|$ where $|X|$ is used to denote the number of leaves in the tree X.

For each inner node, v, each of the incidence oriented edges, see Fig. 2, induces oriented quartets, and the number of oriented quartets induced by the node is

$$\binom{|A|}{2}\cdot|B|\cdot|C|+\binom{|B|}{2}\cdot|A|\cdot|C|+\binom{|C|}{2}\cdot|A|\cdot|B| \quad (1)$$

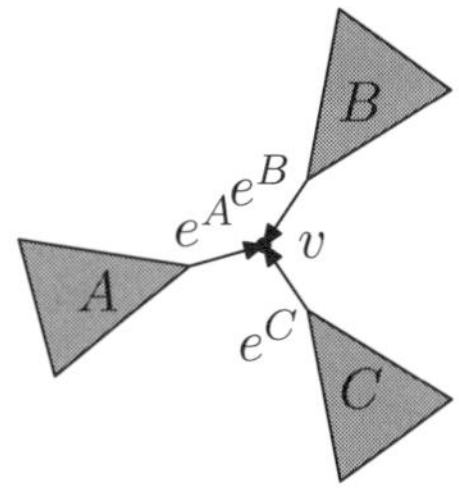

Figure 2. The inner node v induces oriented quartets through each of the three incident edges.

When comparing two trees, T_1 and T_2, we can consider a pair of inner nodes, v_1 in T_1 and v_2 in T_2, with sub-trees A_1, B_1, C_1 and A_2, B_2, C_2 and incident edges e_1^A, e_1^B, e_1^C and e_2^A, e_2^B, e_2^C, respectively, as in Fig. 2. The shared oriented quartets induced by edges e_1^A and e_2^A are the oriented quartets $aa' \to xy$ where $a, a' \in A_1 \cap A_2$ and x and y are found in different subtrees in front of e_1^A and e_2^A, respectively. The number of shared oriented quartets induced by e_1^A and e_2^A, count(e_1^A, e_2^A) is then:

$$\binom{|A_1 \cap A_2|}{2} \cdot \Big(|B_1 \cap B_2| \cdot |C_1 \cap C_2| + |B_1 \cap C_2| \cdot |C_1 \cap B_2|\Big) \tag{2}$$

where $|X_1 \cap Y_2|$, with a slight abuse of notation, denote the number of shared leaves in sub-trees X_1 and Y_2. The number of shared oriented quartets for nodes v_1 and v_2 are the sum of the shared quartets of their incident edges:

$$\text{count}(v_1, v_2) = \sum_{X\in\{A,B,C\}} \sum_{Y\in\{A,B,C\}} \text{count}(e_1^X, e_2^Y) \tag{3}$$

and the total count of shared quartets is the sum of counts for all pairs of nodes v_1 and v_2.

We can express (2) slightly differently: Let node v_1 be an inner node in T_1 with associated sub-trees A, B and C as in Fig. 2. We can then colour all leaves in A with the colour $\mathcal{A}$, the leaves in B with the colour $\mathcal{B}$, and the leaves in C with the colour $\mathcal{C}$. The leaves in T_2 can be coloured with the same colours as in T_1, and for any inner node, v_2 in T_2, the sub-trees will then contain a number of leaves of each colour, see Fig. 3. The size of the intersections of leaves in the sub-trees in T_1 and T_2, respectively, is now captured by the colouring. The oriented quartets induced by both e^A and e^1 (see Fig. 3) are the quartets $aa' \to bc$ where a and a' are coloured $\mathcal{A}$ and present in the first subtree of v_2 and one of b, c is coloured $\mathcal{B}$ and the other $\mathcal{C}$ and present in the two other subtrees respectively. Let $\mathbf{a}(i), \mathbf{b}(i), \mathbf{c}(i)$, $i = 1, 2, 3$ denote the number of leaves with colour $\mathcal{A}$, $\mathcal{B}$, and $\mathcal{C}$ in the three sub-trees of v_2. The "colouring analogue" to (2) is then

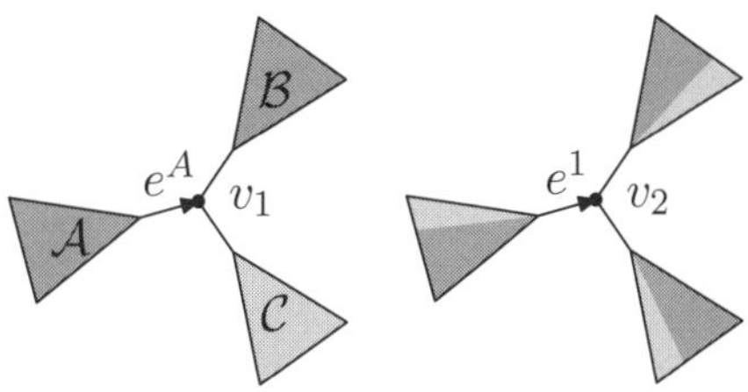

Figure 3. The coloured leaves for an inner node, v_1 in T_1 and the same colouring for inner node v_2, in T_2.

$$\text{count}(e^A, e^1) = \binom{\mathbf{a}(1)}{2} \cdot \Big(\mathbf{b}(2)\cdot\mathbf{c}(3) + \mathbf{b}(3)\cdot\mathbf{c}(2)\Big) \tag{4}$$

Similar expressions can be derived for the remaining pairs of edges.

A naïve approach simply colours the leaves with colours $\mathcal{A}$, $\mathcal{B}$, and $\mathcal{C}$, for each node in T_1, and then, for each node in T_2, counts the numbers $\mathbf{a}(i), \mathbf{b}(i), \mathbf{c}(i)$ and evaluates the sum in (3). The time complexity of this is the time it takes for the colouring of leaves according to the inner node in T_1 plus the time it takes to count $\mathbf{a}(i), \mathbf{b}(i), \mathbf{c}(i)$ and evaluate the sums in T_2. Since the colouring is a simple matter of traversing T_1, and assuming we connect the leaves of T_1 with the corresponding leaves in T_2 with pointers, the colouring takes time $O(n)$ for each node of T_1 giving a total colouring time of $O(n^2)$. If we root T_2 in an arbitrary leaf, we can count $\mathbf{a}(i), \mathbf{b}(i), \mathbf{c}(i)$ in a depth-first traversal: we annotate each edge with the number of leaves below it with colour $\mathcal{A}$, $\mathcal{B}$ and $\mathcal{C}$, respectively, and for a node v_2, the counts out of each child-edge is immediately available from this annotation. The counts on the edge to the root can be obtained from the sum of the counts on the children-edges if we remember the total number of leaves in each colour. This way we can count and then evaluate the sum for all nodes in time $O(n)$ for each colouring, with a total counting time of $O(n^2)$, giving all in all a running time of $O(n^2)$.

The algorithm in Brodal et al.[10] improves on this using two tricks: it reuses parts of the colouring when processing the nodes of T_1 and uses a "smaller part" trick to reduce the number of re-colourings, and it uses a hierarchical decomposition tree to update the counts in T_2 in a clever way thus reducing the counting time. As a preprocessing step, the first tree is rooted in an arbitrary leaf, r, and traversed depth first to calculate the size of each sub-tree. This size is stored in the root of each sub-tree, such that any inner node, v, in constant time can know which of its children is the root of the smaller and the larger sub-tree (with an arbitrary but fixed resolution in case of ties). When counting, initially, r is coloured $\mathcal{C}$ and all other leaves are coloured $\mathcal{A}$. The tree is then traversed, starting from r, with the following invariant: before processing node v, all leaves in the subtree rooted in v are coloured $\mathcal{A}$ and all remaining leaves are coloured $\mathcal{C}$; after the processing, all leaves in v's subtree are coloured $\mathcal{C}$. The processing of node v consists of first colouring the smaller subtree of v $\mathcal{B}$ and then counting the number of shared quartets induced by v. Then the smaller subtree is recoloured $\mathcal{C}$ allowing a calculation of the quartets shared in the larger subtree—since all the leaves in the larger subtree have colour $\mathcal{A}$ and the rest of the leaves now have the colour $\mathcal{C}$. After this, the leaves in the larger subtree have, according to the invariant, been coloured $\mathcal{C}$. The leaves in the smaller subtree are then recoloured $\mathcal{A}$ thus enabling a recursive count of the shared quartets in the smaller sub-tree. The number of shared quartets in v's subtree is the sum of the three counts. For the time usage on colouring, observe that each leaf is only coloured initially and then only when it is part of a smaller tree. Since each leaf can at most be part of $O(\log n)$ smaller trees, the total colouring time is in $O(n \log n)$.

The hierarchical decomposition trick is a balanced tree-structure that enables constant time counting, but has an updating cost associated with re-colourings: updating the colour of l leaves takes $O(l + l \log \frac{n}{l})$ time. Ultimately this yields the $O(n \log^2 n)$ running time of the algorithm. The details of this tree-structure is not important for the algorithm we develop in this paper—it can be reused in this algorithm in a straight forward manner—we will not go into details about them here but instead refer to Brodal et al.[10]

3. The All Pairs Quartet Distance

We consider four algorithms for calculating the all-pairs quartet distance between k trees. The two simplest of these simply performs $O(k^2)$ single comparisons between two trees using time $O(n^2)$ and $O(n \log^2 n)$ per comparison, respectively. These are the algorithms we will attempt to improve upon by exploiting similarity between the compared trees to obtain a heuristic speedup. The use of similarity utilizes a *directed acyclic graph* (DAG) representation of the set of trees. Given our set of k trees, we can root all trees in the same arbitrary leaf, i.e. a leaf label is chosen arbitrarily and all trees are rooted in the leaf with that label. Given this rooting, we can define a representation of the trees as a DAG, D, satisfying:

(1) D has k nodes with in-degree 0: $r_1, r_2, \ldots, r_k$, one for each root in the k trees; we will refer to these as the *roots*.
(2) D has $n - 1$ nodes with out-degree 0, one for each of the non-root leaves in the trees; these DAG leaves are labelled with the $n - 1$ non-root leaf labels; these are the *leaves*.
(3) For each tree T there is an embedding of T in D, in the sense that there is a mapping $e_i : v(T) \to v(D)$ such that the root of T is mapped to the root in D representing T, r_i, the leaves of T are mapped to the leaves of D with the same labels, and for each edge in T, (v, v'), the image of e_i, $(e_i(v), e_i(v'))$, is an edge in D.
(4) For any non-root nodes v_1 and v_2 from trees T_1 and T_2, respectively, if the tree rooted in v_1 is isomorphic to the tree rooted in v_2, then v_1 and v_2 are represented by the same node in D, $e_1(v_1) = e_2(v_2)$.

Conditions 1–3 ensures that each of the k trees are represented in D, in the sense that tree T_i can be extracted by following the directed edges from r_i to the leaves. Condition 4 ensures that this is the minimal such DAG. Condition 1 is implied by the remaining if no two trees in the set are isomorphic. Since isomorphic trees will have distance 0, and this isomorphism is easy to check for when building the DAG as described below, we assume that no trees are isomorphic and thus that merging the trees will automatically ensure condition 1.

We can build the DAG iteratively by merging one tree at a time into D. Initially, we set $D := T_1$, then from $i = 2$ to $i = k$ we merge T_i into D using a depth first traversal of T_i, for each node inserting a corresponding node in the DAG if it isn't already present. Since the set of leaves is the same for all trees, each leaf node can be considered already merged into D. For each inner node v, with children u and w, recursively merge the subtrees of u and w, obtaining the mapped DAG nodes $e(u)$ and $e(w)$. If these share a parent in D, v is mapped to that parent, otherwise a new node $e(v)$ is inserted in the DAG. The test for a shared parent of two nodes can be made efficiently by keeping a table mapping pairs of DAG nodes to their shared ancestor—two DAG nodes can never share more than one parent, since such two parents would be merged together. This table can be kept updated by inserting $(e(u), e(w)) \mapsto e(v)$ whenever we insert $e(v)$ as above. Note that each inner DAG node has out-degree 2.

We observe that if two tree nodes map to the same DAG node, they induce the same oriented quartets: if $e_i(v_i) = e_j(v_j)$ for inner nodes v_i in T_i and v_j in T_j, then, by condition 3, the subtrees of v_i and v_j are isomorphic, i.e. one child of v_i is isomorphic with one of v_j and the other child with the other of v_j. Thus, a colouring as on the left in Fig. 3, if the isomorphic child-trees are coloured the same colour, result in the same leaf-colourings in the two trees T_i and T_j. Counting the oriented quartets in the DAG, as opposed to the trees, we reduce the amount of work necessary, since merged nodes (tree nodes mapped to the same DAG node) need only be processed once, not once for each tree.

Calculating the pairwise quartet distance between a set of trees $S = \{T_1, T_2, \ldots, T_k\}$ and a single tree T can be done by first merging the trees in S into a DAG D. We then preprocess D depth-first for each of its k roots calculating sizes of subtrees, such that each inner node v in D will know which of its two children is root in the smaller and the larger subtree. This enables us to use the same "smaller part" trick when colouring the subtrees of D as is used when comparing two trees. The algorithm starts by colouring all roots of D using $\mathcal{C}$. Now, for each root r_i in D, the algorithm colours all leafs in the embedded tree rooted at r_i, T_i, using the colour $\mathcal{A}$ and traverses T_i recursively, following the colouring schema and invariants used in the one-against-one tree comparison. However, whenever we process a node in the DAG, we store the sum of the count for that node plus the count for all nodes below it (this is done recursively). Since all counts for that node will be the same, regardless of which root we are currently processing, we can reuse the count if we see the node again when processing later trees. The counting as such can run exactly as described in the previous section, except that when a previously counted node v is met, we simply reuse the count and colour the subtree rooted at v with $\mathcal{C}$ in accordance with the stated invariant. It is still possible to use a hierarchical decomposition of the tree T to speed up the counting. Since we only recursively count each node in D once, not each time a node appears in one of the trees in S, we potentially save a significant amount of time if the trees are closely related and thus D is comparably small.

When calculating the all-pairs distance between k trees, this approach will still have an $O(k^2 t)$ worst case time bound. In practice though, especially for highly similar trees, we suspect that this algorithm will outperform the naive approach. In the case of similar trees, that is comparing $S = \{T_1, T_1, \ldots, T_1\}$ of k identical trees and a tree T the time used is $O(kn)$ for building D, $O(t)$ for comparing the first embedded tree of D with T and $O(kn)$ for comparing and recolouring the rest of the trees embedded in D with T (this covers the cost of constructing and updating a hierarchical decomposition tree aswell). As we do this k times in order to get the all-pairs distance we obtain a total running time of $O(k(kn+t))$ which is superior to the naive approach when n is a true lower bound for t, i.e. $t \in \omega(n)$. We expect that nontrivial comparisons will be somewhere in between the stated bounds.

Calculating the distance between all pairs of trees, or more generally the pairwise distances between two sets of trees, $S = \{T_1, T_2, \ldots, T_k\}$ and $S' = \{T'_1, T'_2, \ldots, T'_{k'}\}$ might clearly be done, as stated above: for each $T_i \in S$ calculate the k' distances between T_i and S' using the algorithm described above. Merging both sets into DAGs, however, will let us exploit similar tree structure in both sets.

Let D and D' be the DAG representations of S and S', respectively. Finding the distances between all pairs of trees can be done by counting, for each tree T_i embedded in D, the number of shared quartets between T_i and each tree in S'. This amounts to running through D, colouring recursively as stated above, and counting how many quartets are shared between $v \in v(D)$ and each T' in S'. Storing this set of counts in v, lets us process each v only once. Given a colouring, calculating how many quartets, for each T' embedded in D', is compatible with this colouring, might reuse the technique from above. Running through D' recursively, root for root, storing the number of quartets compatible with the current colouring in the subtree rooted at v' for each v' in D', enables reusing this number when v' is once again processed. The worst case time for this approach remains $O(k^2n^2)$. If we consider identical trees, the time used in calculating the all-pairs quartet distance between such is $O(n^2 + k)$ for comparing the first tree with all others and $O(k(k + n))$ for comparing and colouring the remaining pairs. In total, this is $O(n^2 + k^2)$. In real life examples, we suspect a running time somewhere between these bounds.

4. Experimental Results

The new algorithms are expected to obtain significant speedups when comparing multiple trees of high similarity. In this section we report on experiments to validate this. An essential assumption for the expected speedup of using DAGs is that the trees share much structure, i.e. that the number of nodes in the DAGs are significantly smaller than the number of nodes in all pairs of trees. To evaluate this, we calculated the sizes of the DAGs obtained by merging the trees, see Fig. 4. The results of these experiments are very encouraging. The trees were obtained using the following method: First, to construct similar trees of size n (trees with n leaves), a phylogenetic tree of size n is constructed (using the tool *r8s*[11]). This tree is then used in simulating evolution along its phylogeny, hence evolving n sequences of a given length m (using the tool *Seq-Gen*[12]). The distances between these sequences give rise to an $n \times n$ matrix of distances (calculated using *ednadist*). This matrix, when neighbour-joined (done using *QuickJoin*[13]), will result in a tree of size n.

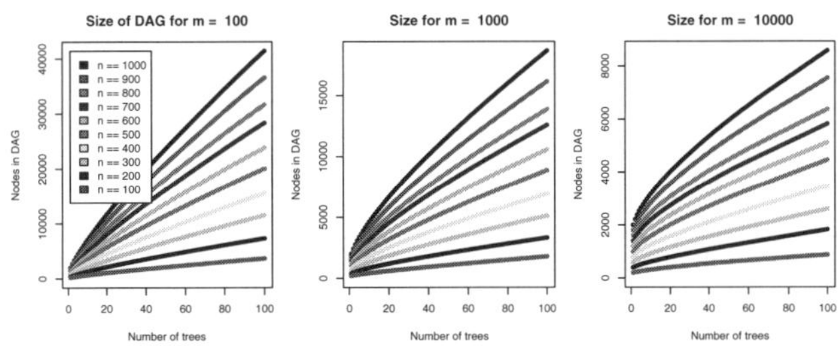

Figure 4. Sizes of DAGs constructed. As predicted, as m grows, the size of the DAG is reduced, and as shown the number of nodes in the DAG is considerably lower than the number of nodes in the sets of trees. (By construction, leaves in a DAG are shared, meaning that the number of nodes in a DAG will be approximately less than half the number of nodes in the corresponding trees.) Note that for $k = 100$, $n = 1000$ and $m = 10000$, the collection of trees contain $100 \cdot 1998$ nodes, whereas the average size of the DAG is 8588. This is $\frac{8588}{100 \cdot 1998} \approx \frac{1}{23}$ the size of the corresponding set of trees or just $\frac{8588}{1998} \approx 4.3$ times the size of one single tree.

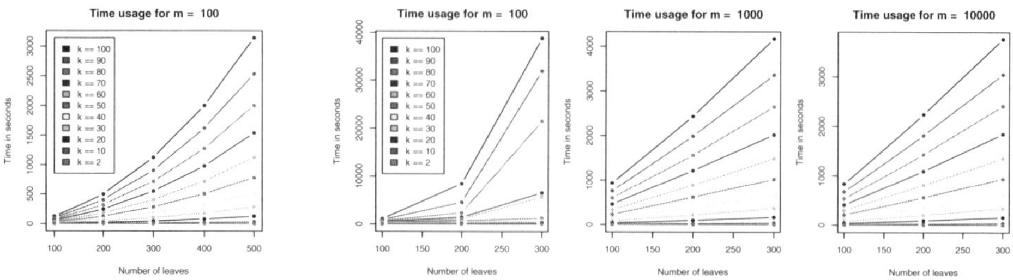

(a) $O(k^2n^2)$ (b) $O(k^2 \cdot n \log^2 n)$

Figure 5. The performance of the two $O(k^2t)$ algorithms. *(a)* Time used in tree against tree comparison, i.e. applying the $O(n^2)$ one against one algorithm for each pair of trees taken from the k trees. As this algorithm is independent of the similarity of the trees (the size of m) only $m = 100$ is included. This experiment has only been repeated six times with $n \in \{100, 200, 300, 400, 500\}$ due to its quite time consuming nature and the above mentioned independence of similarity. For k and n the time used is seen to be close to $1.23 \cdot 10^{-6} \cdot k^2 \cdot n^2$ seconds. This method seems most appropriate for fewer smaller trees of little similarity due to its $O(k^2n^2)$ time complexity. *(b)* Time used in tree against hierarchical decomposition tree comparison. This experiment consists of constructing hierarchical decomposition trees for each of the k trees and running the original $O(n \log^2 n)$ algorithm for each possible pair of tree and hierarchical decomposition tree. This experiment has only been repeated six times with values of $n \in \{100, 200, 300\}$, because of its time consumption. (We note that the graph considering $m = 100$ is not as regular as the others. This might be due to the relatively few experiments run.)

This process is repeated (using the same phylogenetic tree) k times to obtain k trees. The length of the simulated sequences, m, influence the similarity of the trees: as m grows, the sequences evolved, will contain more of the information contained in the original phylogenetic tree, and ultimately the k trees developed will be nearer the original phylogenetic tree and thus the pairwise similarity of the k trees constructed will grow as m grows. For each $n \in \{100, 200, \ldots, 1000\}$, $k = 100$ trees of sequence lengths $m \in \{100, 1000, 10000\}$ were constructed. Construction of data was repeated 30 times. Experiments were run on these 30 sets of data except where otherwise stated.[a]

To have a baseline to compare our new algorithms with, we then timed the two tree-against-tree algorithms. First, we compared the simple $O(n^2)$ algorithm with the more complex $O(n \log^2 n)$ algorithm, both implemented in *QDist*.[8] Since the $O(n \log^2 n)$ time complexity is achieved through a number of complex polynomials in the nodes of the hierarchical decomposition tree, we expect it to have a significant overhead compared to the much simpler $O(n^2)$ algorithm, which is verified in Fig. 6. Observe that for $n = 100$ the simpler algorithm is 8 times faster than the more complex. For $n = 1000$ this factor has shrunk to 2 and for $n \geq 2000$ the complex algorithm is preferable.

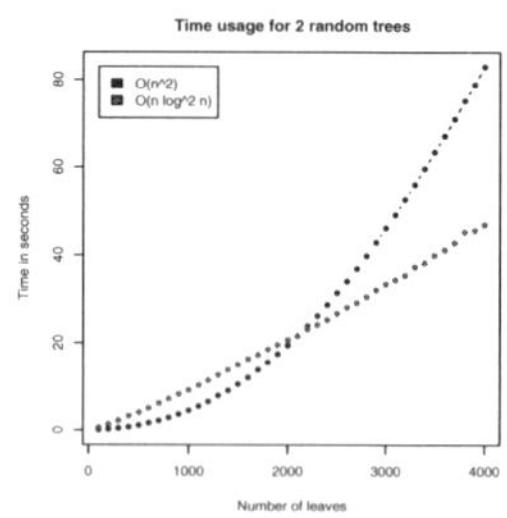

Figure 6. Time used in calculating the quartet distance between two trees. Random trees were used as none of the algorithms should be favoured. The graphs show the average time of 30 runs for each size.

This is not the full story, however. The $O(n \log^2 n)$ al-

[a]6 identical, Intel Pentium 4 1.80 GHz, 512 MB, dedicated machines were used for the experiments.

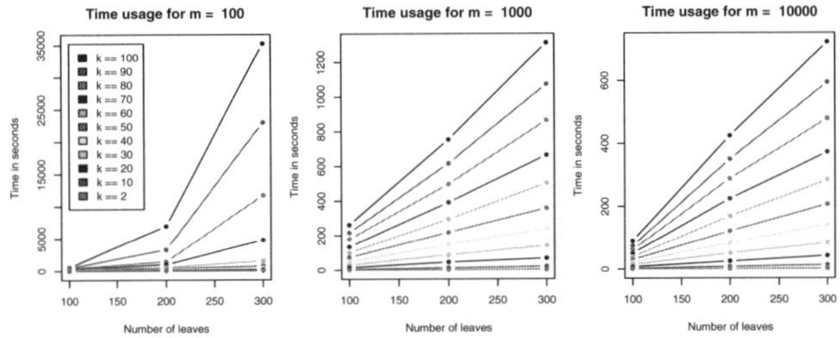

(a) DAG against hierarchical decomposition tree.

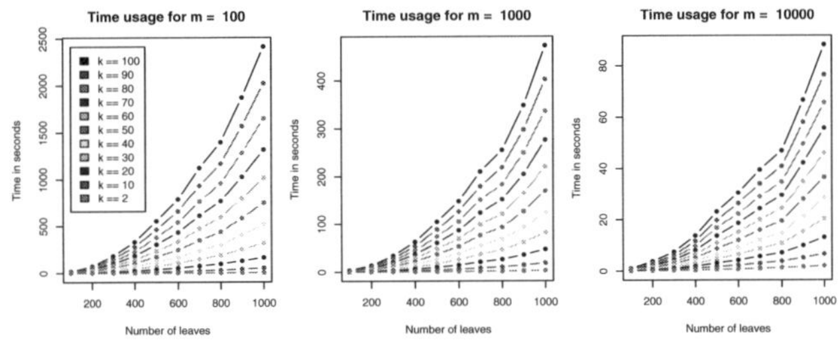

(b) DAG against DAG.

Figure 7. The performance of the two DAG-based algorithms. *(a)* The DAG-against-tree algorithm. Here we construct a DAG D on top of the k trees as well as constructing a set S of hierarchical decomposition trees, each constructed from one of the k trees. Calculating the k^2 quartet distances is accomplished by comparing each $H \in S$ against D. Each experiment is only repeated six times due to the long running times. The graphs show that as m and k grows the algorithm "speeds up" and in the extreme case of $m = 10000$ and $k = 100$ the algorithm is approximately 50 times faster than the same algorithm run on $m = 100$ and $k = 100$. *(b)* The DAG-against-DAG algorithm. Here we construct a DAG D on top of the k trees and comparing D against a copy of itself. The experiments show that DAG against DAG is the fastest way of calculating the quartet distances (for the values of k, n and m examined) and that the speedup gained in comparison with the naive tree against tree algorithm is in the vicinity of a factor of 130 for $k = 100$, $n = 500$ and $m = 10000$.

gorithm, as implemented in the tool *QDist* caches information about polynomial manipulation when handling the hierarchical decomposition tree[8] and is not independent of m when computing all pair-wise distances. It is therefore conceivable that the more complex algorithm is superior for comparing all pair-wise distances for smaller trees than this. Figure 5, however, indicates that the simpler, but $O(k^2 n^2)$, is faster than the more complex, but $O(k^2 \cdot n \log^2 n)$, algorithm for tree sizes up to at least 500 leaves.

Based on these experiments, we expect the DAG-against-DAG algorithm to outperform the DAG-against-tree algorithm for similar sizes of trees, and indeed, as shown in Fig. 7, this is the case. The DAG-against-tree algorithm (Fig. 7(a)) is clearly superior to the tree against hierarchical decomposition algorithm in all cases, and as m grows this superiority becomes more and more apparent, but compared with the DAG-against-DAG algorithm (Fig. 7(b)) it performs rather poorly (notice that the number of leaves for Fig. 7(b) grows from 100 to 1000 while for Fig. 7(a) it grows only to 500). The DAG-gainst-DAG algorithm, however, achieves a speedup factor around 130, compared to the simple (but fastest) tree-against-tree algorithm for $k = 100$, $n = 500$ and $m = 10000$.

5. Conclusions

We have presented two new algorithms for computing the quartet distance between all pairs of a set of trees. The algorithms exploit shared tree structure to speed up the computation by merging the set into a DAG and memorize counts in each node in the DAG. Our experiments show that the algorithms developed works well in practise (on simulated, but realistic trees), with a speedup factor of around 130 in the base case. Calculating the quartet distances between trees with different leaves is not trivially done using the above construction. Two such trees might be compared by removing any leaf not in both trees prior to calculation or by using a fourth colour. When comparing many such trees it is apparent that the DAG approach is not applicable since, assuming we adopt the same mapping of tree nodes to DAG nodes as used above, two tree nodes mapping to the same DAG node does not necessarily induce the same set of quartets and therefore it is not clear how to reuse the count of nodes in the DAG. This might be interesting looking further in to.

References

1. J. Felsenstein. *Inferring Phylogenies*. Sinauer Associates Inc., 2004.
2. B.L. Allen and M. Steel. Subtree transfer operations and their induced metrics on evolutionary trees. *Annals of Combinatorics*, 5:1–13, 2001.
3. G. Estabrook, F. McMorris, and C. Meacham. Comparison of undirected phylogenetic trees based on subtrees of four evolutionary units. *Syst. Zool.*, 34:193–200, 1985.
4. D. F. Robinson and L. R. Foulds. Comparison of weighted labelled trees. In *Combinatorial mathematics, VI*, Lecture Notes in Mathematics, pages 119–126. 1979.
5. D. F. Robinson and L. R. Foulds. Comparison of phylogenetic trees. *Mathematical Biosciences*, 53:131–147, 1981.
6. M. S. Waterman and T. F. Smith. On the similarity of dendrograms. *Journal of Theoretical Biology*, 73:789–800, 1978.
7. G.S. Brodal, R. Fagerberg, and C.N. S. Pedersen. Computing the quartet distance between evolutionary trees in time $O(n \log n)$. *Algorithmica*, 38:377–395, 2003.
8. T. Mailund and C.N.S. Pedersen. QDist—Quartet distance between evolutionary trees. *Bioinformatics*, 20(10):1636–1637, 2004.
9. C. Christiansen, T. Mailund, C.N.S. Pedersen, and M. Randers. Algorithms for computing the quartet distance between trees of arbitrary degree. In *Proc. of WABI*, volume 3692 of *LNBI*, pages 77–88. Springer-Verlag, 2005.
10. G.S. Brodal, R. Fagerberg, and C.N.S. Pedersen. Computing the quartet distance between evolutionary trees. In *Proc. of ISAAC*, pages 731–742, 2001.
11. M.J. Sanderson. r8s: inferring absolute rates of molecular evolution, divergence times in the absence of a molecular clock. *Bioinformatics*, 19(2):301–302, 2003.
12. A. Rambaut and N.C. Grassly. Seq-gen: an application for the monte carlo simulation of dna sequence evolution along phylogenetic trees. *Computer Applications in the Biosciences*, 13(3):235–238, 1997.
13. T. Mailund and C.N.S. Pedersen. QuickJoin—fast neighbour-joining tree reconstruction. *Bioinformatics*, 20(17):3261–3262, 2004. ISSN 1367-4803.

COMPUTING THE QUARTET DISTANCE BETWEEN EVOLUTIONARY TREES OF BOUNDED DEGREE

M. STISSING, C. N. S. PEDERSEN, T. MAILUND* AND G. S. BRODAL

Bioinformatics Research Center, and Dept. of Computer Science, University of Aarhus, Denmark

R. FAGERBERG

Dept. of Mathematics and Computer Science, University of Southern Denmark, Denmark

We present an algorithm for calculating the quartet distance between two evolutionary trees of bounded degree on a common set of n species. The previous best algorithm has running time $O(d^2n^2)$ when considering trees, where no node is of more than degree d. The algorithm developed herein has running time $O(d^9 n \log n))$ which makes it the first algorithm for computing the quartet distance between non-binary trees which has a sub-quadratic worst case running time.

1. Introduction

The evolutionary relationship between a set of species is conveniently described as a tree, where the leaves represent the species and the inner nodes speciation events. Using different biological data, or different methods of inferring such trees (see e.g. Felsenstein[1] for an overview) can yield different inferred trees for the same set of species, and to study such differences in a systematic manner, one must be able to quantify such differences using well-defined and efficient methods. Several distance measures have been proposed,[2–6] each having different properties and reflecting different aspects of biology.

This paper concerns efficient computation of the quartet distance,[6] a distance measure with several attractive properties.[7,8] For an evolutionary tree, the *quartet topology* of four species is determined by the minimal topological subtree containing the four species. The four possible quartet topologies of four species are shown in Fig. 1. The three leftmost of these we denote *butterfly* quartets, the rightmost is a *star* quartet. Given two evolutionary trees on the same set of n species, the *quartet distance* between them is the number of sets of four species for which the quartet topologies differ in the two trees.

For binary trees, the fastest method for computing the quartet distance between two trees runs in $O(n \log n)$[9], but for trees of arbitrary degree, the fastest algorithms run in $O(n^3)$ (independent of the maximal degree) or $O(n^2d^2)$ (where d is the maximal degree in the tree)[10]. This paper focuses on trees where each inner node v has degree at most d, where d is a fixed constant. We develop an $O(d^9 n \log n)$ time and $O(d^8 n)$ space algorithm for

*Current affiliation: Dept. of Statistics, University of Oxford, UK

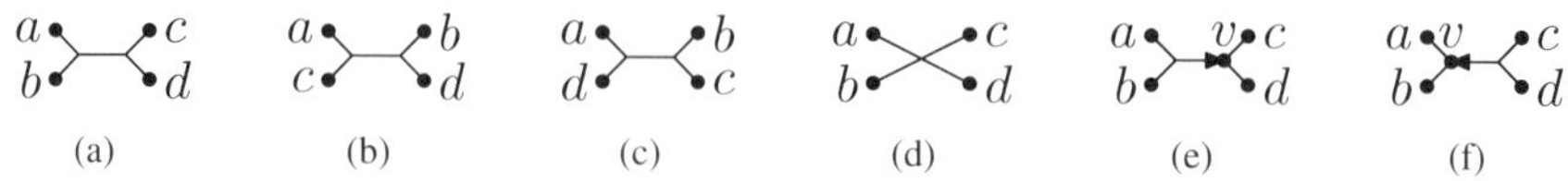

Figure 1. Figures (a)–(d) show the four possible quartet topologies of species a, b, c, and d. Figures (e) and (f) show the two ordered butterfly quartet topologies induced by the butterfly quartet topology in (a).

computing the quartet distance between such two trees, based on the algorithm in Brodal et al.[9] This is the first algorithm for computing the quartet distance between non-binary trees with a sub-quadratic worst case running time. In Brodal et al.[9] the quartet distance was calculated as $\binom{n}{4}$ minus the number of shared quartets. We will adopt this approach, focusing on calculating shared quartets, noting that in our setting trees might include star quartets. We first consider calculating the number of shared butterfly quartets between two trees, and then extend the algorithm into calculating shared star quartets as well.

2. Terminology

An *evolutionary tree* is an unrooted tree where any node, v, is either a leaf or an inner node of degree d_v, where $3 \leq d_v \leq d$. Leaves are uniquely labeled by the elements of a set S of species, where $|S| = n$. For an evolutionary tree T, the *quartet topology* of a set $\{a, b, c, d\} \subseteq S$ of four species is the topological subtree of T induced by these species. The possible quartet topologies for species a, b, c, d are shown in Fig. 1. An evolutionary tree with n leaves gives rise to $\binom{n}{4}$ different quartet topologies. Butterfly quartet topologies are a pairing of the four species into two pairs, defined, see Fig. 1, by letting a and b be a pair if the path from a to b doesn't meet the path from c to d. We view the (butterfly) quartet topology of a four-set of species $\{a, b, c, d\}$ as two *oriented quartet topologies*[9], given by the two possible orientations of the middle edge of the topology, see Fig. 1. An oriented quartet topology is thus an ordered pair of two-sets, e.g. $(\{a, b\}, \{c, d\})$. The number of oriented quartet topologies of a tree is twice the number of unoriented quartet topologies. In the rest of this paper, until Sect. 6 we by *quartet* consider an oriented quartet topology and use the notation $ab|cd$ for $(\{a, b\}, \{c, d\})$.

Let Q be the set of all possible quartets of S. Let $Q_T \subset Q$ denote the set of quartets in an evolutionary tree T. We will *associate* quartets of Q to inner nodes v of T, such that $ab|cd$ is associated to v if v is the node where the paths from c to a and d to a meet (see Fig. 1, right hand side). In the terminology of Christiansen et al.[10] these are all the quartets claimed by edges pointing to v. By Q_v we denote the set of all quartets associated to v. Having the trees incident to v, $T_1, T_2, \ldots, T_{d_v}$, see Fig. 2,

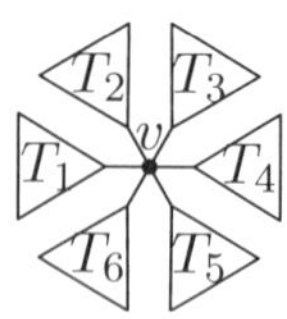

Figure 2. An inner node $v \in T$ with incident subtrees $T_1, \ldots, T_6$

a quartet $ab|cd$ is associated to v if and only if a and b are in the same subtree and c and d are in two different subtrees. The total number of quartets associated to v, $|Q_v|$ is then $\sum_i \sum_{j \neq i} \sum_{\substack{k \neq i \\ k > j}} \binom{|T_i|}{2} |T_j||T_k|$ where i,j, and k is in the interval $1 \ldots d_v$, and $|T|$ denotes the number of leaves in T and denote this the *size* of T. The main strategy of finding the shared quartets between two trees, T and T', is, for each v in T, to count how many of the quartets associated with v are also quartets of T' and calculate the sum over all v, $\sum_{v \in T} |Q_v \cap Q_{T'}|$. Doing this we will relate quartets to a colouring, using the d colours $\mathcal{A}, \mathcal{B}_1, \mathcal{B}_2, \ldots, \mathcal{B}_{d-2}, \mathcal{C}$, of the elements in S. For an internal node v in T, we will say that S is coloured *according* to v if all leaves in each subtree incident to v is coloured using one colour and no other subtree has its leaves coloured this colour. Having a colouring of S and a quartet $ab|cd$, we say that the quartet is *compatible* with the colouring if c and d have different colours and a and b have a third colour. These, almost identical, definitions gives us the following lemma, similar to Brodal et al.[9], Lemma 1.

Lemma 2.1. *When S is coloured according to a choice of v in T, the set of possible quartets compatible with the colouring is exactly the set Q_v of quartets associated with v.*

Consequently, if S is coloured according to v in T, the quartets in $Q_{T'}$ compatible with this colouring are exactly the quartets associated with v that are also quartets of T'. The algorithm will, for each v in T, ensure a colouring according to v and then count the number of quartets of T' compatible with this colouring. In order to do this colouring, we will maintain pointers between elements of S and the leaves of T and T' and vice versa.

3. The Basic Algorithm — $O(d^9 n \log^2 n)$

In this section we expand the idea given above into an algorithm for calculating the shared quartets between T and T' with running time $O(d^9 n \log^2 n)$. The algorithm colours S according to nodes v (using the procedure `colourLeaves`(U, $\mathcal{X}$), which colours all leaves in U with the colour $\mathcal{X}$) and uses a hierarchical decomposition tree $H_{T'}$ in counting the number of quartets in T' compatible with this colouring, `shared`(v, T'). The hierarchical decomposition tree, described in detail in Sect. 5, enables a change of colour of k leaves in time $O(d^9(k + k \log \frac{n}{k}))$ and achieves $O(1)$ time for calculating `shared`(v, T'). The hierarchical decomposition tree $H_{T'}$ is constructable in $O(d^8 n)$ time and $O(d^8 n)$ space.

A pseudocode version of the algorithm is given in Alg. 1. The algorithm assumes T has been rooted in an arbitrary leaf. Let $|v|$ denote the number of leaves in the subtree rooted at v, and call this the *size* of v. A simple traversal lets us annotate each node v such that it knows its largest child, $Large(v)$—where in case of a tie we arbitrarily select one—and which of its children are not the largest, $Small_i(v)$. Let $Small_i(v)$ denote the i'th smallest subtree, with respect to the number of leaves in this subtree. Prior to the first call of the algorithm, the root of T is coloured $\mathcal{C}$ and all (other) leaves are coloured $\mathcal{A}$. The algorithm is initially called with the single child of the root of T. The algorithm recurses through the entire tree, summing the number of shared quartets between v and T', $\sum_{v' \in T'} |Q_v \cap Q_{v'}|$, for each v, ultimately calculating $|Q_T \cap Q_{T'}|$. The algorithm colours the leaves according to v and then counts the number of shared quartets. It then recurses, first on the largest child of v, $Large(v)$ and then on the smaller children of v, $Small_i(v)$. Before recursing on a node v the algorithm ensures that all leaves below v are coloured $\mathcal{A}$. Returning from the recursion, the algorithm ensures that any leaf below v is coloured $\mathcal{C}$.

We see that the algorithm colours a leaf only when this leaf is in a smaller subtree, $Small_i(v)$, of some v on which `count`(v) is invoked. As v is at least twice the size

Algorithm 1 `count`(v, T') - *count number of shared butterfly quartets between the subtree rooted at v and T'*

Require: v a non root node of T, all leaves below v is coloured $\mathcal{A}$, all leaves not in v coloured $\mathcal{C}$.
Ensure: Res is the no. of quartets shared between nodes in v and T'. All leaves in v are coloured $\mathcal{C}$.

```
if v is a leaf then
    colourLeaves(v, C)
    Res ← 0
else
    Res ← 0
    for all Small_i(v) do
        colourLeaves(Small_i(v), B_i)
    Res ← Res + shared(v, T')
    for all Small_i(v) do
        colourLeaves(Small_i(v), C)
    Res ← Res + count(Large(v))
    for all Small_i(v) do
        colourLeaves(Small_i(v), A)
        Res ← Res + count(Small_i(v))
return Res
```

of any $Small_i(v)$, any leaf can at most be coloured $O(\log n)$ times. As the hierarchical decomposition tree enables the change of colour of k leaves in time $O(d^9(k + k\log\frac{n}{k})) \subseteq O(d^9 k \log n)$, we can charge this by letting each colouring of a leaf be of $O(d^9 \log n)$ cost. The entire algorithm, as the colouring is the predominant time consuming factor, is then of time $O(d^9 n \log^2 n)$. The space used is dominated by the space used by the hierarchical decomposition tree, which is $O(d^8 n)$ cf. Sect. 5.

4. The Improved Algorithm — $O(d^9 n \log n)$

The analysis of the basic algorithm above shows that if any node v "uses" time $O(d^9 \log n \cdot \sum_i |Small_i(v)|)$ then the entire algorithm uses time $O(d^9 n \log^2 n)$. This is often referred to as the smaller-half trick:

Lemma 4.1. (Smaller-half trick) *If any inner node v supplies a term $c_v = c \cdot \sum_i |Small_i(v)|$ and any leaf a term $c_v = 0$, then the sum over all nodes $\sum_v c_v \leq c \cdot n \log n$.*

This is easily proved by induction. As an instance of this, the analysis above used $c = d^9 \log n$. The improved algorithm below, uses an extended smaller-half trick which is also easily proved by induction.

Lemma 4.2. (Extended smaller-half trick) *In a rooted tree, if any inner node v supplies a term $c_v = c \cdot \sum_i |Small_i(v)| \log\left(\frac{|v|}{|Small_i(v)|}\right)$ and any leaf a term $c_v = 0$, then the sum over all nodes $\sum_v c_v \leq c \cdot n \log n$.*

The main observation in achieving the improved algorithm comes from noting that, whenever the basic algorithm `count`(v) is called, all leaves outside the subtree rooted at v will have the colour $\mathcal{C}$ and these leaves will not change their colour while `count`(v) is being processed. This, of course, also applies to the leaves of T' coloured $\mathcal{C}$. We will therefore, in certain cases, construct a compact representation of T', by "contracting" nodes of T' coloured $\mathcal{C}$. We will consider any constructed T' as having an associated hierar-

Algorithm 2 `fastCount`(v, T') – *count number of shared butterfly quartets between the subtree rooted at v and T'*

Require: v a non root node of T, all leaves in v coloured $\mathcal{A}$, all leaves not in v coloured $\mathcal{C}$.
Ensure: Res equals the number of quartets shared between v and T'. All leaves in v are coloured $\mathcal{C}$.

```
Res ← 0
if v is a leaf then
    colourLeaves(v, C, T')
else
    for all Small_i(v) do
        colourLeaves(Small_i(v), B_i, T')
    Res ← Res + shared(v, T')
    for all Small_i(v) do
        colourLeaves(Small_i(v), C, T')
    for all Small_i(v) do
        T'_i ← contract(Small_i(v), extract(Small_i(v), T'))
    if |T'| > 5|Large(v)| then
        T' ← contract(Large(v), T')
    Res ← Res + fastCount(Large(v), T')
    for all Small_i(v) do
        colourLeaves(Small_i(v), A, T'_i)
        Res ← Res + fastCount(Small_i(v), T'_i)
return Res
```

chical decomposition tree $H_{T'}$, see below. A pseudocode version of the improved algorithm is given in Alg. 2. If we ensure that T' (and thus $H_{T'}$) is of size $O(|v|)$ whenever `fastCount`(v, T') is processed, we know that k leaves can have their colour updated in time $O(d^9(k + k\log\frac{|v|}{k}))$. The extended smaller-half trick then ensures that the total time spent colouring is $O(d^9 n\log n)$.

The algorithm resembles the basic algorithm except for `contract`(U, Y) and `extract`(U, Y), the details of which are given in Sect. 5. For the analysis of the improved algorithm it suffices to note that `contract`(U, Y) makes a compact representation of Y by contracting anything in Y except the leaves in U. This yields a tree with no more than $4|U|$ nodes in time $O(d^9|Y|)$. Likewise `extract`(U, Y) makes a compact representation of Y. This representation also lets the leaves of U in Y remain intact. All other nodes are contracted. The leaves of the arising tree are (implicitly) coloured $\mathcal{C}$. The operation `extract`(U, Y) completes in $O(d^9|U|\log\frac{|Y|}{|U|})$ time and yields a tree of size $O(d|U|\log\frac{|Y|}{|U|})$. When constructing such a new tree, as a result of `contract`(U, Y), we will update the pointers of S to point to the leaves of the newly created tree. This manipulation of S enables the colouring of leaves in the newly created trees.

Regarding correctness, assuming the leaves in v are coloured $\mathcal{A}$ and the leaves outside v are coloured $\mathcal{C}$ when `fastCount`(v, T') is called, the algorithm will, as the basic algorithm, ensure a colouring according to v prior to the call `shared`(v, T'). Furthermore, before recursing on $Small_i(v)$ (or $Large(v)$) the algorithm ensures that the tree used in the recursion is coloured such that all leaves in $Small_i(v)$ ($Large(v)$) are coloured $\mathcal{A}$ and the leaves outside are coloured $\mathcal{C}$. The correctness of the algorithm follows from the correctness of the basic algorithm.

For time complexity, we see that $|T'|$ is of size $O(|v|)$ when `fastCount`(v, T') is called. This implies that the trees T'_i are each of size $O(|Small_i(v)|)$. The time used in constructing these is $O(d^9\sum_i |Small_i(v)|\log\frac{|v|}{|Small_i(v)|})$, i.e. construction time is dominated by the time taken colouring the leaves `colourLeaves`($Small_i(v),\ \mathcal{B}_i, T'$). We note that each $H_{T_i'}$ is constructable in time $O(d^8|T_i'|)$, see Sect. 5, i.e. is dominated by the time used obtaining T_i' by contraction. Contracting the larger part of T', `contract`($Large(v), T'$), completes in time $O(d^9|T'|)$ and yields a tree of size at most $4|Large(v)|$ (see Lemma 5.5 below). The total time spent on repeatedly contracting larger parts of T', as we only do this when $5 \cdot |Large(v)| \leq |T'|$ is thus bounded by the sum of the geometric series $\frac{4}{5}^k$ times d^9. This implies that the time spent contracting T' is linear in the initial size of T' (times d^9), i.e. the time spent is bounded by the time constructing T', the time used by `contract`(`extract`($Small_i(v), T'$)). Ultimately this implies that the algorithm completes in $O(d^9 n \log n)$ time.

Regarding the space used by the algorithm, we see that the only additional space it consumes is when creating T_i''s (and corresponding $H_{T_i'}$'s) at each node $v \in T$; in total no more than the maximal space used on any root-to-leaf path P_j in T, i.e. $O(d^8 \max_j \sum_{v \in P_j} \sum_i |Small_i(v)|)$. Consider a path P_j, there will be a number of nodes v, where both v and $Small_i(v)$ are on the path. The total space consumed by all such v is no more than $d^8 \frac{d-1}{d} n \sum_i \frac{1}{2^i} \in O(d^8 n)$, that is we store at most $\frac{d-1}{d}$ parts of what is left, i.e. all the smaller children, and as we know $Small_i(v)$ is on the path, we can cut of at least half. The "rest" of the path consists of pairs v and $Large(v)$. For each such pair we consume $d^8 \sum_i |Small_i(v)|$ space, we might think of this as marking the leaves in each of the $Small_i(v)$. However as no other pair v, $Large(v)$ can mark an already marked leaf, we conclude that these pairs consume $O(d^8 n)$ space. In total $O(d^8 n)$ space is used.

5. Hierarchical Decomposition Tree

The algorithms developed uses the hierarchical decomposition tree heavily. The data structure can, in constant time, calculate the number of quartets in an evolutionary tree T compatible with the current colouring of S. The data structure allows a change of the colour of k elements of S in time $O(d^9(k + k \log \frac{n}{k}))$ where n is the number of leaves in T. In the following we describe how to build and update such a tree inspired by the approach in Brodal et al.[9]

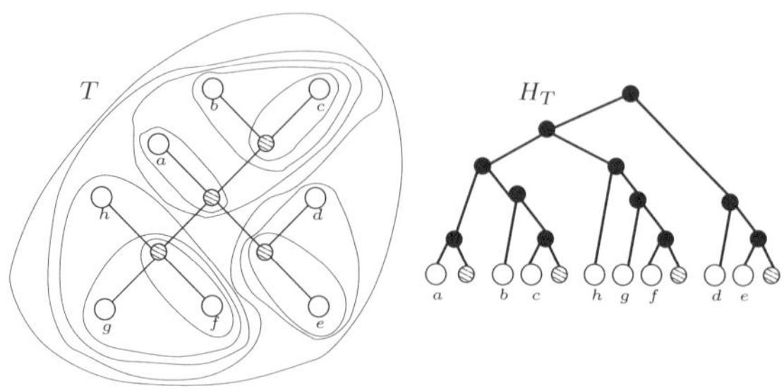

Figure 3. A tree T and a hierarchical decomposition of this tree. H_T is the hierarchical decomposition tree corresponding to the shown hierarchical decomposition of T.

The *hierarchical decomposition* of T is based on the notion of components. A component C of T is a connected subset of nodes in T. An *external edge* of C is connecting a node in C with a node outside of C. The number of external edges is the *degree* of C. We will allow two types of components: (1) *Simple components* containing a single node of T, see Fig. 4(a), 4(b). (2) *Composite components* composing two other components, where both of these are of degree two 4(c) or at least one of these are of degree one, see Fig. 4(d), 4(e).

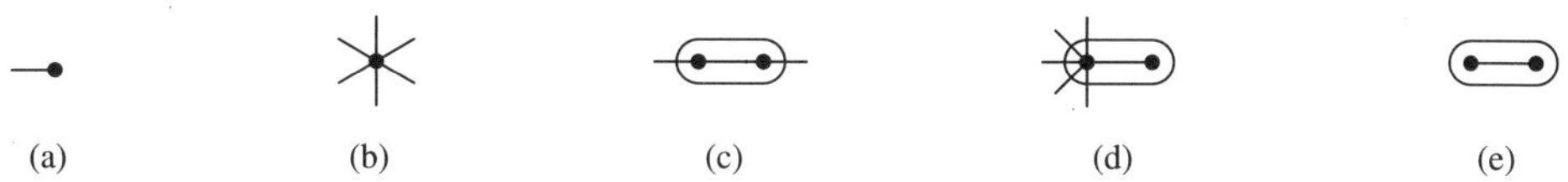

Figure 4. Possible components: (a), (b) Simple components, a leaf and an inner node component respectively. (c) – (e) Composite components: (c) Composing two components of degree two. (d) Composing a component of degree d_C with a component of degree one. (e) Composing two degree one components – as seen, a special case of (d).

Letting each node of T be a component by itself, a hierarchical decomposition of T is a set of components created by repeatedly composing these. Note that the degree of a composite component will be no more than the maximum degree of the components it is composed of. In decomposing T, note that, the current set of components form a tree, hence there will always be at least one component of degree 1, and we can therefore always continue composing until we are left with a component containing all simple components of T.

Having a hierarchical decomposition of T including a component containing all simple components of T, we might in a natural way view this as a tree. A hierarchical decomposition tree H_T for T is a rooted binary tree with leaves corresponding to simple components of T and inner nodes corresponding to composite components (components in a hierarchical decomposition of T), see Fig. 3. An inner node v, with children v' and v'', corresponds to the component C arising when the two components C' and C'', corresponding to the children of the node, are composed. The root, r, corresponds to a component containing all simple components of T. In this sense many hierarchical decomposition trees exist. We will show how to construct a *locally-balanced* hierarchical decomposition tree. A rooted binary tree with n nodes is *c-locally-balanced* if for all nodes v in the tree, the height of the subtree rooted at v is at most $c \cdot (1 + \log |v|)$, where $|v|$ is the number of leaves in the subtree and the height is the maximal number of edges on any root-to-leaf path. The following lemma is an extension of Brodal et al.[9], Lemma 3.

Lemma 5.1. *For any unrooted tree T with n nodes of degree at most d, a $6d$-locally balanced hierarchical decomposition tree H_T can be computed in time $O(dn)$.*

The following lemma from Brodal et al.[9] bounds the number of nodes on k root-to-leaf paths in a hierarchical decomposition tree.

Lemma 5.2. *The union of k root-to-leaf paths in a c-locally balanced rooted binary tree with n leaves contains at most $k(3 + 4c) + 2ck \log \frac{n}{k}$ nodes.*

Having an evolutionary tree T with n leaves and the associated hierarchical decomposition tree H_T we want to count the number of quartets in T compatible with the current colouring of S in constant time. Further, when k elements of S change their colour, we should handle this update in time $O(d^9(k + k \log \frac{n}{k}))$.

We will associate functions and vectors to the nodes of H_T. At any node v, having the associated component C, in H_T the vector $\mathbf{c} = (c_1, c_2, \ldots, c_d)$ stored holds the number of leaves contained in C of colours $\mathcal{A}, \mathcal{B}_1, \mathcal{B}_2, \ldots, \mathcal{B}_{d-2}$ and $\mathcal{C}$ respectively. If C is of degree

d_C, the function F stored at v, is a function of d_C vector variables. The function counts the number of quartets associated to any node in C compatible with the current colouring of S. This implies that the function stored at the root of H_T counts the total number of quartets in T compatible with the current colouring of S. Furthermore, since the component associated to the root of H_T has 0 external edges, the function stored here is a constant. The elements $\mathbf{c}^{\mathbf{i}}_{i'}$ of the vector variables $\mathbf{c}^{\mathbf{i}}$ of F correspond to the number of leaves coloured with the i''th colour in the component incident to the i'th external edge of C.

First we describe how to associate the vectors and functions to the leaves of H_T, that is the simple components of T. If v has an associated component of degree 1, i.e. it represents a leaf l in T, having the colour $\mathcal{A}, \mathcal{B}_1, \mathcal{B}_2, \ldots, \mathcal{B}_{d-2}$ or $\mathcal{C}$ the vector stored at v is $(1, 0, \ldots, 0, 0)$,$(0, 1, \ldots, 0, 0)$,...,$(0, 0, \ldots, 1, 0)$ or $(0, 0, \ldots, 0, 1)$ respectively. Since the number of quartets associated to l is 0, the function stored at v is identically zero: $F(\mathbf{c}^{\mathbf{1}}) = 0$. Otherwise, if v, with associated component C of degree d_C, represents an internal node u in T the tuple stored here is $(0, 0, \ldots, 0, 0)$ as the component contains no leaves of any colour. The function F stored here, counts the number of quartets associated to u compatible with the colouring of S. Recall that a quartet $ab|cd$ associated to u has a and b in the same subtree incident to u and c and d in two different subtrees. Further, if $ab|cd$ is compatible with the colouring of S, a and b have the same colour and c and d have two different colours. F is then:

$$F(\mathbf{c}^{\mathbf{1}}, \mathbf{c}^{\mathbf{2}}, \ldots, \mathbf{c}^{\mathbf{d_C}}) = \sum_{i}^{d_C} \sum_{j \neq i}^{d_C} \sum_{\substack{k \neq i \\ k > j}}^{d_C} \sum_{i'}^{d} \sum_{j' \neq i'}^{d} \sum_{\substack{k' \neq i' \\ k' \neq j'}}^{d} \binom{\mathbf{c}^{\mathbf{i}}_{i'}}{2} \mathbf{c}^{\mathbf{j}}_{j'} \mathbf{c}^{\mathbf{k}}_{k'}$$

We now turn to the tuples and functions associated to the inner nodes of H_T. The inner node v, with children v' and v'', will store the vector $\mathbf{c}' + \mathbf{c}''$, assuming v' and v'' store the vector $\mathbf{c}'$ and $\mathbf{c}''$ respectively. Letting F' and F'' be the functions stored at v' and v'', we express F stored at v. Let C be the component corresponding to v, likewise for C' and C''. If both C' and C'' are degree 2 components (Fig. 4(c)) we construct F as $F(\mathbf{c}^{\mathbf{1}}, \mathbf{c}^{\mathbf{2}}) = F'(\mathbf{c}^{\mathbf{1}}, \mathbf{c}^{\mathbf{2}} + \mathbf{c}'') + F''(\mathbf{c}^{\mathbf{1}} + \mathbf{c}', \mathbf{c}^{\mathbf{2}})$, assuming the second external edge of C' is the first external edge of C'' and the first external edge of C' is the first external edge of C and the second external edge of C'' is the second external edge of C (other edge "numberings" are handled similarly). If C' is a component of degree $d_{C'} \geq 2$ and C'' a component of degree 1 (Fig. 4(d)), we construct F, this time assuming the $d_{C'}$'th external edge of C' is the first (and only) external edge of C'', the d_C external edges of C correspond to the d_C first external edges of C': $F(\mathbf{c}^{\mathbf{1}}, \mathbf{c}^{\mathbf{2}}, \ldots, \mathbf{c}^{\mathbf{d_C}}) = F'(\mathbf{c}^{\mathbf{1}}, \mathbf{c}^{\mathbf{2}}, \ldots, \mathbf{c}^{\mathbf{d_C}}, \mathbf{c}'') + F''(\mathbf{c}^{\mathbf{1}} + \mathbf{c}^{\mathbf{2}} + \ldots + \mathbf{c}^{\mathbf{d_C}} + \mathbf{c}')$. As a special case of the above, if both C' and C'' are of degree 1 (Fig. 4(d)), we note that F is a constant: $F = F'(\mathbf{c}'') + F''(\mathbf{c}')$. If C is a simple component, F is a polynomial of degree at most 4 with no more than $d \cdot d_C \leq d^2$ variables. By induction in the way F's are constructed, this is then seen to hold for any component. At any node v we observe that the F (and $\mathbf{c}$) to be stored is constructable in $O(d^8)$ time. This implies the following lemma, similar to Brodal et al.[9], Lemma 5:

Lemma 5.3. *The tree H_T can be decorated with the information described above in time and space $O(d^8 n)$.*

The following lemma, similar to Brodal et al.[9], Lemma 6, arises as a consequence of Lem. 5.2 and the fact that the decoration stored at a node v in H_T is constructable in $O(d^8)$ time, given the decoration at its children.

Lemma 5.4. *The decoration of H_T can be updated in $O(d^9(k + k \log \frac{n}{k}))$ time when the colour of k elements in S changes.*

The above results imply the running time of the basic algorithm. We now turn to the details of `contract` and `extract` used in the improved algorithm. The procedure `contract`(U, Y) yields a tree Y' of size $O(|U|)$ in time $O(d^9|Y|)$ letting the leaves present in U remain untouched in Y'. This is accomplished by copying Y and contracting edges corresponding to legal compositions. This way Y' contains nodes corresponding to simple or composite components. The functions and vectors stored at these components is inherited by the nodes they correspond to. Y''s edges are a subset of the edges of Y, namely the edges not contracted. The following lemma, an extension of Brodal et al.[9], Lemma 4), ensures that Y' has no more than $4|U|$ nodes, and that each of the leaves in U is a leaf in Y'.

Lemma 5.5. *Let T be an unrooted tree with n nodes of degree at most d, and let $k \geq 0$ leaves be marked as* non-contractible. *In $O(dn)$ time a decomposition of T into at most $4k$ components can be computed such that each marked leaf is a component by itself.*

Creating the information to be stored at the nodes of Y' uses $O(d^8)$ time per contraction made, that is `contract`(U, Y) completes in time $O(d^9|Y|)$. We can calculate $H_{Y'}$, in the time stated, as each node of Y' has an associated vector and function. Likewise `extract`(U, Y) yields a contracted tree Y' of size $O(d|U| \log \frac{|Y|}{|U|})$ in time $O(d^9|U| \log \frac{|Y|}{|U|})$. This is achieved by using the hierarchical decomposition tree H_Y. We mark the internal nodes of H_Y on the $|U|$ root-to-leaf paths leading to the leaves in U. Doing this bottom-up, one leaf at a time, we can stop marking when an already marked node is encountered. Lem. 5.2 then bounds the number of marked nodes. Removing all these marked nodes yields a set of subtrees of H_Y. The root nodes of these subtrees correspond to components in Y. We let these root nodes be the nodes of Y'. Having the external edges of each of the components corresponding to the nodes of Y' we connect such two nodes if they share an external edge. This can be done in time linear in the number of edges, assuming that the edges are labelled. The leaves in U are also leaves in Y'.

In order to consider all leaves in Y coloured $\mathcal{C}$ in Y' we let the nodes of H_Y store another vector $\mathbf{c}_\mathcal{C}$ and function $F_\mathcal{C}$. These are defined equivalently to $\mathbf{c}$ and F with the exception that they assume that all leaves in the associated component are coloured $\mathcal{C}$. These can be constructed once and for all when H_Y is constructed. We let $\mathbf{c}_\mathcal{C}$ and $F_\mathcal{C}$ be the information stored at the nodes of Y'. We note that we use $O(d^8)$ time, copying information, per node in the extraction.

6. Calculating Shared Star Quartets

The last step is the calculation of shared star quartets between T and T'. We adopt the notion of associated and compatible from butterfly quartets. We can, in the same way as above, construct polynomials G counting the number of star quartets associated with simple components of T' and compatible with the current colouring of S. As there are no star quartets associated to a leaf of T', $G(\mathbf{c^1}) = 0$. At internal nodes of T':

$$G(\mathbf{c^1}, \mathbf{c^2}, \ldots, \mathbf{c^{d_C}}) = \sum_{i}^{d_C} \sum_{j>i}^{d_C} \sum_{k>j}^{d_C} \sum_{l>k}^{d_C} \sum_{i'}^{d} \sum_{j' \neq i'}^{d} \sum_{\substack{k' \neq i' \\ k' \neq j'}}^{d} \sum_{\substack{l' \neq i' \\ l' \neq j' \\ l' \neq k'}}^{d} \mathbf{c}^{\mathbf{i}}_{i'} \mathbf{c}^{\mathbf{j}}_{j'} \mathbf{c}^{\mathbf{k}}_{k'} \mathbf{c}^{\mathbf{l}}_{l'}$$

The construction of G's at internal nodes of the hierarchical decomposition tree corresponds to the construction of F's at these nodes. We note that G is itself a polynomial of degree 4 with no more than d^2 variables, i.e. it can be stored and manipulated in $O(d^8)$ space and time. We conclude that we can extend both the basic and the improved algorithm, by associating G's to the nodes of the trees, into counting shared star quartets as well as the shared butterfly quartets. This enables the calculation of the quartet distance between T and T'.

References

1. J. Felsenstein. *Inferring Phylogenies*. Sinauer Associates Inc., 2004.
2. D. F. Robinson and L. R. Foulds. Comparison of weighted labelled trees. In *Combinatorial mathematics, VI (Proc. 6th Austral. Conf)*, Lecture Notes in Mathematics, pages 119–126. Springer, 1979.
3. M. S. Waterman and T. F. Smith. On the similarity of dendrograms. *Journal of Theoretical Biology*, 73:789–800, 1978.
4. B. L. Allen and M. Steel. Subtree transfer operations and their induced metrics on evolutionary trees. *Annals of Combinatorics*, 5:1–13, 2001.
5. D. F. Robinson and L. R. Foulds. Comparison of phylogenetic trees. *Mathematical Biosciences*, 53:131–147, 1981.
6. G. Estabrook, F. McMorris, and C. Meacham. Comparison of undirected phylogenetic trees based on subtrees of four evolutionary units. *Syst. Zool.*, 34:193–200, 1985.
7. D. Bryant, J. Tsang, P. E. Kearney, and M. Li. Computing the quartet distance between evolutionary trees. In *Proceedings of the 11th Annual Symposium on Discrete Algorithms (SODA)*, pages 285–286, 2000.
8. M. Steel and D. Penny. Distribution of tree comparison metrics–some new results. *Syst. Biol.*, 42(2):126–141, 1993.
9. G. S. Brodal, R. Fagerberg, and C. N.S̃. Pedersen. Computing the quartet distance between evolutionary trees in time $O(n \log n)$. *Algorithmica*, 38:377–395, 2003.
10. C. Christiansen, T. Mailund, C. N. S. Pedersen, and M. Randers. Computing the quartet distance between trees of arbitrary degree. In R. Casadio and G. Myers, editors, *WABI*, volume 3692 of *LNCS*, pages 77–88. Springer, 2005. ISBN 3-540-29008-7.

A GLOBAL MAXIMUM LIKELIHOOD SUPER-QUARTET PHYLOGENY METHOD

P. WANG, B. B. ZHOU, M. TARAENEH, D. CHU, C. WANG and A. Y. ZOMAYA

School of Information Technologies, University of Sydney
NSW 2006, Australia

R. P. BRENT

Mathematical Science Institute, Australian National University
Canberra, ACT 0200, Australia

Extending the idea of our previous algorithm [17, 18] we developed a new sequential quartet-based phylogenetic tree construction method. This new algorithm reconstructs the phylogenetic tree iteratively by examining at each merge step every possible super-quartet which is formed by four subtrees instead of simple quartet in our previous algorithm. Because our new algorithm evaluates super-quartet trees, each of which may consist of more than four molecular sequences, it can effectively alleviate a traditional, but important problem of quartet errors encountered in the quartet-based methods. Experiment results show that our newly proposed algorithm is capable of achieving very high accuracy and solid consistency in reconstructing the phylogenetic trees on different sets of synthetic DNA data under various evolution circumstances.

1 Introduction

For systematic biology, evolutionary history is one of the most important topics. Therefore reconstruction of phylogenetic trees from molecular sequences has strong research significance. The quartet-based approach is one of the primary methods for phylogeny reconstruction. The basic idea is to construct a tree based on the topological properties of a set of four molecular sequences (or quartets). The advantages of the quartet-based method are that theoretically it guarantees a one-to-one correspondence between a tree topology and a set of quartet trees and if the tree topology for each individual quartet can be correctly identified, the entire evolutionary tree for a given problem can be reconstructed in polynomial time. The main disadvantage, however, is that it can be very difficult to obtain correctly resolved quartet trees using any existing methods [1, 8]. This quartet error problem greatly hinders the quartet-based approach from a wide application.

Previously we developed a quartet-based algorithm for the reconstruction of evolutionary trees [17, 18, 19]. Instead of constructing only one tree as output, this algorithm constructs a limited number of trees for a given set of DNA or protein sequences. Experimental results showed that the probability for the correct phylogenetic tree to be included in this small number of trees is very high. When we selected just a few best ones (say three) from these trees under maximum likelihood (ML) criterion, extensive tests using synthetic test data sets showed that the algorithm outperforms many known algorithms for phylogeny reconstruction in terms of tree topology [19]. One problem associated with the original form of this algorithm is that under certain circumstances it does not perform well when reconstructing only a single tree as output without the selection stage using ML. Though the method can tolerate quartet errors better than many other quartet-based algorithms, the quality of the generated tree still depends heavily on the quality of quartet trees.

In this paper we introduce a new algorithm. This algorithm is similar to our previous quartet-based algorithm. However, it reconstructs a tree based on an idea of taking more than four taxa into quartet topological estimates, which is called "super-quartet" in this paper. The main reason of using super-quartets is to alleviate the problem of quartet errors. Inherited from previous algorithm, this new algorithm maintains the theoretical advantage of one-to-one mapping of the tree and corresponding (super-) quartet weights. The main difference is that this new method iteratively merges taxa on super-quartet weights calculated by log ML values (with a probabilistic normalization). Since a super-quartet consists of more than four molecular sequences, it is expected that the quality of super-quartet weights are higher than that of basic quartet (four sequences), hence improving the accuracy of the generated tree. Our experiments confirm this and the results demonstrate that this new algorithm is able to achieve better accuracy, when reconstructing only a single tree, than many ML-based algorithms including the very popular PHYML [7] in the terms of phylogenetic tree topology.

The paper first reviews our previous algorithm and its associated problem in reconstructing of a single tree in Section 2. The super-quartet idea and our new phylogenetic reconstruction algorithm are then introduced in Section 3. We present some experiment results in Section 4. Finally the conclusions and the future work are discussed in Section 5.

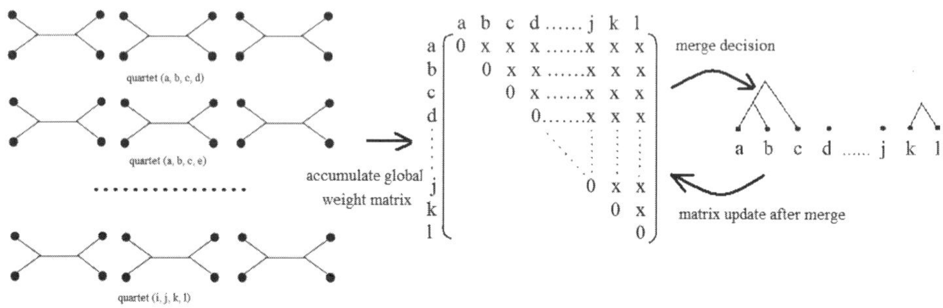

Figure 1. The three-stage procedure of previous algorithm

2 The Previous Algorithm and Problem

Our previous quartet-based algorithm rebuilds the evolutionary tree using quartet weights [17, 18]. As shown in Figure 1, the algorithm is a three-stage procedure: 1. generate all the quartets and calculate the quartet weights; 2. accumulate weights to a global quartet weight matrix; 3. iteratively merge subtrees using this matrix. The idea of quartet weights was first introduced in [5] and then extended and used in a tree-puzzling algorithm [16]. Each quartet is associated with three topologies and their normalized weights. In the ideal case when all quartets are correctly and fully resolved, there is a one-to-one correspondence between this matrix and the true tree.

In order to overcome the quartet inconsistency problem, the previous algorithm also deploys two additional mechanisms: the first is to generate more than one phylogenetic tree (a small number) when merging ambiguity occurs; the other is to update the corresponding global quartet weights to theoretical values. Theoretical values are obtained by assuming the two subtrees that have been merged in a heuristic merging step are real neighbors in the true tree. Previous experimental results [18] demonstrated that the probability for the correct tree to be included in the small set of generated trees is very high. As shown in Figure 2, the algorithm is able to achieve better results than PHYML when a limited number of trees are constructed. (The reason we use PHYML as benchmark to compare our method is because it is one of the most popular packages used by biologists, and its accuracy in building the tree is among the highest several methods up to now.) However, PHYML achieves better accuracy than our previous algorithm when the algorithm is limited to generate only a single tree. Though our previous algorithm is able to tolerate the quartet errors better than other quartet-based algorithms, the schemes used are still not sufficiently good enough to compensate the quartet error problem. It is necessary to find more vigorous mechanisms to deal with the quartet error problem.

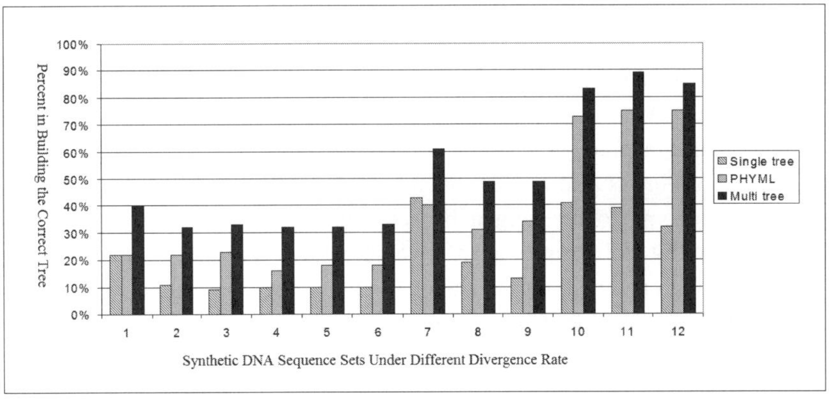

Figure 2. Experimental comparison of PHYML and previous algorithm (both on constructing single tree and multiple 5 trees) under the same condition as in experiment section.

3 The New Super-Quartet Algorithm

Phylogeny inference for only four taxa is often considered hard and the result unreliable. This is because sampling of taxa prior to phylogenetic tree reconstruction strongly influences the accuracy. If one seeks to establish the phylogenetic relationship between four groups of taxa by using a single representative for each of these groupings, the result generally depends on which representatives are selected [1, 13]. Although our previous algorithm builds the tree on a global view of quartet relationship, it is still limited to examination of the topological relationship on the basis of four taxa only.

The basic idea of quartet is to represent the topological relation of four taxa, through three possible binary (quartet) trees. In the extension of this idea we place a subtree rather than a single taxon on each vertex of a quartet binary tree, as shown in Figure 3. The weight of a super-quartet is measured using the same procedure for quartet weight calculation [9], by firstly calculating the maximum likelihood values for each possible super-quartet tree out of three possible topologies and then transforming these likelihood values into super-quartet weights. Bayes theorem with a uniform prior for all three possible trees is used for such transformation. Since each vertex may contain more than one taxon, we expect the super-quartet weights are more reliable than weights of simple quartets.

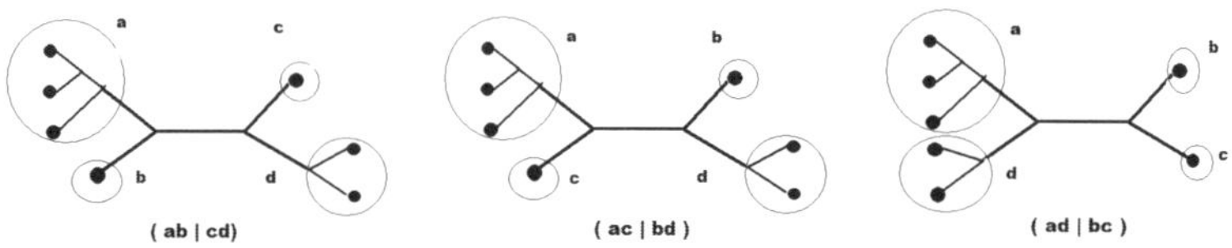

Figure 3. super-quartet trees for taxon sets of {a, b, c, d}.

After the weights for all possible super-quartets are calculated, the super-quartet weight matrix can be generated the same way as the simple quartet weight matrix [17, 18, 19] with each row or column corresponding to a subtree rather than a single taxon. Our new super-quartet method shares the same theoretical property of one-to-one tree topology and matrix mapping [17, 18].

The new algorithm is also an iterative merge algorithm; it makes decision on which two subtrees are going to be merged by selecting the pair of subtrees which shows highest probability by global super-quartet weight matrix. The entry values in the weight matrix are the agglomerated normalization weights for corresponding subtrees at a particular merge step. The metric of making the merge decision is to evaluate how close the super-quartet weight is to the theoretical value. This is easily implemented by:

$$C_{ij} = M_{ij} / T_{ij}, \qquad (1)$$

where C_{ij} is called "confidence value", and T_{ij} is the theoretical value for the current subtrees to be merged and can be calculated by:

$$T_{ij} = \binom{n_k - 2}{2}. \qquad (2)$$

where n_k represents the current global super-quartet weight matrix dimension.

The structure of our super-quartet algorithm is given below:

1. set $n_k = n$. (Initially every taxon represents itself as a subtree)
2. number the subtrees from 1 to n_k.

3. calculate the likelihood values of all possible super-quartet trees and the associated weights.
4. update the weight matrix. (Each subtree represented by one number corresponding to a particular row or column.)
5. for each pair of subtrees calculate the confidence value (using the entry value against a desired one) to determine how likely the pair is to be merged directly.
6. choose the pair of subtrees that has the highest confidence value and merge them into a bigger subtree.
7. reduce n_k by one.
8. if $n_k > 3$, go back to *step 2*; otherwise merge the remaining three subtrees into one final tree.

A simple example of one merge step using our algorithm is illustrated in Figure 4. In this example we assume the taxon 3 and 4 have been merged in previous step, and the current global super-quartet weight matrix is of size 6x6 since previous merging step reduced the matrix size from the original 7x7. The algorithm calculates the confidence value and find that subtree pair 3 and 5 has the highest value. These two subtrees are merged into one larger subtree. The algorithm then re-calculates the likelihood values of super-quartet trees and the size of the global super-quartet weight matrix is reduced to 5x5. This heuristic merge process continues until the whole tree is constructed.

4 Experiments

The experiments are carried out using the synthetic data sets from *Montpellier Laboratory of Computer Science, Robotics, and Microelectronics (www.lirmm.fr)*. The data sets, each consists of 12 taxa, are generated using six model trees. Three model trees are molecular clock-like, while the other three present varying substitution rates among lineages. With these model trees under various evolutionary conditions, the test data sets of DNA sequences, each being of length 300, are generated using Seq-Gen. The reason we select these synthetic data as benchmark is because for real DNA data it is very hard to clarify the correctness of the constructed tree. On the other hand, these synthetic data sets are very comprehensive in evolutionary conditions, both with molecular clock and without, both balanced and unbalanced.

We present in Figure 5 our experimental results and compare them with those obtained using PHYML in terms of the percentage of correctly constructed phylogenetic trees. For our algorithm, HKY model as evolutionary model, transition to transversion rate of one, nucleotide frequency all at 25%, uniform model of rate heterogeneity are selected as parameters to perform the experiments. We run PHYML (version 2.44) on the same data set using the same parameters to compare the results. The trees generated from both algorithms are compared with the true trees using the Robinson and Foulds topological distance (RF) method [14].

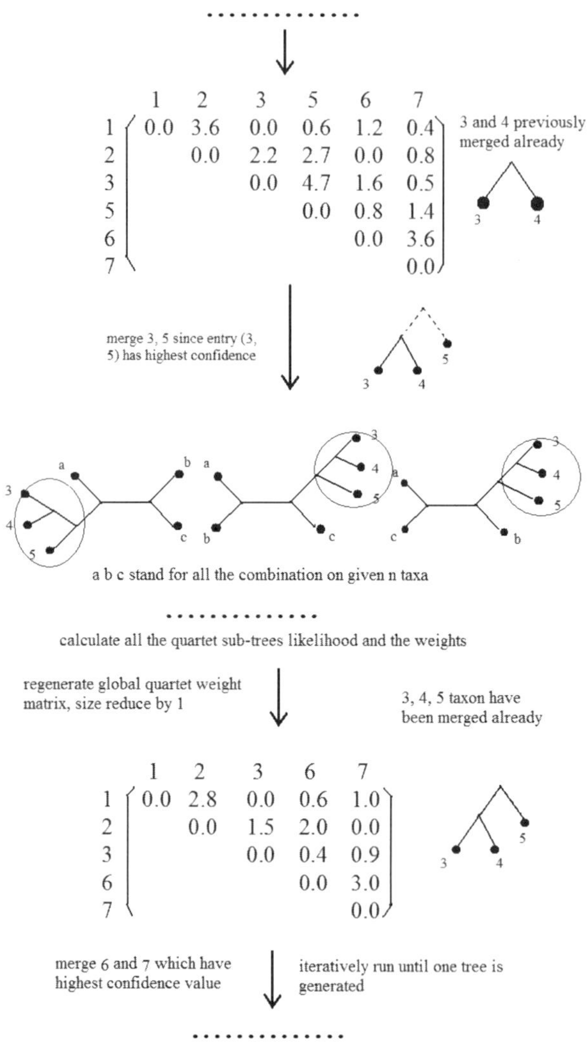

Figure 4. The algorithm iterative step procedure illustration

The results demonstrate that our new global ML super-quartet algorithm outperforms PHYML in most circumstances. First of all, our algorithm made great improvement for data sets without molecular clock: our method constructs the correct tree at much higher frequency than PHYML. This can be seen clearly from Figure 5 (b) and (c) for the experiment data sets without molecular clock (right 3 columns in the figure). These sets of data represent the most common evolution circumstances for phylogenetic study and the percentage of correctly constructed trees using our algorithm is nearly 15% higher than that of PHYML. Secondly on data set with large variation (MD around 2.0) with or

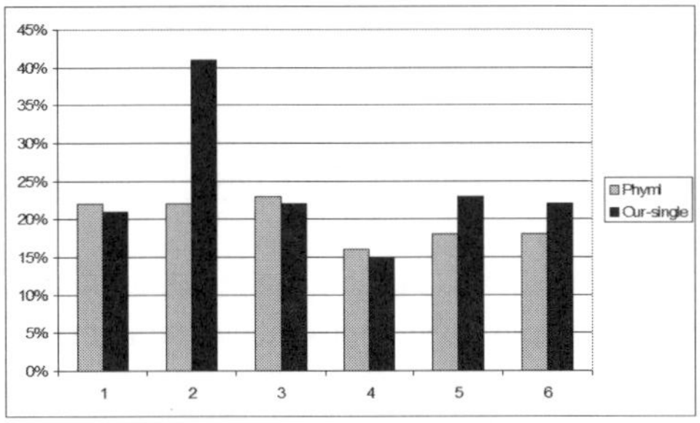

(a) MD = 0.1 experiment results

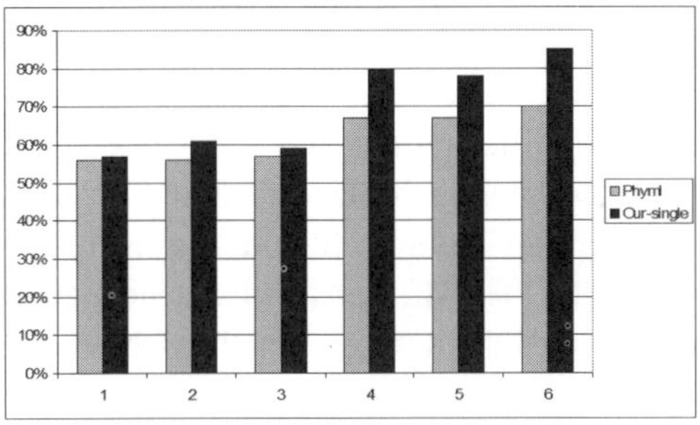

(b) MD = 0.3 experiment results

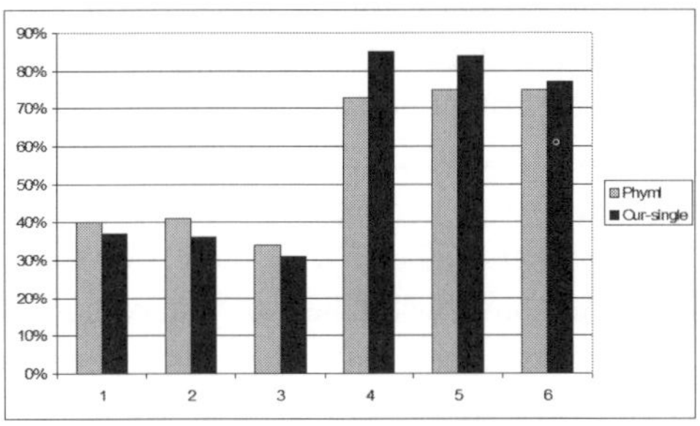

(c) MD = 1.0 experiment results

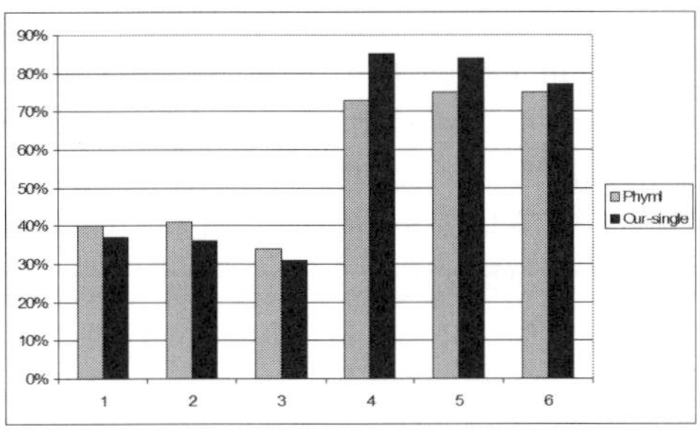

(d) MD = 2.0 experiment results

Figure 5. Results on synthetic data. X axis: different DNA set; Y axis: percent in building correct tree, figures are rounded to the nearest integers.

without molecular clock, our algorithm nearly doubles the percentage of constructing the correct tree compared with PHYML. This is another significant improvement. Thirdly for the data sets with molecular clock, as shown in Figure 5 (left 3 columns), our super-quartet method is able to reconstruct the true phylogenetic tree with slightly better accuracy on the average than PHYML.

The reasons why we use PHYML as the benchmark to compare our algorithm lie in two aspects. Firstly PHYML is one of the most accurate algorithms among existing ML methods [10, 11, 12], and it is one of the most popular and used packages. Secondly PHYML works by perturbing one initial tree (generated by BIONJ). The perturbation is carried out to optimize the ML value by swapping the subtrees connected at each vertex of internal branch, i.e., it evaluates the three possible topologies of the tree by interchanging the four subtrees at vertices of an internal branch. The most important common characteristic of PHYML and our new method is that these two methods all construct the tree through examining the taxon subtree neighborhood relationship on quartet-like binary trees under maximum likelihood. The main difference of PHYML and our super-quartet method is that PHYML starts with a generated tree while our algorithm starts from single taxa and builds the tree sequentially. Obviously our method examines the subtree relations from a much more aggressive way than PHYML, i.e., PHYML uses an initial *N*-sequence tree and changes the tree topology on *N-3* branches, while our method may examine all the subtrees combinations on previous heuristic (*N* is the number of taxa). Our algorithm may have two advantages. Firstly at each merge step we calculate all the likelihood values of all possible super-quartets and take a global view by accumulating all these super-quartet weights to examine every neighborhood relationship of subtrees. The second advantage is our super-quartet approach inherits the theoretical mapping advantage of the quartet method. With these two theoretical advantages, our super-quartet method is able to converge on the global maximum with higher probability. This can be seen from our experiment results that our super-quartet method is able to achieve higher accuracy than PHYML.

One disadvantage of our algorithm is that it takes $O(n^5)$ steps to complete and thus more expensive than PHYML which only takes $O(n^3)$ steps. The main computation cost for our super-quartet algorithm lies in two parts, i.e., the computation of confidence values for every merge step and the calculation of likelihood values of super-quartet trees. The likelihood value re-calculation is the most expensive part. To reduce the computational cost we may introduce a threshold. The likelihood value re-calculation takes place in a merge step only when the confidence value for the merged pair is below this threshold. Another feasible way in reduction of the computational cost is to incorporate the idea of out-group sequences. When there are ambiguities on which pair of subtrees should be merged, we may pick up only a few other subtrees which are "out-groups" of those which are currently considered to be merged, use one out-group subtree at a time which those considered to be merged to form a super-quartet and calculate its likelihood of three possible trees. Since we do not re-calculate likelihood values for all possible super-quartets from the total number of super-trees, the computational cost can thus be significantly reduced.

5 Conclusion and Future Work

In this paper, we proposed one super-quartet phylogenetic tree reconstruction algorithm. This new algorithm extends our previous quartet-based algorithm and employs an iterative super-quartet approach to enhance the algorithm accuracy. We presented our experiment results and compared them with those obtained using PHYML, one of the most accurate ML algorithms. The experimental results demonstrate that our new algorithm can achieve better results than PHYML. With super-quartets and global quartet weights mechanism, our new algorithm is able to effectively alleviate the problem of quartet errors encountered in traditional quartet-based methods. However our algorithm is computationally more expensive than other methods due to super-quartet weight re-calculation. In the paper we proposed several methods to reduce the total computational cost.

Even though our super-quartets approach is able to achieve very high accuracy, there is still no guarantee it can avoid local maxima. Possible extensions are to develop several mechanisms to deal with those critical merge steps when the ambiguity occurs. One possible extension is to build multiple trees as output in case of possible local maxima convergence.

References

1. J. Adachi and M. Hasegawa, Instability of quartet analyses of molecular sequence data by the maximum likelihood method: the cetacean/artiodactyla relationships. *Cladistics,* Vol. 5, pp.164-166, 1999.
2. J. Felsenstein, Evolutionary trees from DNA sequences: a maximum likelihood approach. *J. Mol Evol.,* 17(6):368-76, 1981.
3. J. Felsenstein, The evolutionary advantage of recombination. II. Individual selection for recombination. *Genetics,* 83(4):845-59, 1976.
4. J. Felsenstein, PHYLIP (phylogeny inference package), *version 3.6a2. Distributed by the author, Department of Genetics, Univ. Washington, Seattle*, 1993.
5. W. M. Fitch, A non-sequential method for constructing trees and hierarchical classifications. *J. Mol. Evol.,* 18, pp. 30-37, 1981.
6. O. Gascuel, BIONJ: An improved version of the NJ algorithm based on a simple model of sequence data. *Mol. Biol. Evol.,* 14:685– 695, 1997.
7. S. Guindon and O. Gascuel, A simple, fast and accurate algorithm to estimate large phylogenies by maximum likelihood. *Syst. Biol.,* 52, pp. 696-704, 2003.
8. T. Jiang, P. E. Kearney and M. Li, Orchestrating quartets: Approximation and data correction. *Proceedings of the 39th IEEE Symposium on Foundations of Computer Science,* pp.416-425, 1998.
9. K. Nieselt-Struwe and A. von Haeseler, Quartet-mapping, a generalization of the likelihood-mapping procedure. *Mol. Biol. Evol.,* 18(7), pp.1204-1219, 2001.

10. G. Olsen, H. Matsuda, R. Hagstrom, and R. Overbeek. FastDNAml: A tool for construction of phylogenetic trees of DNA sequences using maximum likelihood. *Comput. Appl. Biosci.,* 10:41– 48, 1994.
11. S. Ota and W.-H. Li. NJML: A hybrid algorithm for the neighbor-joining and maximum-likelihood methods. *Mol. Biol. Evol.,* 17:1401–1409, 2000.
12. S. Ota and W.-H. Li. NJML+: An extension of the NJML method to handle protein sequence data and computer software implementation. *Mol. Biol. Evol.,* 18:1983–1992, 2001.
13. H. Philippe and E. Douzery, The pitfalls of molecular phylogeny based on four species, as illustrated by the Cetacea/ArtiodacKuhner. *J. Mammal. Evol,.* 2: 133–152, 1994.
14. D. R. Robinson and L. R. Foulds, An optimal way to compare additive trees using circular orders. *J. Comp. Biol.*, pp.731-744, 1981.
15. V. Ranwez and O. Gascuel, Quartet-based phylogenetic inference: Improvements and limits. *Mol. Biol. Evol.*, 18(6), pp.1103-111, 2001.
16. K. Strimmer and A. von Haeseler, Quartet puzzling: A quartet maximum-likelihood method for reconstructing tree topologies. *Mol. Biol. Evol.*, 13(7), pp.964-969, 1996.
17. B. B. Zhou, M. Tarawneh, C. Wang, D. Chu, A. Y. Zomaya and R. P. Brent. A novel quartet-based method for phylogenetic inference. *Proceedings of IEEE 5th Symposium on bioinformatics and bioengineering*, pp.32-39, 2005.
18. B. B. Zhou, M. Tarawneh, D. Chu, P. Wang, C. Wang, A. Zomaya, R. Brent, Evidence of Multiple Maximum Likelihood Points for a Phylogenetic Tree. *Proceedings of IEEE 6th Symposium on bioinformatics and bioengineering, Wasington D.C*, 2006.
19. B. B. Zhou, M. Tarawneh, D. Chu, P. Wang, C. Wang, A. Y. Zomaya, and R.P. Brent, On a New Quartet-based Phylogeny Reconstruction Algorithm. *Proceedings of the 2006 International Conference on Bioinformatics and Computational Biology, Las Vegas*, 2006.

A RANDOMIZED ALGORITHM FOR COMPARING SETS OF PHYLOGENETIC TREES

SEUNG-JIN SUL AND TIFFANI L. WILLIAMS

Department of Computer Science
Texas A&M University
College Station, TX 77843-3112 USA
E-mail: {sulsj,tlw}@cs.tamu.edu

Phylogenetic analysis often produce a large number of candidate evolutionary trees, each a hypothesis of the "true" tree. Post-processing techniques such as strict consensus trees are widely used to summarize the evolutionary relationships into a single tree. However, valuable information is lost during the summarization process. A more elementary step is to produce estimates of the topological differences that exist among all pairs of trees. We design a new randomized algorithm, called *Hash-RF*, that computes the all-to-all Robinson-Foulds (RF) distance—the most common distance metric for comparing two phylogenetic trees. Our approach uses a hash table to organize the bipartitions of a tree, and a universal hashing function makes our algorithm randomized. We compare the performance of our Hash-RF algorithm to PAUP*'s implementation of computing the all-to-all RF distance matrix. Our experiments focus on the algorithmic performance of comparing sets of biological trees, where the size of each tree ranged from 500 to 2,000 taxa and the collection of trees varied from 200 to 1,000 trees. Our experimental results clearly show that our Hash-RF algorithm is up to 500 times faster than PAUP*'s approach. Thus, Hash-RF provides an efficient alternative to a single tree summary of a collection of trees and potentially gives researchers the ability to explore their data in new and interesting ways.

1. Introduction

The objective of a phylogenetic analysis is to infer the evolutionary relationships for a given set of organisms (or taxa). Since the true evolutionary history is unknown, many phylogenetic techniques use stochastic search algorithms to solve NP-hard optimization criteria such as maximum likelihood and maximum parsimony. Under these criteria, trees that have better scores are believed to be better approximations of the truth. A typical phylogenetic search results in t trees (i.e., hundreds to thousands of trees can be found), each representing a hypothesis of the "true" tree. Afterwards, post-processing techniques often use consensus methods to transform the rich output of a phylogenetic heuristic into a single summary tree [2]. Yet, much information is lost by summarizing the evolutionary relationships between the t trees into a single consensus tree [7,14].

Given a set of t input trees, we design a randomized hash-based algorithm, called *Hash-RF*, that outputs a $t \times t$ matrix representing the topological distances between every pair of trees. The $t \times t$ distance matrix provides a more information-rich approach for summarizing t trees. The most popular distance measure used to compare two trees is the Robinson-Foulds (RF) distance [12]. Under RF, the distance between two trees is based on the edges (or bipartitions) they share. It is a widely used measure and can be computed in $O(n)$ time using Day's algorithm [5], where n is the number of taxa. Very few algorithms have been designed specifically to compute the all-to-all RF distance. Notable exceptions include PAUP* [15], Phylip [6], and Split-Dist [9a]. Pattengale and Moret provide an approximation algorithm [10], which provides with high probability a $(1+\epsilon)$ approximation of the true RF distance matrix. Given that Pattengale and Moret's approach provides an approximation of the RF distance, we do not compare our approach to their algorithm.

Our experimental results compare the performance on biological trees of our Hash-RF algorithm to the all-to-all RF algorithm embodied in PAUP*, a widely-used commercial application for inferring and interpreting phylogenetic trees. Here, n ranges from 500 to 2,000 taxa and t varies from 200 to 1,000 trees. The results clearly demonstrate that our approach outperforms PAUP*, where greater performance is achieved with increasing values of n and t. On the largest dataset ($n = 2,000$ and $t = 1,000$), our algorithm is 500 times faster than PAUP*. We also compared our approach to Phylip and Split-Dist, but Phylip is tremendously slow even on our smallest dataset. Performance comparisons with Split-Dist followed the same trends as those shown with PAUP* (not shown). Thus, our Hash-RF algorithm provides an efficient alternative to consensus approaches for summarizing a large collection of trees.

2. Background

2.1. *Phylogenetic trees*

The leaves of an evolutionary tree are always labeled with the taxa, and permuting the labels on a tree with fixed topology generally produces a different evolutionary tree. Internal nodes—the hypothetical ancestors—are generally unlabeled. Phylogenies may be rooted or unrooted, and edges may be weighted or unweighted. Order is unimportant. For example, for a node in a rooted tree, swapping the left and the right child does not change the tree.

It is useful to represent evolutionary trees in terms of *bipartitions*. Removing an edge e from a tree separates the leaves on one side from the leaves on the other. The division of the leaves into two subsets is the bipartition B_i associated with edge e_i. In Figure 1, T_2 has two bipartitions: $AB|CDE$ and $ABD|CE$. An evolutionary tree is uniquely and completely defined by its set of $O(n)$ bipartitions. For ease of computation, many algorithms represent each bipartition as a bit-string. At each internal node, those taxa whose subset includes a specified taxon are represented by the bit value '0'. For example, in Figure 1, those taxa that are in the subset of taxon A, are labeled '0'. All other taxa are labeled '1'. Thus, the bipartition, $AB|CDE$ is represented as the bit-string 00111 and $ABD|CE$ is represented as 00101.

[a] Actually, these approaches compute the symmetric distance between two trees. Dividing the symmetric distance by two easily converts it into the RF distance.

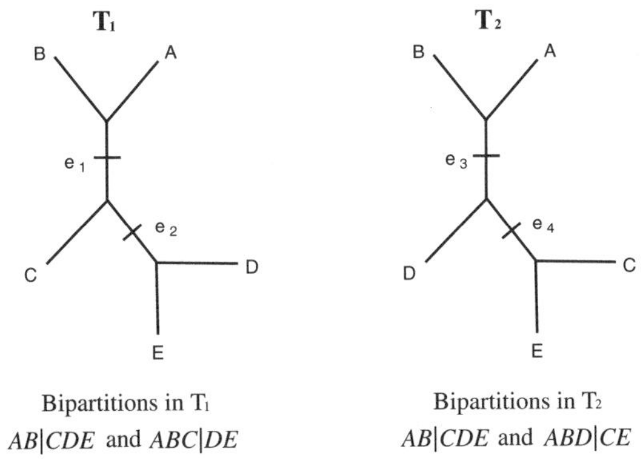

Figure 1. Two phylogenetic trees, T_1 and T_2, and their respective bipartitions. Internal edges are represented by the labels e_1 through e_4.

2.2. *Robinson-Foulds distance*

The Robinson-Foulds (RF) distance between two trees is the number of bipartitions that differ between them. Let $\Sigma(T)$ be the set of bipartitions defined by all edges in tree T. The RF distance between trees T_1 and T_2 is defined as:

$$d_{RF}(T_1, T_2) = \frac{|\Sigma(T_1) - \Sigma(T_2)| + |\Sigma(T_2) - \Sigma(T_1)|}{2} \tag{1}$$

Figure 1 depicts how the RF distance between two trees is calculated. Trees T_1 and T_2 consist of five taxa, and each tree has two non-trivial bipartitions (or internal edges). In this example, the trees are binary and thus you can find $d_{RF}(T_1, T_2)$ is same with $d_{RF}(T_2, T_1)$. By equation (1), the RF distance between the two trees in Figure 1 is equal to 1.

2.3. *Hashing*

A set abstract data type (set ADT) is an abstract data type that maintains a set S under the following three operations:

(1) Insert(x): Add the key x to the set.
(2) Delete(x): Remove the key x from the set.
(3) Search(x): Determine if the key x is contained in the set, and if so, return x.

Hash tables are the most practical and widely used methods of implementing the set ADT and perform the three set ADT operations in $O(1)$ expected time.

The main idea behind all *hash table* implementations is to store a set of $k = |S|$ elements in an array (the hash table) of length $m \geq k$. Hence, we require a function that maps any element x (also called the *hash key*) to an array location. This function is called a *hash function h* and the value $h(x)$ is called the *hash value* of x. That is, the element x gets stored at the array location $H[h(x)]$. Given two distinct elements x_1 and x_2, a *collision* occurs if $h(x_1) = h(x_2)$. Ideally, one would be interested in a perfect hash function, which guarantees no collisions. However, this is only possible when the set of keys are known *a priori* (e.g., compiler keywords). Thus, most hash table implementations must explicitly handle collisions—especially since the performance of the underlying implementation is dependent upon the operations used to resolve the collision.

3. The Hash-RF Algorithm

We were inspired by the work of Amenta et al. [1] to use a hash table as a mechanism for organizing tree bipartitions. Although their algorithm computes a majority consensus tree, we incorporate many of their ideas into our approach. Our Hash-RF algorithm consists of two major steps. The first step requires collecting all of the bipartitions from the evolutionary trees and hashing them into the hash table. Once all of the bipartitions are hashed into the table, the pairwise RF distances can be computed quite quickly. Algorithm 1 presents our hash-based approach. In the following subsections, we explain each of these major steps in detail.

3.1. *Populating the hash table*

Figure 2 provides an overview of the steps required in placing a tree's bipartitions into the hash table. As each input tree, T_i is traversed in post-order, each of its bipartitions is fed through two hash functions, h_1 and h_2. Hash function h_1 is used to generate the location needed for storing a bipartition in the hash table, H. h_2 is responsible for creating a unique identifier for each unique bipartition. For each bipartition, its associated hash table record contains its bipartition ID (BID) along with the index of the tree from where it originated.

For every hash function h, there exists bad sets S, where all distinct keys hash to the same address. To get around this difficulty, we need a collection of hash functions from which we can choose the one that works well for S. Even better would be a collection of hash functions such that, for any given S, most of the hash functions work well. Then, we could randomly pick one of the functions and have a good chance of it working well.

We employ the use of universal hash functions, where $A = (a_1, ..., a_n)$ is a list of random integers in $(0, ..., m_1 - 1)$ and $B = (b_1, ..., b_n)$ is a bipartition. Our h_1 and h_2 hash functions are defined as follows:

$$h_1(B) = \sum b_i a_i \bmod m_1 \tag{2}$$

$$h_2(B) = \sum b_i a_i \bmod m_2 \tag{3}$$

Using these universal hash functions, the probability that any two distinct bipartitions B_i and B_j collide (i.e., $h_1(B_i) = h_1(B_j)$) is $\frac{1}{m_1}$ [3,4]. We call this a Type II collision. (Collisions are described in more detail in the following subsection.) If we choose $m_1 > tn$, the expected number of Type II collisions is $O(tn)$. A double collision (i.e., $h_1(B_i) = h_1(B_j)$ and $h_2(B_i) = h_2(B_j)$) occurs with probability $\frac{1}{m_1 m_2}$. Since the size of m_2 has no impact on the hash table size, m_2 can be made arbitrarily large to avoid double collisions with high probability. We provide more detail on how we detect double collisions (i.e., Type III collisions) below.

Algorithm 1 The Hash-RF algorithm

Require: A set of $T_1, T_2, ..., T_t$ binary trees
for $i = 1$ to t **do**
  Traverse tree T_i in post order
  for all bipartition $B_j \in T_i$ **do**
    Determine collision type at hash table $H[h_1(B_j)]$.
    if Type I collision **then**
      Increment count at KeyMap[B_j]
      Insert $h_2(B_j)$ and tree index i into $H[h_1(B_j)]$
    else if Type III collision **then**
      Terminate and restart algorithm
    else
      Insert $h_1(B_j)$, $h_2(B_j)$ and B_j into KeyMap[B_j]
      Increment count at KeyMap[B_j]
      Insert $h_2(B_j)$ and tree index i into $H[h_1(B_j)]$
    end if
  end for
end for
for all KeyMap[i].count ≥ 2 **do**
  Retrieve the linked list l_i of tree ids (TIDs) from $H[i]$
  for all TID pairs $j, k \in l_i$ **do**
    if BID(j) = BID(k) **then**
      Increment SIM[j][k]
    end if
  end for
end for
for all $i < j \leq t$ **do**
  $RF[i][j] = (n-3) - SIM[i][j]$
end for

Table 1. Collision types in the Hash-RF algorithm.

Collision Type	$B_i = B_j$?	$h_1(B_i) = h_1(B_j)$?	$h_2(B_i) = h_2(B_j)$?
Type I	Yes	Yes	Yes
Type II	No	Yes	No
Type III	No	Yes	Yes

3.2. *Handling collisions*

Given two bipartitions B_i and B_j, there are three types of collisions in the algorithm. Table 1 provides a summary of the different collision types. The first collision type, which we call Type I, occurs as a result of identical bipartitions B_i and B_j appearing in two different trees. Hence, the record for each of these bipartitions at $h_1(B_i)$ will differ in the tree index part of their hash record. In a standard hash implementation, collisions occur

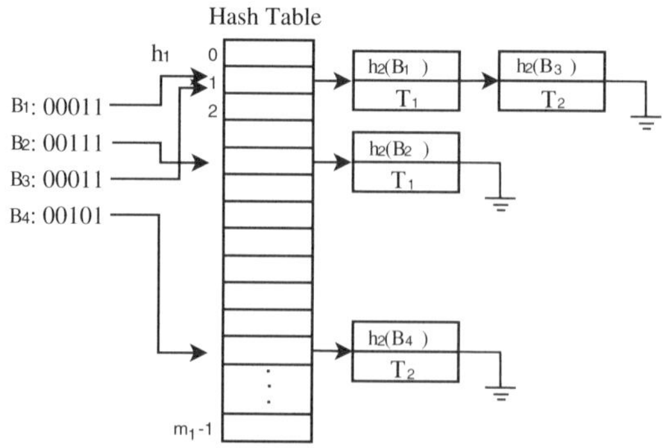

Figure 2. Populating the hash table with the bipartitions from trees T_1 and T_2, which are shown in Fig. 1. Bipartitions B_1 and B_2 define T_1, and B_3 and B_4 are from T_2. Each bipartition is fed to the hash functions h_1 and h_2. For each bipartition, its associated hash table record contains its bipartition ID (BID) along with the associated tree index (TID).

between two different keys hashing to the same location. For our implementation, we keep track of all trees that contain bipartition B_i in order to compute the all-to-all RF distance. Trees that contain bipartition B_i are chained together at location $h_1(B_i)$. Therefore, we consider this situation a collision in our algorithm.

We use an additional data structure, called a *KeyMap*, which is a `map` container from the C++ Standard Template Library, for collision detection. The KeyMap table is used to store key/value pairs, where the keys are logically maintained in sorted order. Each unique bipartition from the set of t trees is given an entry in the KeyMap table. Our KeyMap table contains four fields for each unique bipartition B_i: $h_1(B_i)$, $h_2(B_i)$, B_i, and the frequency of B_i. To detect if B_i causes a Type I collision at $h_1(B_i)$, we search for $h_1(B_1)$ in the KeyMap table. If an entry is found, a collision has occurred. If the bipartition at this location is equal to B_i, we have a Type I collision. Otherwise, if no entry for KeyMap[B_i] is found, B_i is a new bipartition, and a new entry is created in the KeyMap table.

For a Type II collision, $h_1(B_i) = h_1(B_j)$ and $h_2(B_i) \neq h_2(B_j)$. Hash function h_2 is used to generate a bipartition identifier (BID), which attempts to distinguish B_i from the other unique bipartitions. Let B_j represent the bipartition field of the entry KeyMap[$h_1(B_i)$]. If $h_2(B_i) \neq h_2(B_j)$, then a Type II collision has occurred. Hence, two different bipartitions hash to the same location in the hash table. (We note that this is the standard definition of collision in hash table implementations.) Otherwise, there is a double collision (or Type III collision), that is, the bipartitions B_i and B_j are different, but they have the same h_1 and h_2 values. In our algorithm, this is a critical collision, and the algorithm must be restarted with a different set of random integers for the set A.

3.3. *Computing the pairwise RF distances*

Once all of the bipartitions are organized in the hash table, then the RF distance can be calculated. First, we search the KeyMap table to identify bipartitions that have occurred two or more times. Bipartitions that have occurred can be ignored when computing the

RF distance matrix. We update the similarity matrix (SIM) for all pairs of trees in the linked list at $H[i]$. Suppose the linked list consists of trees T_1, T_3, and T_{11}. The Hash-RF algorithm uses their tree ids or indexes (TIDs) 1, 3, and 11 to update the similarity matrix. Here, we increment $SIM[1,3], SIM[1,11]$, and $SIM[3,11]$ by one. We perform the above operations for all of the bipartitions in the hash table. Since our algorithm assumes binary trees, we subtract the similarity matrix entries from $(n-3)$ to obtain the all-pairs RF distance (see lines 25–27 of Algorithm 1). Furthermore, we only compute the upper diagonal of the RF matrix.

3.4. *Analysis*

In our algorithm, the hash table must be populated with nt bipartitions. Hence, this stage of our algorithm requires $O(nt \log nt)$ time since each bipartition must be processed by the KeyMap table to detect the collision type. However, the distribution of the bipartitions in the hash table is responsible for the running time involved in calculating the RF distance matrix. The best case running time of $O(nt \log nt + t^2)$ arises when each hash location has one record, which occurs when there are no bipartitions shared among the input trees. The worst case occurs when each of the nt bipartitions hash to the same location i. Here, the size of the linked list at location i will be nt, which requires $O(n^2t^2)$ time to compute the RF distance matrix.

Our worst case performance matches that of a brute-force all-pairs RF algorithm. Consider computing the RF distance between trees T_i and T_j. We compare each edge in T_i with $O(n)$ edges in T_j. Hence, the RF distance between T_i and T_j requires $O(n^2)$ time. Using this algorithm, we can compute the all-pairs RF matrix in $O(n^2t^2)$ time. Although there is no documentation describing the tree distance algorithm in PAUP*, we suspect it is using the above brute-force algorithm.

4. Our Collection of Biological Trees

Since the performance of our algorithm is dependent upon the distribution of the nt bipartitions, our experiments consider the behavior of our Hash-RF algorithm between the best and worst running time bounds. Our experimental approach is to explore the performance of the algorithm on biological trees produced from a phylogenetic search. Since phylogenetic search techniques operate within a defined neighborhood of the search space, the resulting output tree tends to share many bipartitions among the t trees.

The biological trees used in this study were obtained by running the *Recursive-Iterative DCM3 (Rec-I-DCM3)* algorithm [13], one of the best algorithms for obtaining maximum parsimony trees. We used the following molecular datasets to obtain phylogenetic trees from a Rec-I-DCM3 search: (1) a set of 500 aligned *rbc*L DNA sequences (1,398 sites) [11]; (2) a set of 1,127 aligned large subunit ribosomal RNA sequences (1,078 sites) obtained from the Ribosomal rRNA database [16]; and (3) a set of 2,000 aligned Eukaryotic sRNA sequences (1,251 sites) obtained from the Gutell Lab at the Institute for Cellular and Molecular Biology, The University of Texas at Austin. Thus, n ranged from 500 to 2,000 taxa.

For each of the above datasets, a single run of the Rec-I-DCM3 algorithm produced 1,000 trees (i.e., the Rec-I-DCM3 was run for 1,000 iterations). From these 1,000 trees, we created five sets consisting of 200, 400, 600, 800, and 1,000 trees. Hence, t ranged from 200 to 1,000 trees. Overall, we performed five runs of the Rec-I-DCM3 algorithm on each of the biomolecular datasets leading to 75 different collections of biological trees. Since there are five sets of trees for each pairing of n and t, our experimental results show the average performance of the algorithms on the five tree collections for each pair of n and t.

5. Experimental Results

We ran a series of experiments to study the performance of Hash-RF and PAUP* on the collection of biological trees described in the previous section. All experiments were run on an Intel Pentium platform with 3.0GHz dual-core processors and a total of 2GB of memory. Since our biological trees are binary, we can simply compute the upper triangle of the RF distance matrix as shown in Algorithm 1. However, since PAUP* computes the full matrix, our algorithm does the same to ensure a fair comparison of the algorithms.

Figure 3 compares the performance of the Hash-RF algorithm with PAUP* in terms of running time and speedup. The Hash-RF algorithm requires a minimum of 1.58 seconds on the smallest dataset (n = 500, t = 200) and a maximum of 93.65 seconds on the largest dataset (n = 2, 000 and $t = 1,000$). For the same datasets, PAUP* requires 31.09 seconds and 13.94 hours, respectively. Hence, on our largest dataset, the Hash-RF algorithm is over 500 times faster than PAUP*'s all-to-all RF distance algorithm. Moreover, the results demonstrate that even greater speedups can be expected with larger collections of biological trees.

For $n = 2,000$, Table 2 provides information on the number of hash locations where $|l_i| = t$, which implies that bipartition B_i is shared among the t trees. A large number of locations with $|l_i| = t$ in the hash table will result in a slowdown in the performance of the Hash-RF algorithm since it will require $O(|l_i|^2)$ time to process the linked list at location i. The number of edges in an unrooted binary tree with n taxa is $n - 3$, which also represents the maximum number of distinct bipartitions that can be shared across the t trees. Shared bipartitions among the t trees compose the strict consensus tree. Table 2 shows that the average number of bipartitions shared among the trees range from 43.77% to 45.62%, when $n = 2,000$. When n = 500, the resolution of the strict consensus tree ranges from 68.21% to 71.23% (not shown). So, even under diverse conditions of overlap among the bipartitions, the Hash-RF algorithm performs quite well in comparison to PAUP*.

6. Conclusions and Future Work

Phylogenetic search methods can produce large numbers of candidate trees as approximations to the "true" evolutionary tree. Such trees provide a powerful data-mining opportunity for examining the evolutionary relationships that exist among them. Post-processing methods such as strict consensus trees are the most common methods for providing a single tree that summarizes the information contained in the candidate trees. We advocate a more

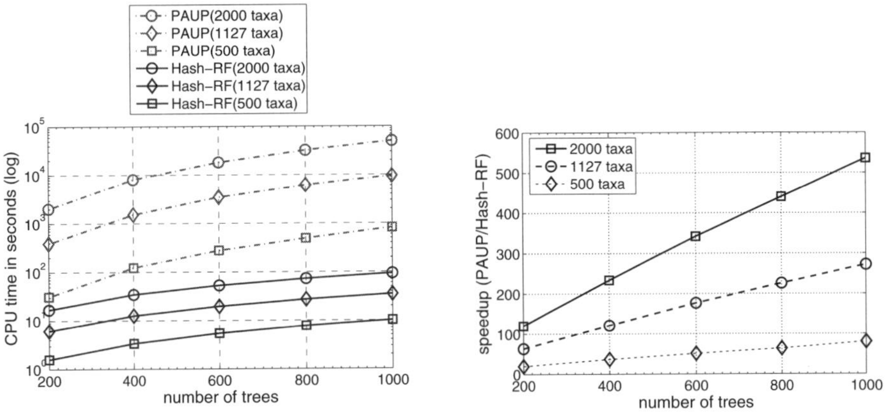

(a) CPU time *(b) Speedup*

Figure 3. Performance of Hash-RF and PAUP* on the collection of biological trees. (a) provides the running time required by the algorithms to compute the all-to-all RF distance matrix for various values of n and t. (b) shows the speedup of the Hash-RF approach over PAUP*.

Table 2. For n = 2,000, the number of identical bipartitions shared between the t trees and resulting resolution of the strict consensus tree.

t	# hash locations where $\lvert l_i \rvert = t$	strict consensus tree resolution (%)
200	911	45.62
400	908	45.47
600	884	44.27
800	882	44.17
1,000	874	43.77

information-rich approach to analyzing the trees returned from a phylogenetic search. As a step in this direction, we present a fast randomized algorithm that calculates the Robinson-Foulds (RF) distance between each pair of evolutionary trees that in the best and worst case require $O(nt \log nt + t^2)$ and $O(n^2t^2)$ running times, respectively.

Our experiments explore the behavior of our approach within these two boundaries. We compared the performance of our Hash-RF algorithm to PAUP*—a popular, commercially-available software package for inferring and interpreting phylogenetic trees—on large collections of biological trees. Our Hash-RF algorithm is up to 500 times faster than PAUP*'s approach. We also compared our approach to Phylip and Split-Dist, but Phylip is extremely slow and Split-Dist performed similarly to PAUP* (not shown). The experiments with biological trees share between 43.77% and 71.23% of their bipartitions among the t trees. Given the diverse distributions of bipartition sharing among the biological trees, the results clearly demonstrate the performance gain achieved by using a hash-based approach for computing the RF distance between each pair of trees. Moreover, fast algorithms such as Hash-RF will enable users to perform interactive analyses of large tree collections in such applications as Mesquite [8].

Our work can be extended in many different directions. One immediate source of improvement would be a better mechanism for detecting collisions in our Hash-RF algorithm. Additional experiments will include randomly-generated trees (i.e, to control the degree of bipartition sharing among the t trees) and larger tree collections. Using Day's $O(n)$ algorithm to compute the RF distance between two trees, it is possible theoretically to compute the all-pairs RF distance in $O(nt^2)$ time. We plan on implementing Day's algorithm and comparing its performance in practice to our Hash-RF approach. Finally, we are extending our algorithm for use with multifurcating trees.

References

1. N. Amenta, F. Clarke, and K. S. John. A linear-time majority tree algorithm. In *Workshop on Algorithms in Bioinformatics*, volume 2168 of *Lecture Notes in Computer Science*, pages 216–227, 2003.
2. D. Bryant. A classification of consensus methods for phylogenetics. In M. Janowitz, F. Lapointe, F. McMorris, B. Mirkin, and F. Roberts, editors, *Bioconsensus*, volume 61 of *DIMACS: Series in Discrete Mathematics and Theoretical Computer Science*, pages 163–184. American Mathematical Society, DIMACS, 2003.
3. J. L. Carter and M. N. Wegman. Universal classes of hash functions. *Journal of Computer and Systems Sciences*, 18(2):143–154, 1979.
4. T. H. Cormen, C. E. Leiserson, R. L. Rivest, and C. Stein. *Introduction to algorithms*. MIT Press, Inc., 2001.
5. W. H. E. Day. Optimal algorithms for comparing trees with labeled leaves. *Journal Of Classification*, 2:7–28, 1985.
6. J. Felsenstein. Phylogenetic inference package (PHYLIP), version 3.2. *Cladistics*, 5:164–166, 89.
7. D. M. Hillis, T. A. Heath, and K. S. John. Analysis and visualization of tree space. *Syst. Biol*, 54(3):471–482, 2004.
8. W. P. Maddison and D. R. Maddison. Mesquite: a modular system for evolutionary analyses. Version 1.11, 2006. http://mesquiteproject.org.
9. T. Mailund. SplitDist—calculating split-distances for sets of trees. Available from http://www.daimi.au.dk/ mailund/split-dist.html.
10. N. D. Pattengale and B. M. E. Moret. A sublinear-time randomized approximation scheme for the robinson-foulds metric". In *Proc. 10th Int'l Conf. on Research in Comput. Molecular Biol. (RECOMB'06)*, volume 3909 of *Lecture Notes in Computer Science*, pages 221–230, 2006.
11. K. Rice, M. Donoghue, and R. Olmstead. Analyzing large datasets: *rbcL* 500 revisited. *Systematic Biology*, 46(3):554–563, 1997.
12. D. F. Robinson and L. R. Foulds. Comparison of phylogenetic trees. *Mathematical Biosciences*, 53:131–147, 1981.
13. U. Roshan, B. M. E. Moret, T. L. Williams, and T. Warnow. Rec-I-DCM3: a fast algorithmic techniques for reconstructing large phylogenetic trees. In *Proc. IEEE Computer Society Bioinformatics Conference (CSB 2004)*, pages 98–109. IEEE Press, 2004.
14. C. Stockham, L. S. Wang, and T. Warnow. Statistically based postprocessing of phylogenetic analysis by clustering. In *Proceedings of 10th Int'l Conf. on Intelligent Systems for Molecular Biology (ISMB'02)*, pages 285–293, 2002.
15. D. L. Swofford. PAUP*: Phylogenetic analysis using parsimony (and other methods), 2002. Sinauer Associates, Underland, Massachusetts, Version 4.0.
16. J. Wuyts, Y. V. de Peer, T. Winkelmans, and R. D. Wachter. The European database on small subunit ribosomal RNA. *Nucleic Acids Research*, 30:183–185, 2002.

PROTEIN STRUCTURE-STRUCTURE ALIGNMENT WITH DISCRETE FRÉCHET DISTANCE

MINGHUI JIANG*

Department of Computer Science, Utah State University,
Logan, UT 84322-4205, USA
Email: mjiang@cc.usu.edu

YING XU†

Department of Biochemistry and Molecular Biology, University of Georgia,
Athens, GA 30602-7229, USA
Email: xyn@bmb.uga.edu

BINHAI ZHU

Department of Computer Science, Montana State University,
Bozeman, MT 59717-3880, USA
Email: bhz@cs.montana.edu

Matching two geometric objects in 2D and 3D spaces is a central problem in computer vision, pattern recognition and protein structure prediction. In particular, the problem of aligning two polygonal chains under translation and rotation to minimize their distance has been studied using various distance measures. It is well known that the Hausdorff distance is useful for matching two point sets, and that the Fréchet distance is a superior measure for matching two polygonal chains. The discrete Fréchet distance closely approximates the (continuous) Fréchet distance, and is a natural measure for the geometric similarity of the folded 3D structures of bio-molecules such as proteins. In this paper, we present new algorithms for matching two polygonal chains in 2D to minimize their discrete Fréchet distance under translation and rotation, and an effective heuristic for matching two polygonal chains in 3D. We also describe our empirical results on the application of the discrete Fréchet distance to the protein structure-structure alignment.

1. Introduction

Matching two geometric objects in 2D and 3D spaces is a central problem in computer vision, pattern recognition and protein structure prediction. A lot of research has been done in this aspect using various distance measures. One of the most popular distance measures is the Hausdorff distance $d_{\mathcal{H}}$. For arbitrary bounded sets $A, B \subseteq \mathbb{R}^2$, it is defined

*Supported by USU research funds A13501 and A14766.
†Supported in part by National Science Foundation (NSF/DBI-0354771, NSF/ITR-IIS-0407204) and by a Distinguished Cancer Scholar grant from the Georgia Cancer Coalition.

as follows:

$$d_{\mathcal{H}}(A,B) = \max\left(\sup_{a\in A}\inf_{b\in B} dist(a,b), \sup_{b\in B}\inf_{a\in A} dist(a,b)\right).$$

where $dist$ is the underlying metric in the plane, for example the Euclidean metric. Given two point sets with m and n points respectively in the plane, their minimum Hausdorff distance under translation can be computed in $O(mn(m+n)\alpha(mn)\log(mn))$ time [12] and, when both translation and rotation are allowed, in $O((m+n)^6\log(mn))$ time [11]. Given two polygonal chains with m and n vertices respectively in the plane, their minimum Hausdorff distance under translation can be computed in $O((mn)^2\log^3(mn))$ time [6] and, when both translation and rotation are allowed, in $O((mn)^4(m+n)\log(m+n))$ time [2].

The Hausdorff distance is a good measure for the similarity of point sets, but it is inadequate for the similarity of polygonal chains; one can easily come up with examples of two polygonal chains with a small Hausdorff distance but drastically different geometric shapes. Alt and Godau [3] proposed to use the Fréchet distance to measure the similarity of two polygonal chains. The Fréchet distance $\delta_{\mathcal{F}}$ between two parametric curves $f:[0,1]\to\mathbb{R}^2$ and $g:[0,1]\to\mathbb{R}^2$ is defined as follows:

$$\delta_{\mathcal{F}}(f,g) = \inf_{\alpha,\beta}\max_{s\in[0,1]} dist(f(\alpha(s)), g(\beta(s))),$$

where α and β range over all continuous non-decreasing real functions with $\alpha(0)=\beta(0)=0$ and $\alpha(1)=\beta(1)=1$. Imagine that a person and a dog walk along two different paths while connected by a leash; they always move forward, though at different paces. The minimum possible length of the leash is the Fréchet distance between the two paths. Given two polygonal chains with m and n vertices respectively in the plane, their Fréchet distance at fixed positions can be computed in $O(mn\log(m+n))$ time [4]; their minimum Fréchet distance under translation can be computed in $O((mn)^3(m+n)^2\log(m+n))$ time [5] and, when both translation and rotation are allowed, in $O((m+n)^{11}\log(m+n))$ time [17].

The Fréchet distance is a superior measure for the similarity of polygonal curves, but it is very difficult to handle. Eiter and Mannila [9] introduced the discrete Fréchet distance as a close approximation of the (continuous) Fréchet distance. We now review their definition of the discrete Fréchet distance using our notations (but with exactly the same idea).

Definition 1.1. Given a polygonal chain $P=\langle p_1,\ldots,p_n\rangle$ of n vertices, a k**-walk** along P partitions the vertices of P into k disjoint non-empty subsets $\{P_i\}_{i=1..k}$ such that $P_i=\langle p_{n_{i-1}+1},\ldots,p_{n_i}\rangle$ and $0=n_0<n_1<\cdots<n_k=n$.

Given two polygonal chains $A=\langle a_1,\ldots,a_m\rangle$ and $B=\langle b_1,\ldots,b_n\rangle$, a **paired walk** along A and B is a k-walk $\{A_i\}_{i=1..k}$ along A and a k-walk $\{B_i\}_{i=1..k}$ along B for some k, such that, for $1\le i\le k$, either $|A_i|=1$ or $|B_i|=1$ (that is, either A_i or B_i contains exactly one vertex). The **cost** of a paired walk $W=\{(A_i,B_i)\}$ along two chains A and B is

$$d_{\mathcal{F}}^W(A,B) = \max_i \max_{(a,b)\in A_i\times B_i} dist(a,b).$$

The **discrete Fréchet distance** between two polygonal chains A and B is

$$d_{\mathcal{F}}(A, B) = \min_{W} d_{\mathcal{F}}^{W}(A, B).$$

The paired walk that achieves the discrete Fréchet distance between two polygonal chains A and B is called the **Fréchet alignment** of A and B.

Let's consider again the scenario in which the person walks along A and the dog along B. Intuitively, the definition of the paired walk is based on three cases:

(1) $|B_i| > |A_i| = 1$: the person stays and the dog moves forward;
(2) $|A_i| > |B_i| = 1$: the person moves forward and the dog stays;
(3) $|A_i| = |B_i| = 1$: both the person and the dog move forward.

The following figure shows the relationship between discrete and continuous Fréchet distances. In Figure 1 (I), we have two polygonal chains $\langle a, b\rangle$ and $\langle c, d, e\rangle$; their continuous Fréchet distance is the distance from d to the segment $\overline{ab}$, that is, $dist(d, o)$. The discrete Fréchet distance is $dist(d, b)$. As we can see from the figure, the discrete Fréchet distance could be arbitrarily larger than the continuous distance. On the other hand, if we put enough sample points on the two polygonal chains, then the resulting discrete Fréchet distance, that is, $dist(d, f)$ in Figure 1 (II), closely approximates $dist(d, o)$.

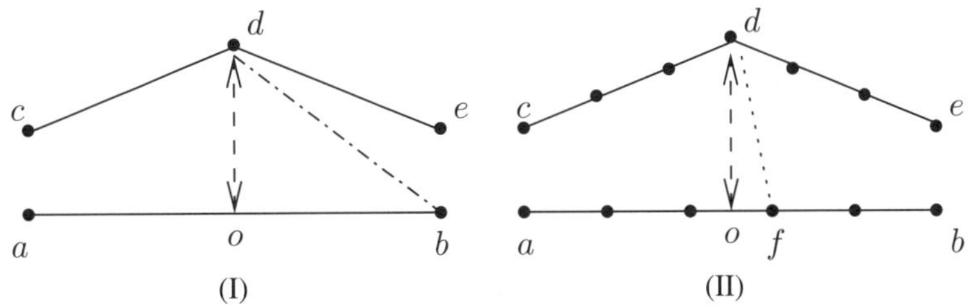

Figure 1. The relationship between discrete and continuous Fréchet distances.

Given two polygonal chains of m and n vertices respectively, their discrete Fréchet distance can be computed in $O(mn)$ time by a dynamic programming algorithm [9]. We now describe our algorithm based on the same idea.

Given two polygonal chains $A = \langle a_1, \ldots, a_m\rangle$ and $B = \langle b_1, \ldots, b_n\rangle$, and their two subchains $A[1..i] = \langle a_1, \ldots, a_i\rangle$ and $B[1..j] = \langle b_1, \ldots, b_j\rangle$, let $d_<(i, j)$ (respectively, $d_>(i, j)$) denote the discrete Fréchet distance between $A[1..i]$ and $B[1..j]$ such that a_i (respectively, b_j) belongs to a single-vertex subset in the paired walk, and define $d(i, j) = \min\{d_<(i, j), d_>(i, j)\}$. The discrete Fréchet distance $d_{\mathcal{F}}(A, B) = \min\{d_<(m, n), d_>(m, n)\}$ can be computed in $O(mn)$ time with the base conditions

$$d_<(i, 0) = d_<(0, j) = 0 \quad \text{and} \quad d_>(i, 0) = d_>(0, j) = 0 \quad \text{and} \quad d(i, 0) = d(0, j) = 0,$$

and the recurrences

$$d_<(i,j) = \max \begin{cases} dist(a_i, b_j) \\ \min\{d(i-1, j-1), d(i, j-1)\} \end{cases}$$

$$d_>(i,j) = \max \begin{cases} dist(a_i, b_j) \\ \min\{d(i-1, j-1), d(i-1, j)\} \end{cases}$$

$$d(i,j) = \min\{d_<(i,j), d_>(i,j)\}.$$

In this paper, we present new algorithms that compute the minimum discrete Fréchet distance of two polygonal chains in the plane under translation in $O((mn)^3 \log(m+n))$ time and, when both rotation and translation are allowed, in $O((mn)^4 \log(m+n))$ time. These bounds are two or three orders of magnitude smaller than the corresponding best bounds [5,17] using the continuous Fréchet distance measure.

Our interest in matching two polygonal chains in 2D and 3D spaces is motivated by the application of protein structure-structure alignment. The discrete Fréchet distance is a very natural measure in this application because a protein can be viewed essentially as a chain of discrete amino acids in 3D. We design a heuristic method for aligning two polygonal chains in 3D based on the intuition behind our theoretical results for the 2D case, and use it to measure the geometric similarity of protein tertiary structures with real protein data drawn from the Protein Data Bank (PDB) hosted at `http://www.rcsb.org/pdb/`.

The paper is organized as follows. In Section 2, we present our algorithms for matching two polygonal chains in 2D under translation and rotation. In Section 3, we describe our heuristic method for matching two polygonal chains in 3D under translation and rotation, and present our empirical results on protein structure-structure alignment with the discrete Fréchet distance. In Section 4, we conclude the paper.

2. Matching 2D Polygonal Chains Under Translation and Rotation

Definition 2.1. *(Optimization Problem)* Given two polygonal chains A and B, a transformation class T, and a distance measure d, find a transformation $\tau \in T$ such that $d(A, \tau(B))$ is minimized.

Definition 2.2. *(Decision Problem)* Given two polygonal chains A and B, a transformation class T, a distance measure d, and a real number $\epsilon \geq 0$, decide whether there is a transformation $\tau \in T$ such that $d(A, \tau(B)) \leq \epsilon$.

Observation 2.1. Given two polygonal chains A and B, if there is a transformation τ such that $d_{\mathcal{F}}(A, \tau(B)) = \epsilon$, then there are a vertex $a \in A$ and a vertex $b \in B$ such that $dist(a, \tau(b)) = \epsilon$.

2.1. *Matching Under Translation*

We first consider the transformation class T_t of all translations.

Lemma 2.1. *Given two 2D polygonal chains A and B, if there is a translation $\tau \in T_t$ such that $d_{\mathcal{F}}(A, \tau(B)) = \epsilon > 0$, then one of the following four cases is true:*

(1) *there are a vertex* $a \in A$ *and a vertex* $b \in B$ *such that, for any translation* $\tau' \in T_t$, $dist(a, \tau'(b)) = \epsilon \implies d_{\mathcal{F}}(A, \tau'(B)) \le \epsilon$;

(2) *there are two vertices* $a, c \in A$, *a vertex* $b \in B$, *and a translation* $\tau' \in T_t$ *such that* $dist(a, \tau'(b)) = dist(c, \tau'(b)) = \epsilon$ *and* $d_{\mathcal{F}}(A, \tau'(B)) \le \epsilon$;

(3) *there are a vertex* $a \in A$, *two vertices* $b, d \in B$, *and a translation* $\tau' \in T_t$ *such that* $dist(a, \tau'(b)) = dist(a, \tau'(d)) = \epsilon$ *and* $d_{\mathcal{F}}(A, \tau'(B)) \le \epsilon$;

(4) *there are two vertices* $a, c \in A$, *two vertices* $b, d \in B$, *and a translation* $\tau' \in T_t$ *such that* $\overrightarrow{ac} \neq \overrightarrow{bd}$ *(that is, either* $|ac| \neq |bd|$, *or* $\overrightarrow{ac}$ *and* $\overrightarrow{bd}$ *have different directions),* $dist(a, \tau'(b)) = dist(c, \tau'(d)) = \epsilon$, *and* $d_{\mathcal{F}}(A, \tau'(B)) \le \epsilon$.

Proof. Let $a \in A$ and $b \in B$ be the two vertices such that $dist(a, \tau(b)) = \epsilon$, the existence of which is guaranteed by Observation 2.1. Let $W = \{(A_i, B_i)\}$ be the Fréchet alignment of A and $\tau(B)$ such that $d_{\mathcal{F}}^W(A, \tau(B)) = \epsilon$. We translate B with τ' (starting at τ) such that the distance between the two vertices a and b remains at exactly ϵ, that is, $dist(a, \tau'(b)) = \epsilon$. We consider the distance $d_{\mathcal{F}}^W(A, \tau'(B)) = \max_i \max_{(p,q) \in A_i \times B_i} dist(p, \tau'(q))$ as τ' changes continuously.

As τ' changes continuously, $\tau'(b)$ rotates around a in a circle of radius ϵ. If $d_{\mathcal{F}}^W(A, \tau'(B))$ always remains at ϵ, we have case 1; otherwise, there are two vertices $c \in A_i$ and $d \in B_i$ for some i such that the distance $dist(c, \tau'(d))$ crosses the threshold ϵ. We cannot have both $a = c$ and $b = d$ because the distance $dist(a, \tau'(b))$ always remains at ϵ; for the same reason, we cannot have $\overrightarrow{ac} = \overrightarrow{bd}$. There are three possible cases: if $a \neq c$ and $b = d$, we have case 2; if $a = c$ and $b \neq d$, we have case 3; if $a \neq c$ and $b \neq d$, we have case 4. □

The previous lemma implies the following algorithm that checks the four cases:

(1) For every two vertices $a \in A$ and $b \in B$, compute an arbitrary translation τ' such that $dist(a, \tau'(b)) = \epsilon$, and check whether $d_{\mathcal{F}}(A, \tau'(B)) \le \epsilon$.

(2) For every three vertices $a, c \in A$ and $b \in B$, compute all possible translations τ' such that $dist(a, \tau'(b)) = dist(c, \tau'(b)) = \epsilon$, and check whether $d_{\mathcal{F}}(A, \tau'(B)) \le \epsilon$.

(3) For every three vertices $a \in A$ and $b, d \in B$, compute all possible translations τ' such that $dist(a, \tau'(b)) = dist(a, \tau'(d)) = \epsilon$, and check whether $d_{\mathcal{F}}(A, \tau'(B)) \le \epsilon$.

(4) For every four vertices $a, c \in A$ and $b, d \in B$ such that $\overrightarrow{ac} \neq \overrightarrow{bd}$, compute all possible translations τ' such that $dist(a, \tau'(b)) = dist(c, \tau'(d)) = \epsilon$, and check whether $d_{\mathcal{F}}(A, \tau'(B)) \le \epsilon$.

The algorithm answers yes if it finds at least one translation τ' such that $d_{\mathcal{F}}(A, \tau'(B)) \le \epsilon$; otherwise, it answers no. As we can see from the following lemma, this algorithm solves the decision problem.

Lemma 2.2. *If there is translation* τ' *such that* $d_{\mathcal{F}}(A, \tau'(B)) = \epsilon'$, *then, for any distance* $\epsilon \ge \epsilon'$, *there exists a translation* τ *such that* $d_{\mathcal{F}}(A, \tau(B)) = \epsilon$.

Proof. As we translate B from $\tau'(B)$ to infinity, the discrete Fréchet distance between A and the translated B changes continuously (since it is a composite function based on the continuous Euclidean distance functions) from $d_{\mathcal{F}}(A, \tau'(B)) = \epsilon'$ to infinity. The continuity implies that, for any $\epsilon \geq \epsilon'$, there exists a translation τ such that $d_{\mathcal{F}}(A, \tau(B)) = \epsilon$. □

We now analyze the algorithm. In cases 2 and 3, given two points p and q such that $p \neq q$, the two equations $dist(x, p) = \epsilon$ and $dist(x, q) = \epsilon$ together determine x (there are at most two solutions for x) since the 2D point x has two variable components. In case 4, given two points p and q, and a vector $\vec{v} \neq \overrightarrow{pq}$, the two equations $dist(x, p) = \epsilon$ and $dist(x + \vec{v}, q) = \epsilon$ are independent and determine x (there are at most a constant number of solutions for x). Given a translation τ', to check whether $d_{\mathcal{F}}(A, \tau'(B)) \leq \epsilon$ takes $O(mn)$ time. The overall time complexity is $O(mn \cdot m^2n^2) = O(m^3n^3)$.

With binary search, our algorithm for the decision problem implies an $O(m^3n^3 \log(1/\epsilon))$ time $1 + \epsilon$ approximation for the optimization problem; with parametric search (applying Cole's sorting trick [5,8]), it implies an $O(m^3n^3 \log(m + n))$ time exact algorithm. We have the following theorem.

Theorem 2.1. *For minimizing the discrete Fréchet distance between two 2D polygonal chains under translation, we have an $O(m^3n^3 \log(1/\epsilon))$ time $1 + \epsilon$ approximation algorithm and an $O(m^3n^3 \log(m + n))$ time exact algorithm.*

2.2. *Matching Under Translation and Rotation*

We next consider the transformation class T_{tr} that includes both translations and rotations.

Lemma 2.3. *Given two 2D polygonal chains A and B, if there is a transformation $\tau \in T_{tr}$ such that $d_{\mathcal{F}}(A, \tau(B)) = \epsilon > 0$, then one of the following seven cases is true:*

(1) *there are a vertex $a \in A$ and a vertex $b \in B$ such that, for any transformation $\tau' \in T_{tr}$, $dist(a, \tau'(b)) = \epsilon \implies d_{\mathcal{F}}(A, \tau'(B)) \leq \epsilon$;*
(2) *there are two vertices $a, c \in A$ and two vertices $b, d \in B$ such that, for any transformation $\tau' \in T_{tr}$, $dist(a, \tau'(b)) = dist(c, \tau'(d)) = \epsilon \implies d_{\mathcal{F}}(A, \tau'(B)) \leq \epsilon$;*
(3) *there are two vertices $a, c \in A$, three vertices $b, d, f \in B$, and a transformation $\tau' \in T_{tr}$ such that $dist(a, \tau'(b)) = dist(c, \tau'(d)) = dist(c, \tau'(f)) = \epsilon$ and $d_{\mathcal{F}}(A, \tau'(B)) \leq \epsilon$.*
(4) *there are three vertices $a, c, e \in A$, two vertices $b, d \in B$, and a transformation $\tau' \in T_{tr}$ such that $dist(a, \tau'(b)) = dist(c, \tau'(d)) = dist(e, \tau'(d)) = \epsilon$ and $d_{\mathcal{F}}(A, \tau'(B)) \leq \epsilon$.*
(5) *there are three vertices $a, c, e \in A$ and three vertices $b, d, f \in B$ ($\triangle ace$ and $\triangle bdf$ are not congruent), and a transformation $\tau' \in T_{tr}$ such that $dist(a, \tau'(b)) = dist(c, \tau'(d)) = dist(e, \tau'(f)) = \epsilon$, and $d_{\mathcal{F}}(A, \tau'(B)) \leq \epsilon$.*
(6) *there are three vertices $a, c, e \in A$ and three vertices $b, d, f \in B$ ($\triangle ace$ and $\triangle bdf$ are congruent), and a transformation $\tau' \in T_{tr}$ such that the two triangles*

$\triangle ace$ and $\tau'(\triangle bdf)$ are not parallel (their corresponding edges are not parallel), $dist(a, \tau'(b)) = dist(c, \tau'(d)) = dist(e, \tau'(f)) = \epsilon$, and $d_{\mathcal{F}}(A, \tau'(B)) \leq \epsilon$.

(7) *there are three vertices $a, c, e \in A$ and three vertices $b, d, f \in B$ ($\triangle ace$ and $\triangle bdf$ are congruent) such that, for any transformation $\tau' \in T_{tr}$, if $\triangle ace$ and $\tau'(\triangle bdf)$ are parallel, and if $dist(a, \tau'(b)) = dist(c, \tau'(d)) = dist(e, \tau'(f)) = \epsilon$, then $d_{\mathcal{F}}(A, \tau'(B)) \leq \epsilon$.*

Proof. Let $a \in A$ and $b \in B$ be the two vertices such that $dist(a, \tau(b)) = \epsilon$, the existence of which is guaranteed by Observation 2.1. Let $W = \{(A_i, B_i)\}$ be the Fréchet alignment of A and $\tau(B)$ such that $d^W_{\mathcal{F}}(A, \tau(B)) = \epsilon$. Without loss of generality, we assume that $a \in A_i$, $b \in B_i$, and b is the only vertex in B_i.

Starting with $\tau' = \tau$, we rotate B around the vertex b. During the rotation, the distance between the two vertices a and b remains at exactly ϵ, that is, $dist(a, \tau'(b)) = \epsilon$. If $d^W_{\mathcal{F}}(A, \tau'(B))$ always remains at ϵ, we have case 1.

Otherwise, there are two vertices $c \in A_j$ and $d \in B_j$ for some j such that the distance $dist(c, \tau'(d))$ crosses the threshold ϵ. We must have $i \neq j$ because b is the only vertex in B_i and the positions of the vertices in A_i are fixed as we rotate B around b. It follows that $a \neq c$ and $b \neq d$. Now, we continue to transform B while keeping the two constraints $dist(a, \tau'(b)) = \epsilon$ and $dist(c, \tau'(d)) = \epsilon$ satisfied. If $d^W_{\mathcal{F}}(A, \tau'(B))$ always remains at ϵ, we have case 2.

Otherwise, there are two vertices $e \in A_k$ and $f \in B_k$ for some k such that the distance $dist(e, \tau'(f))$ crosses the threshold ϵ. We must have $k \neq i$ for the same reason that $j \neq i$. We consider two possibilities: either $k = j$ or $k \neq j$.

If $k = j$, then we must have either $e = c$ or $f = d$ because either A_j or B_j contains a single vertex. We cannot have both $e = c$ and $f = d$ because we keep the constraint $dist(c, \tau'(d)) = \epsilon$ satisfied during the transformation. If $e = c$, we have case 3; if $f = d$, we have case 4.

If $k \neq j$, then we consider the two triangles $\triangle ace$ and $\tau'(\triangle bdf)$.

(1) If they are not congruent, we have case 5.
(2) If they are congruent by not parallel, we have case 6.
(3) If they are both congruent and parallel, then we translate B continuously while keeping the three constraints $dist(a, \tau'(b)) = dist(c, \tau'(d)) = dist(e, \tau'(f)) = \epsilon$ satisfied. During the translation, we either encounter another pair of vertices e' and f' whose distance crosses the threshold ϵ or not. If we encounter e' and f', then the two triangles $\triangle ace'$ and $\triangle bdf'$ must not be congruent, and we have case 5; otherwise, we have case 7. □

As before, the previous lemma implies an algorithm for the decision problem. We now analyze the running time. In cases 1, 2, and 7, we only need to find one transformation τ'. In cases 3 and 4, there are at most four transformations for τ'. In case 5, the transformation for τ' can be specified by six variables: the x and y coordinates of the three vertices b, d, and f; we also have six constraints for the lengths of the six segments ab, cd, ef, bd, df, and bf. Each constraint is specified by a quadratic equation. There are at most a constant number of solutions for these equations.

In case 6, we have two congruent triangles $\triangle ace$ and $\triangle b'd'f'$ ($\triangle b'd'f' = \tau'(\triangle bdf)$). If the two triangles have the same enclosing circle, then there are at most two transformations such that $|ab'| = |cd'| = |ef'| = \epsilon$. If the two triangles do not have the same enclosing circle, then we can always translate $\triangle ace$ to $\triangle a'c'e'$ such that $\triangle a'c'e'$ and $\triangle b'd'f'$ have the same enclosing circle, then rotate $\triangle a'c'e'$ to $\triangle b'd'f'$. We have $|ab'| = |cd'| = |ef'| = \epsilon > 0$, $|a'b'| = |c'd'| = |e'f'| = x > 0$ (since they are not parallel), and

$$\overrightarrow{ab'} = \overrightarrow{a'b'} + \vec{v}, \qquad \overrightarrow{cd'} = \overrightarrow{c'd'} + \vec{v}, \qquad \overrightarrow{ef'} = \overrightarrow{e'f'} + \vec{v}.$$

Given a fixed vector $\vec{v}$, the equation $\vec{w} = \vec{u} + \vec{v}$, subject to the two constraints $|\vec{w}| = \epsilon > 0$ and $|\vec{u}| = x > 0$, has at most two solutions for $\vec{w}$ and $\vec{u}$. On the other hand, the three vectors $\overrightarrow{a'b'}$, $\overrightarrow{c'd'}$, and $\overrightarrow{e'f'}$ are distinct, which is a contradiction. Therefore, the two triangles $\triangle ace$ and $\triangle b'd'f'$ must have the same enclosing circle.

Theorem 2.2. *For minimizing the discrete Fréchet distance between two 2D polygonal chains under translation and rotation, we have an $O(m^4n^4 \log(1/\epsilon))$ time $1 + \epsilon$ approximation algorithm and an $O(m^4n^4 \log(m + n))$ time exact algorithm.*

3. Protein Structure-Structure Alignment

The discrete Fréchet distance between two polygonal chains is a natural measure for comparing the geometric similarity of protein tertiary structures because the alpha-carbon atoms along the backbone of a protein essentially forms a 3D polygonal chain.

Generalizing the theoretical results in the previous section, it is possible to match two polygonal chains with m and n vertices in 3D in roughly $O((mn)^5)$ time under both translation and rotation. However, this would be too slow for our target application of protein structure-structure alignment, where a typical protein corresponds to a 3D polygonal chain with 300–500 amino acids. Instead of an exact algorithm, we propose an intuitive heuristic and present our empirical results showing its effectiveness in matching two similar polygonal chains.

3.1. *A Heuristic for Matching 3D Polygonal Chains Under Translation and Rotation*

Given a 3D chain C of n vertices, the coordinates of each vertex c_i of C can be represented by a 3D vector $\vec{c_i}$. The *center* c of the chain C corresponds to the vector $\vec{c} = \frac{\sum_i \vec{c_i}}{n}$. We observe that, given two polygonal chains $A = \langle a_1, \ldots, a_m \rangle$ and $B = \langle b_1, \ldots, b_n \rangle$, if $d_{\mathcal{F}}(A, B) = \epsilon$, then we must have both $dist(a_1, b_1) \leq \epsilon$ and $dist(a_m, b_n) \leq \epsilon$. If ϵ is smaller than half the minimum distance between two consecutive vertices in either A or B, then the Fréchet alignment of A and B must contain only one-to-one matches between vertices of A and B. That is, we must have $m = n$ and, for $1 \leq i \leq n$, $dist(a_i, b_i) \leq \epsilon$. It follows that $dist(a, b) \leq \epsilon$, where a and b are the centers of A and B, respectively.

The observation above suggests that we can use the three points, the two end-vertices and the center, as the reference points [1] for each chain. For two polygonal chains with a small discrete Fréchet distance, their corresponding reference points must be close. In general, the position and orientation of each polygonal chain is determined by the positions of its three reference points. We have the following heuristic for matching A and B under translation and rotation:

(1) Translate B such that the center a of A and the center b of B coincide.
(2) Rotate B around b such that the two triangles $\triangle a a_1 a_m$ and $\triangle b b_1 b_n$ are co-planar and such that the two vectors $\frac{\vec{a_1}+\vec{a_m}}{2} - \vec{a}$ and $\frac{\vec{b_1}+\vec{b_n}}{2} - \vec{b}$ have the same direction.
(3) Rotate B for a small angle around the axis through its two randomly chosen vertices. If this does not decrease the discrete Fréchet distance between A and B, rotate back.
(4) Repeat the previous tuning step for a number of times.

3.2. *The Experiment*

We implemented our protein structure-structure alignment heuristic and a protein visualization software [a] in Java. The experiment was conducted on an Apple iMac with a 2GHz PowerPC G5 processor and 2GB DDR SDRAM memory running Mac OS 10.4.3 and Java 1.4.2.

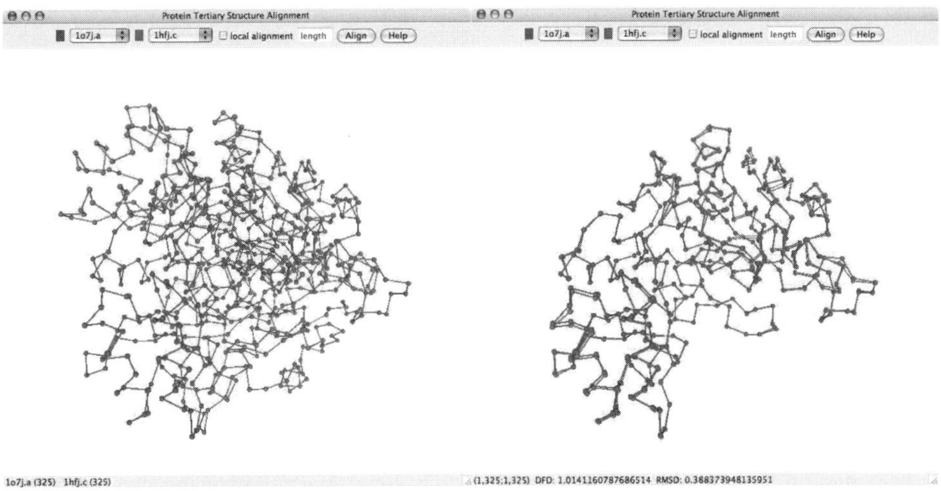

Figure 2. The alignment of 1o7j.a and 1hfj.c by our heuristic.

In the experiment, we align the protein chain 1o7j.a (PDB ID 1o7j; chain A) with seven other protein chains 1hfj.c, 1qd1.b, 1toh, 4eca.c, 1d9q.d, 4eca.b, and 4eca.d. Each of these

[a] The program is hosted on the web at `http://www.cs.usu.edu/~mjiang/frechet.html`.

eight chains contains exactly 325 vertices, where each vertex represents an alpha-carbon atom on the protein backbone. When the number of tuning steps is set to 20, our program takes less than one second to align two chains of lengths 325 on our test machine. Figure 2 shows two screenshots of our program, before and after aligning the two protein chains 1o7j.a and 1hfj.c.

We compare our heuristic with ProteinDBS [16], an online protein database search engine hosted at `http://proteindbs.rnet.missouri.edu/` that supports protein structure-structure alignment. ProteinDBS uses computer vision techniques to align two protein chains based on the two-dimensional distance matrix generated from the 3D coordinates of the alpha-carbon atoms on the protein backbones. The two chains 1o7j.a and 1hfj.c are examples given in the ProteinDBS paper [16]. According to the query result from the ProteinDBS website, the seven chains (1hfj.c, 1qd1.b, 1toh, 4eca.c, 1d9q.d, 4eca.b, 4eca.d) have global tertiary structures most similar to 1o7j.a.

Table 1. The characteristics of the seven chains with the highest similarity ranking by ProteinDBS.

Protein Chain	Alignment Length	RMSD (in angstrom)	Discrete Fréchet Distance (in angstrom)
1hfj.c	325	0.27	1.01
1qd1.b	85	2.81	22.90
1toh	55	2.91	35.09
4eca.c	317	1.10	6.01
1d9q.d	81	2.88	22.18
4eca.b	317	1.09	5.76
4eca.d	318	1.45	5.92

By comparing the image patterns in the distance matrices instead of aligning the tertiary structures geometrically, ProteinDBS is very efficient but not so accurate. We refer to Table 1, which lists the characteristics of the alignments generated by ProteinDBS. The three protein chains, 1qd1.b, 1toh, and 1d9q.d, have global tertiary structures dissimilar to that of the chain 1o7j.a, but they are incorrectly ranked among the top by ProteinDBS. The discrete Fréchet distances of these chains and the query chain computed by our heuristic correctly identify the three dissimilar protein chains.

4. Conclusion

In this paper, we present the first algorithms for matching two polygonal chains in 2D to minimize their discrete Fréchet distance under translation and rotation. Our algorithms are two or three orders of magnitude faster than the fastest algorithms using the continuous Fréchet distance, and can be readily generalized to higher dimensions.

The discrete Fréchet distance is a natural measure for comparing the folded 3D structures of bio-molecules such as proteins. Our experiment shows that our heuristic for aligning protein tertiary structures using the discrete Fréchet distance is more accurate than ProteinDBS's structure aligning algorithm, which is based on computer vision techniques. We are currently conducting more empirical studies and refining our protein structure-structure alignment algorithm with additional ideas from some other popular algorithms such as the

Combinatorial Extension (CE) Method [15] hosted at `http://cl.sdsc.edu/`. We see great potential for using the discrete Fréchet distance in the local alignment [10], the feature identification, and the consensus shape construction [7] of multiple proteins.

References

1. O. Aichholzer, H. Alt, and G. Rote. Matching shapes with a reference point. *International Journal of Computational Geometry & Applications*, 7(4):349–363, 1997.
2. H. Alt, B. Behrends, and J. Blömer. Approximate matching of polygonal shapes (extended abstract). In *Proceedings of the 7th Annual Symposium on Computational Geometry (SoCG'91)*, pages 186–193, 1991.
3. H. Alt and M. Godau. Measuring the resemblance of polygonal curves. In *Proceedings of the 8th Annual Symposium on Computational Geometry (SoCG'92)*, pages 102–109, 1992.
4. H. Alt and M. Godau. Computing the Fréchet distance between two polygonal curves. *International Journal of Computational Geometry & Applications*, 5:75–91, 1995.
5. H. Alt, C. Knauer, and C. Wenk. Matching polygonal curves with respect to the Fréchet distance. In *Proceedings of the 18th Annual Symposium on Theoretical Aspects of Computer Science (STACS'01)*, pages 63–74, 2001.
6. P.K. Agarwal, M. Sharir, and S. Toledo. Applications of parametric search in geometric optimization. In *Proceedings of the 3rd Annual ACM-SIAM Symposium on Discrete Algorithms (SODA'92)*, pages 72–82, 1992.
7. L.P. Chew and K. Kedem. Finding the consensus shape of a protein family. In *Proceedings of the 18th Annual Symposium on Computational Geometry (SoCG'02)*, pages 64–73, 2002.
8. R. Cole. Slowing down sorting networks to obtain faster sorting algorithms. *Journal of the ACM*, 34:200–208, 1987.
9. T. Eiter and H. Mannila. Computing discrete Fréchet distance. Technical Report CD-TR 94/64, Information Systems Department, Technical University of Vienna, 1994.
10. D. Gusfield. *Algorithms on Strings, Trees, and Sequences: Computer Science and Computational Biology*. Cambridge University Press, 1997.
11. D.P. Huttenlocher, K. Kedem, and J.M. Kleinberg. On dynamic Voronoi diagrams and the minimum Hausdorff distance for point sets under Euclidean motion in the plane. In *Proceedings of the 8th Annual Symposium on Computational Geometry (SoCG'92)*, pages 110–119, 1992.
12. D.P. Huttenlocher, K. Kedem, and M. Sharir. The upper envelope of Voronoi surfaces and its applications. In *Proceedings of the 7th Annual Symposium on Computational Geometry (SoCG'91)*, pages 194–203, 1991.
13. P. Indyk. Approximate nearest neighbor algorithms for Fréchet distance via product metrics. In *Proceedings of the 18th Annual Symposium on Computational Geometry (SoCG'02)*, pages 102–106, 2002.
14. A. Mosig and M. Clausen. Approximately matching polygonal curves with respect to the Fréchet distance. *Computational Geometry: Theory and Applications*, 30(2):113–127, 2005.
15. I.N. Shindyalov and P.E. Bourne. Protein structure alignment by incremental combinatorial extension (CE) of the optimal path. *Protein Engineering*, 11(9):739–747, 1998.
16. C.-R. Shyu, P.-H. Chi, G. Scott, and D. Xu. ProteinDBS: a real-time retrieval system for protein structure comparison. *Nucleic Acids Research*, 32:W572–575, 2004.
17. C. Wenk. *Shape Matching in Higher Dimensions*. PhD thesis, Freie Universitaet Berlin, 2002.

DERIVING PROTEIN STRUCTURE TOPOLOGY FROM THE HELIX SKELETON IN LOW RESOLUTION DENSITY MAP USING ROSETTA

YONGGANG LU, JING HE[1]

Dept. Computer Science, New Mexico State University
Las Cruces, NM 88003-8001, USA

CHARLIE E. M. STRAUSS

Bioscience Division, M888, Los Alamos National Laboratory
Los Alamos, NM 87545, USA

Electron cryo-microscopy (cryo-EM) is an experimental technique to determine the 3-dimensional structure for large protein complexes. Currently this technique is able to generate protein density maps at 6 to 9 Å resolution. Although secondary structures such as α-helix and β-sheet can be visualized from these maps, there is no mature approach to deduce their tertiary topology, the linear order of the secondary structures on the sequence. The problem is challenging because given N secondary structure elements, the number of possible orders is $(2^N)*N!$. We have developed a method to predict the topology of the secondary structures using *ab initio* structure prediction. The Rosetta structure prediction algorithm was used to make purely sequence based structure predictions for the protein. We produced 1000 of these *ab initio* models, and then screened the models produced by Rosetta for agreement with the helix skeleton derived from the density map. The method was benchmarked on 60 mainly alpha helical proteins, finding that for about 3/4 of all the proteins, the majority of the helices in the skeleton were correctly assigned by one of the top 10 suggested topologies from the method, while for about 1/3 of all the proteins the best topology assignment without errors was ranked the first. This approach also provides an estimate of the sequence alignment of the skeleton. For most of those true-positive assignments, the alignment was accurate to within +/- 2 amino acids in the sequence.

1. Introduction

Electron cryo-microscopy (cryo-EM) is an attractive method for structure determination because it can work with proteins that are poorly soluble or otherwise fail to crystallize, and is amenable to the structure determination of large protein complexes as well [1-7]. Although the cryo-EM can generate structures in the form of electron density maps at 6 to 9 Å resolution, it is currently not sufficient to determine the atomic structure directly since the side-chains cannot be resolved at this low-to-intermediate resolution [7, 8]. However, the location of secondary structures (SS), such as helices and β-sheets, can be visually and computationally identified [8-12]. It has been an emerging question about how to combine the low resolution density map with structure prediction techniques in order to derive the 3-dimensional structure of the protein [7, 8, 13]. The identified SS are often the major components of a protein and they form its skeleton (Fig. 1). Although the

---

This work is supported by LANL-NMSU MOU, NSF-HRD-0420407, LANL-LDRD-DR and DOE: Genomics: GTL Carbon sequestration.

[1]Email address of the corresponding author: jinghe@cs.nmsu.edu

skeleton contains the geometrical location of SS, it does not provide the information about where the SS are aligned with the protein sequence. The order of the SS with respect to the sequence, the topology, is also unknown. In this paper, we used helix skeleton, which is composed of the helices computationally identified by our software HelixTracer [12]. Given a protein density map at 6-10 Å resolution, HelixTracer can output the locations of helices represented by their central axial lines which can be potentially curved (Fig. 1).

Rosetta is one of the most successful *ab initio* structure prediction methods [14-18]. Unlike comparative modeling, *ab initio* methods do not require a structural homology to start with. For small protein domains, Rosetta is frequently able to produce low resolution models with correct topologies spanning the majority of the protein sequence. Previous work has shown that Rosetta is useful in refining NMR data [15, 18].

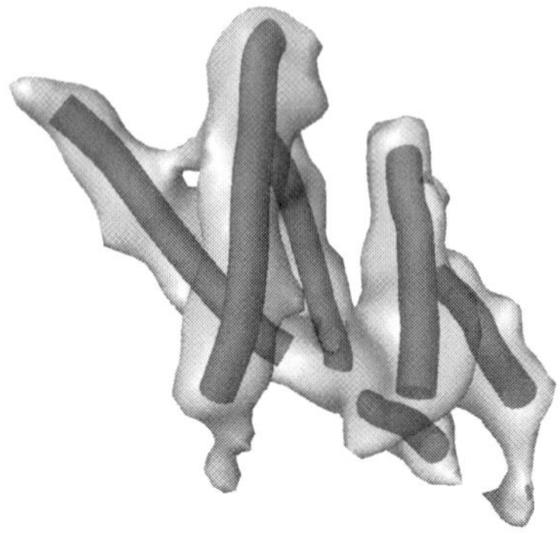

Fig. 1. Surface representation of the simulated density map (green) and the helix skeleton (purple cylinder-like sticks) found by HelixTracer for protein 1abv

We have developed a method to derive the topology of a protein by combining its helix skeleton information with the predicted models obtained from Rosetta. Our method involves two components: MatchHelices and consensus analysis.

There were two impetuses to develop MatchHelices. First, we were not aware of an existing efficient approach to matching predicted model structures to a skeleton without any sequence information. Second, it is a waypoint towards fully integrating density maps into constrained *ab initio* modeling that rapidly reduces the search space. An alternative approach for comparison is used in "Foldhunter" from EMAN [9, 19] which uses correlation of density maps to align the skeleton with a structure. This approach is slow because of the grid search for translation and rotation parameters. Since MatchHelices only searches through the possible orientations suggested by the helix skeleton, the computation is significantly less. For a typical alignment between two structures of 100 amino acids, Foldhunter needs several minutes while MatchHelices takes a few seconds on a 3GHz machine.

2. Methods

2.1 The Overall Approach

Given a protein sequence, Rosetta can generate protein-like conformations known as decoys. The decoys are predicted possible conformations of the protein. Different decoys may have quite different conformations. The idea is to use helix skeleton to group the decoys and to derive

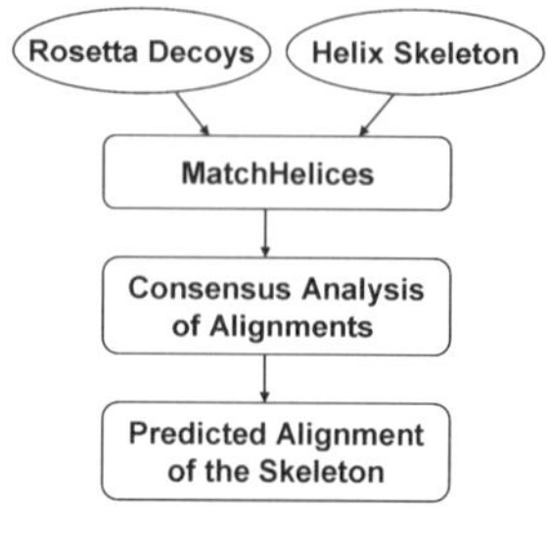

Fig. 2. The overall approach

the topology from each group. The overall procedure is shown in Fig. 2. It was tested on 60 mainly alpha-helical single domain proteins ranging in size of 50 to 150 residues. For each protein, Rosetta was used to generate 1000 decoys, and "pdb2mrc" was used to generate its density map at 8Å resolution [19]. Then the helix skeleton was identified from the density map by HelixTracer [12], and all the decoys were aligned with the skeleton by MatchHelices. Finally a consensus analysis was employed to identify the ten most popular alignments of the skeleton with the sequence.

2.2 MatchHelices

MatchHelices is a method to align a decoy with a helix skeleton (Fig. 3). The input is composed of a decoy and the helix skeleton found by HelixTracer (step 1). MatchHelices is a greedy and iterative method which first constructs seed alignment and then refines the alignment. A seed alignment is an initial trial of alignment that satisfies the following two criteria. The first requires the alignment of the two mass centers, one from the decoy and the other from the density map (step 3). The second requires that a pair of helices, one from the decoy and the other from the skeleton (selected in step 2), is positioned in a way so that the two helices are as close as possible while keeping the mass centers aligned. The second requirement is satisfied by a rotation of the decoy model around its mass center to maximally superimpose the two helices in a pair (step 4). The seed alignment is then refined by allowing certain level of mismatch between the mass centers. Since the seed alignment roughly positions the two helices in a pair, the corresponding points can be assigned between the two helices. If the C_α atoms of the helix in the decoy are within 5 Å distance from the helix axis in the skeleton, they are assigned to the nearest corresponding points on the skeleton (step 5). By doing this, a set of corresponding points are determined between the two helices in a pair. During the refinement step, the decoy is rotated and translated to minimize the RMS deviation between the corresponding points of a pair of helices (step 6). The step 5 and step 6 are repeated iteratively finding atoms within the cut-off and re-superposition to convergence. In step 7, the alignment score is calculated and the best alignment is updated. The alignment score is an ad hoc combination of the number of overlapped C-alpha atoms and the number of helices

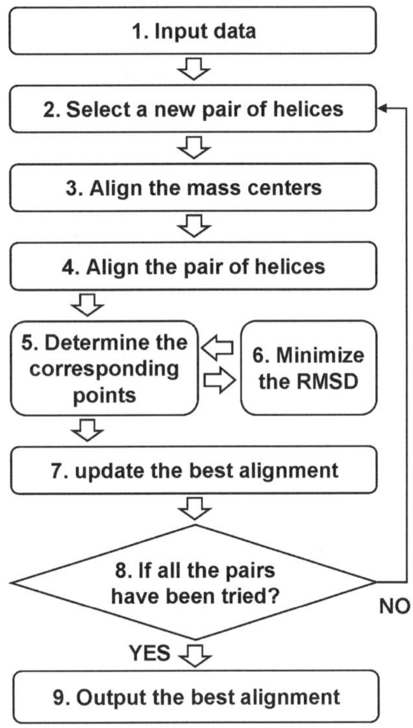

Fig. 3. The flowchart of MatchHelices

matched. The process (step 2 to step 7) is repeated for all the permutations of helix pairs and their relative directionalities to determine the best alignment between the decoy and the skeleton. In case of ambiguity, in which different parts of a decoy helix overlap two or more skeleton helices, the assignment is made according to the center residue of the helical segment of the model. Two example decoys aligned to the skeleton by MatchHelices are shown in Fig. 4.

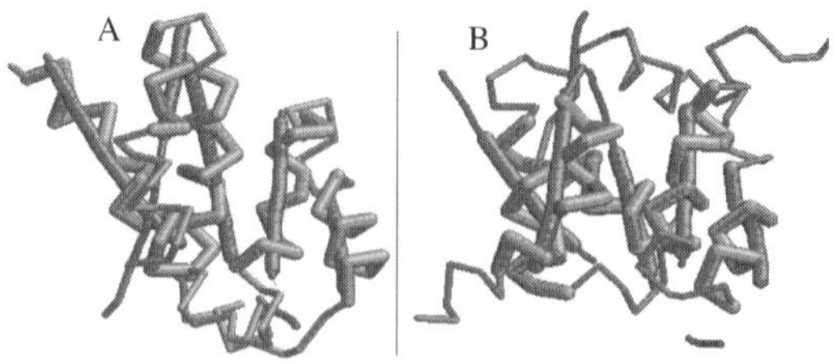

Fig. 4. Examples of MatchHelices Results: Rosetta decoy backbone (blue) and it's superposition with the helix skeleton (red) for protein 1abv. The thick regions show where the superposition criteria are satisfied. A. Corresponds to one of the decoys in first ranked topology prediction (a good prediction) in Fig. 6C, and B. corresponds to second ranked (a poor prediction) in Fig. 6C.

Since we are hunting for partial alignments over subsets of the skeleton, it is in principle possible that there could be multiple alignments of the skeleton to a given decoy that are equally good. In practice, several attributes of our approach seem to avoid this issue. First, because we require the decoy center of mass to be concentric with the density center, this forces an overall overlap beyond just the helices we are pairing, and breaks most degenerate cases. Second, HelixTracer produces skeletons with curved segments, not straight lines, and this too breaks the degeneracy. Third, we simply discard skeleton matches that are insufficiently complex (e.g. if only a single segment is matched while there are more than two helix segments in the skeleton). Lastly, even if this does occur in a particular decoy, we use many decoys and form a consensus.

2.3 Consensus Analysis

After all the decoys have been aligned with the helix skeleton, they are clustered based on where the skeleton helices are aligned on the protein sequence (Fig. 5). We used the secondary structure locations ("estimated helix positions") generated by Rosetta as a guide to assist the clustering process. For each residue, if more than half of the Rosetta decoys assign helix secondary structure to this residue, the residue is labeled to be within an estimated helix (indicated by H's below the protein sequence in Fig. 5). Within the regions of the sequence where the estimated helices reside, the nature of the alignment is examined. Two alignments are grouped into the same cluster if the skeleton helices they align roughly have the same location on the sequence. Particularly, each center amino acid of the skeleton helix has to be within the corresponding "estimated" helix. Therefore, the decoys in the same cluster are those with the same number of aligned skeleton helices. Moreover, their order, their directions on the sequence, and the corresponding "estimated" helices must be the same. For example, in Fig. 5, "Decoy A", "Decoy B" and "Decoy C" are grouped into one cluster while "Decoy E" and "Decoy F" are grouped into another cluster. "Decoy D" has only 3 skeleton helices aligned, so it

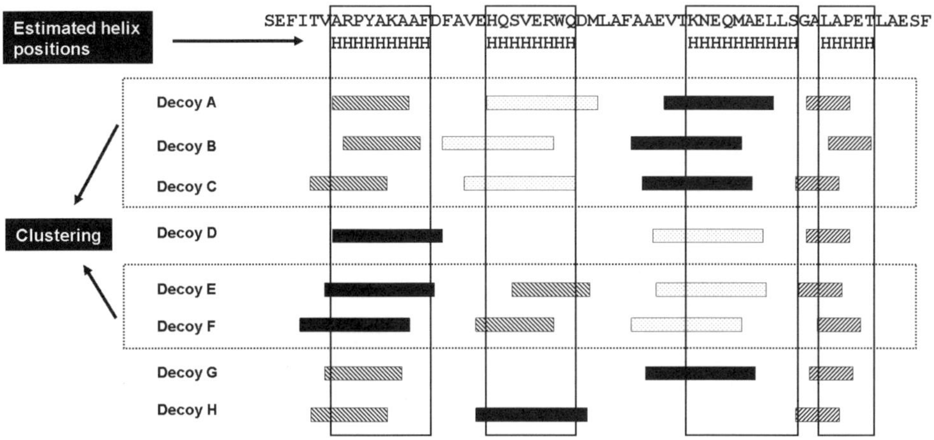

Fig. 5. Illustration of the clustering process of the decoy alignments. Horizontal bars following the label "Decoy A"..."Decoy H" are the alignment results of different decoys. Different patterns of the bars represent different helices in the skeleton.

cannot be grouped into the same cluster with "Decoy E" and "Decoy F". Although "Decoy G" and "Decoy H" have the same set of skeleton helices aligned to the sequence, they cannot be grouped together since the center of the skeleton helix shown as the black bar is aligned to two different estimated helices in the two alignments. After the clustering, the resulting clusters are ranked by the number of decoys which they contain.

3. Results and Discussion

Fig. 6 shows the top ranked four topologies for protein 1abv. The top ranked topology correctly predicted the order of the six skeleton helices identified by HelixTracer (Fig. 6C). Besides the correct topology, our method also roughly aligned the skeleton helices to their correct locations on the sequence for the top ranked result (Fig. 6C). Each topology is derived from a cluster of decoys (Fig. 6C). The two decoys fitted to the skeleton of protein 1abv in Fig. 4 are the members of the two clusters corresponding to the first and the second topologies diagrammed in Fig. 6C. It can be seen in Fig. 6C that the topology inferred from the alignment in Fig. 4A is the correct one and the topology inferred from Fig. 4B is incorrect.

In the case of 1abv, HelixTracer correctly identified all of the six helices. But this is not always the case. Sometimes it misses or over-predicts some of the helices. Therefore, there could be an error in the result of HelixTracer. Similarly, there is often an error in the predicted model of Rosetta. However, the error sometimes can be partially compensated in the consensus analysis step, which will be shown later in this section.

We tested our method on 60 mainly alpha helical proteins. In Table 1, we only listed 44 of them. These 44 proteins have majority of the skeleton helices correctly assigned by one of the top 10 predicted topologies, judging by the column "Correct TH" and the column "Assigned Helices" (Table 1). From the last column "Alignment Offsets", we can see that for most of the alignments with the sequence, the offsets of the centers are within

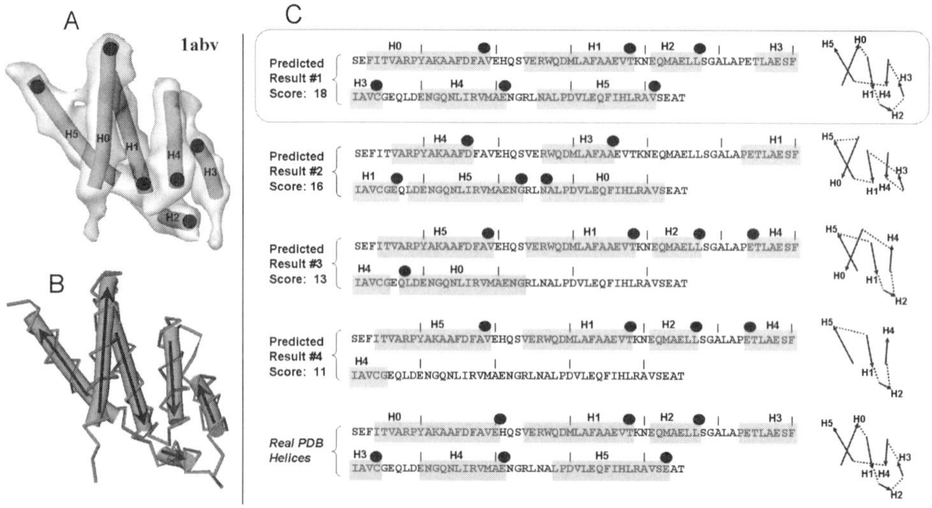

Fig. 6. Predicted sequence alignment of the skeleton for protein 1abv.
A: Simulated density of protein 1abv at 8Å resolution, with inset helix skeleton (purple sticks).
B: The helix skeleton (purple sticks) overlapped on the backbone of the protein from the PDB.
C: Top 4 predicted topologies and the true topology of 1abv. Green: predicted sequence alignments of the helices. Purple: the true sequence alignment of the helices. Right: diagrams of the topologies. Blue dots indicate true C-termini of helices labeled in A and C.

+/-2 residues. From the column "Rank", it is noticeable that for 17 proteins, about 1/3 of the total proteins tested, the best assignment is ranked the first. Column "Cluster Size" lists the number of the decoys in the cluster that was used to produce the correct assignment. In general, a larger value in the "Cluster Size" column makes the result more reliable. However, for some proteins, the "Cluster Size" value is very small and it still produces good results. It can also be noticed that when the value of the "Total Clusters" is larger, the value of the "Cluster Size" is usually smaller. This is reasonable, because the number of decoys is the same for each protein, and fewer decoys will usually be in a cluster if more clusters are formed.

The other 16 proteins (with PDB ID 1ag2_, 1bgf_, 1bo9A, 1d8bA, 1eo0A, 1ey1A, 1jhgA, 1jli_, 1jw2A, 1k04A, 1klxA, 1koyA, 1l9lA, 1lre_, 1lriA and 1qqvA) failed to generate correct topologies from the top 10 predictions. These 16 proteins are not listed in the table. Although for the 16 proteins we didn't get satisfied results in the top ten topologies, we noticed that a subset of the assignments of the skeleton helices may still be correct. They were not qualified to be "good results" because we used a very strict criterion that all of the assignments in a result including the topological ordering of the helical segments and their directionalities should be correct. So it is possible that useful information still can be acquired from the top 10 results for the 16 proteins.

In some cases only a subset of the skeleton helices were aligned (see Fig. 6C), thus there could be more than one result having correct topologies but using different helical subsets within the top 10 predictions. In such cases, we only listed the one that has the most skeleton helices involved in the assignment.

Table 1. Results of the 44 out of 60 proteins which have the majority of the helices in the skeleton correctly assigned by one of the top 10 suggested topologies

Protein ID	Total Residues	Possible Clusters[a]	Total Clusters[b]	Rank[c]	Cluster Size[d]	PDB Helices[e]	Total TH[f]	Correct TH[g]	**Assigned Helices[h]**	Alignment Offsets[i]
1a43_	72	62700	315	3rd	23	6	5	5	**4**	{0,2.5,0,0.5 }
1a6s_	87	360	213	9th	19	4	3	3	**3**	{0,2.5,1.5 }
1a7w_	68	138	80	3rd	56	3	3	3	**2**	{0,0.5 }
1abv_	105	291792	611	1st	18	6	6	6	**6**	{1,0,0,1,0,2 }
1bby_	69	138	71	1st	346	3	3	3	**3**	{0.5,0.5,0 }
1bkrA	108	378640	607	2nd	5	8	5	5	**3**	{1,0.5,0 }
1c3yA	108	62700	670	1st	12	6	5	5	**4**	{0.5,0.5,2,0 }
1daqA	71	36	18	4th	95	3	2	2	**1**	{1 }
1dgnA	89	1044204	578	8th	8	7	6	6	**4**	{1,0,0,0.5 }
1dk8A	147	1432180	725	3rd	5	10	5	5	**3**	{1,0,0.5 }
1dlwA	116	39040	334	1st	34	8	4	4	**4**	{0.5,0,0.5,1.5 }
1doqA	69	4360	87	1st	234	5	4	4	**4**	{0,0,1,1 }
1dp3A	55	138	35	1st	194	3	3	3	**2**	{1,0 }
1du2A	76	64	61	7th	36	4	2	2	**2**	{0,1 }
1dxsA	57	4360	234	5th	21	5	4	4	**4**	{0,1,0,2.5 }
1ef4A	55	138	47	1st	518	3	3	3	**3**	{0.5,0,2.5 }
1eyhA	144	1.69E+08	823	1st	15	10	7	7	**4**	{0,0.5,1,4.5 }
1f68A	103	62700	427	2nd	44	6	5	5	**3**	{0.5,0.5,0.5 }
1f6vA	91	1356	235	7th	21	6	3	3	**2**	{0.5,2.5 }
1fe5A	118	750	182	7th	16	5	3	3	**2**	{1.5,2 }
1g03A	134	291792	635	5th	8	6	6	6	**3**	{2,1,5.5 }
1g7dA	106	19090	338	2nd	53	5	5	5	**5**	{1,1,1.5,0,1.5 }
1gab_	53	138	116	4th	57	3	3	3	**3**	{1.5,0.5,0.5 }
1gd6A	119	166390	347	1st	14	7	5	4	**3**	{0.5,0.5,1 }
1gxgA	85	360	149	2nd	62	4	3	3	**3**	{1,1,0.5 }
1gxqA	105	1472	95	1st	68	4	4	4	**3**	{0.5,0.5,1 }
1hb6A	86	4360	235	1st	35	5	4	4	**4**	{1,1,0,0.5 }
1hbkA	89	1472	218	3rd	17	4	4	4	**4**	{0,1,0,0.5 }
1hdj_	77	360	76	1st	532	4	3	3	**3**	{1.5,1,2 }
1hp8_	68	360	57	1st	253	4	3	3	**2**	{1.5,1.5 }
1ig6A	107	166390	517	7th	5	7	5	5	**5**	{1,0,2,2,9 }
1iizA	120	62700	313	1st	36	6	5	3	**3**	{0.5,1.5,2 }
1iygA	133	1044204	501	6th	18	7	6	5	**4**	{1,0,0,0.5 }
1ji8A	111	166390	316	7th	17	7	5	5	**5**	{1,2,2,0,0.5 }
1kr7A	110	1044204	602	6th	9	7	6	5	**4**	{0.5,0,0,0.5 }
1myo_	118	19748176	816	2nd	6	8	7	7	**4**	{0,0,1,1 }
1ngr_	85	1044204	752	8th	4	6	7	6	**3**	{0,1,0.5 }
1nkl_	78	19090	303	6th	17	5	5	5	**3**	{0.5,0.5,4 }
1pru_	56	138	62	1st	320	3	3	3	**3**	{0.5,0,0 }
1utg_	70	1472	111	1st	79	4	4	4	**4**	{0,1,0.5,1 }
2asr_	142	19090	447	7th	4	5	5	5	**3**	{3,0.5,0.5 }
2end_	137	138	204	8th	14	3	3	3	**2**	{1,5 }
2lisA	131	1044204	695	7th	3	6	7	5	**3**	{2.5,0.5,2 }
2mhr_	118	4360	258	1st	84	5	4	4	**4**	{4,0.5,1,0.5 }

[a] The total number of all possible clusters in the search space.
[b] The total number of clusters produced in the consensus analysis step.
[c] The rank of the cluster from which the correct topology assignment was produced.
[d] The size (the number of the decoy members in the cluster) of the cluster from which the correct topology assignment was produced.
[e] The total number of helices in the crystal structure.
[f] The total number of helices (in the skeleton) identified by HelixTracer.
[g] The number of correctly identified helices (in the skeleton) by HelixTracer.
[h] The number of helices (in the skeleton) assigned in the correct topology assignment.
[i] The offsets of the centers of helices (in the skeleton) from the actual positions in the alignment corresponding to the correct topology assignment. The offsets are separated by commas.

The column "Possible Clusters" lists the total number of all possible clusters in the search space, N_P, which can be calculated by:

$$N_p = \sum_{k=1}^{\min(Ht,Hp)} C_{Ht}^k C_{Hp}^k \left(k! \times 2^k\right)$$

in which Ht is the total number of helices identified by HelixTracer, and Hp is the total number of helices in the crystal structure.

It can be seen from the table that for protein "1iizA" with 6 helices, HelixTracer only correctly identified 3 of them, with 3 helices missed and 2 helices over-predicted. With so many errors contained in the input, the correct topology was still found and it was ranked the first in the final results. Other examples for which HelixTracer over-predicted helices include protein "igd6A", "1iygA", "1kr7A", "1ngr_" and "2lisA". So our method can sometimes compensate the errors produced in the skeleton identification step.

4. Conclusion

This work is a preliminary study to see if we can quickly collapse the factorial complexity of the topology assignment and sequence alignment problem to a small number of possibilities that include an assured true positive. The results showed that our method was capable of assign majorities of the helices in the skeleton correctly within top 10 assignments for most of the proteins tested. For about 1/3 of all the proteins, the best result with the correct topology was ranked the first. Our method also showed robustness in working with bad inputs, namely the false positive and false negative identifications of the true helices by HelixTracer. The predicted topologies for the helix skeleton can be very helpful for structure determination with cryo-EM method. And it also can become the basis of a more careful constrained search using Rosetta.

Previously Rosetta has been used to refine NMR data [15, 18]. Here we are using it only as a consensus screen and we are not (as yet) incorporating the cryo-EM data into the prediction algorithm as restraints as it was done for NMR. Our previous work with NMR data showed that only a few constraints are needed to achieve very high accuracy, however when false positive constraints are included in a constraint set, the prediction quality rapidly deteriorates. Thus our screening protocol was biased away from making complete assignments of all the SS elements, and towards predictions with few false positives over the majority of the SS elements at the top of its ranked list.

References

1. Chiu, W., M. L. Baker, W. Jiang, and Z. H. Zhou. 2002. Deriving folds of macromolecular complexes through electron cryomicroscopy and bioinformatics approaches. Curr Opin Struct Biol 12: 263-9.
2. Topf, M., and A. Sali. 2005. Combining electron microscopy and comparative protein structure modeling. Curr Opin Struct Biol 15: 578-85.

3. Yonekura, K., S. Maki-Yonekura, and K. Namba. 2005. Building the atomic model for the bacterial flagellar filament by electron cryomicroscopy and image analysis. Structure 13: 407-12.
4. Zhou, Z. H., M. Dougherty, J. Jakana, J. He, F. J. Rixon, and W. Chiu. 2000. Seeing the herpesvirus capsid at 8.5 A. Science 288: 877-80.
5. Zhou, Z. H., M. L. Baker, W. Jiang, M. Dougherty, J. Jakana, G. Dong, G. Lu, and W. Chiu. 2001. Electron cryomicroscopy and bioinformatics suggest protein fold models for rice dwarf virus. Nat Struct Biol 8: 868-73.
6. Chiu, W., M. L. Baker, W. Jiang, M. Dougherty, and M. F. Schmid. 2005. Electron cryomicroscopy of biological machines at subnanometer resolution. Structure 13: 363-72.
7. Topf, M., M. L. Baker, B. John, W. Chiu, and A. Sali. 2005. Structural characterization of components of protein assemblies by comparative modeling and electron cryo-microscopy. J Struct Biol 149: 191-203.
8. He, J., Y. Lu, and E. Pontelli. 2004. A Parallel Algorithm for Helix Mapping between 3-D and 1-D Protein Structure using the Length Constraints. Lecture Notes in Computer Science 3358: 746-756.
9. Jiang, W., M. L. Baker, S. J. Ludtke, and W. Chiu. 2001. Bridging the information gap: computational tools for intermediate resolution structure interpretation. J Mol Biol 308: 1033-44.
10. Kong, Y., and J. Ma. 2003. A structural-informatics approach for mining beta-sheets: locating sheets in intermediate-resolution density maps. J Mol Biol 332: 399-413.
11. Kong, Y., X. Zhang, T. S. Baker, and J. Ma. 2004. A Structural-informatics approach for tracing beta-sheets: building pseudo-C(alpha) traces for beta-strands in intermediate-resolution density maps. J Mol Biol 339: 117-30.
12. Del Palu, A., J. He, E. Pontelli, and Y. Lu. 2006. (accepted) Identification of Alpha-Helices from Low Resolution Protein Density Maps. CSB Computational Systems Bioinformatics.
13. Wu, Y., M. Chen, M. Lu, Q. Wang, and J. Ma. 2005. Determining protein topology from skeletons of secondary structures. J Mol Biol 350: 571-86.
14. Rohl, C. A., C. E. Strauss, K. M. Misura, and D. Baker. 2004. Protein structure prediction using Rosetta. Methods Enzymol 383: 66-93.
15. Kim, D. E., D. Chivian, and D. Baker. 2004. Protein structure prediction and analysis using the Robetta server. Nucleic Acids Res 32: W526-31.
16. Simons, K. T., I. Ruczinski, C. Kooperberg, B. A. Fox, C. Bystroff, and D. Baker. 1999. Improved recognition of native-like protein structures using a combination of sequence-dependent and sequence-independent features of proteins. Proteins 34: 82-95.
17. Simons, K. T., C. Kooperberg, E. Huang, and D. Baker. 1997. Assembly of protein tertiary structures from fragments with similar local sequences using simulated annealing and Bayesian scoring functions. J Mol Biol 268: 209-25.
18. Bowers, P. M., C. E. Strauss, and D. Baker. 2000. De novo protein structure determination using sparse NMR data. J Biomol NMR 18: 311-8.
19. Ludtke, S. J., P. R. Baldwin, and W. Chiu. 1999. EMAN: semiautomated software for high-resolution single-particle reconstructions. J Struct Biol 128: 82-97.

FITTING PROTEIN CHAINS TO CUBIC LATTICE IS NP-COMPLETE

JÁN MAŇUCH

School of Computing Science,
Simon Fraser University,
Burnaby, BC, V5A 1S6, Canada
E-mail: jmanuch@sfu.ca

DAYA RAM GAUR

Department of Math and Computer Science,
University of Lethbridge,
Lethbridge, AB, T1K 3M4, Canada
E-mail: gaur@cs.uleth.ca

It is known that folding a protein chain into the cubic lattice is an NP-complete problem. We consider a seemingly easier problem, given a 3D fold of a protein chain (coordinates of its C_α atoms), we want to find the closest lattice approximation of this fold. This problem has been studied under names such as "lattice approximation of a protein chain", "the protein chain fitting problem" and "building protein lattice models". We show that this problem is NP-complete for the cubic lattice with side 3.8Å and the coordinate root mean-square deviation.

1. Introduction

A protein is a linear sequence of amino acids which when placed into a solvent forms a 3D structure (fold). One of the main problems in proteomics is to computationally determine the structure of a protein given a sequence of amino acids. This problem appears extremely hard even when very simplified models are considered. For instance in Dill's HP-model[6], it is assumed that the "centers" of amino acids (C_α atoms) of the protein structure occupy vertices of a given lattice. A fold of a protein then can be represented as a path in the lattice. The second simplification, is the energy function of the fold. Instead of considering all forces affecting the folding process, only hydrophobic interactions between amino acids neighboring in the lattice are considered. It was shown in Refs. 1 and 4 that protein folding is NP-complete even in this simplified model.

Even though protein folding in lattice models is NP-complete, it is more computationally feasible than in the general non-lattice models as the lattice significantly limits the degree of freedom. In fact, lattice models are widely used in investigation of folding kinetics and thermodynamics[5,7] and for computer investigation of protein folding[15,20]. However, even if the optimal (native) fold in a certain lattice model is found it could be quite far from the real fold of the protein. Identifying lattice models which have a potential to produce folds close to real 3D structures is an important question in structural proteomics studied in handful of papers[3,11,10,19,18,16,17,12,14] to cite a few.

To measure the accuracy of representation of lattice models, the following procedure is commonly used: (1) select a test set of proteins with known 3D structure (for instance from PDB[2]); (2) find the closest lattice representation of each protein minimizing the overall distance of the lattice representation to the exact structure, as measured by the coordinate root mean squared deviation (c-RMS) or by the distance rms deviation (d-RMS); (3) calculate average of the c-RMS (d-RMS) values over all proteins in the test set. The crucial part of this procedure is the computation of the closest lattice representation (of the chain) of a given protein structure. This problem is also referred to as "the discretization of a protein backbone", "lattice approximation of 3D structure of a chain molecule", "constructing lattice models of protein chains", "modeling protein structures on a lattice", "discrete state model fitting to X-ray structures" or "fitting of a protein chain to a lattice". In this paper, we call the problem *protein chain lattice fitting problem* (the PCLF problem). Also note that an algorithm for the PCLF problem is an essential part of genetic protein folding algorithm, cf. Ref. 17.

The first algorithm for the PCLF problem proposed in Ref. 3 enumerates all possible conformations and picks the best one. Dynamic programming based algorithms were presented in several papers, cf. Refs. 19, 18 and 17. A greedy approach keeping about 500 "best" lattice folds was used in Ref. 16 and another greedy approach was used in Ref. 14. A completely different approach using the self-consistent mean field theory was presented in Ref. 12. All these algorithms either exhaustively enumerate all conformations, which can be applied only on very short proteins, or produce approximate solutions. It is questionable how reliable is the comparison of accuracy of various lattices based on an approximate algorithm. A chosen approximation algorithm might have a better approximation ratio for certain types of lattices which would consequently show higher accuracy for those lattice than other ones. Therefore, it would be highly desirable to develop a fast (polynomial) and exact algorithm solving the protein chain fitting problem.

In this paper, we show that the protein chain lattice fitting problem is NP-complete for the cubic lattice with side 3.8Å and c-RMS deviation. Although this result does not immediately imply that the problem is intractable for other lattices as well, it would be very unlikely if it is not the case.

2. Formalization of the Problem

Given is a protein as a sequence of 3D points and a regular lattice embedded into space (each lattice vertex is a 3D point). The goal is to map every point of the protein to a lattice point such that consecutive protein points are mapped to lattices points connected by an edge, the mapping is injective and the "distance" between the sequences of protein points and mapped points is minimized. The following properties of proteins whose structure is available in PDB[2] are well known:

- the distances between consecutive points of a protein vary very little from 3.8Å;
- the distances between non-consecutive points of a protein is at least 3.8Å.

We will assume that the lengths of the edges of the lattice are equal to 3.8Å. We can then easily scale the whole setting so that the distances between consecutive protein points and between two neighboring lattice points are one. Hence, we can formalize the protein chain fitting problem as follows:

Protein Chain Lattice Fitting (PCLF) Problem
Instance: Equilateral lattice L with side 1, a sequence of points $p = p_1, \ldots, p_n$ such that

(P1) $d(p_i, p_{i+1}) = 1$, for every $1 \leq i \leq n$, and
(P2) $d(p_i, p_j) \geq 1$ for every $2 \leq i + 1 < j \leq n$,

a distance measure α on the sequence of points, and a number K.
Question: Is there a path $l = l_1, \ldots, l_n$ in L such that $\alpha(p, l) \leq K$?

$\alpha(p, l)$ represents the quality of the lattice representation of a given protein structure. Two most common distance measures used to measure this quality are the *coordinate root mean square deviation* (c-RMS) and the *distance root mean square deviation* (d-RMS), defined as follows

$$\text{c-RMS}(p, l) = \sqrt{\frac{\sum_{i=1}^{n} d^2(p_i, l_i)}{n}}, \quad \text{d-RMS}(p, l) = \sqrt{\frac{\sum_{1 \leq i < j \leq n} [d(p_i, p_j) - d(l_i, l_j)]^2}{n(n-1)/2}}.$$

We show that the PCLF problem is NP-complete for the cubic lattice and the c-RMS measure. We would like to point out that the assumption that $l = l_1, \ldots, l_n$ is a *path* in the lattice is a crucial assumption. In fact it can be proved that if l is a *walk* in the lattice, the problem can be solved in polynomial time[9]. We use a reduction from a special variant of the satisfiability problem shown to be NP-complete in Ref. 13.

Var-linked Planar 3-SAT (VLP-3-SAT)
Instance: A formula Φ with a set C of clauses over a set X of variables in a conjunctive normal form such that:

(S1) Every clause contains at most three variables.
(S2) The set X of variables allows a linear ordering, $x_1, \ldots, x_n$ such that the graph $G_\Phi = (C \cup X, \{xc;\ x \in c \in C \text{ or } \neg x \in c \in C\} \cup \{x_i x_{i+1};\ i = 1, \ldots, n\})$ is planar (here $x_{n+1} = x_1$).

Question: Is Φ satisfiable?

Note that for a different variant of planar 3-SAT, in which clauses are linked in a circled chain instead of variables (clause-linked), it was shown in Ref. 8 that it can be assumed that each variable occurs in exactly three clauses, once negated and twice positive. A similar assumption for the var-linked version of the planar 3-SAT problem simplifies our proof. We provide the justification for this assumption in the next paragraph.

Consider an instance of the VLP-3-SAT problem. It is possible to draw G_Φ in a way that all the variables in X lie on one vertical line, and clauses lie either to the left or to the right of this line and connect directly to the variables without crossing the vertical line.

Henceforth, we distinguish between left and right clauses. Now, similar to the construction in Ref. 8, we replace every variable x which does not have exactly 3 occurrences such that at least one of them is positive and one negative, with variables $l_1, \ldots, l_k$ and $r_1, \ldots, r_m$, where x was connected to left clauses $L_1, \ldots, L_k$ and right clauses $R_1, \ldots, R_m$. Obviously, we can assume that $k + m \geq 2$. Figure 1 shows the connections before and after the replacement. Note that we have introduced new clauses $(l_i \vee \neg l_{i+1})$, $(l_k \vee \neg r_m)$, $(r_{i+1} \vee \neg r_i)$ and $(r_1 \vee \neg l_1)$, which guarantees that all new the variables have the same value in any satisfiable assignment. Therefore, the new 3-SAT formula is satisfiable if and only if the original one was, and every variable has at least one positive and at least one negative occurrence. Finally, replacing the variables with negative occurrences by their negations we obtain a var-linked planar 3-SAT formula satisfying the following condition:

(S3) Every variable occurs in exactly three clauses, once negated and twice positive.

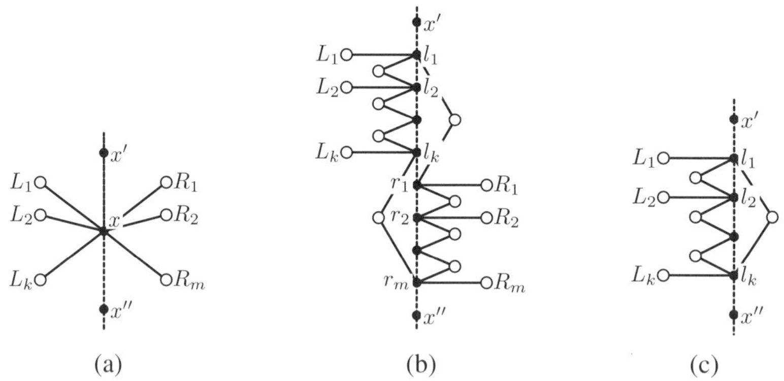

Figure 1. Replacement of variable x with sequence of variables: (a) the original configuration; (b) a new configuration in case $k, m > 0$; (c) a new configuration in case $m = 0$.

3. NP-completeness of the PCLF Problem for the Cubic Lattice and the c-RMS Measure

To prove the NP-completeness of the PCLF problem, we consider special instances of the problem. Next we give a simple lower bound on the c-RMS of protein sequence p and its lattice approximation l. For every p_i, let $L(p_i)$ denote the set of lattice vertices which are the closest to p. Let d_i be the distance of p_i from $L(p_i)$, i.e., $d_i = d(p_i, q)$, where $q \in L(p_i)$. Let $D = \sqrt{\frac{\sum_{i=1}^{n} d_i^2}{n}}$. Obviously, for any lattice path l, c-RMS$(p, l) \geq D$. In the special instances we set K to D, i.e., we get an affirmative answer to the instance if and only if there is a path l in the lattice such that $l_i \in L(p_i)$. We show that this happens if and only if the formula Φ for which we built the instance is satisfiable. Since the cubic lattice $L_\square$ is equilateral, we can assume that the vertices of $L_\square$ are all integral points, i.e., $V(L_\square) = \{[x, y, z];\ x, y, z \in \mathbb{Z}\}$ and $E(L_\square) = \{([x, y, z], [x', y', z']);\ |x - x'| + |y -$

$y'|+|z-z'|=1\}$. A point p for which $L(p)$ is not a singleton will be called a *flexible* point. Obviously, flexible points will play an important role in the construction. It is sufficient to prove the following lemma.

Lemma 3.1. *It is NP-complete to decide whether for a given sequence of points* $p = p_1, \ldots, p_n$ *there is a path* $l = l_1, \ldots, l_n$ *in* $L_\Box$ *such that for every* $i = 1, \ldots, n$, $l_i \in L(p_i)$.

Proof. We establish the NP-completeness by a reduction from the VLP-3-SAT problem. Let Φ be a formula and G_Φ a planar drawing of Φ such that every variable occurs twice positive and once negated (cf. the definition of VLP-3-SAT and the discussion after the definition). We construct p in two phases. In the first phase, we construct several subsequences of p lying either in or close to (within distance 3 from) the plane $z = 0$, each ending in $L_\Box$-points. These sub-sequences closely follow the planar drawing G_Φ. In the second phase, these subsequence are connected to one sequence p using only $L_\Box$-points not lying in the plane $z = 0$. Obviously, this can be done without any problem and in any solution they have to be mapped to the same points, therefore we omit the explicit description of the second phase.

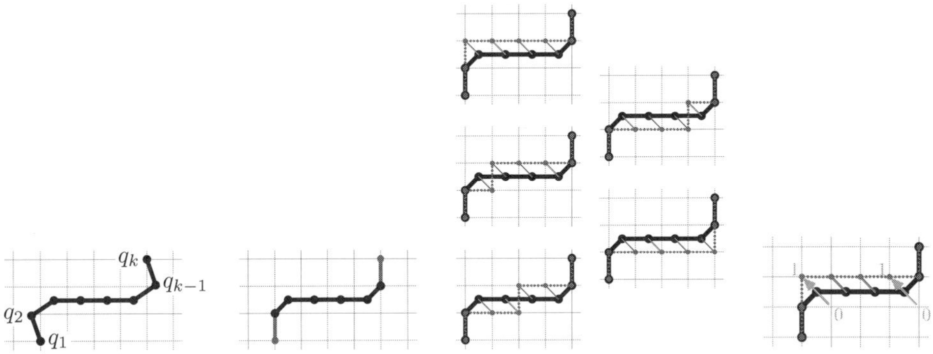

Figure 2. Illustration of a "wire": (a) a real subsequence (a distance between two consecutive points is 1); (b) a schematic drawing of subsequence (the gray edges will be usually omitted); (c) 5 possible lattice approximations of the subsequence (depicted with dotted lines); (d) pulling the wire on the left hand side forces a unique state of the wire.

The first basic building block of the construction is a "*wire*", cf. Figure 2(a). The end points of the wire, q_1 and q_k are lattice points, as required. The middle points $q_3, \ldots, q_{k-2}$ are flexible points, each having four choices for the closest lattice point, since they lie in the centers of square faces of plane $z = 0$, i.e., in the set $L_{1/2} = [1/2, 1/2, 0] + V(L_\Box)$. The second and the penultimate points, q_2 and q_{k-1}, lie neither in $L_\Box$ nor in $L_{1/2}$, as they connect lattice points to flexible points. We will call such points *connecting points*. Even though they do not lie directly in the lattice, their closest lattice points sets $L(q_2)$ and $L(q_{k-1})$ are singletons, i.e., in every solution, they are uniquely mapped to the lattice points. Therefore, in our schematic drawing, the end points of the wire are usually omitted and the connecting points are shifted to the lattice points to which they are uniquely mapped, cf. Figure 2(b).

Next we show that the connecting points satisfying conditions (P1) and (P2) do exist. Wlog assume that $q_1 = [1, 0]$ and the leftmost flexible point is $[5/2, 3/2]$ then $q_2 = [3/4 + \sqrt{15}/20, 5/4 - 3\sqrt{15}/20] = [0.9436, 0.669]$. It can be verified that q_2 obeys the distance constraints (P1) and (P2).

Figure 2(c) shows all the possible ways a "wire" can be mapped to the lattice points sequence forming a path in $L_\square$. An important property of a wire is that forcing the first (leftmost) flexible point (by some other gadget of the construction) to position 1, as depicted in Figure 2(d), yields a unique state of the wire, and most importantly, makes the last flexible point to map into position 1 as well (which can affect state of another gadget on this end of wire). One can imagine that the wire is sending a signal from left to the right (or symmetrically, from right to the left). More formally, consider two boolean variables s and t. Let $s = 1$ iff the first flexible point of the wire is in position 1 and let $t = 1$ iff the last flexible point is in position 1. Then in every solution, we have $s \implies t$.

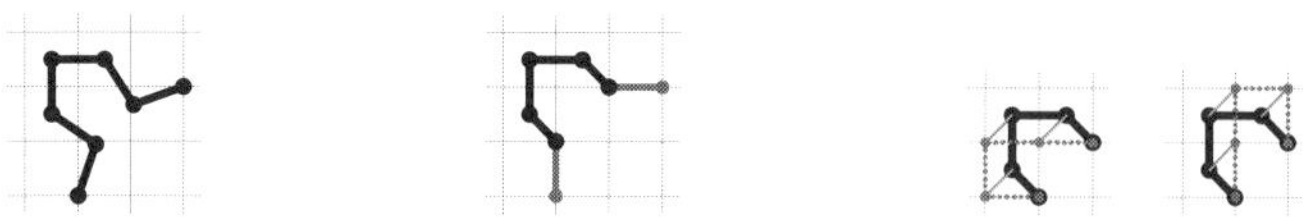

(a) (b) (c)

Figure 3. Illustration of a "flipper": (a) a real subsequence; (b) a schematic drawing (the gray edges will be usually omitted); (c) two possible states of the subsequence.

The second basic building block is a "*flipper*", cf. Figure 3(a). Note that the distance between two connecting points of the subsequence is $\frac{3}{\sqrt{2}} - \frac{\sqrt{6}}{\sqrt{5}} > 1$, i.e., condition (P2) is satisfied. In the schematic drawing we again omit the end points and shift the second and the second last point to positions where they are uniquely mapped, cf. Figure 3(b). Figure 3(c) shows all the possible states for a flipper.

To model the graph G_Φ we replace each vertex (clause) $c \in C$ by a "clause gadget" and every vertex (variable) $x \in X$ with a "variable gadget". We will use two different types of gadgets for clauses depending on the number of literals, and several different types of gadgets for variables depending on occurrences of variables in the formula Φ (positive or negative, in left or right clauses, as well as their relative order). The variable gadgets are placed vertically on top of each other and there are no connections between them as in G_Φ. For each edge between a clause and a variable, we have a wire connecting the corresponding clause and variable gadgets. It is not always possible to draw G_Φ in a way that all clause-variable edges are horizontal. Therefore, we need a wire which bends and correctly sends a signal from one end to the other. Such wires can be constructed using only two bends. Figure 4 shows how using two flippers we can achieve exactly this. Consider Figure 4(a), if the top part of the wire is forced to be in state 1 (pulled to the top) then the bottom part of is also pulled. If the top part of the wire is in state 0, then the top flipper is in $\sqsupset$ state, this forces the bottom flipper to be in $\sqcup$ state, this forces the bottom part of the wire to be in state 0.

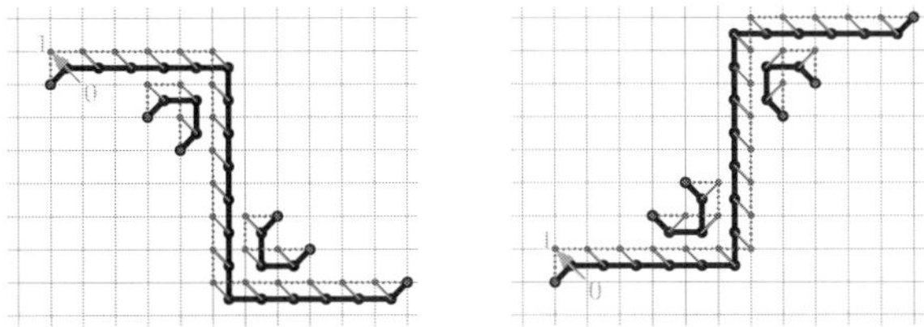

(a) (b)

Figure 4. Illustration of a wire which is shifted from one horizontal line to another: (a) down shift; (b) up shift. The dotted line shows the unique state when the first flexible point is forced to position 1.

Consider a left clause $c \in C$ (for right clauses, we would use a symmetrical design). First, assume that the clause contains two literals. The gadget for such a clause is depicted in Figure 5. The point q, common to the two wires which appears as a lattice point in the schematic drawings (a)–(c) has to be shifted off the lattice (by 1/4 in both coordinates) as shown in Figure 5(d), otherwise the connecting points p and r would be closer to each other than 1 which would violate condition (P2). Enumerating all the possibilities (using a computer program), we found out that there are 42 states for this gadget, 2 states with the upper wire pulled (the blue arrow is in position 1) and the lower wire not pulled (position 0), cf. Figure 5(a) for one of them, 2 states with the reversed situation of the wires, cf. Figure 5(b), and the remaining have both wires are pulled, cf. Figure 5(c) for instance. If s and t are boolean variables representing the position of the last flexible point of the wires, then in any solution, we have $s \vee t$ is satisfiable, and vice versa, for any values of s and t such that $s \vee t$ is satisfied, there is a solution with these values. Consider the case when the gadget in Figure 5 has both the top and the bottom wires in state 0 (not pulled). The state of the bottom wire forces the bottom flipper to be in $\sqcup$ state, and the flipper in the middle to be in $\sqcap$ state. The state of the top wire forces the top flipper to be in $\sqsubset$ state, but point q can be occupied by only one flipper, this leads to a contradiction. Hence, both the wires cannot be in 0 state.

Next, assume that the clause contains three literals. We want to design a gadget with 3 wires coming out of it such that it allows all and only those states in which at least one of the wires is pulled. It seems hard to design such a planar gadget, therefore we use a 3D-version of the flipper depicted in Figure 6(a). Note that we again need two connecting points q_2 and q_3 on left end of the sequence. Points q_2 and q_3 are exactly above each other at distance 1. If q_2 were placed on a lattice vertex, it would be too close to the point q_7 on the other end of the subsequence. Figure 6(b) shows all possible states of the flipper. Observe that in each state, the mapped sequence goes through exactly one of the points p, q, r. Two of them, p and q lie directly in the plane $z = 0$. For point r, we use one vertically placed flipper to transfer information that r is occupied to plane $z = 0$, cf. Figure 7(a–b).

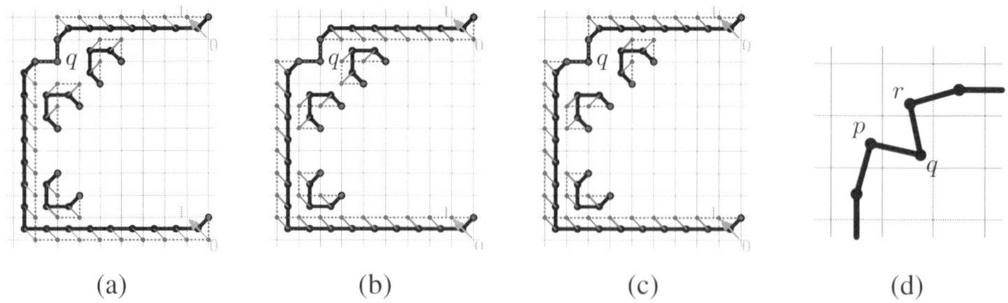

Figure 5. Illustration of a 2-clause gadget: (a) a state when the upper wire is pulled; (b) a state when the lower wire is pulled; (c) a state when both wires are pulled; (d) the real sequence around the point where the two wires meet (upper left corner). The two wires coming out of it (on the right) would end in corresponding variable gadgets (after possible shifts).

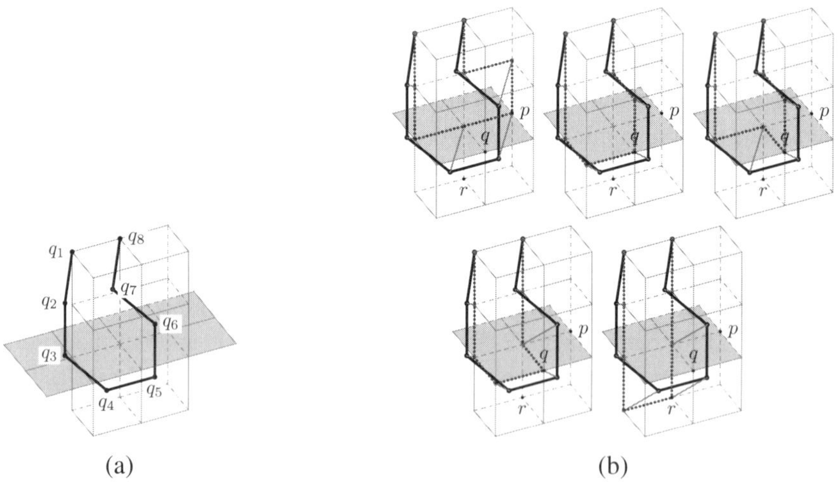

Figure 6. Illustration of a 3D-flipper: (a) a real subsequence; (b) 5 possible states of the 3D-flipper. The grey area indicates the plane $z = 0$ where the ordinary gadgets are placed.

Figure 7(a) shows the unique state of the vertical flipper in the case when the 3D-flipper is in the state using point r. Obviously, in this case points r_1 and r_2, both lying in the plane $z = 0$ are used. In any other state of the 3D-flipper, the vertical flipper can be in either of the two states. Figure 7(b) shows an example, when it is in the state not using points r_1 and r_2. Figure 7(c) shows five possible situations in the plane $z = 0$ caused by the non-planar part of the 3-clause gadget. Observe that for any of the points p, q and r_2, there is a solution such that this point is occupied while the other two are not. Given 3 wires, and using several flippers it is possible to ensure that at least one of the wires is pulled, cf. Figure 7(d). If s_1, s_2, s_3 are boolean variables representing positions of the rightmost flexible points of the three wires, then in any solution, we have $s_1 \vee s_2 \vee s_3$ is satisfiable, and vice versa, for any values of s_1, s_2, s_3 such that $s_1 \vee s_2 \vee s_3$ is satisfied, there is a solution with these values. Consider the case when all the three wires are in state 0, then the flippers are forced to occupy points p, q, r_2 but at least one of these points is needed by the non-planar part of the 3-clause gadget, a contradiction. All the other states are valid, and again have been checked using a computer program.

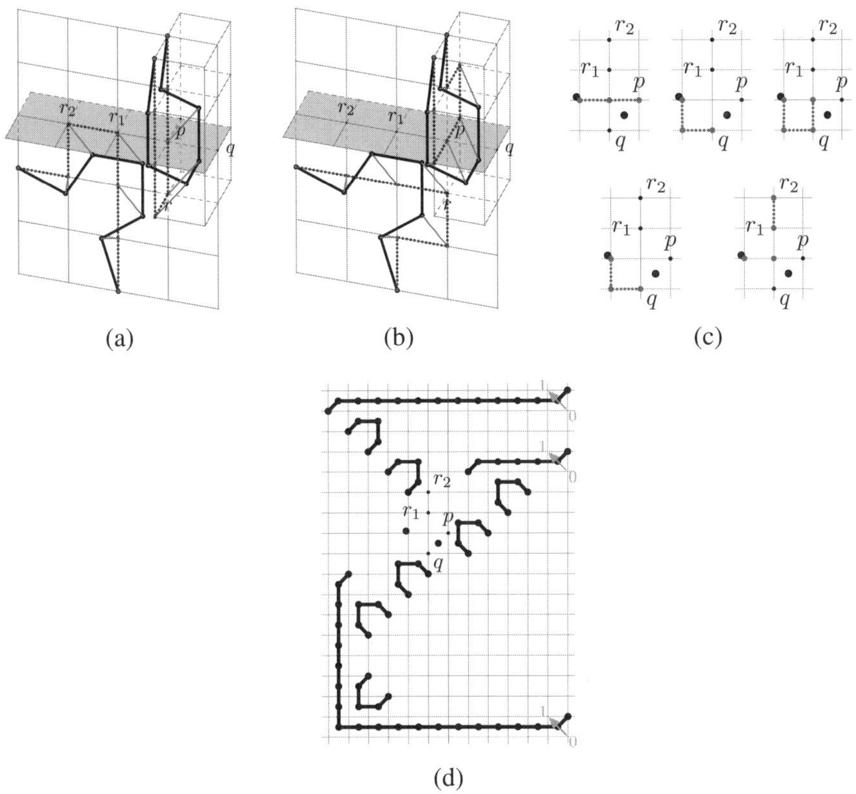

Figure 7. Illustration of a 3-clause gadget: (a–b) the part of the gadget outside of the plane $z = 0$; (c) five possible situations in the plane $z = 0$ caused by the non-planar part of the gadget (two larger black dots mark the places where the subsequence of the 3D-flipper crosses the plane $z = 0$); (d) the planar part of the gadget.

Finally, we construct the variable gadgets. Recall that each $x \in X$ occurs in exactly 3 clauses, twice positive and once negated. Figure 8 shows all the cases (up to symmetries) of how the neighborhood of a variable x occurring in 3 clauses looks like in a planar drawing of G_Φ.

All variable gadgets are variations of the gadget depicted in Figure 9(a). There are 2439 states for this gadget. Let s_1, s_2, s_3, s_4 be boolean variables whose values depend on the positions of flexible points at the end of wires as depicted in the figure. If s_1 is in state 1, the flippers force s_2 into state 0. Symmetrically, if s_4 is in state 1, then the flippers force s_3 to be in state 0. Also note that if s_1 is in state 1 then s_3 is in state 0, and symmetrically if s_2 is in state 1 then s_4 is in state 0. Therefore in any solution for the gadget we have that the formula $(s_1 \vee s_4) \oplus (s_2 \vee s_3)$ is satisfied, and vice versa, for any satisfiable assignment

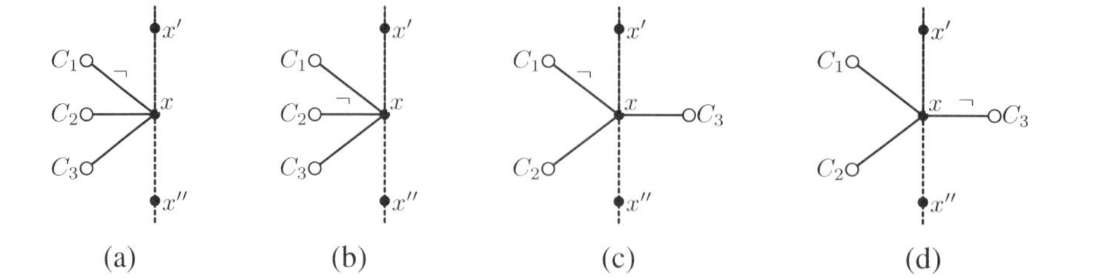

Figure 8. Neighborhood of a variable $x \in X$ in a planar drawing of G_Φ. For negated occurrence we put the symbol ¬ on the edge connecting the clause and x.

of the formula there is a valid state for the gadget. Once again, all the valid states for the variable gadget have been verified using a computer program.

Now, consider the occurrence of some variable x. For every positive occurrence of x in a clause $c \in C$, connect the wire coming out of the gadget of c to the end of wire marked with s_1 or s_4 (depending on whether it is a left or right clause), and for every negative occurrence, to the end marked with s_2 or s_3, cf. Figure 9(b). Obviously, this strategy can be directly applied in the case (c) depicted in Figure 8. Note that one wire coming out of the variable gadget stays unused, and remains unconnected to any clause.

Given a solution to the PCLF problem, we determine value for variable x in a satisfiable assignment as follows: if $s_1 = 1$ or $s_4 = 1$ then $x = 1$; if $s_2 = 1$ or $s_3 = 1$ then $x = 0$; otherwise (all s_1, s_2, s_3, s_4 have 0 values), the value for x can be chosen arbitrarily. Note that this assignment to variables X based on any solution to the PCLF problem, guarantees that every clause is satisfied. Indeed, in every clause gadget atleast one of the wires is in state 1 (pulled) this correponds to a literal in the corresponding clause to be true (by construction).

On the other hand, for every satisfiable assignment to variables in X of the formula Φ, set each clause gadget to the state in which they are pulling each wire which corresponds to a literal satisfied by the assignment, and set each variable gadget to the state in which $s_1 = s_4 = 1$ if the corresponding variable x has value 1, or $s_2 = s_3 = 1$ otherwise.

Finally, for the remaining cases of the neighborhood of x, cf. Figure 8(a)(b)(d), we need to bend one wire from the right hand side of the configuration to the left hand side. For the case (a), the complete variable gadget is depicted in Figure 10. For other two cases, the construction is analogous.

It follows by the construction that the formula Φ is satisfiable if and only if there exists a solution to the constructed PCLF problem. It is also clear that the construction can be done in polynomial time and space. Therefore, the PCLF problem is NP-complete. □

4. Conclusions

We have proved that the protein chain lattice fitting problem is NP-complete for the cubic lattice with side 3.8Å and the coordinate root mean square deviation (c-RMS) as the distance measure. From the theoretical point of view it would be very interesting to further investigate the complexity for the 2D square lattice with side 3.8Å. The proof of NP-

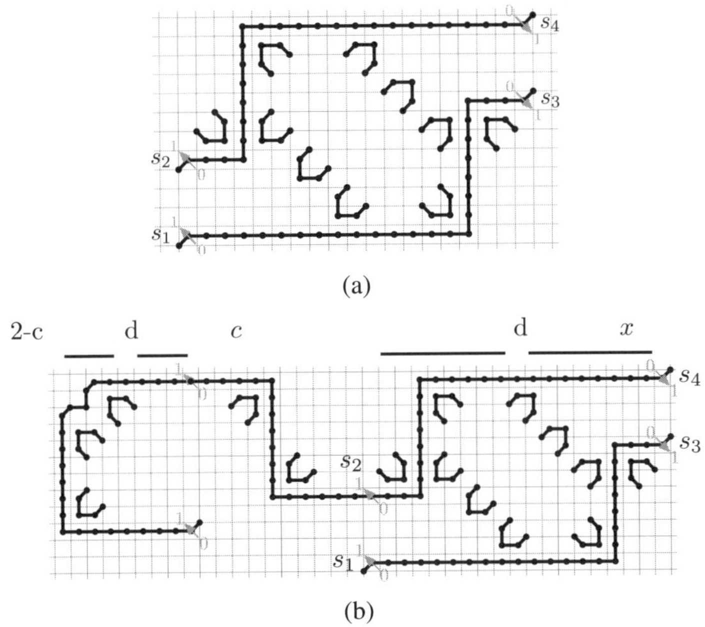

Figure 9. (a) A common substructure to all variable gadgets. (b) Example of a connection between a left 2-clause gadget for $c = \neg x \vee y$ and the variable gadget for x.

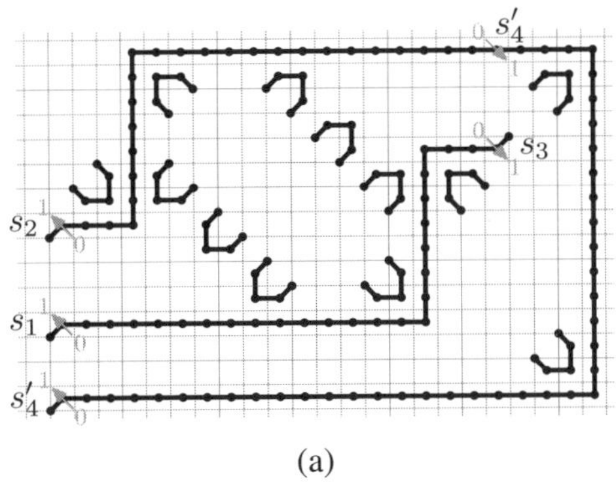

Figure 10. A variant of the variable gadget used in the case (a) depicted in Figure 8. Note that if the wire is pulled at position s'_4, it is also pulled at position s_4.

completeness presented in this paper mostly uses one plane of the 3D cubic lattice ($z = 0$) and is based on the planar 3-SAT problem. However, it cannot be applied directly to the square lattice for two reasons: (1) we were unable to design the 3-clause gadget without using the third dimension; (2) connecting the gadgets into one protein string without using the third dimension seems to be a nontrivial task.

From the practical point view, the questions whether our result applies to different types of lattices or cubic lattices with sides different from 3.8Å or d-RMS used as the distance measure between two 3D structures are more important. It can be shown that a greedy algorithm can perform arbitrary bad for constructed sequences of points (although, the performance on proteins from PDB is better). It would be interesting to study whether the existing DP-based algorithms have bounded performance ratio, or to design a new algorithm with constant performance ratio.

References

1. B. Berger and T. Leighton. Protein folding in the hydrophobic-hydrophilic (HP) model is NP-complete. *J. Comp. Biol.*, 5:27–40, 1998.
2. H. Berman, J. Westbrook, Z. Feng, G. Gilliland, T. Bhat, H. Weissig, I. Shindyalov, and P. Bourne. The protein data bank. *Nucleic Acids Res*, 28(1):235–242, 2000.
3. D. G. Covell and R. L. Jernigan. Conformations of folded proteins in restricted spaces. *Biochemistry*, 29:3287–3294, 1990.
4. P. Crescenzi, D. Goldman, C. Papadimitriou, A. Piccolboni, and M. Yannakakis. On the complexity of protein folding. In *Proc. of STOC'98*, pages 597–603, 1998.
5. H. DA and L. M. Exploring conformational space with a simple lattice model for protein structure. *J. Mol. Biol.*, 243(4):668–682, 1994.
6. K. Dill. Theory for the folding and stability of globular proteins. *Biochemistry*, 24(6):1501–1509, 1985.
7. K. A. Dill, S. Bromberg, K. Yue, K. M. Fiebig, D. P. Yee, P. D. Thomas, and H. S. Chan. Principles of protein folding: A perspective from simple exact models. *Protein Science*, 4:561–602, 1995.
8. M. Fellows, J. Kratochvíl, M. Middendorf, and F. Pfeiffer. The complexity of induced minors and related problems. *Algorithmica*, 13(3):266–282, 1995.
9. D. Gaur, J. Maňuch, and D. Thomas. Polynomial algorithm for aligning proteins to cubic lattice as a walk (manuscript).
10. A. Godzik, A. Kolinski, and J. Skolnick. Lattice representations of globular proteins: How good are they? *J. Comp. Chem.*, 14:1194–1202, 1993.
11. D. A. Hinds and M. Levitt. A lattice model for protein structure prediction at low resolution. *Proc. Natl. Acad. Sci. USA*, 89:2536–2540, 1992.
12. P. Koehl and M. Delarue. Building protein lattice models using self-consistent mean field theory. *J. Chem. Physics*, 108(22):9540–9549, 1998.
13. D. Lichtenstein. Planar formulae and their uses. *SIAM Journal of Computing*, 11(2):329–343, 1982.
14. C. Mead, J. Maňuch, X. Huang, B. Bhattacharyya, L. Stacho, and A. Gupta. Investigating lattice structure for inverse protein folding (abstract). *FEBS Journal*, 272 (s1):4739_1_380, 2005.
15. K. P and L. M. A brighter future for protein structure prediction. *Nat. Struct. Biol.*, 6(2):108–111, 1999.
16. B. H. Park and M. Levitt. The complexity and accuracy of discrete state models of protein structure. *J. Mol. Biol.*, 249:493–507, 1995.
17. A. A. Rabow and H. A. Sheraga. Improved genetic algorithm for the protein folding problem by use of a Cartesian combination operator. *Protein Science*, 5:1800–1815, 1996.
18. B. A. Reva, D. S. Rykunov, A. J. Olson, and A. V. Finkelstein. Constructing lattice models of protein chains with side groups. *Journal of Comp. Biology*, 2(4):527–535, 1995.
19. D. S. Rykunov, B. A. Reva, and A. V. Finkelstein. Accurate general method for lattice approximation of three-dimensional structure of a chain molecule. *Proteins: Structure, Function and Genetics*, 22:100–109, 1995.
20. Y. Xia, E. Huang, M. Levitt, and R. Samudrala. Ab initio construction of protein tertiary sctructures using a hierarchical approach. *J. Mol. Biol.*, 300:171–185, 2000.

INFERRING A CHEMICAL STRUCTURE FROM A FEATURE VECTOR BASED ON FREQUENCY OF LABELED PATHS AND SMALL FRAGMENTS

TATSUYA AKUTSU *

Bioinformatics Center, Institute for Chemical Research, Kyoto University
Gokasho, Uji, Kyoto 611-0011, Japan
E-mail: takutsu@kuicr.kyoto-u.ac.jp

DAIJI FUKAGAWA

National Institute of Informatics
Chiyoda-ku, Tokyo 101-8430, Japan
E-mail: daiji@nii.ac.jp

This paper proposes algorithms for inferring a chemical structure from a feature vector based on frequency of labeled paths and small fragments, where this inference problem has a potential application to drug design. In this paper, chemical structures are modeled as trees or tree-like structures. It is shown that the inference problems for these kinds of structures can be solved in polynomial time using dynamic programming-based algorithms. Since these algorithms are not practical, a branch-and-bound type algorithm is also proposed. The result of computational experiment suggests that the algorithm can solve the inference problem in a few or few-tens of seconds for moderate size chemical compounds.

1. Introduction

Drug design is one of the important targets of bioinformatics. For designing new drugs, classification of chemical compounds is important and thus a lot of studies have been done. Recently, *kernel methods* have been applied to classification of chemical compounds[1,2,3,4]. In most of these approaches, chemical compounds are mapped to *feature vectors* (i.e., vectors of reals) and then *support vector machines* (SVMs)[5] are employed to learn classification rules. Though several methods have been proposed, feature vectors based on *frequency of small fragments*[1,2] or *frequency of labeled paths*[3,4] are widely used, where other chemical properties such as molecular weights, partial charges and logP are sometimes combined with these, and weights/probabilities are sometimes put on paths/fragments.

On the other hand, a new approach was recently proposed for designing and/or optimizing objects using kernel methods[6,7]. In this approach, a desired object is computed as a point in the feature space using suitable objective function and optimization technique and then the point is mapped back to the input space, where this mapped back object is

*Work partially supported by Grants-in-Aid "Systems Genomics" and #16300092 from MEXT, Japan, and by the Kayamori Foundation of Information Science Advancement.

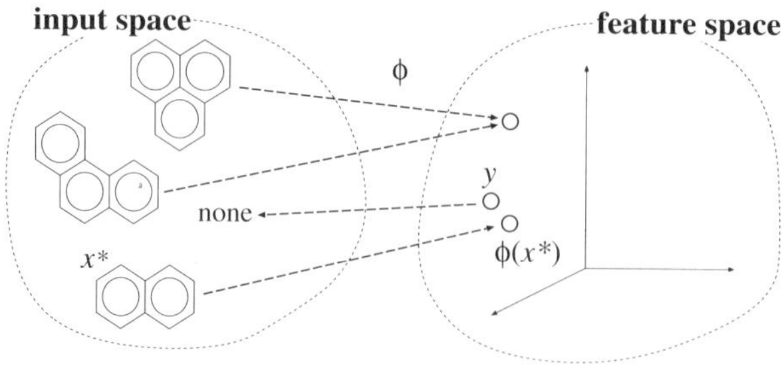

Figure 1. Inference of a chemical compound from a feature vector. Multiple compounds may be mapped to the same point in a feature space.

called a *pre-image*. Let ϕ be a mapping from an input space to a feature space. Then, the problem is, given a point y in the feature space, to find a pre-image x in the input space such that $y = \phi(x)$. It should be noted that ϕ is not necessarily injective or surjective. If ϕ is not surjective, we need to compute the approximate pre-image x^* defined by $x^* = \arg\min_x \ dist(y, \phi(x))$ (see Fig. 1).

The pre-image problem has potential application to drug design[7] by using a suitable objective function reflecting desired properties. For example, suppose that we have two chemical compounds x and y having different functions and we want to design a new chemical compound having both functions. In such a case, we may develop a new compound by computing a pre-image of the middle point of feature vectors corresponding to x and y (i.e., a pre-image of $(\phi(x) + \phi(y))/2$). Though this example is too simple, several ways can be considered for applications of the pre-image problem to drug design and other bioinformatics problems.

There are several related works. Bakir, Weston and Scölkopf proposed a method to find pre-images in a general setting by using Kernel Principal Component Analysis and regression[6]. Bakir, Zien and Tsuda developed a stochastic search algorithm to find pre-images for graphs[7]. However, the pre-image problems are not studied from a computational viewpoint. Graphical degree sequence problems[8], graph inference from walks[9] and the graph reconstruction problem[10] are related to the pre-image problem for graphs. However, results on these problems are not directly applicable to the pre-image problem.

In our previous works[11,12], we studied a theoretical aspect of the pre-image problem on graphs. In Ref. 11, we studied the problem of inferring a graph from the numbers of occurrences of vertex-labeled paths. We showed that this problem can be solved in polynomial time of the size of an output graph if graphs are trees of bounded degree and the lengths of given paths are bounded by a constant, whereas this problem is NP-hard even for planar graphs of bounded degree. In Ref. 12, these results were further improved. We showed that the inference problem can be solved in polynomial time if graphs are outerplanar of bounded degree and bounded face size and the lengths of given paths are bounded by a constant, whereas this problem is NP-hard even for trees of bounded degree if the lengths of paths are not bounded.

In this paper, we extend our previous algorithms so that constraints on valences of atoms are taken into account. Moreover, we modify and extend these so that feature vectors based on frequencies of small fragments can be treated. These modifications are important because major feature vectors are based on either frequencies of small fragment structures[1,2] or frequencies of labeled paths[3,4]. The modified algorithms have another application: *elucidation of chemical structures* from mass/NMR spectra data. This elucidation problem has a long history and many methods have been developed[13,14]. However, to our knowledge, no polynomial time algorithm was known for the problem. A more important result of this paper is that a heuristic algorithm is developed for inference of tree-like chemical structures. It works within a few or few-tens of seconds for inference of moderate size chemical compounds with tree or tree-like structures.

2. Problem Definitions

First, we review the definition of the problem on inference of graph from path frequency[11]. Let $G(V, E)$ be an undirected vertex-labeled connected graph and Σ be a set of vertex labels, where all results can be modified for including edge labels. Since we are considering chemical structures, we reasonably assume in this paper that the maximum degree of vertices and the size of Σ are bounded by constants. Let $\Sigma^{\leq k}$ be the set of label sequences (i.e., the set of strings) over Σ whose lengths are between 1 and k. Let $l(v)$ be the label of vertex v. For a path $P = (v_0, \ldots, v_h)$ of G, $l(P)$ denotes the label sequence of P. For graph G and label sequence t, $occ(t, G)$ denotes the number of paths P in G such that $l(P) = t$. Then, the feature vector $\boldsymbol{f}_K(G)$ of level K for $G(V, E)$ is defined by $\boldsymbol{f}_K(G) = (occ(t, G))_{t \in \Sigma^{\leq K+1}}$. See Fig. 2 for an example. It should be noted that the size (i.e., number of vertices) n of the original graph can be obtained from $\boldsymbol{f}_K(G)$. In this paper, we assume for simplicity that *tottering paths* (paths for which there exists some i such that $v_i = v_{i+2}$) are not counted in feature vectors.

G(V,E)

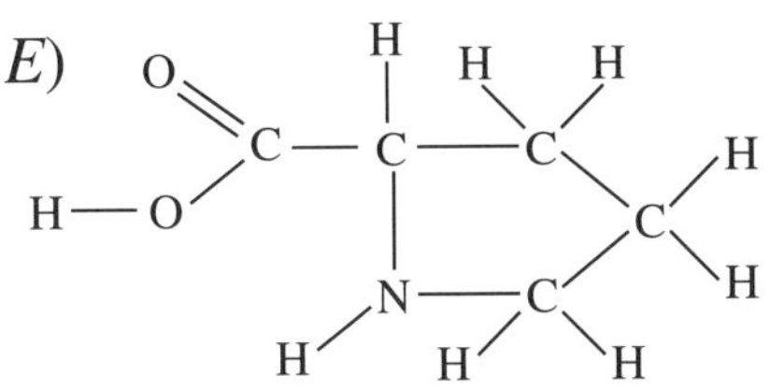

$f_1(G)$

C	N	O	H	CC	CN	CO	CH	NC	NN	• • •	OH	• • •
5	1	2	9	8	2	2	7	2	0	• • •	1	• • •

$f_F(G)$

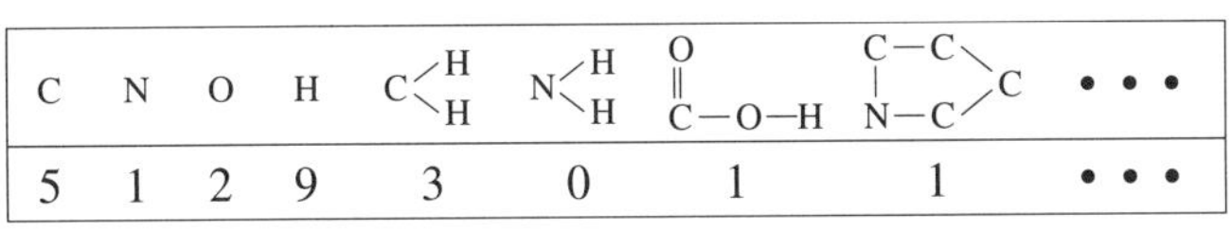

Figure 2. Examples of feature vectors $\boldsymbol{f}_K(G)$ for GIPF/CIPF ($K = 1$) and $\boldsymbol{f}_{\mathcal{F}}(G)$ for CIFF.

Graph Inference from Path Frequency (GIPF)[11] Given a feature vector $\boldsymbol{v}$ of level K, output a graph $G(V, E)$ satisfying $\boldsymbol{f}_K(G) = \boldsymbol{v}$. If there does not exist such $G(V, E)$, output "no solution".

For the case of "no solution", we can consider the problem (**GIPF-M**) of finding $G(V, E)$ which minimizes the L_1 distance between $\boldsymbol{v}$ and $\boldsymbol{f}_K(G)$ (see also Fig. 1)[11].

We sometimes omit K from $\boldsymbol{f}_K(G)$ if K is obvious or is not relevant. We may also use $\boldsymbol{f}$ to denote a feature vector if G and K are not relevant. For a vector $\boldsymbol{v}$, $(\boldsymbol{v})_i$ denotes the i-th element of $\boldsymbol{v}$ (i.e., the value of i-th coordinate of $\boldsymbol{v}$). For vectors $\boldsymbol{v}$ and $\boldsymbol{u}$, $\boldsymbol{v} \preceq \boldsymbol{u}$ means that $(\boldsymbol{v})_i \leq (\boldsymbol{u})_i$ for all i.

In order to treat chemical compounds, constraints on *valences* of atoms must be taken into account. For example, a carbon atom can be connected to at most four other atoms. If double bounds are used, it can be connected to at most two other atoms. Let Σ be the set of atom types. For each $a \in \Sigma$, the maximum valence $val(a)$ is associated. We also assume that each edge e has multiplicity $m(e)$ where $m(e)$ is usually 1 (single bond), 2 (double bond) and 3 (triple bond). When treating aromatic structures, aromatic bond may be modeled as an edge with multiplicity 1.5. In this paper, we use "degree" to mean the number of edges connected to a vertex and "valence" to mean the sum of multiplicities of edges connected to a vertex, respectively. Then, we define chemical compound inference problem from path frequency as follows:

Chemical compound Inference from Path Frequency (CIPF) Given a feature vector $\boldsymbol{v}$ of level K, output a graph $G(V, E)$ satisfying $\boldsymbol{f}_K(G) = \boldsymbol{v}$ and $\sum_{w:\{v,w\}\in E} m(\{v, w\}) \leq val(l(v))$ for all $v \in V$. If there does not exist such $G(V, E)$, output "no solution".

For the case of "no solution", **CIPF-M** is defined in the same way as in GIPF-M. Next, we define pre-image problems for feature vectors based on frequencies of fragments. Let $\mathcal{F} = \{F_1, \ldots, F_M\}$ be a set of graphs (chemical substructures) satisfying the valence conditions. Since information on the number of occurrences of each atom type is usually included in feature vectors, we assume that a graph consisting of each single atom is contained in $\mathcal{F}$. We also assume that the size of each F_i is bounded by a constant K because small fragments are usually employed. Let $occ(F_i, G)$ denote the number of subgraphs of G that are isomorphic to F_i, where we assume that subgraphs consisting of the same vertices are counted only once for each F_i. Then, a feature vector $\boldsymbol{f}_{\mathcal{F}}(G)$ for G is defined by $\boldsymbol{f}_{\mathcal{F}}(G) = (occ(F_i, G))_{F_i \in \mathcal{F}}$. The pre-iamge problem from a feature vector based on fragments is defined as below (see also Fig. 2).

Chemical compound Inference from Fragment Frequency (CIFF) Given a feature vector $\boldsymbol{v}$ based on a set of fragments $\mathcal{F}$, output a graph $G(V, E)$ satisfying $\boldsymbol{f}_{\mathcal{F}}(G) = \boldsymbol{v}$ and $\sum_{w:\{v,w\}\in E} m(\{v, w\}) \leq val(l(v))$ for all $v \in V$. If there does not exist such $G(V, E)$, output "no solution".

CIFF-M is defined in the same way as in GIPF-M and CIPF-M. In the case of elucidation of chemical structures from mass/NMR spectra[13], upper and lower bounds of the number of occurrences of each fragment are specified. Let $\boldsymbol{ub}$ and $\boldsymbol{lb}$ be vectors corresponding to upper and lower bounds, respectively. Then, $CIULF$ is defined as:

Chemical compound Inference from Upper and Lower bounds of frequencies of Fragments (CIULF) Given feature vectors $\boldsymbol{ub}$ and $\boldsymbol{lb}$ based on a set of fragments $\mathcal{F}$, output a graph $G(V, E)$ satisfying $\boldsymbol{lb} \preceq \boldsymbol{f}_{\mathcal{F}}(G) \preceq \boldsymbol{ub}$ and $\sum_{w:\{v,w\}\in E} m(\{v,w\}) \leq val(l(v))$ for all $v \in V$. If there does not exist such $G(V, E)$, output "no solution".

It is worthy to note that CIFF is clearly a subproblem of CIULF. It should also be noted that CIPF is a subproblem of CIFF because each labeled path can be treated as a fragment.

3. Dynamic Programming Algorithms

In this section, we extend our previous algorithms[11,12] for CIPF, CIFF and CIULF.

3.1. *A Basic Algorithm for CIPF*

As in Ref. 11 we begin with a very simple case: we consider inference of chemical compounds with tree structures from a feature vector of level 1 (i.e., $K = 1$). For simplicity, we assume that only two kinds of atoms N and H, and single bonds (i.e., edges with multiplicity 1) can appear in chemical compounds. In this case, a feature vector for tree T has the following form: $\boldsymbol{f}_1(T) = (n_N, n_H, n_{NN}, n_{NH}, n_{HN}, n_{HH})$, where n_x denotes the number of atoms of type x and n_{xy} denotes the number of occurrences of a labeled path of (x, y).

We construct the dynamic programming table $D(\ldots)$ defined by

$$D(n_{N1}, n_{N2}, n_{N3}, n_H, n_{NN}, n_{NH}, n_{HN}) = \begin{cases} 1, & \text{if there exists a chemical compound (tree) } T \text{ such that} \\ & \quad \boldsymbol{f}_1(T) = (n_{N1} + n_{N2} + n_{N3}, n_H, n_{NN}, n_{NH}, n_{HN}, 0), \\ & \quad \text{the number of nitrogen atoms with degree 1 is } n_{N1}, \\ & \quad \text{the number of nitrogen atoms with degree 2 is } n_{N2}, \\ & \quad \text{and the number of nitrogen atoms with degree 3 is } n_{N3}, \\ 0, & \text{otherwise.} \end{cases}$$

It should be noted that we ignore chemical compound of H_2 here and thus n_{HH} should be always 0. This table can be constructed by a dynamic programming procedure based the following recursion where the initialization part is straight-forward.

$$\begin{aligned} D(n_{N1}, n_{N2}, n_{N3}, n_H, n_{NN}, n_{NH}, n_{HN}) = 1 \ \textbf{iff.} \\ & D(n_{N1}, n_{N2} - 1, n_{N3}, n_H, n_{NN} - 2, n_{NH}, n_{HN}) = 1 \quad \textbf{or} \\ & D(n_{N1} - 1, n_{N2} + 1, n_{N3} - 1, n_H, n_{NN} - 2, n_{NH}, n_{HN}) = 1 \quad \textbf{or} \\ & D(n_{N1} + 1, n_{N2} - 1, n_{N3}, n_H - 1, n_{NN}, n_{NH} - 1, n_{HN} - 1) = 1 \quad \textbf{or} \\ & D(n_{N1}, n_{N2} + 1, n_{N3} - 1, n_H - 1, n_{NN}, n_{NH} - 1, n_{HN} - 1) = 1. \end{aligned}$$

The correctness of the algorithm follows from the fact that any tree can be constructed incrementally by adding a vertex (leaf) one by one. Since the value of each element of the feature vector is $O(n)$, the table size is $O(n^7)$ and thus the computation time is $O(n^7)$. Since it is straight-forward to extend this algorithm for a fixed number of atom types and a fixed number of bond types, we have:

Theorem 3.1. *CIPF for trees is solved in polynomial time in n for $K = 1$.*

As in Ref. 11, we can modify the algorithm for CIPF-M (since we only need to examine polynomial number of items in the DP table).

Corollary 3.1. *CIPF-M for trees is solved in polynomial time in n for $K = 1$.*

Recently, Nagamochi[15] developed a much more efficient algorithm for GIPF/CIPF for general graphs and $K = 1$. But, his algorithm can not be extended for cases of $K > 1$, which are much more important. Our algorithm can be extended for cases of $K > 1$ as to be shown below, and the idea in our algorithm is also used in a practical algorithm in Sec. 4.

3.2. *Algorithm for CIPF*

In this subsection, we show that CIPF can be solved in polynomial time for fixed K. For that purpose, we modify our previous algorithm[11] for GIPF as below, where details are omitted here. As in the original algorithm, we maintain the current degrees and valences of vertices of subtrees. When adding a new leaf u to an existing vertex w in a subtree, we check the constraint on valences and update information about degrees and valences. Clearly, this part can be done in constant time per addition of a leaf and thus does not affect the order of the time complexity.

Proposition 3.1. *CIPF (and CIPF-M) for trees can be solved in polynomial time in n if K and Σ are fixed.*

3.3. *Algorithms for CIFF and CIULF*

We develop algorithms for CIFF and CIULF by modifying the algorithm for GIPF. Since CIULF is more general than CIFF, we only consider CIULF here. We modify the table $D(\boldsymbol{v}, \boldsymbol{e}, d)$ in Ref. 11 as follows. Let $\boldsymbol{h} = (h_1, h_2, \ldots, h_M)$ be a vector of non-negative integers, where h_i corresponds to the number of occurrences of fragment F_i. Then, we define the table $D'(\boldsymbol{h}, \boldsymbol{e}, d)$ by

$$D'(\boldsymbol{h}, \boldsymbol{e}, d) = 1 \text{ iff. there exists a tree } T \text{ such that } \boldsymbol{f}_{\mathcal{F}}(T) = \boldsymbol{h}, \boldsymbol{g}_K(T) = \boldsymbol{e}, \text{ and } d(T) = d.$$

Based on this table, we can develop a dynamic programming algorithm, where details are omitted here.

Theorem 3.2. *CICF and CIULF (and CICF-M) for trees of bounded degree can be solved in polynomial time in n if K, M and Σ are fixed.*

As a negative result, it was shown[12] that GIPF (a subproblem of CIFF) is NP-hard for trees of bounded degree if K is not fixed. This suggests that the time complexity increases non-polynomially in K. Here, we show another hardness result for CIULF (proof omitted), which suggests that the time complexity increases non-polynomially in M.

Theorem 3.3. *CIULF can not be solved in polynomial time unless P=NP even if the maximum degree is bounded by 2, where the size and number of fragments are not bounded by a constant.*

3.4. *Extensions to Outerplanar Graphs*

Though many chemical compounds have tree structures, many other chemical compounds have rings such as benzene rings. Therefore, it is desirable to develop algorithms for more general structures than trees. In the case of GIPF, the algorithm for trees[11] was extended for outerplanar graphs[12]. The same technique can also be applied to CICF and CIFF.

Theorem 3.4. *CIPF, CIPF-M, CICF, CICF-M and CIULF for chemical compounds with outerplanar structures can be solved in polynomial time in n if K, M and Σ are fixed and the number of edges of each face is bounded by a constant.*

4. A Branch-and-Bound Type Algorithm

Though the algorithms in the previous section work in polynomial time, these are not practical. Thus, we develop a branch-and-bound type algorithm (called BB-CIPF) for CIPF, where this algorithm can be modified for inferring more general classes of chemical compounds and/or for feature vectors based on frequency of small fragments.

Before presenting BB-CIPF, we need several definitions. Let $\boldsymbol{f}^{target}$ be the given feature vector for which a pre-image should be computed. Since information on paths of length 0 is included in $\boldsymbol{f}^{target}$, we know the number of occurrences of atom types in the pre-image of $\boldsymbol{f}^{target}$. Let $atomset(\boldsymbol{f})$ be the multi-set of atom types in the pre-image of a feature vector $\boldsymbol{f}$. Let $ATOMBONDPAIRS$ be a set of possible atom-bond pairs. For example, if we only consider C, N, O, H and do not consider aromatic bonds, the set is defined as $\{(\mathrm{C},1),\ (\mathrm{C},2),\ (\mathrm{C},3),\ (\mathrm{N},1),\ (\mathrm{N},2),\ (\mathrm{N},3),\ (\mathrm{O},1),\ (\mathrm{O},2),\ (\mathrm{H},1)\}$. It should be noted that $(\mathrm{C},4)$ is not included since it is not necessary.

The basic idea of BB-CIPF is similar to that of the algorithm in Sec. 3.1: beginning from a small tree, a leaf is added one by one. Though trees are not explicitly constructed in Sec. 3.1, BB-CIPF maintains trees.

When adding a leaf u, BB-CIPF basically examines all combinations of an atom-bond pair (a, b) and a vertex w in the current tree. However, we do not need to examine the following cases, where T^{cur} be the current tree and T^{next} be the next candidate tree obtained by adding a leaf to T^{cur}:

(i) Addition of a leaf with atom label a violates the condition on the numbers of atoms,

(ii) Connection of a leaf to $w \in T^{cur}$ by bond type b violates the condition on the valence of w,

(iii) Connection of a leaf to $w \in T^{cur}$ violates the condition on feature vectors (i.e., $\boldsymbol{f}(T^{next}) \preceq \boldsymbol{f}(T^{target})$ must hold since T^{next} must be a subgraph of T^{target}).

Therefore, we do not examine T^{next} further in these cases. These conditions significantly contribute to reducing the search space and are implemented in BB-CIPF.

BB-CIPF employs a kind of distance defined by:

$$dist(\boldsymbol{f}, \boldsymbol{f}^{target}) = \begin{cases} \infty, & \text{if } (\boldsymbol{f}^{target})_i < (\boldsymbol{f})_i \text{ for some } i, \\ \sum_i c^{k(i)}((\boldsymbol{f}^{target})_i - (\boldsymbol{f})_i), & \text{otherwise.} \end{cases}$$

where summation is taken over all elements (i.e., dimensions) of feature vectors, c is a constant (currently, $c = 10$) and $k(i)$ denotes the length of a path corresponding to the i-th element of a feature vector. Though we use the word "distance" for the sake of convenience, this measure is not symmetric and thus does not satisfy the conditions on usual distances. The meaning of weighting factor $c^{k(i)}$ is that priorities are put on longer paths when calculating distances. The pseudocode of BB-CIPF is given below.

Procedure $BB - CIPF(n, \boldsymbol{f}^{target})$
  Let T^{cur} be an initial tree constructed from a longest path appearing in $\boldsymbol{f}^{target}$;
  Compute feature vector $\boldsymbol{f}^{cur}$ from T^{cur};
  if $DFS - CIPF(T^{cur}, \boldsymbol{f}^{cur}, n, \boldsymbol{f}^{target})$=false **then** output "no solution";

Procedure $DFS - CIPF(T^{cur}, \boldsymbol{f}^{cur}, n, \boldsymbol{f}^{target})$
  if $|V(T^{cur})| = n$ **then**
    if $\boldsymbol{f}^{cur} = \boldsymbol{f}^{target}$ **then** output T^{cur}; **return** true;
    else return false;
  for $(a, b) \in ATOMBONDPAIRS$ **do**
    $L \leftarrow \emptyset$;
    if $\{l(v)|v \in V(T^{cur})\} \cup \{a\} \not\subseteq atomset(\boldsymbol{f}^{target})$;   (set means multiset)
    then continue (i.e., examine the next pair in $ATOMBONDPAIRS$);
    for all $w \in V(T^{cur})$ **do**
      Let T^{next} be a tree got by connecting new leaf u with label a to w by bond b;
      if w does not satisfy the valence constraint **then continue**;
      Compute $\boldsymbol{f}^{next}$ from T^{next} and $\boldsymbol{f}^{cur}$;
      $dist^{next} \leftarrow dist(\boldsymbol{f}^{next}, \boldsymbol{f}^{target})$;
      if $dist^{next} \neq \infty$ **then** Add $(T^{next}, \boldsymbol{f}^{next}, dist^{next})$ to L;
    while L is not empty **do**
      Remove $(T^{next}, \boldsymbol{f}^{next}, dist^{next})$ from L such that $dist^{next}$ is the minimum;
      if $DFS - CIPF(T^{next}, \boldsymbol{f}^{next}, T^{target}, \boldsymbol{f}^{target})$ =true **then return** true;
  return false;

The details of the implemented code are slightly different from the above. The followings are the main differences: (i) Hydrogen atoms are added at the last stage of the search procedure: hydrogen atoms are added only if the frequencies of the other atoms are the same as those in the target feature vector. (ii) L does not contain T^{next} and $\boldsymbol{f}^{next}$ (in order to save memory space). These are reconstructed just before the recursive call of DFS-CIPF. (iii) When calculating $\boldsymbol{f}^{next}$ from T^{next} and $\boldsymbol{f}^{cur}$, paths beginning from and ending at a new leaf are only computed. (iv) Benzene rings can be added as if these were leaves, where structural information on benzene is utilized for calculating feature vectors. (v) A benzene ring is given as an initial structure when a compound is small and contains a benzene ring.

It should be noted that *BB-CIPF finds an exact solution* (i.e., an exact pre-image of a given feature vector) if it exists. BB-CIPF may be modified so that it can find a kind of approximate pre-image or it can enumerate all possible pre-images.

5. Computational Experiment

We performed computational experiment on BB-CIPF in order to evaluate practical computation time. We used a PC cluster with Intel Xeon 2.8GHz CPUs working under the LINUX operating system where only one CPU was used per execution.

We examined several chemical structures by varying K. As mentioned before, BB-CIPF can handle chemical compounds with tree structures where benzene rings can also appear in structures. In the experiment, a target feature vector is computed from a target chemical compound and is given as an input for BB-CIPF (a target chemical compound is not given to BB-CIPF). Then, BB-CIPF computes a chemical structure whose feature vector coincides with the target feature vector. We examined 8 chemical compounds with $K = 1, 2, 3, 4$. CPU times are shown in Table 1. CPU time is shown with underline if the same structure as the target compound was obtained. N/A means that search did not succeed in 10 minutes.

Table 1. Computation time of BB-CIPF for various chemical compounds.

Name	Chemical Formula	CPU time (sec.)			
		$K = 1$	$K = 2$	$K = 3$	$K = 4$
Methionine	$C_5H_{11}NO_2S$	9.08	0.16	0.019	0.002
Phenylalanine	$C_9H_{11}NO_2$	0.020	0.010	0.010	0.014
Arginine	$C_6H_{14}N_4O_2$	N/A	500.0	19.9	1.51
Aspirin	$C_9H_8O_4$	0.060	0.001	0.002	0.003
2-Ethylhexyl phthalate	$C_{16}H_{22}O_4$	N/A	4.29	6.04	7.88
Etidocaine	$C_{17}H_{28}N_2$	N/A	N/A	N/A	0.470
Esatenolol	$C_{14}H_{22}N_2O_3$	N/A	N/A	25.6	1.46
Trimethobenzamide	$C_{21}H_{28}N_2O_5$	N/A	N/A	N/A	30.7

It is seen from the table that the computation time decreases as K increases in general. It is reasonable that pruning operations are effectively performed if longer paths are employed. It is also seen that the same structures as the target ones are inferred when larger K is used. From this table, it is suggested that the algorithm can output a solution for moderate size chemical structures (e.g., the number of carbon atoms is less than 20) if K is 3 or 4. Though further improvements are required, the result of computational experiment suggests that it is possible to solve the pre-image problem practically.

The current implementation of BB-CIPF can only output one solution. But, there are many possible solutions especially when K is small. Therefore, it is important future work to modify BB-CIPF so that it can efficiently output all possible solutions. It is also important future work to develop a method to select the best solution from the possible solutions.

References

1. E. Byvatov, U. Fechner, J. Sadowski and G. Schneider. Comparison of support vector machine and artificial neural network systems for drug/nondrug classification. *Journal of Chemical Information and Computer Sciences*, 43:1882–1889, 2003.
2. M. Deshpande, M. Kuramochi, N. Wale and G. Karypis. Frequent substructure-based approaches for classifying chemical compounds. *IEEE Trans. Knowledge and Data Engineering*, 17:1036–1050, 2005.
3. H. Kashima, K. Tsuda and A. Inokuchi. Marginalized kernels between labeled graphs. *Proc. 20th Int. Conf. Machine Learning*, 321–328, 2003.
4. P. Mahé, N. Ueda, T. Akutsu, J-L. Perret and J-P. Vert. Graph kernels for molecular structure-activity relationship analysis with support vector machines. *Journal of Chemical Information and Modeling*, 45:939–951, 2005.
5. N. Cristianini and J. Shawe-Taylor. *An Introduction to Support Vector Machines and Other Kernel-based Learning Methods*. Cambridge Univ. Press, 2000.
6. G. H. Bakir, J. Weston and B. Schölkopf. Learning to find pre-images. *Advances in Neural Information Processing Systems*, 16:449–456, 2003.
7. G. H. Bakir, A. Zien and K. Tsuda. Learning to find graph pre-images. *Lecture Notes in Computer Science*, 3175:253–261, 2004.
8. T. Asano. An $O(n \log \log n)$ time algorithm for constructing a graph of maximum connectivity with prescribed degrees. *Journal of Computer and System Sciences*, 51:503–510, 1995.
9. O. Maruyama and S. Miyano. Inferring a tree from walks. *Theoretical Computer Science*, 161:289–300, 1996.
10. J. Lauri and R. Scapellato. *Topics in Graph Automorphisms and Reconstruction*. Cambridge Univ. Press, 2003.
11. T. Akutsu and D. Fukagawa. Inferring a graph from path frequency. *Lecture Notes in Computer Science*, 3537:371–392, 2005.
12. T. Akutsu and D. Fukagawa. On inference of a chemical structure from path frequency. *Proc. BIOINFO 2005*, 96–100, 2005.
13. K. Funatsu and S. Sasaki. Recent advances in the automated structure elucidation system, CHEMICS. Utilization of two-dimensional NMR spectral information and development of peripheral functions for examination of candidates. *Journal of Chemical Information and Computer Sciences*, 36:190–204, 1996.
14. A. Korytko, K-P. Schulz, M. S. Madison and M. E. Munk. HOUDINI: a new approach to computer-based structure generation. *Journal of Chemical Information and Computer Sciences*, 43:1434–1446, 2003.
15. H. Nagamochi. A detachment algorithm for inferring a graph from path frequency. *Lecture Notes in Computer Science*, 4112:274–283, 2006.

EXACT AND HEURISTIC APPROACHES FOR IDENTIFYING DISEASE-ASSOCIATED SNP MOTIFS

GAOFENG HUANG and PETER JEAVONS

Computing Laboratory, University of Oxford

{Gaofeng.Huang, Peter.Jeavons}@comlab.ox.ac.uk

DOMINIC KWIATKOWSKI

Wellcome Trust Centre for Human Genetics, University of Oxford

Dominic.Kwiatkowski@paediatrics.oxford.ac.uk

A Single Nucleotide Polymorphism (SNP) is a small DNA variation which occurs naturally between different individuals of the same species. Some combinations of SNPs in the human genome are known to increase the risk of certain complex genetic diseases. This paper formulates the problem of identifying such disease-associated SNP motifs as a combinatorial optimization problem and shows it to be $\mathcal{NP}$-hard. Both exact and heuristic approaches for this problem are developed and tested on simulated data and real clinical data. Computational results are given to demonstrate that these approaches are sufficiently effective to support ongoing biological research.

1. Introduction

The DNA sequences of different individuals within the same species are highly conserved but not exactly the same. One common type of variation is called a Single Nucleotide Polymorphism (SNP), where a single position within a DNA sequence is altered from one nucleotide base to another. A general research question within the field of human genetics is to ask whether a disease of interest is related to the occurrence of (unknown but) particular SNPs within the human genome. This question is hard to answer in most cases due to the fact that the frequency of SNPs occurring in the human genome is estimated to be 1 SNP per 300 bases, which means that there are about 10 million common SNPs to be investigated. It is currently technically and economically impossible to screen the whole genome of all patients.

Fortunately, the phenomenon of Linkage Disequilibrium (LD) makes some progress possible. Two SNPs are said to be in high Linkage Disequilibrium if their respective alleles are not randomly associated with each other. This allows us to investigate only some SNPs and infer others. To calculate the LD value between two SNP loci, several quantitative LD measures have been proposed, including D' [4], r^2 [3] and λ [9]. The International HapMap Project (http://www.hapmap.org/), which aims to produce a map of all the common SNPs in the human genome, was started in October 2002.

Within this framework there are currently two different typical scenarios for finding the positions of disease-associated SNPs depending on how much *a priori* knowledge we have about these positions. In the first scenario, where we know nothing about where these SNPs might be, we have to do a genome-wide scan which samples a small proportion of the SNPs from the whole genome and outputs the estimated region which is most likely to contain disease-associated SNPs. This problem has been the subject of extensive research by statisticians [5,6,7,8]. Some research has also been done by computer scientists, including HPM (Haplotype Pattern Mining) [11], which has had some success. Many of these methods are statistically powerful; they can often locate candidate disease-associated SNPs with a resolution of about 50-100Kb (where a Kb is 1000 nucleotide bases).

In the second scenario, we assume that the disease-associated SNPs we wish to identify are already known to lie within some small region of the genome (50-100Kb). This is the scenario that our biological collaborators are currently investigating. Even in this case it is still not trivial to determine whether particular SNPs are "significantly" associated with the disease. One reason is that complex genetic diseases are influenced by both genetic and environmental factors (e.g. lifestyle). The "causative" SNPs only affect the *risk* of getting the disease in question. Another difficulty is due to the *combinatorial* nature of disease association with SNPs, which has recently been observed in biological research. If we only examine each single SNP locus, we might not find any significant association between each individual SNP and the disease. However, a particular combination of several SNPs leads to high disease risk. One possible explanation is that this particular combination of several SNPs might be in high LD with another SNP, which is the real causative SNP (but has not been measured - remember that it is not realistic at present to screen a patient for all possible SNPs, even in a restricted region).

In this paper, we will focus on the second scenario in the context of a classical clinical case-control study: given several candidate SNP loci within a small region, and the SNP data observed from both *cases* (patients) and *controls* (healthy individuals), the problem is to identify the most significant SNP *combination* (motif) that associates with the disease. Our biological collaborators currently perform this task manually, and our aim is to develop efficient automated tools to do this job.

The rest of the paper is organized as follows. In Section 2, we formulate the problem as a combinatorial optimization problem. In Section 3 we show this problem to be $\mathcal{NP}$-hard, and develop both exact and heuristic approaches. Computational results are presented in Section 4 for both simulated data and real data. In the final section, we summarise our conclusions and discuss directions for future work.

2. Problem Formulation

Throughout this paper the input SNP data for an individual will be represented in a standard way by a vector of m values over the alphabet $\{0, 1, 2\}$. The reason this alphabet is usually adopted by biologists is that almost all (99%) of the SNPs in the human genome are bi-alleles, for example `A/T`. The standard practice is to use "1" to denote the major allele (i.e. the one with the highest frequency in the human population) and "2" to denote the minor allele. Some allele values may be missing (unknown) due to experimental reasons, and these are denoted with "0".

The SNP data from n_p patients (cases) and n_h healthy individuals (controls) will be represented by two matrices $M^{(p)}_{n_p \times m}$ and $M^{(h)}_{n_h \times m}$. Each row of such a matrix represents the data for a single individual for the m SNPs under consideration. The vector in row i of matrix $M^{(x)}$, i.e. the data for individual i, will be denoted $M^{(x)}_i$.

A SNP *motif* is an expression of the form "`--11---2-`", where "-" means "don't care". A convenient sparse representation for such a motif is to use two vectors, a position vector $P = (p_1, p_2, ..., p_k)$, where $1 \leq p_1 < p_2 < ... < p_k \leq m$ and a data vector $D = (d_1, d_2, ..., d_k)$, where each $d_i \in \{1, 2\}$. These 2 vectors specify the motif by requiring that the allele in each position p_i should be d_i; for example, the motif "`--11---2-`" is represented as $P = (3, 4, 8), D = (1, 1, 2)$.

We now define a matching function between a motif (P, D) and an individual data vector $M^{(x)}_i$, as follows:

$$Match((P,D), M^{(x)}_i) = \begin{cases} 1, & \text{if } \forall k, M^{(x)}_{i,p_k} = 0 \; or \; M^{(x)}_{i,p_k} = d_k; \\ 0, & \text{otherwise.} \end{cases}$$

Using this function, the number of cases (a) and controls (b) that match a motif (P, D) is given by the formula $a = \sum_{i=1}^{n_p} Match((P, D), M^{(p)}_i)$ and $b = \sum_{i=1}^{n_h} Match((P, D), M^{(h)}_i)$ respectively.

	#Match	#non-Match	Total
#Cases	a	$c = n_p - a$	n_p
#Controls	b	$d = n_h - b$	n_h
Total	$e = a + b$	$f = n - e$	$n = n_h + n_p$

Figure 1. 2×2 contingency table

To measure the significance of a motif, a standard method is to put these numbers into a 2×2 contingency table (see Figure 1) and perform a standard chi-squared test. A convenient formula for calculating the value of the chi-squared statistic is:

$$\chi^2(a, b) = \frac{n(ad - bc)^2}{efn_pn_h} = \frac{n(an_h - bn_p)^2}{(a+b)(n-a-b)n_pn_h}. \tag{1}$$

Our search problem can now be formulated as finding a motif (P^*, D^*) which yields the maximal possible value for χ^2, given the data matrices $M^{(p)}_{n_p \times m}$ and $M^{(h)}_{n_h \times m}$. This problem will be called the SNP motif identification problem.

We believe that this formulation of the problem captures essential aspects of current biological research in this area, including the fact that SNP motifs may be only weakly associated with the occurrence of disease, and yet may be biologically significant. By seeking to maximise a statistical measure of association our formulation is also able to cope with noisy data and missing values.

3. Methods

3.1. $\mathcal{NP}$-hardness

In this section, we will show that the SNP motif identification problem is $\mathcal{NP}$-hard, by constructing a reduction from the standard MAX-SAT problem. The MAX-SAT (Maximum Satisfiability) problem is a well-known $\mathcal{NP}$-hard problem [2], which can be stated as follows. Let $x_1, x_2, ..., x_m$ be m boolean variables. A clause C_i is a disjunction of $|C_i|$ literals, i.e. $C_i = \bigvee_{j=1}^{|Ci|} l_{i,j}$, where each literal $l_{i,j}$ is either of the form x_j or $\bar{x}_j$. Given a conjunction of n clauses $\bigwedge_{i=1}^{n} C_i$, the MAX-SAT problem is to find an assignment of boolean values to the variables $x_1, x_2, ..., x_m$, such that the number of satisfied (true) clauses is maximized.

To reduce MAX-SAT to our SNP motif identification problem we proceed as follows. As a first step, given any instance of MAX-SAT, we construct a corresponding instance of the SNP motif identification problem by the following procedure:
1. Let the m boolean variables of the MAX-SAT instance correspond to m SNPs.
2. Construct the matrix $M^{(h)}$ by transforming each clause C_i to the vector $M_i^{(h)}$:

$$M_{i,j}^{(h)} = \begin{cases} 1 & \text{if } x_j \text{ occurs in clause } C_i; \\ 2 & \text{if } \bar{x}_j \text{ occurs in clause } C_i; \\ 0 & \text{otherwise.} \end{cases}$$

3. Let the matrix $M^{(p)}$ consist of a single row containing only zeros.

We will now show that an optimal solution to this artificial instance always corresponds to an optimal solution to the original MAX-SAT instance. Since the matrix $M^{(p)}$ only contains a single line with all zeros, which will match any motif, we will always have the number of matches $a = 1$ for any motif. Therefore, the objective function (Equation (1)) becomes $\chi^2(1, b) = \dfrac{(n_h - b)}{(1 + b)n_h}$, which is a decreasing function with increasing b. This means that an optimal solution to the artificial instance is one which minimizes b, i.e. a motif (P^*, D^*) which matches the least number of lines in $M^{(h)}$. Note that we can assume that $P^* = (1, 2, 3, ..., m)$ (otherwise, we could add some extra position, which would not increase b). We can transform (P^*, D^*) to a solution X^* of the original MAX-SAT instance by setting $x_j^* = true$ if $d_j^* = 2$, and $x_j^* = false$ if $d_j^* = 1$.

If the motif (P^*, D^*) does not match line i in $M^{(h)}$, then there is some position j such that $M_{i,j}^{(h)} \neq 0$ and $M_{i,j}^{(h)} \neq d_j^*$. There are two cases:

1. $M_{i,j}^{(h)} = 1$ **and** $d_j^* = 2$. In this case $M_{i,j}^{(h)} = 1$ means that clause C_i contains x_j while $d_j^* = 2$ means $x_j^* = true$, so clause C_i is satisfied.

2. $M_{i,j}^{(h)} = 2$ **and** $d_j^* = 1$. In this case $M_{i,j}^{(h)} = 2$ means that clause C_i contains $\bar{x}_j$ while $d_j^* = 1$ means $x_j^* = false$, so clause C_i is again satisfied.

Hence, if motif (P^*, D^*) does **not** match line i in $M^{(h)}$, then clause C_i is satisfied by assignment X^*. A similar argument shows that the converse is also true. Since (P^*, D^*) is a motif which matches the least number of lines in $M^{(h)}$ (minimizes b), this means that the number of lines that motif (P^*, D^*) does *not* match is maximized. Therefore, the number of clauses that assignment X^* satisfies is maximized, i.e. X^* is an optimal solution for the original MAX-SAT problem.

In this way, we have reduced the MAX-SAT problem to our problem of finding a motif (P^*, D^*) with the maximal value for χ^2. Since MAX-SAT is known to be $\mathcal{NP}$-hard, it follows that the SNP motif identification problem is also $\mathcal{NP}$-hard.

3.2. *Exact Algorithm*

A straight-forward exhaustive search algorithm needs to explore $O(3^m)$ motifs, and for each motif it takes $O((n_p + n_h)m)$ time to test matching and hence compute a and b. In this section, we will develop an effective exact algorithm using a branch-and-prune tree search technique, which dramatically reduces the search space.

3.2.1. *Search Tree Representation.*

We will search for the motif (P, D) in a sequence of steps. At step j, we enumerate the possible choices of p_j and d_j.

Each node at level j in the tree determines a motif (P, D), $P = (p_1, p_2, ..., p_j)$, $D = (d_1, d_2, ..., d_j)$, where the values for each p_i and d_i can be retrieved by tracing the path from the root node to the given node. To speed up the calculation of χ^2 objective function and allow the search tree to be pruned, we will associate each node with a triple (j, S^+, S^-), where S^+ is the set of lines in $M^{(p)}$ which match motif (P, D), i.e. $S^+ = \{i \mid Match((P, D), M_i^{(p)}) = 1\}$ and $S^- = \{i \mid Match((P, D), M_i^{(h)}) = 1\}$.

The root node is associated with the triple $(j = 0, S^+ = \{1, 2, ..., n_p\}, S^- = \{1, 2, ..., n_h\})$. Given a node with associated triple (j, S^+, S^-), its child node along the branch (p_{j+1}, d_{j+1}), has associated triple $(j + 1, S'^+, S'^-)$, where

$$S'^+ = \{i \mid i \in S^+ \text{ and } M^{(p)}_{i,p_{j+1}} = d_{j+1}\}, \text{ and} \tag{2}$$

$$S'^- = \{i \mid i \in S^- \text{ and } M^{(h)}_{i,p_{j+1}} = d_{j+1}\}. \tag{3}$$

Hence, generating a new node with its associated triple, can be done in $O(n_p + n_h)$ time. Since $a = |S'^+|$ and $b = |S'^-|$, we can calculate the χ^2 objective function for that node in $O(n_p + n_h)$ as well. Hence by storing the information in the triples at each node we have reduced the time complexity of the calculations at each node from $O((n_p + n_h)m)$ to $O(n_p + n_h)$.

3.2.2. *Branching Strategy and Pruning Rules.*

We use a Depth-First-Search strategy to minimise the storage requirements. Two pruning rules are applied to reduce the search space:

1. Absorption Rule : When we generate a child node of a node with triple (j, S^+, S^-) following branch $(p_{j+1} = p^*, d_{j+1} = d^*)$, as shown in Figure 2, if the child node has the associated triple $(j + 1, S'^+, S'^-)$, where $S'^+ = S^+$ and $S'^- = S^-$, then this child node can be "absorbed", and therefore the whole subtree T' rooted at the child node can be pruned, as we will now explain.

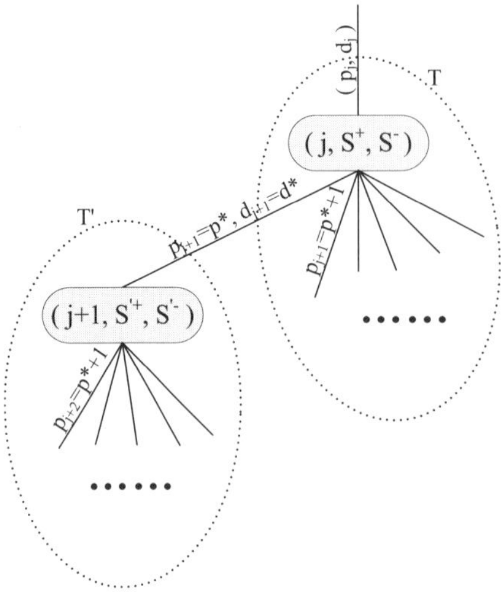

Figure 2. Absorption Rule

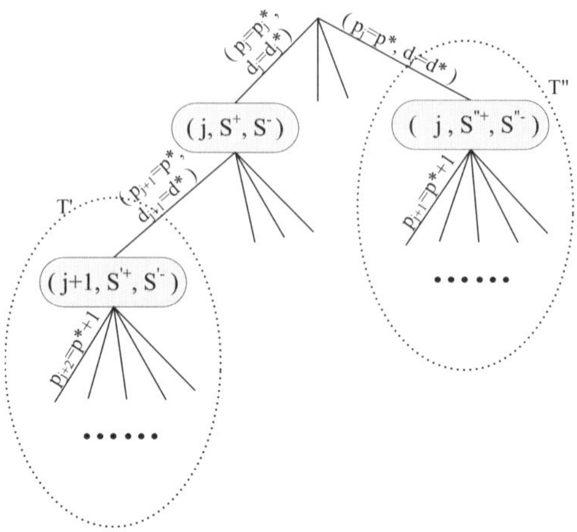

Figure 3. Sibling Rule

Recall the calculation of S'^{+} and S'^{-} in Equations (2) and (3). If $S'^{+} = S^{+}$ and $S'^{-} = S^{-}$, then we have that $\forall i \in S^{+}, M_{i,p^*}^{(p)} = d^*$ and $\forall i \in S^{-}, M_{i,p^*}^{(h)} = d^*$. In other words, for both cases and controls, all the remaining individuals have the same allele d^* in the SNP locus p^*, so adding position p^* to the motif does not provide any more information. This implies that subtree T' and subtree T in Figure 2 are identical, so we can prune subtree T' and explore subtree T only.

2. Sibling Rule : Suppose that we generate a child node of a node with associated triple (j, S^{+}, S^{-}), following branch $(p_{j+1} = p^*, d_{j+1} = d^*)$, and assume that the original node has a sibling node which follows branch $(p_j = p^*, d_j = d^*)$ from its parent, as shown in Figure 3. If the child node has associated triple (j, S'^{+}, S'^{-}), and the sibling node has associated triple (j, S''^{+}, S''^{-}), where $S''^{+} = S'^{+}$ and $S''^{-} = S'^{-}$, then the child node can be pruned, as we will now explain.

It follows from their relative positions in the search tree that the child node and the sibling node only differ in one SNP locus: p_j^*. The equations $S''^{+} = S'^{+}$ and $S''^{-} = S'^{-}$ imply that adding position p_j^* does not provide more information. In other words, for any node in the subtree T'' obtained by following the path $(p_j = p^*, d_j = d^*), (p_{j+1} = p_{j+1}^*, d_{j+1} = d_{j+1}^*), ..., (p_k = p_k^*, d_k = d_k^*)$, there exists a corresponding node in the subtree T', matching the same cases and controls, which is obtained by following the path $(p_j = p_j^*, d_j = d_j^*), (p_{j+1} = p^*, d_{j+1} = d^*), (p_{j+2} = p_{j+1}^*, d_{j+2} = d_{j+1}^*), ..., (p_{k+1} = p_k^*, d_{k+1} = d_k^*)$, and vice versa. Hence, subtree T' and subtree T'' are identical, and we only need to explore one of them.

3.3. *Heuristic Approach*

Our heuristic approach includes two stages: a search stage and a refinement stage.

The search stage is adapted from the exact search algorithm described in the previous section by extending the absorption rule. In Figure 2, a child node with triple $(j+1, S'^{+}, S'^{-})$ is absorbed by its parent node with triple (j, S^{+}, S^{-}) if $S'^{+} = S^{+}$ and $S'^{-} = S^{-}$, because adding position p^* does not provide more information. Now we extend this rule: a child node with triple $(j+1, S'^{+}, S'^{-})$ will be absorbed by its parent node with triple (j, S^{+}, S^{-}) if S'^{+} is *highly similar* to S^{+} and S'^{-} is *highly similar* to S^{-}. Keeping in mind that $S'^{+} \subset S^{+}$ and $S'^{-} \subset S^{-}$, the similarity conditions can be formulated as follows: $|S'^{+}| \geq 0.95\,|S^{+}|$ and $|S'^{-}| \geq 0.95\,|S^{-}|$.

Using this rule, when we generate the children of a node we prune those branches which do not provide "enough" new information. This pruning is very effective because it makes use of the statistical relationship between different SNP loci, i.e. Linkage Disequilibrium.

In the refinement stage, all motifs in the candidate list obtained from the search stage are refined by a local search procedure, and the most significant motif obtained is returned as the final result.

4. Results

We have tested our algorithms on both simulated and real clinical data. All experiments were performed on an IBM Pentium 1.5GHz laptop with 512MB of memory. We set the time limit of running each single testcase to be 5 minutes.

4.1. *Simulated Data with Realistic Linkage Disequilibrium*

At present the best model for Linkage Disequilibrium between SNP loci is still unclear, but it is unreasonable to simply generate random SNP data without attempting to model the Linkage Disequilibrium. To provide a suitable test set for our computational tools we therefore had to develop a novel way to generate simulated data with realistic Linkage Disequilibrium between SNP at different loci. The procedure is briefly described as follows:

Step 1: Obtain raw data from the HapMap Project[10]. The HapMap raw data (release 14) contains 60 unrelated Caucasian (CEU) individuals, 60 Yoruba individuals, 45 unrelated individuals from Tokyo (Japan) and 45 unrelated individuals from Beijing (China). We randomly select $m+1$ consecutive SNP loci.

Step 2: Estimate haplotype frequency using a popular program SNPHAP [1]

Step 3: Generate a population of 100000 individuals according to the frequency table obtained from SNPHAP.

Step 4: Simulate an unknown disease and remove the causative SNP. We first randomly assume one SNP locus $p*$ among the $m+1$ loci to be the disease causative SNP. Then we assign a disease risk $r_1 = 0.001$ and $r_2 = 0.01$ respectively to allele "1" and "2" in SNP locus $p*$, and randomly simulate whether each individual gets the disease or not according to that risk. Finally we remove the column of the disease causative SNP, locus $p*$, and only use the remaining m SNP loci.

Step 5: Simulate a clinical scenario. A classical case-control study samples the same number of cases and controls. Here we consider all individuals in our simulated population sample who get the disease to be cases. Then we randomly sample the same number of healthy individuals from the whole population to be controls.

4.1.1. *Exact Search vs. Heuristic*

We first carried out experiments with the number of SNPs, m, set to 7 different values ranging from 20 to 50. For each value of m, we did 100 simulations to generate 100 testcases from 10 different genomic regions. Both exact and heuristic approaches were then tested on these $7 \times 100 = 700$ testcases. The results are reported in Table 1. Each row presents the average result over the 100 testcases with the same m. The first column is the average number of nodes that the exact search algorithm explores. The second column is the average CPU time that the exact search algorithm takes. The third column is the number of testcases solved within 5 minutes. Similarly, the fourth, fifth and sixth columns are for the heuristic approach. The last column shows for how many testcases the solution that the heuristic returns is optimal, i.e. the same solution as the exact algorithm returns.

Table 1. Results for Simulated Data ($7 \times 100 = 700$ testcases in total)

	Exact Search			Heuristic			
m	average nodes	average time(sec)	#solved	average nodes	average time(sec)	#solved	#optimal
20	16988.0	0.62	100	7110.3	0.24	100	100
25	47166.3	3.34	100	16533.1	0.88	100	100
30	161205.8	21.54	97	41637.1	4.13	100	≥97
35	-	-	85	77899.7	11.23	100	≥85
40	-	-	-	124126.4	29.16	100	-
45	-	-	-	187257.0	67.38	97	-
50	-	-	-	-	-	83	-

These results show that our exact algorithm using a branch-and-prune strategy is considerably more efficient than a brute-force approach: the number of nodes explored is much smaller than 3^m, as shown in Table 1, because a large number are pruned. The results also indicate that our heuristic method can provide high quality motifs using less time. In fact, for all of the 382 testcases that we can compare, the motifs that the heuristic method provides are always optimal.

4.1.2. *Are the Motifs Found Real Signals?*

More specifically, what if the disease is not associated with any SNP? Under our simulation framework, We simulated this "no association" scenario by setting disease risk $r_1 = r_2 = 0.005$. We did 100 simulations to generate 100 testcases with $m = 40$ SNPs to test our heuristic approach. We compare the results with the 100 testcases ($m = 40$) from previous experiments. Figure 4 shows the histogram for both scenarios (wider bars for the association scenario and narrower bars for non-association). The x-axis represents the χ^2 value of the motif we found, and the y-axis represents the number of testcases.

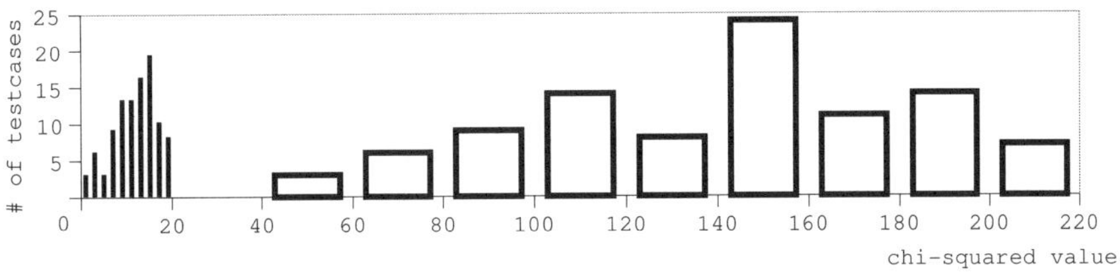

Figure 4. Histogram (Non-association vs. association)

Figure 4 shows that if the disease is not associated with any SNP, then the χ^2 value of the motif that our algorithm returns is very low (less than 20). On the other hand, if the disease has a causative SNP, even though this causative SNP is not directly observed, our algorithm finds a motif with much higher χ^2 value (above 40). This indicates that if our algorithm find motifs with high χ^2 value, then the motifs are very likely to be real biological signals.

4.2. *Clinical Data*

Finally, we obtained a real clinical testcase from the biological laboratory. (Due to privacy policy and copyright restrictions, we cannot give full details of the clinical background to this dataset.) This clinical testcase contains $m = 19$ SNP loci, $n_p = 1362$ patients and $n_h = 914$ controls. Table 2 shows the computational results. As mentioned before, currently the identification of disease-associated SNP motifs is done manually by experienced biologists. The first row of Table 2 shows the result that was obtained in the biological laboratory using this manual process.

As shown in Table 2, the motifs obtained by our methods are very similar to those obtained by the current labour-intensive manual process. For this particular dataset the resulting χ^2 value (22.22) is not high enough to conclude with confidence that the SNP motif is significantly associated with the disease. However, we have successfully automated the process of finding the best possible motif, and this approach can now be used to support ongoing biological research.

Table 2. Real Clinical Data

	Motif	#cases	#controls	χ^2 value
Manual Result	11-2----2--------2-	93	113	19.69
Exact Search	11-2----2-11-----21	89	112	22.22
Heuristic	11-2----2-11-----21	89	112	22.22

5. Conclusions and Future Work

In this paper, we studied the problem of identifying disease-associated SNP motifs. We formulated it as a combinatorial optimization problem and showed it to be $\mathcal{NP}$-hard. Both exact and heuristic approaches for this problem were developed and tested on both simulated and real data. The results demonstrate that these computational approaches can support ongoing biological research.

For simplicity of problem description in this paper we haven't made the distinction between "haplotype" data and "genotype" data. In fact, our approach deals with SNP haplotype data. To infer haplotype data from genotype data is another very active research topic. In this paper, we have used the SNPHAP program [1] as a preprocessor to obtain estimated haplotype data. Our plan for the next stage of this research is to develop algorithms which can deal directly with the unphased genotype SNP data to identify significant SNP motifs.

References

1. D. Clayton. SNPHAP : A program for estimating frequencies of large haplotypes of SNPs. `http://www-gene.cimr.cam.ac.uk/clayton/software/snphap.txt`, 2002.
2. M. R. Garey and D. S. Johnson. *Computers and intractability: A guide to the theory of NP-completeness.* W.H. Freeman and Company, New York, 1979.
3. W. G. Hill and A. Robertson. Linkage disequilibrium in finite populations. *Theoret. Appl. Genet.*, 38:226–231, 1968.
4. R. C. Lewontin. The interaction of selection and linkage. I. general considerations; heterotic models. *Genetics*, 49:49–67, 1964.
5. J S Liu, C Sabatti, J Teng, B J B Keats, and N Risch. Bayesian analysis of haplotypes for linkage disequilibrium mapping. *Genome Research*, 11:1716–1724, 2001.
6. X. Lu, T. Niu, and Jun S. Liu. Haplotype information and linkage disequilibrium mapping for single nucleotide polymorphisms. *Genome Research*, 13:2112–2117, 2003.
7. M. S. McPeek and A. Strahs. Assessment of linkage disequilibrium by the decay of haplotype sharing, with application to fine-scale genetic mapping. *Am. J. Hum. Genet.*, 65:858–875, 1999.
8. A P Morris, J C Whittaker, and D J Balding. Fine-scale mapping of disease loci via shattered coalescent modeling of genealogies. *Am. J. Hum. Genet.*, 70:686–707, 2002.
9. J. D. Terwilliger. A powerful likelihood method for the analysis of linkage disequilibrium between trait loci and one or more polymorphic marker loci. *Am. J. Human Genet.*, 56:777–787, 1995.
10. The International HapMap Consortium. The International HapMap Project. *Nature*, 426:789–796, December 2003.
11. Hannu T. T. Toivonen, P. Onkamo, K. Vasko, V. Ollikainen, P. Sevon, H. Mannila, Mathias Herr, and Juha Kere. Data mining applied to linkage disequilibrium mapping. *American Journal of Human Genetics*, 67:133–145, 2000.

GENOTYPE-BASED CASE-CONTROL ANALYSIS, VIOLATION OF HARDY-WEINBERG EQUILIBRIUM, AND PHASE DIAGRAMS

YOUNG JU SUH

BK21 Research Division of Medicine and Department of Preventive Medicine, College of Medicine, Ewha Womans University, Seoul, Korea. E-mail: ysprite@hotmail.com

WENTIAN LI

The Robert S. Boas Center for Genomics and Human Genetics, Feinstein Institute for Medical Research, North Shore LIJ Health System, Manhasset, NY 11030, USA
E-mail: wli@nslij-genetics.org

We study in detail a particular statistical method in genetic case-control analysis, labeled "genotype-based association", in which the two test results from assuming dominant and recessive model are combined in one optimal output. This method differs both from the allele-based association which artificially doubles the sample size, and the direct χ^2 test on 3-by-2 contingency table which may overestimate the degree of freedom. We conclude that the comparative advantage (or disadvantage) of the genotype-based test over the allele-based test mainly depends on two parameters, the allele frequency difference δ and the Hardy-Weinberg disequilibrium coefficient difference δ_ϵ. Six different situations, called "phases", characterized by the two X^2 test statistics in allele-based and genotype-based test, are well separated in the phase diagram parameterized by δ and δ_ϵ. For two major groups of phases, a single parameter $\theta = tan^{-1}(\delta/\delta_\epsilon)$ is able to achieves an almost perfect phase separation. We also applied the analytic result to several types of disease models. It is shown that for dominant and additive models, genotype-based tests are favored over allele-based tests.

1. Introduction

Genetic association analysis is a major tool in mapping human disease genes[16,7,11]. A simple association study is the case-control analysis, in which individuals with and without disease are collected (roughly the equal number of sample per group for an optimal design), DNA samples extracted and genetic markers typed. The prototype of a genetic marker is the two-allele single-nucleotide-polymorphism (SNP)[4]. If the two alleles are A and a, there three possible genotypes: AA, Aa, aa, consisting of the maternally-derived and paternally-derived copy of an allele. The three genotype frequencies are calculated in case (disease) and control (normal) group, and a strong contrast of the two sets of genotype frequencies can be used to indicate an association between that marker and the disease.

The statistical analysis in an association study seems to be simple – mostly the standard Pearson's χ^2 test in categorical analysis[1], there are nevertheless subtle differences among various approaches. Some people use the 2×3 genotype count table to carry about test with χ^2 distribution of $df = 2$ degrees of freedom[6]. This method may overestimate the degree of freedom if the Hardy-Weinberg equilibrium holds true. Other people use the allele-based test, where each person contributes two allele counts, and the allele frequency is compared in a 2×2 allele count table. This approach artificially doubles the sample sizes without a theoretical justification[17]. A third approach, what we called "genotype-based" case-control association analysis, remains faithful to the sample size, while does not overestimate the degrees of freedom.

A genotype-based analysis can be simply summarized here. Two Pearson's χ^2 tests are carried out on two 2×2 count tables: the first is constructed by combining the AA and Aa genotype counts and keeping the aa genotype column, and the second by combining the Aa and aa genotype counts. If the marker happens to be the disease gene and A is the mutant allele (a is the wild type allele), then the first table is consistent with a dominant disease model, whereas second a recessive disease model. The two χ^2 tests lead to two p-values, and the smallest one (the more significant one) is chosen as the final test result.

Genotype-based analysis has been used in practice many times[20,18,9], without a particular name, and without a theoretical study. In this article, we will take a deeper look of the genotype-based analysis. We will show that the justification of using genotype-based tests is intrinsically related to the Hardy-Weinberg disequilibrium, but there are more than just a non-zero Hardy-Weinberg disequilibrium coefficient that is important.

The article is organized as follows: we first show that there is no advantage in using genotype-based test if there is no Hardy-Weinberg disequilibrium; we then examine the situation with Hardy-Weinberg disequilibrium, and use the two parameters, the allele frequency difference and the difference of two Hardy-Weinberg disequilibrium coefficients, to construct a phase diagram; the phase diagram is further simplified by using just one parameter; our analytic result is illustrated by a real example from the study of rheumatoid arthritis; we apply the formula to different models; and finally future works are discussed.

2. No advantage for genotype-based analysis if Hardy-Weinberg equilibrium holds true exactly

In an ideal situation, we assume N case samples and N control samples, and the A allele frequency in case and control groups is p_1 and p_2 ($q_1 = 1 - p_1, q_2 = 1 - p_2$). On average (or in the asymptotic limit), the allele and genotype counts are listed in Table 1 where the Hardy-Weinberg equilibrium (HWE) is assumed.

For a $\{N_{ij}\}$ $(i, j = 1, 2)$ 2-by-2 contingency table, the Pearson's (O-E)^2/E (O for observed count, and E for expected count) test statistic is:

$$X^2 = \frac{(N_{11}N_{22} - N_{12}N_{21})^2(N_{11} + N_{12} + N_{21} + N_{22})}{(N_{11} + N_{12})(N_{21} + N_{22})(N_{11} + N_{21})(N_{12} + N_{22})} \tag{1}$$

Using the table elements in Table 1, we can derive

$$\begin{aligned}
X^2_{\text{allele}} &= \frac{(2N)^4(p_1q_2 - p_2q_1)^2 4N}{(2N)^4(p_1 + p_2)(q_1 + q_2)} = \frac{4N(p_1 - p_2)^2}{(p_1 + p_2)(q_1 + q_2)} \\
X^2_{\text{dom}} &= \frac{N^4[(p_1^2 + 2p_1q_1)q_2^2 - (p_2^2 + 2p_2q_2)q_1^2]^2 2N}{N^4(p_1^2 + 2p_1q_1 + p_2^2 + 2p_2q_2)(q_1^2 + q_2^2)} = \frac{2N(p_1 - p_2)^2(q_1 + q_2)^2}{(2 - q_1^2 - q_2^2)(q_1^2 + q_2^2)} \\
X^2_{\text{rec}} &= \frac{N^4[(q_1^2 + 2p_1q_1)p_2^2 - (q_2^2 + 2p_2q_2)p_1^2]^2 2N}{N^4(q_1^2 + 2p_1q_1 + q_2^2 + 2p_2q_2)(p_1^2 + p_2^2)} = \frac{2N(q_1 - q_2)^2(p_1 + p_2)^2}{(2 - p_1^2 - p_2^2)(p_1^2 + p_2^2)}
\end{aligned} \tag{2}$$

Table 1. Count tables for genotype-based analysis under HWE

	A	a	$AA + Aa$	aa	AA	$AA + Aa$
	allele count		dominant model		recessive model	
case	$2\,Np_1$	$2\,Nq_1$	$N(p_1^2 + 2p_1q_1)$	Nq_1^2	Np_1^2	$N(2p_1q_1 + q_1^2)$
control	$2\,Np_2$	$2\,Nq_2$	$N(p_2^2 + 2p_2q_2)$	Nq_2^2	Np_2^2	$N(2p_2q_2 + q_2^2)$

To further simplify the notation, let's denote $\delta \equiv p_1 - p_2$ as the allele frequency difference, $\overline{p} \equiv (p_1 + p_2)/2$ as the averaged A allele frequency across groups, and the averages of the squared terms $\overline{p^2} \equiv (p_1^2 + p_2^2)/2$ ($\overline{q}$ and $\overline{q^2}$ are defined similarly). Then Eq.(2) becomes:

$$X^2_{\text{allele}} = \frac{N\delta^2}{\overline{p} \cdot \overline{q}},$$
$$X^2_{\text{dom}} = \frac{2N\delta^2\overline{q}^2}{\overline{q^2}(1 - \overline{q^2})},$$
$$X^2_{\text{rec}} = \frac{2N\delta^2\overline{p}^2}{\overline{p^2}(1 - \overline{p^2})}. \tag{3}$$

Since the genotype-based test is determined by the maximum value among X^2_{dom} and X^2_{rec}, we would like to prove an inequality between X^2_{allele} and $\max(X^2_{\text{dom}}, X^2_{\text{rec}})$.

Towards this aim, we first compare X^2_{allele} and X^2_{dom}. Due to the following two inequalities:

$$\overline{q^2} = \frac{2q_1^2 + 2q_2^2}{4} \geq \frac{2q_1^2 + 2q_2^2 - (q_1 - q_2)^2}{4} = \frac{(q_1 + q_2)^2}{4} = \overline{q}^2$$
$$2\overline{p} \cdot \overline{q} = 2\overline{q} - 2\overline{q}^2 = q_1 + q_2 - (q_1q_2 + \overline{q^2}) = 1 - \overline{q^2} - (1 - q_1)(1 - q_2) \leq 1 - \overline{q^2},$$

we have

$$\frac{\overline{q^2}}{2\overline{p} \cdot \overline{q}} \geq \frac{\overline{q}^2}{1 - \overline{q^2}}, \tag{4}$$

which leads to $X^2_{\text{allele}} \geq X^2_{\text{dom}}$. The similar approach shows that $\overline{p^2} \geq \overline{p}^2$ and $2\overline{p} \cdot \overline{q} \leq 1 - \overline{p^2}$, which leads to $X^2_{\text{allele}} \geq X^2_{\text{rec}}$.

With the proof that $X^2_{\text{allele}} \geq \max(X^2_{\text{dom}}, X^2_{\text{rec}})$, we have shown that allele-based X^2 (p-value) is always larger (smaller) than the genotype-based X^2 (p-value). In other words, if HWE holds exactly true, there is no need to carry out a genotype-based association analysis. To certain extend, this result is not surprising since allele-based test utlizes twice the number of samples as the genotype-based test, even though the latter has one advantage of testing multiple (two) disease models. Clearly, the increase in sample size more than compensates the advantage of testing multiple models, when HWE is true.

3. Adding violation of Hardy-Weinberg equilibrium

The result in the previous section actually does not disapprove the genotype-based association, since HWE in real data is often violated, even if it is not significantly violated.

Table 2. Count tables for genotype-based analysis under HWD

	A	a	$AA+Aa$	aa	AA	$AA+Aa$
	allele count		dominant model		recessive model	
case	$2\,Np_1$	$2\,Nq_1$	$N(p_1^2+2p_1q_1-\epsilon_1)$	$N(q_1^2+\epsilon_1)$	$N(p_1^2+\epsilon_1)$	$N(2p_1q_1+q_1^2-\epsilon_1)$
control	$2\,Np_2$	$2\,Nq_2$	$N(p_2^2+2p_2q_2-\epsilon_2)$	$N(q_2^2+\epsilon_2)$	$N(p_2^2+\epsilon_2)$	$N(2p_2q_2+q_2^2-\epsilon_2)$

To characterize a realistic genotype count table, one more parameter besides the allele frequency is needed: the Hardy-Weinberg disequilibrium coefficient (HWDc)[19]. The HWDc ϵ is defined as[19] $\epsilon = p_{AA} - p_A^2 = p_{aa} - p_a^2 = -(p_{Aa} - 2p_ap_A)/2 = p_{aa}p_{AA} - p_{Aa}^2/4$. For case and control groups, two HWDc's are used ϵ_1 and ϵ_2. The three count tables under HWD are now parameterized in Table 2.

Applying the definition of X^2 in Eq.(1) to the count tables in Table 2 (note that the allele counts are not affected by HWD), we have

$$
\begin{aligned}
X^2_{\text{allele}} &= \frac{4N(p_1-p_2)^2}{(p_1+p_2)(q_1+q_2)} \\
X^2_{\text{dom,HWD}} &= \frac{N^4[(p_1^2+2p_1q_1-\epsilon_1)(q_2^2+\epsilon_2)-(p_2^2+2p_2q_2-\epsilon_2)(q_1^2+\epsilon_1)]^2 2N}{N^4(p_1^2+2p_1q_1+p_2^2+2p_2q_2-\epsilon_1-\epsilon_2)(q_1^2+q_2^2+\epsilon_1+\epsilon_2)} \\
&= \frac{2N[(p_1-p_2)(q_1+q_2)-(\epsilon_1-\epsilon_2)]^2}{(2-q_1^2-q_2^2-\epsilon_1-\epsilon_2)(q_1^2+q_2^2+\epsilon_1+\epsilon_2)} \\
X^2_{\text{rec,HWD}} &= \frac{N^4[(q_1^2+2p_1q_1-\epsilon_1)(p_2^2+\epsilon_2)-(q_2^2+2p_2q_2-\epsilon_2)(p_1^2+\epsilon_1)]^2 2N}{N^4(q_1^2+2p_1q_1+q_2^2+2p_2q_2-\epsilon_1-\epsilon_2)(p_1^2+p_2^2+\epsilon_1+\epsilon_2)} \\
&= \frac{2N[(q_1-q_2)(p_1+p_2)-(\epsilon_1-\epsilon_2)]^2}{(2-p_1^2-p_2^2-\epsilon_1-\epsilon_2)(p_1^2+p_2^2+\epsilon_1+\epsilon_2)} \qquad (5)
\end{aligned}
$$

Again shorthand notations are introduced: $\delta_\epsilon \equiv \epsilon_1 - \epsilon_2$, and $\overline{\epsilon} \equiv (\epsilon_1+\epsilon_2)/2$. Eq.(5) is rewritten as

$$
\begin{aligned}
X^2_{\text{allele}} &= \frac{N\delta^2}{\overline{p}\cdot\overline{q}}, \\
X^2_{\text{dom,HWD}} &= \frac{2N(\delta\overline{q}-\frac{\delta_\epsilon}{2})^2}{(\overline{q^2}+\overline{\epsilon})(1-\overline{q^2}-\overline{\epsilon})} \\
X^2_{\text{rec,HWD}} &= \frac{2N(\delta\overline{p}-\frac{\delta_\epsilon}{2})^2}{(\overline{p^2}+\overline{\epsilon})(1-\overline{p^2}-\overline{\epsilon})} \qquad (6)
\end{aligned}
$$

From Eq.(6), it is not clear whether X^2_{allele} is still larger than $X^2_{\text{dom,HWD}}$ and $X^2_{\text{rec,HWD}}$. Systematic scanning of the 4-parameter space $(p_1, p_2, \epsilon_1, \epsilon_2)$ would offer a solution, but the result cannot be displayed on a 2-dimensional space. In the following, we simplify the display of the "phase diagram" by using only two (or one) parameters.

4. Phase diagram with one and two parameters

The term "phase diagram" is borrowed from the field of statistical physics[12]. In a typical diagram used in statistical or chemical physics, phases (e.g. solid, liquid and gas) as well as phase boundaries (e.g. melting line) are displayed as a function of physical quantities such as temperature and pressure. Phase transition occurs at phase boundaries. For our topic, a phase indicates, for example, whether allele-based or genotype-based test leads to a higher X^2 value; or it can indicate whether or not the X^2 value leads to a statistically significant result (e.g. p-value < 0.05). The quantities chosen to mimic temperature or pressure for our topic should highlight the phase separation and phase transitions.

Eq.(6) provides us a hint that the allele frequency difference in two groups, δ, and the HWDc difference, δ_ϵ, could be good quantities for phase separation. First of all, δ directly controls the magnitude of X^2, so it should separate "significant phases" from "insignificant phases". Secondly, the relative magnitude and sign of δ and δ_ϵ seems to control the difference between X^2_{allele} and $X^2_{\text{dom,HWD}}$ or $X^2_{\text{rec,HWD}}$, so it should be a good quantity to separate "favoring-allele-based-test phase" (when $X^2_{\text{allele}} > \max(X^2_{\text{dom,HWD}}, X^2_{\text{rec,HWD}})$) and "favoring-genotype-based-test phase" (when $X^2_{\text{allele}} < \max(X^2_{\text{dom,HWD}}, X^2_{\text{rec,HWD}})$).

We carried out the following simulation to construct the phase diagram: 5000 replicates of case-control datasets with 100 cases and 100 controls (in another simulation, the sample size is 1000 per group); For each replicate, the three genotypes are randomly chosen, then the allele frequency and Hardy-Weinberg disequilibrium coefficient were determined. Fig.1 shows the simulation result parameterized by δ_ϵ (x-axis) and δ (y-axis). Six phases (labeled I-VI) are illustrated using 6 different colors, within the two larger categories:

- Favoring genotype-based tests (crosses in Fig.1)
  - I. p-values for both genotype- and allele-based tests are < 0.05 (red)
  - II. p-values for both genotype- and allele-based tests are > 0.05 (yellow)
  - III. p-value for genotype-based test is < 0.05, that for allele-based test is > 0.05 (pink)
- Favoring allele-based tests (circles in Fig.1)
  - IV. p-values for both genotype- and allele-based tests are < 0.05 (purple)
  - V. p-values for both genotype- and allele-based tests are > 0.05 (blue)
  - VI. p-value for allele-based test is < 0.05, that for genotype-based test is > 0.05 (green)

As can be seen from Fig.1, the two parameters, δ and δ_ϵ does a pretty good job in separating six different phases, although minor overlap between phases occurs. The overall performance of δ and δ_ϵ as phase parameters is satisfactory.

As expected, the magnitude of p-values is mainly controlled by the y-axis. Smaller allele frequency differences (smaller δ's) result in non-significant p-values, and significant results are located far away from the $\delta = 0$ line. On the other hand, the δ_ϵ mainly controls whether allele-based or genotype-based test is more significant. However, δ_ϵ itself is not enough: it acts jointly with δ to achieve the phase separation: for genotype-based test to have a smaller p-value than the allele-based test and both are smaller than 0.05 (red points in Fig.1), δ_ϵ tends to have the different sign as that of δ.

The effect of sample size on the phase diagram can be examined by comparing Fig.1(A) and Fig.1(B). Phases II, III, V, VI all shrink in area simply because a larger sample size is

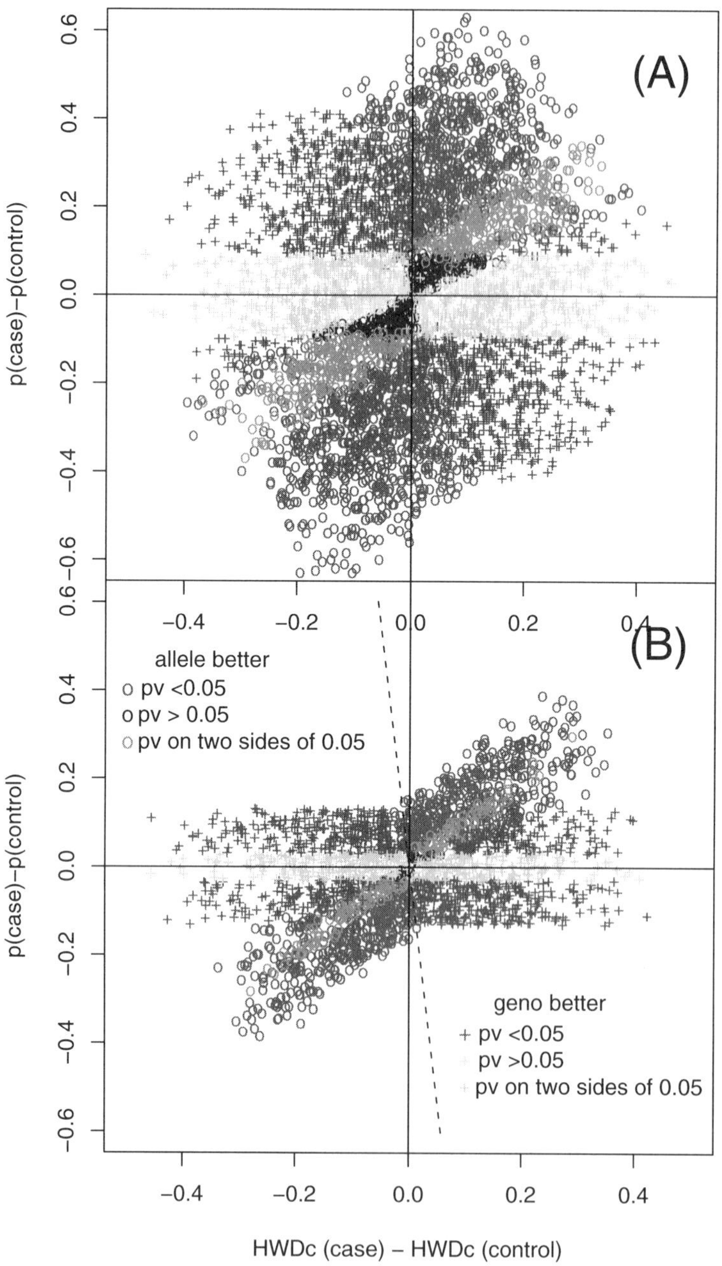

Figure 1. The phase diagram parameterized by $\delta_\epsilon = \epsilon_1 - \epsilon_2$ (x-axis) and $\delta = p_1 - p_2$ (y-axis), where p is the allele frequency for A and ϵ is the Hardy-Weinberg disequilibrium coefficient, determined by a numerical simulation. (A) 100 samples per group with 5000 replicates (5000 points in the plot); (B) 1000 samples per group with 5000 replicates. Six phases are marked: I. p-value for genotype-based test is smaller than that for allele-based test (and both p-values are smaller than 0.05) (red cross); II. similar to I, but both p-values are larger than 0.05 (yellow cross); III. similar to I, but one p-value is smaller than 0.05 and another larger than 0.05 (pink cross); IV. p-value for allele-based test is smaller than that for genotype-based test (and both p-values are smaller than 0.05) (purple circle); V. similar to IV, but both p-values are larger than 0.05 (blue circle); VI. similar to V, but one p-value is smaller than 0.05 and another larger than 0.05 (green circle).

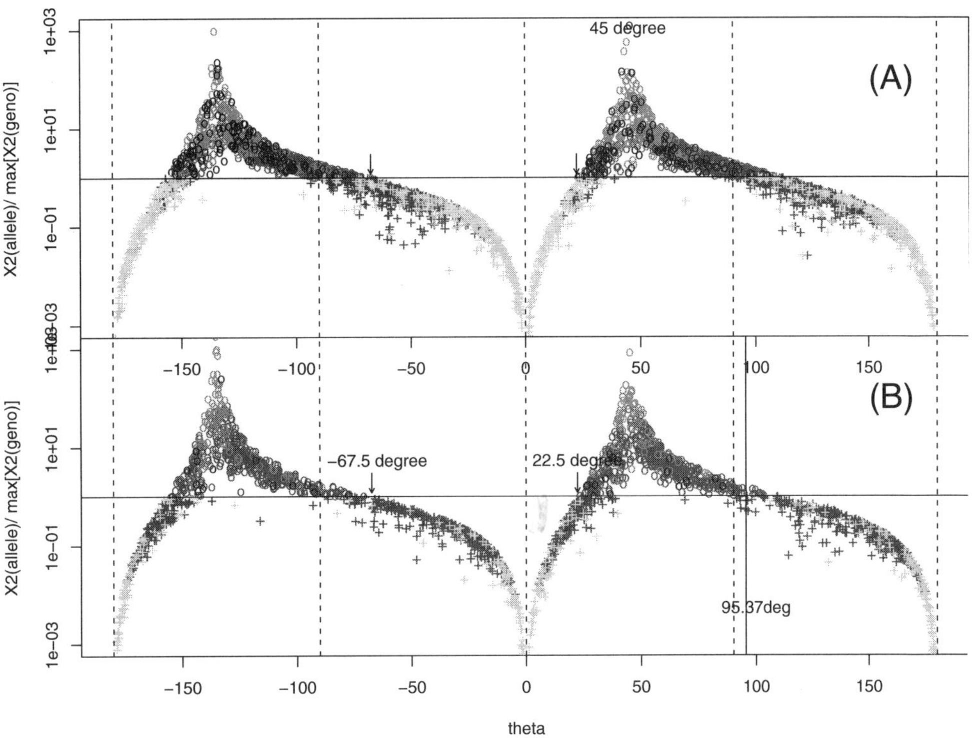

Figure 2. The X^2 ratio $\lambda = X^2_{\text{allele}} / \max(X^2_{\text{rec}}, X^2_{\text{dom}})$ as a function of the parameter $\theta = tan^{-1}(\delta/\delta_\epsilon)$. The y-axis is in a logarithm scale. The same color code for the six phases as used in Fig.1 is also used here. For phases that favor the genotype-based test, $\lambda < 1$; for those favoring allele-based test, $\lambda > 1$. (A) 100 samples per group with 5000 replicates; (B) 1000 samples per group with 5000 replicates.

more likely to lead to a p-value < 0.05 replicate. The relative location of different phases in Fig.1 remains the same.

If we focus on the two major categories (phases I,II,III versus phases IV,V,VI), we notice that the phase boundaries are radiuses. The observation led to the following phase diagram by using a single parameter $\theta = tan^{-1}(y/x) = tan^{-1}(\delta/\delta_\epsilon)$, i.e., the angle between a radius and the x-axis. To measure the relative advantage (disadvantage) of allele-based test over genotype-based test, we use the ratio of two X^2's: $\lambda = X^2_{\text{allele}} / \max(X^2_{\text{rec}}, X^2_{\text{dom}})$. Fig.2 shows λ as a function of θ, using the simulation result in Fig.1 (100 samples per group and 1000 samples per group) and the same color code for six phases.

Fig.2 shows nicely that within the range of $-3\pi/8 < \theta < \pi/8$ ($-67.5^o < \theta < 22.5^o$, or $-2.414 < \delta/\delta_\epsilon < 0.414$), the genotype-based test is favored over the allele-based test. Overlap of phases still occurs in Fig.2, e.g., for θ's slightly below $-3\pi/8$ or slightly above

Table 3. Count tables of marker genotype for a SNP within the gene PTPN22

	TT	TC	CC	total	p_T	ϵ
case	16	245	677	938	.147655	-0.004744
control	12	221	1168	1401	.087438	+0.000920
difference					.060217	-.005664

$\pi/8$. The allele-based test is much better than the genotype-based test when $\theta = \pi/4$ (45°, or $\delta = \delta_\epsilon$). On the other hand, the genotype-based test is much better than the allele-based test when $\theta = 0$ (or $\delta = 0$), though the test result is not significant (phase III). From both Fig.1 and Fig.2, we can see that $\theta = \pi/4$ and $\theta = 0$ line is mainly related to phase VI and phase III.

The sample size per group does not affect the phase boundary between the two major categories, though it does affect phases within a major category. This observation can be understood theoretically by the formula of X^2's in Eq.(6): the relative magnitude between X^2_{allele} and $X^2_{\text{dom,HWD}}$ or $X^2_{\text{rec,HWD}}$ is independent of N as it is canceled out.

5. Illustration by a real dataset

The genotype counts of a missense SNP in gene PTPN22 in Rheumatoid Arthritis samples and in control samples are listed in Table 3 (combining the "discovery" dataset and the "single sib" option in the "replication" dataset in Ref. 3). Our formula predicts that $\theta = tan^{-1}(0.147655 - 0.087438)/(-0.004744 - 0.000920) = tan^{-1}(-0.060217/0.005664) = 95.37^o$. This θ line is marked both in Fig.1 and Fig.2, and is right at the phase boundary. Our calculation predicts that the allele-based test and genotype-based test should lead to similar result.[a] Indeed, X^2_{allele}=41.10, $X^2_{\text{dom,HWD}}$=42.26, $X^2_{\text{rec,HWD}}$=3.43, and the difference between the allele-based and genotype-based tests is very small.

6. Hardy-Weinberg disequilibrium in the patient population given a disease model

In the population of patients (case group), a SNP marker within the disease gene or in linkage disequilibrium with the disease usually violates the Hardy-Weinberg equilibrium. This fact has been used in the proposal of using HWD in case samples to map the disease gene[8]. The HWD coefficient in the case group can be calculated if the disease model is given[10], which is reproduced here. Assuming the penetrance for AA, Aa, aa genotypes to be $f_{\text{AA}}, f_{\text{Aa}}, f_{\text{aa}}$, the disease prevalence is $K = f_{\text{AA}}p_1^2 + f_{\text{Aa}}2p_1q_1 + f_{\text{aa}}q_1^2$, and the genotype frequencies for the case group are (using the Bayes' theorem):

$$p_{\text{AA,aff}} = \frac{f_{\text{AA}}p_1^2}{K}, \qquad p_{\text{Aa,aff}} = \frac{f_{\text{Aa}}2p_1q_1}{K}, \qquad p_{\text{aa,aff}} = \frac{f_{\text{aa}}q_1^2}{K}. \tag{7}$$

[a]One difference however is that the theoretical calculation is based on equal number of samples in case and control group. In our example, the sample size in two groups is slightly different.

The HWD coefficient for the case group is then[10]:

$$\epsilon_1 = p_{\mathrm{AA,aff}} \cdot p_{\mathrm{aa,aff}} - \frac{p^2_{\mathrm{Aa,aff}}}{4} = \left(\frac{p_1 q_1}{K}\right)^2 (f_{\mathrm{AA}} f_{\mathrm{aa}} - f^2_{\mathrm{Aa}}), \quad (8)$$

and the HWD coefficient for the control group is assumed to be zero ($\epsilon_2 = 0$).

If the disease model is multiplicative, i.e., $f_{\mathrm{AA}}/f_{\mathrm{Aa}} = f_{\mathrm{Aa}}/f_{\mathrm{aa}}$, there is no HWD in the case group, so HWD can not be used to map the disease gene. With $\delta_\epsilon = 0 - 0 = 0$, from the result in Sec. 2, the allele-based test is favored over the genotype-based test. For dominant models, $f_{\mathrm{AA}} \approx f_{\mathrm{Aa}} = F$, and $\epsilon_1 \propto F(f_{\mathrm{aa}} - F)$. Since we usually assume low phenocopy rate, i.e., $f_{\mathrm{aa}} \approx 0$, the HWDc $\epsilon_1 \propto -F^2$ is negative. If the mutant allele A is enriched in case samples ($\delta = p_1 - p_2 > 0$), with the $\delta_\epsilon < 0$ in dominant models, we conclude that genotype-based test is favored over allele-based tests. For recessive models, $f_{\mathrm{Aa}} \approx f_{\mathrm{aa}} \approx 0$, $\epsilon_1 \propto 0$, so the allele-based test is better. For additive models, $f_{\mathrm{Aa}} = f_{\mathrm{aa}} + \Delta$, $f_{\mathrm{AA}} = f_{\mathrm{aa}} + 2\Delta$, where Δ is the contribution to the penetrance by adding one copy of the mutant allele. The δ_ϵ is equal to $\epsilon_1 \propto (f_{\mathrm{aa}} + 2\Delta) f_{\mathrm{aa}} - (f_{\mathrm{aa}} + \Delta)^2 = -\Delta^2 < 0$. Thus genotype-based test is favored for additive disease models.

7. Discussion and future works

The main point of this article is that genotype-based test may take advantage of certain Hardy-Weinberg disequilibrium in case samples to overcome the advantage of larger sample sizes in allele-based tests. Another advantage of the genotype-based test is that it tests two models and picks the best one. This multiple testing might be corrected by multiplying the p-value by a factor of 2 (Bonferroni corrections), which was not done in this article. Whether correcting multiple testing or not is always under debate[14,2,15], but its effect on our problem is probably to shift the phase boundary slightly.

The X^2 test statistic calculation in this article was all carried out assuming equal number of samples in case and control group. Changing this assumption to unequal number of samples per group is not difficult, but its effect on the conclusion has not been examined.

Here we are addressing the type-I error of the test, the p-value, which is determined by the X^2 test statistic. For type-II error under alternative hypothesis, usually a non-central χ^2 distribution could be used[13]. However, other alternatives to non-central χ^2 distribution to calculate type-II error and the power have been proposed[5].

Acknowledgments

W.L. acknowledges the support from The Robert S. Boas Center for Genomics and Human Genetics at the Feinstein Institute for Medical Research.

References

1. A. Agresti. *Categorical Data Analysis* (Wiley-Interscience, 2002).
2. M. Aickin. Other method for adjustment of multiple testing exists. BMJ, 318:127, 1999.
3. A.B. Begovich, et al., A missense single-nucleotide polymorphism in a gene encoding a protein tyrosine phosphatase (PTPN22) is associated with rheumatoid arthritis. *Am. J. Human Genet.*, 75:330-337, 2004.

4. A.J. Brookes. The essence of SNPs. *Gene*, 234:177-186, 1999.
5. J. Bukszar, E.J. van den Oord. Accurate and efficient power calculations for $2 \times m$ tables in unmatched case-control designs. *Stat. in Med.*, 25:2623-2646, 2006.
6. P.R. Burton, M.D. Tobin, J.K. Happer. Key concepts in genetic epidemiology. *Lancet*, 366:941-951, 2005.
7. H.J. Cordell, D.G. Clayton. Genetic association studies. *Lancet*, 366:1121-1131, 2005.
8. J.N. Feder, et al., A novel MHC class I-like gene is mutated in patients with hereditary haemochromatosis. *Nature Genet.*, 13:399-408, 1996.
9. A.T. Lee, W. Li, A. Liew, C. Bombardier, M. Weisman, E.M. Massarotti, J. Kent, F. Wolfe, A.B. Begovich, P.K. Gregersen. The PTPN22 R620W polymorphism associates with RF positive rheumatoid arthritis in a dose-dependent manner but not with HLA-SE status. *Genes and Immunity*, 6:129-133, 2005.
10. W.C. Lee. Searching for disease-susceptibility loci by testing for Hardy-Weinberg disequilibrium in a gene bank of affected individuals. *Am. J. Epidemiology*, 158:397-400, 2003.
11. W. Li, edited, Bibliography: linkage disequilibrium analysis URL: *http://www.nslij-genetics.org/ld/*
12. E.M. Lifshitz, L.D. Landau. *Statistical Physics: Course of Theoretical Physics, Volume 5*, 3rd edition (Butterworth-Heinemann, 1980).
13. P.B. Patnaik. The non-central χ^2- and F-Distribution and their applications. *Biometrika*, 36:202-232, 1949.
14. T.V. Perneger. What's wrong with Bonferroni adjustments. BMJ, 316:1236-1238, 1998.
15. T.V. Perneger. Adjusting for multiple testing in studies is less important than other concerns. BMJ, 318:1288, 1999.
16. N. Risch, K. Merikangas. The future of genetic studies of complex human diseases. *Science*, 273:1516-1517, 1996.
17. P.D. Sasieni. From genotypes to genes: doubling the sample size. *Biometrics*, 53:1253-1261, 1997.
18. S. Tokuhiro, et al. An intronic SNP in a RUNX1 binding site of SLC22A4, encoding an organic cation transporter, is associated with rheumatoid arthritis. Nature Genet., 35:341-348, 2003.
19. B.S. Weir. *Genetic Data Analysis II* (Sinauer Associates, 1996).
20. R. Yamada, et al., Association between a single-nucleotide polymorphism in the promoter of the human interleukin-3 gene and rheumatoid arthritis in Japanese patients, and maximum-likelihood estimation of combinatorial effect that two genetic loci have on susceptibility to the disease. *Am. J. Hum. Genet.*, 68:674-685, 2001.

A PROBABILISTIC METHOD TO IDENTIFY COMPENSATORY SUBSTITUTIONS FOR PATHOGENIC MUTATIONS

B. C. EASTON AND A. V. ISAEV

Department of Mathematics,
Mathematical Sciences Institute,
Australian National University,
Canberra, ACT 0200, Australia
E-mail: Brett.Easton@maths.anu.edu.au

G. A. HUTTLEY AND P. MAXWELL

Computational Genomics Laboratory,
John Curtin School of Medical Research,
Australian National University,
Canberra, ACT 0200, Australia

Complex systems of interactions govern the structure and function of biomolecules. Mutations that substantially disrupt these interactions are deleterious and should not persist under selection. Yet, several instances have been reported where a variant confirmed as pathogenic in one species is fixed in the orthologs of other species. Here we introduce a novel method for detecting compensatory substitutions for these so-called compensated pathogenic deviations (CPDs), incorporating knowledge of pathogenic variants into a probabilistic method for detecting correlated evolution. The success of this approach is demonstrated for 26 of 31 CPDs observed in mitochondrial transfer RNAs and for one in beta hemoglobin. The detection of multiple compensatory sites is demonstrated for two of these CPDs. The methodology is applicable to comparative sequence data for biomolecules expressed in any alphabet, real or abstract. It provides a widely applicable approach to the prediction of compensatory substitutions for CPDs, avoiding any reliance on rigid non-probabilistic criteria or structural data. The detection of compensatory substitutions that facilitate the substitution of otherwise pathogenic variants offers valuable insight into the molecular constraints imposed on adaptive evolution.

1. Introduction

Recent studies on the adaptive evolution of pathogens with resistance to chemical agents have reported striking examples of correlated mutation. In the absence of the chemical agent, resistance persists in some populations despite evidence that the resistance conferring mutation comes at a fitness cost in such an environment. These studies point to the rapid succession of secondary mutations at other loci to, at least partially, ameliorate this reduction in fitness.[1–7] A similar compensatory process has been observed in the Australian sheep blowfly (*Lucilia cuprina*), with the fitness cost of resistance to an organophosphate insecticide reduced by mutation at a second locus.[8] Further evidence of such dependence has been given by the observation of several biomolecules containing compensated pathogenic deviations (CPDs) – mutations known to be pathogenic in one species (typically humans), but occurring naturally in the orthologs of other species.[9–12] Such variants necessitate the presence of additional compensatory mutations in either the same, or else functionally-related, biomolecules.

These results allude to a key role for compensatory mutations in molecular adaptation. Several methods for detecting these and other instances of correlated evolution have been suggested,[13–24] many introducing novel models of multi-site evolution. With ever increasing efforts made to understand the genetic components of disease, knowledge on pathogenic variants of human sequences has uncovered a new resource of information to such studies.[9] This information was used to develop a list of criteria to identify compensatory sites for several CPDs observed in mammalian protein sequences[9] (similar criteria also suggested by Ref. 10). Another approach specific to RNA molecules validates predictions based on secondary structure, with the associated variations in free energy.[11]

The success of these approaches confirms the value of information on pathogenic variants, but their reliance on rigid criteria and structural data limit their wider utility. We conjectured that prediction could be achieved solely from models of correlated substitution that incorporate knowledge of pathogenic variants. In this paper, we confirm this hypothesis with the identification of compensatory sites for several CPDs observed in mammalian mitochondrial transfer RNAs (mt tRNAs) and for one CPD observed in beta hemoglobin (HBB), through a method supported by a novel model for dependent evolution.

2. Materials and Methods

Calculations were implemented in PyEvolve version 0.86 alpha[25]. The sequence alignments and phylogenies are available from

http://jcsmr.anu.edu.au/org/dmb/compgen/publications.php.

2.1. *Data*

For each mt tRNA we used published[11] alignments for 106 mammalian species, along with secondary structure annotation[26]. We eliminated gaps by removing a combination of columns and sequences. Mitomap[27] provides a list of point mutations that are pathogenic in the human mitochondrial genome. From this we selected 31 mutations – each found to disrupt a Watson-Crick pair in a stem of a mt tRNA, and observed in at least two species (using the above alignments, with gaps removed). Amino acid sequences for mammalian HBB were obtained from Swiss-Prot[28]. These were aligned with ClustalW version 1.8[29] and partial sequences were deleted, leaving 123 sequences. Columns containing gaps were removed. In the human sequence the substitution V20E is documented as pathogenic.[30]

2.2. *Reducing and recoding the data*

We applied the same procedure to each of the 32 CPDs mentioned above, but for explanatory purposes we describe the method for V20E in HBB. Among the 123 full sequences, we observe the pathogenic variant E in six species – *Suncus murinus*, *Ceratotherium simum*,

Rhinoceros unicornis, *Equus hemionus kulan*, *Equus caballus* and *Equus zebra*. For these species we conjecture that some form of compensation has occurred and that the relevant substitutions are expressed in the alignment. By contrast, in the 104 sequences with a V at the site of the CPD we expect such compensatory variants to be less common. For the other species with neither V nor E at the site of the CPD, we have no expectation on the frequency of these variants and therefore removed the sequences from the alignment, leaving 110 sequences. Each site was then recoded according to an abstract two-state alphabet, with variants designated as either potentially compensatory (ρ) or non-compensatory (η). This recoding highlights those transitions that may be important, and ensures the method is applicable to sequence data expressed in any alphabet, real or abstract.

At each site a variant is deemed potentially compensatory if it accompanies the CPD in one or more species, and is not found in *H. sapiens* (since the CPD is pathogenic in this species). For example, at site 50 we see the variants $\{A, N, S, T\}$, with T found in humans and $\{N, S, T\}$ observed in species with the CPD. Therefore at site 50, $\{N, S\}$ are recoded as potentially compensatory and $\{A, T\}$ are recoded as non-compensatory. In *Ovis aries musimon* and *Dasypus novemcinctus* an IUPAC ambiguity was found at some sites. Ambiguities in the raw alignment were left as ambiguities when recoded, unless all of the variants they represented were recoded to a single state (in which case this state was used). The maximum likelihood procedure sums over all possibilities when dealing with ambiguities. For branch length estimation, at site 20 we recoded V as η and E as ρ.

We constructed a rooted phylogeny for the species represented in the alignment by following a selection of the current literature[31–41] and the Tree of Life Web Project (http://tolweb.org/tree). We sampled all lineages descended from the most recent common ancestor of humans and of those species with the CPD. Further to this, we removed one species from each pair of identical (in the recoded alignment) sequences with the same immediate ancestor. These steps were taken to reduce the computational burden.

Since we measure dependence for pairs of sites, and because estimates of branch lengths would be unreliable for such pairs, we obtained the branch lengths from the entire alignment. For this we assumed that sites were independent and identically distributed, with evolution at each site (in the recoded alphabet) occurring under Felsenstein's model[42] (F81). We set the stationary motif probabilities from their relative frequencies in the alignment (but note that we do not use these values later, to measure dependence). Since the dynamics of rate heterogeneity have not been investigated in the abstract alphabet we supposed, for simplicity, that sites evolved under the same constant rate (fixed to unity). The branch lengths were deduced as the set of maximum likelihood parameter estimates. Since F81 is a reversible Markov model, this method does not resolve the position of the root between its two immediate descendants (which is needed for the dependent model).[42] This was input as a free parameter to the dependent model.

2.3. *Identifying compensatory sites*

For each site with a variant recoded as potentially compensatory, we coupled this site to the site of the CPD and measured the level of dependence shown in their evolution. This was done by scaling a likelihood ratio (LR) statistic, calculated as the ratio of the maximum likelihood of the data under a dependent model $L(D)$ with that under a nested independent model $L(I)$, i.e. $LR = \frac{L(D)}{L(I)}$. The highest ranked site was predicted as the most likely location of compensatory substitution. Combining the approach to recoding (and possibly the novel scaling described below) with any procedure capable of detecting dependent evolution in a two-site alignment (with each site expressed in an abstract two-state alphabet) produces a similar methodology. We defer an analysis of such hybrid methods to future studies.

2.3.1. *A model for dependent evolution*

Let $P(t)$ denote the matrix of transition probabilities for a single independently evolving site. For this we use the analogue of F81 applied to our abstract alphabet. That is,

$$P_{ij}(t) = e^{-ut} \cdot \delta_{ij} + (1 - e^{-ut}) \cdot \pi_j, \tag{1}$$

where π_j is the stationary motif probability of j, u is the rate of substitution per unit time (which we henceforth assume is 1), and δ_{ij} is the Kronecker delta function ($\delta_{ij} = 1$ if $i = j$, and 0 otherwise). Denote by α the site of the CPD and by β the potentially compensatory site. We will use superscripts α and β to indicate site-specific values.

To specify evolution at α and β along a tree, we require the joint motif probabilities (that is, the probability of each pair of motifs at the ancestral root) and the subsequent transition probabilities. We denote the former by π_{AB} (A at α, B at β) and the latter by $P_{A\mapsto a,B\mapsto b}(t)$ (AB to ab on a branch of length $t \geq 0$). If α and β evolve independently, then their evolution is given by

$$\begin{aligned} \pi_{AB} &= \pi_A^\alpha \cdot \pi_B^\beta \\ P_{A\mapsto a,B\mapsto b}(t) &= P_{Aa}^\alpha(t) \cdot P_{Bb}^\beta(t), \end{aligned} \tag{2}$$

with the motif probabilities at α and β provided as free parameters, subject to the constraint that they each partition 1. To account for dependence at the ancestral root we instead allow the joint motif probabilities as free parameters, and deduce the motif probabilities at α and β by

$$\pi_A^\alpha = \pi_{A\rho} + \pi_{A\eta} \quad \text{and} \quad \pi_B^\beta = \pi_{\rho B} + \pi_{\eta B}. \tag{3}$$

Similarly, we account for dependence in the transition probabilities by allowing each as a free parameter. This may be formulated equivalently by scaling the transition probabilities of the independent model with free non-negative parameters $C_{AaBb}(t)$, so that

$$P_{A\mapsto a,B\mapsto b}(t) = P_{Aa}^\alpha(t) \cdot P_{Bb}^\beta(t) \cdot C_{AaBb}(t). \tag{4}$$

However, with this high degree of freedom the application of the model to any data will result in severe over-fitting. Accordingly we assume that each $C_{AaBb}(t)$ may be decomposed into the product of two parameters: the first $F_{AB}(t)$, to encapsulate the effect of any interaction between the initial states on the subsequent transitions; and the second $\frac{\pi_{ab}}{\pi_a^\alpha \cdot \pi_b^\beta}$, provided as a measure of the impact of dependence on the frequency of the destination motifs ab. It then follows from the constraint that the transition probabilities partition 1, that

$$F_{AB}(t) = \frac{1}{\sum_{a,b \in \{\rho,\eta\}} P^{\alpha}_{Aa}(t) \cdot P^{\beta}_{Bb}(t) \cdot \frac{\pi_{ab}}{\pi^{\alpha}_{a} \cdot \pi^{\beta}_{b}}}. \quad (5)$$

By Eqns. (3) and (5) the free parameters in the dependent model are just the joint motif probabilities, substantially reducing the possibility of over-fitting. This model of dependent evolution avoids a number of biologically unjustified assumptions that are commonly imposed for computational convenience. It does not assume that the distribution of joint motif probabilities is stationary and as such is non-reversible. The model is also non-Markovian in the sense that we do not impose the constraint that $P(s+t) = P(s) \cdot P(t)$.

2.3.2. *Scaling the LR statistic*

The raw LR statistic for a two-site alignment provides some measure of the extent of dependence governing the evolution of the two sites. It is however sensitive to site-specific factors such as the frequency of motifs at each site. We used instead a scaled statistic, defined as the probability of a lower LR value if we replace the motif at one site in one species with the alternative in the recoded alphabet. As a simple example to illustrate the benefit of scaling, consider the alignments summarized in Table 1 for species related by a tree with infinite branch lengths (this is a theoretical approximation to long branches). An alignment comprised exclusively of the motif pairs $\rho\rho$ and $\eta\eta$ shows perfect dependence. This is observed using scaled dependence values for alignments 1 to 8. If we vary the frequencies of $\rho\rho$ and $\eta\eta$ it seems reasonable to suggest that those in which both are common provide better support for the prediction of dependence. However, all of the values taken by these alignments should exceed the values of alignments in which $\eta\rho$ and/or $\rho\eta$ are also seen (alignments 9 to 13). With scaling this condition is satisfied, but without scaling high frequencies of $\rho\rho$ and $\eta\eta$ obscure the impact of introducing a mismatch pair – alignments 9 to 13 do not consistently rank lower than 1 to 8 when the raw statistic is used.

3. Results and Discussion

3.1. *CPDs in mt tRNAs*

The pathology of any point mutation which disrupts a Watson-Crick pair in a stem can be attributed to the destabilization of this stem. Any compensatory substitution should stabilize the corresponding stem and the success of our method can be measured by its recognition of such substitutions. With the exception of G7497A the method provided sufficient resolution to predict a single site as the most probable location of a compensatory substitution. This highest ranking site (i.e., the predicted compensatory site) for each CPD is shown in Table 2. For 26 of the CPDs the predicted compensatory site (one such site for G7497A) was located in the same stem. In 25 instances this site was found to be complementary to the CPD and among the most frequent substitutions accompanying the CPD at this site, we find a substitution which provides a Watson-Crick interaction in place of the disturbed pair.

Table 1. The impact of scaling.

Alignment	Motif pairs				Scaled	
	$\rho\rho$	$\rho\eta$	$\eta\rho$	$\eta\eta$	Yes	No
1	8	0	0	8	1.0000	11.0904
2	9	0	0	7	1.0000	10.9650
3	10	0	0	6	1.0000	10.5850
4	11	0	0	5	1.0000	9.9374
5	12	0	0	4	1.0000	8.9974
6	13	0	0	3	1.0000	7.7212
7	14	0	0	2	1.0000	6.0283
8	15	0	0	1	1.0000	3.7407
9	8	0	1	7	0.9375	7.9509
10	8	0	2	6	0.8750	6.0863
11	8	0	3	5	0.8125	4.6449
12	8	0	4	4	0.7500	3.4522
13	8	1	1	6	0.8750	4.9547

Note: Above are the dependence values of selected 2-site alignments of 16 taxa related by a star topology with infinite branch lengths; counts of motif pairs are provided in lieu of alignments.

An advantage of our probabilistic approach over criteria-based methods is that we avoid the binary classification of sites as compensatory or otherwise. Instead we have a measure of the likelihood that each site contains a compensatory substitution. As an example we look to G5540A which disrupts a GC pair on the fourth rung of the anticodon stem of mt tRNA Trp. The secondary structure[26] of mt tRNA Trp is given in Figure 1 along with the anticodon stems of the three species in which the mutation is observed. Only in *Ursus americanus* is there compensation at the complementary site 5550. Therefore the CPD must be accompanied by another compensation in *Ceratotherium simum* and *Balaenoptera acutorostrata*. In these species the sequences of the anticodon stem are equivalent and differ from that of *H. sapiens* only by the substitutions A5539G and G5540A, corresponding to the gain and loss of Watson-Crick interactions on rungs three and four respectively. It has been suggested therefore that the presence of a G at 5539 compensates for G5540A.[11] The two highest ranking potential compensatory sites for G5540A were 5550 and 5539 (results not shown), indicating that both compensatory mechanisms were detected.

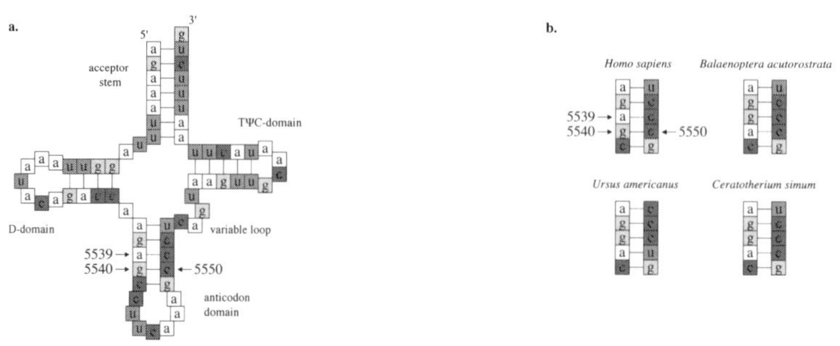

Figure 1. Mitochondrial tRNA Trp. (a) Secondary structure in *H. sapiens*. (b) Anticodon stem in selected species.

Table 2. CPDs in mt tRNAs.

CPD	Frequency		mt tRNA	Potential	Predicted
x site y	x	y		compensatory sites	compensatory site(s)
T582C	92	8	F	25	641**
A606G	25	76	F	34	618**
A608G	99	2	F	13	627
T618C	24	77	F	33	606**
G1606A	61	42	V	39	1665**
C1624T	100	2	V	22	1625
G1642A	19	84	V	37	1604
C3256T	98	4	L1	21	3239**
T3258C	100	2	L1	15	3274**
T3271C	95	7	L1	25	3261**
A3280G	100	2	L1	15	3245
C3303T	82	19	L1	29	3230**
G4298A	90	7	I	14	4297*
G4309A	18	79	I	23	4321**
G5540A	100	3	W	15	5550**
T5628C	86	8	A	29	5620**
G5703A	11	86	N	43	5687**
T5814C	72	3	C	14	5810**
G7497A	4	95	S1	46	7471 & 7503**
T7510C	85	15	S1	34	7451**
T7512C	12	88	S1	44	7449**
A7543G	97	7	D	34	7555**
G8342A	68	17	K	27	8352**
T8355C	67	31	K	35	8339**
T8356C	11	87	K	41	8338**
G8361A	26	70	K	40	8297**
T9997C	87	13	G	30	10051**
G10014A	35	61	G	42	10030**
G12147A	95	2	H	16	12159**
G12183A	83	12	H	36	12197**
A15924G	77	23	T	37	15899

Note: The columns from left to right are: the CPD; the frequency of the motif found in humans and the motif pathogenic to humans, in the alignment with gaps removed (prior to the subsequent recoding and reduction); the mt tRNA containing the CPD; the number of sites identified as potentially compensatory by recoding; and, the highest ranking site(s). Sites are numbered according to Mitomap[27].

* contained in the same stem as the CPD.

** complementary to the CPD.

Multiple compensatory mechanisms were detected also for G4309A. This mutation disrupts a GC pair on the third rung of the TΨC stem in mt tRNA Ile, the secondary structure[26] of which is given in Figure 2. The motif U at the complementary site 4321 restores this rung and 4321 is the highest ranking potential compensatory site. For 11 species containing the CPD however, no such compensation is observed. In these species the sequences for the TΨC stem differ from that of *H. sapiens* only by the substitutions G4309A and A4310G, resulting in the loss and gain of GC pairs on rungs three and four respectively. This suggests that the absence of a GC interaction on the third rung is compensated for in these species by the presence of a GC pair on the fourth rung. Site 4310 is ranked second among the potential compensatory sites (results not shown), with G the single motif identified by recoding as potentially compensatory at this site.

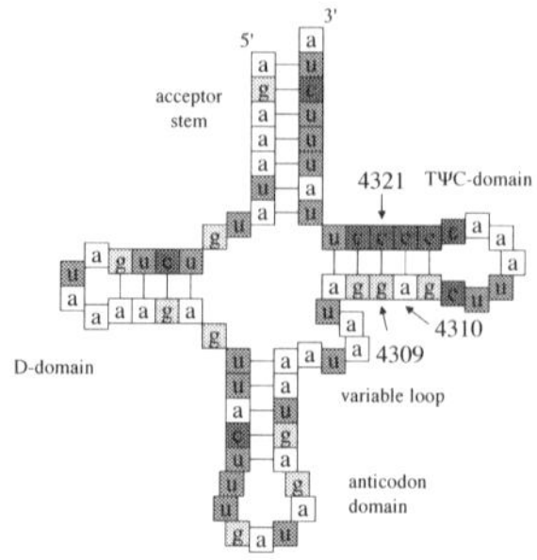

Figure 2. Mitochondrial tRNA Ile. Secondary structure in *H. sapiens*.

3.2. *V20E in HBB*

The method predicted 69 as the most likely location of a compensatory substitution for V20E in HBB. In the alignment with gaps removed the motifs observed at the site-pair (20,69) are $\{(\text{A,D}) : 2, (\text{A,H}) : 3, (\text{A,T}) : 1, (\text{E,H}) : 6, (\text{I,T}) : 1, (\text{I,V}) : 3, (\text{L,T}) : 1, (\text{P,N}) : 1, (\text{Q,N}) : 1, (\text{V,A}) : 5, (\text{V,B}) : 1, (\text{V,D}) : 16, (\text{V,G}) : 25, (\text{V,H}) : 1, (\text{V,N}) : 37, (\text{V,Q}) : 2, (\text{V,S}) : 7, (\text{V,T}) : 10\}$. Removing those species with neither V nor E at 20, this reduces to $\{(\text{E,H}) : 6, (\text{V,A}) : 5, (\text{V,B}) : 1, (\text{V,D}) : 16, (\text{V,G}) : 25, (\text{V,H}) : 1, (\text{V,N}) : 37, (\text{V,Q}) : 2, (\text{V,S}) : 7, (\text{V,T}) : 10\}$ and on recoding 69 it reduces further to $\{(\text{V},\eta) : 103, (\text{V},\rho) : 1, (\text{E},\rho) : 6\}$. This aptly illustrates the benefit of recoding, eliminating the impact on our analysis of any transitions between the eight motifs classified as η. Given this recoded dataset the prediction of 69 as the location of a compensatory substitution for V20E is not surprising. In recoding the only motif identified at 69 as potentially compensatory was H, hence the substitution G69H is the expected compensatory substitution (noting that G is found at 69 in *H. sapiens*). This is consistent with the results of Ref. 9, who confirmed predictions based on non-probabilistic criteria with structural data.

4. Conclusions

A complex system of interactions gives rise to the structure and function of most biomolecules. Variants that disrupt such interactions putatively become fixed in species only where a compensatory mutation occurs. Most methods introduced to detect these compensatory substitutions are limited by their reliance on structural data and/or use of rigid non-probabilistic criteria. The approach presented herein avoids these restrictions and is applicable to comparative sequence data on any single biomolecule, or functionally-related system of biomolecules. That the method is applicable to both nucleotide and amino acid sequences was demonstrated with the detection of compensatory sites for 26 of 31 CPDs contained in mammalian orthologs of mt tRNAs and for V20E in HBB. The detection of secondary compensatory sites was demonstrated for two of these CPDs. To our knowledge, the method is the first of its kind for which success has been verified in applications to both nucleotide and amino acid data.

Acknowledgements

This research used facilities provided by the Australian Partnership for Advanced Computing.

References

1. A. M. Borman, S. Paulous and F. Clavel. Resistance of human immunodeficiency virus type 1 to protease inhibitors: selection of resistance mutations in the presence and absence of the drug. *J. Gen. Virol.*, 77:419–26, 1996.
2. S. J. Schrag and V. Perrot. Reducing antibiotic resistance. *Nature*, 381:120–1, 1996.
3. J. Björkman, D. Hughes and D. I. Andersson. Virulence of antibiotic-resistant *Salmonella typhimurium*. *Proc. Natl. Acad. Sci. USA*, 95:3949–53, 1998.
4. J. Björkman, I. Nagaev, O. G. Berg, D. Hughes et al. Effects of environment on compensatory mutations to ameliorate costs of antibiotic resistance. *Science*, 287:1479–82, 2000.
5. B. R. Levin, V. Perrot and N. Walker. Compensatory mutations, antibiotic resistance and the population genetics of adaptive evolution in bacteria. *Genetics*, 154:985–97, 2000.
6. M. G. Reynolds. Compensatory evolution in rifampin-resistant *Escherichia coli*. *Genetics*, 156:1471–81, 2000.
7. I. Nagaev, J. Björkman, D. I. Andersson and D. Hughes. Biological cost and compensatory evolution in fusidic acid-resistant *Staphylococcus aureus*. *Mol. Microbiol.*, 40:433–9, 2001.
8. P. Batterham, A. G. Davies, A. Y. Game and J. A. McKenzie. Asymmetry–where evolutionary and developmental genetics meet. *Bioessays*, 18:841–5, 1996.
9. A. S. Kondrashov, S. Sunyaev and F. A. Kondrashov. Dobzhansky-Muller incompatibilities in protein evolution. *Proc. Natl. Acad. Sci. USA*, 99:14878–83, 2002.
10. L. Gao and J. Zhang. Why are some human disease-associated mutations fixed in mice? *Trends Genet.*, 19:678–81, 2003.
11. A. D. Kern and F. A. Kondrashov. Mechanisms and convergence of compensatory evolution in mammalian mitochondrial tRNAs. *Nat. Genet.*, 36:1207–12, 2004.
12. R. J. Kulathinal, B. R. Bettencourt and D. L. Hartl. Compensated deleterious mutations in insect genomes. *Science*, 306:1553–4, 2004.
13. W. P. Maddison. A method for testing the correlated evolution of two binary characters: are gains or losses concentrated on certain branches of a phylogenetic tree? *Evolution*, 44:539–557, 1990.
14. M. Pagel. Detecting correlated evolution on phylogenies: a general method for the comparative analysis of discrete characters. *Proc. R. Soc. Lond. B*, 255:37–45, 1994.
15. I. N. Shindyalov, N. A. Kolchanov and C. Sander. Can three-dimensional contacts in protein structures be predicted by analysis of correlated mutations? *Protein Eng.*, 7:349–58, 1994.
16. S. V. Muse. Evolutionary analyses of DNA sequences subject to constraints on secondary structure. *Genetics*, 139:1429–39, 1995.
17. S. W. Lockless and R. Ranganathan. Evolutionarily conserved pathways of energetic connectivity in protein families. *Science*, 286:295–9, 1999.
18. D. D. Pollock, W. R. Taylor and N. Goldman. Coevolving protein residues: maximum likelihood identification and relationship to structure. *J. Mol. Biol.*, 287:187–98, 1999.
19. V. R. Akmaev, S. T. Kelley and G. D. Stormo. Phylogenetically enhanced statistical tools for RNA structure prediction. *Bioinformatics*, 16:501–12, 2000.
20. S. T. Kelley, V. R. Akmaev and G. D. Stormo. Improved statistical methods reveal direct interactions between 16S and 23S rRNA. *Nucleic Acids Res.*, 28:4938–43, 2000.
21. P. Tufféry and P. Darlu. Exploring a phylogenetic approach for the detection of correlated substitutions in proteins. *Mol. Biol. Evol.*, 17:1753–9, 2000.

22. K. R. Wollenberg and W. R. Atchley. Separation of phylogenetic and functional associations in biological sequences by using the parametric bootstrap. *Proc. Natl. Acad. Sci. USA*, 97:3288–91, 2000.
23. M. W. Dimmic, M. J. Hubisz, C. D. Bustamante and R. Nielsen. Detecting coevolving amino acid sites using Bayesian mutational mapping. *Bioinformatics*, 21 Suppl 1:i126–35, 2005.
24. J. Dutheil, T. Pupko, A. Jean-Marie and N. Galtier. A model-based approach for detecting coevolving positions in a molecule. *Mol. Biol. Evol.*, 22:1919–28, 2005.
25. A. Butterfield, V. Vedagiri, E. Lang, C. Lawrence et al. PyEvolve: a toolkit for statistical modelling of molecular evolution. *BMC Bioinformatics*, 5:1, 2004.
26. M. Helm, H. Brule, D. Friede, R. Giege et al. Search for characteristic structural features of mammalian mitochondrial tRNAs. *RNA*, 6:1356–79, 2000.
27. M. C. Brandon, M. T. Lott, K. C. Nguyen, S. Spolim et al. MITOMAP: a human mitochondrial genome database – 2004 update. *Nucleic Acids Res.*, 33:D611–613, 2005.
28. B. Boeckmann, A. Bairoch, R. Apweiler, M.-C. Blatter et al. The SWISS-PROT protein knowledgebase and its supplement TrEMBL in 2003. *Nucleic Acids Res.*, 31:365–70, 2003.
29. J. D. Thompson, D. G. Higgins and T. J. Gibson. CLUSTAL W: improving the sensitivity of progressive multiple sequence alignment through sequence weighting, position-specific gap penalties and weight matrix choice. *Nucleic Acids Res.*, 22:4673–80, 1994.
30. B. Landin, S. Berglund and B. Lindoff. Hb Trollhättan [β 20(B2)Val→Glu] - a new haemoglobin variant with increased oxygen affinity causing erythrocytosis. *Eur. J. Haematol.*, 53:21–5, 1994.
31. M. A. Cronin, R. Stuart, B. J. Pierson and J. C. Patton. K-casein gene phylogeny of higher ruminants (Pecora, Artiodactyla). *Mol. Phylogenet. Evol.*, 6:295–311, 1996.
32. M. Hasegawa, J. Adachi and M. C. Milinkovitch. Novel phylogeny of whales supported by total molecular evidence. *J. Mol. Evol.*, 44:S117–20, 1997.
33. M. Robinson, F. Catzeflis, J. Briolay and D. Mouchiroud. Molecular phylogeny of rodents, with special emphasis on murids: evidence from nuclear gene LCAT. *Mol. Phylogenet. Evol.*, 8:423–34, 1997.
34. O. R. Bininda-Emonds, J. L. Gittleman and A. Purvis. Building large trees by combining phylogenetic information: a complete phylogeny of the extant Carnivora (Mammalia). *Biol. Rev. Camb. Philos. Soc.*, 74:143–75, 1999.
35. D. Huchon, F. M. Catzeflis and E. J. Douzery. Molecular evolution of the nuclear von Willebrand factor gene in mammals and the phylogeny of rodents. *Mol. Biol. Evol.*, 16:577–89, 1999.
36. A. D. Yoder and J. A. Irwin. Phylogeny of the Lemuridae: effects of character and taxon sampling on resolution of species relationships within *Eulemur*. *Cladistics*, 15:351–361, 1999.
37. W. J. Murphy, E. Eizirik, W. E. Johnson, Y. P. Zhang et al. Molecular phylogenetics and the origins of placental mammals. *Nature*, 409:614–8, 2001.
38. H. Amrine-Madsen, M. Scally, M. Westerman, M. J. Stanhope et al. Nuclear gene sequences provide evidence for the monophyly of australidelphian marsupials. *Mol. Phylogenet. Evol.*, 28:186–96, 2003.
39. C. J. Douady and E. J. P. Douzery. Molecular estimation of eulipotyphlan divergence times and the evolution of "Insectivora". *Mol. Phylogenet. Evol.*, 28:285–96, 2003.
40. Y. Murata, M. Nikaido, T. Sasaki, Y. Cao et al. Afrotherian phylogeny as inferred from complete mitochondrial genomes. *Mol. Phylogenet. Evol.*, 28:253–60, 2003.
41. M. A. Nilsson, A. Gullberg, A. E. Spotorno, U. Arnason et al. Radiation of extant marsupials after the K/T boundary: evidence from complete mitochondrial genomes. *J. Mol. Evol.*, 57:S3–S12, 2003.
42. J. Felsenstein. Evolutionary trees from DNA sequences: a maximum likelihood approach. *J. Mol. Evol.*, 17:368–76, 1981.

EXPLORING GENOME REARRANGEMENTS USING VIRTUAL HYBRIDIZATION

M. BELCAID[1], A. BERGERON[2], A. CHATEAU[3], C. CHAUVE[2,4], Y. GINGRAS[2], G. POISSON[1] AND M. VENDETTE[2]

1) Information and Computer Sciences, University of Hawaii at Mānoa, USA
2) Comparative Genomics Laboratory, Université du Québec à Montréal, Canada
3) Laboratoire d'informatique, robotique et microélectronique de Montpellier, France
4) Department of Mathematics, Simon Fraser University, Vancouver, Canada

Genomes evolve with both mutations and large scale events, such as inversions, translocations, duplications and losses, that modify the structure of a set of chromosomes. In order to study these types of large-scale events, the first task is to select, in different genomes, sub-sequences that are considered "equivalent". Many approaches have been used to identify equivalent sequences, either based on biological experiments, gene annotations, or sequence alignments. These techniques suffer from a variety of drawbacks that often result in the impossibility, for independent researchers, to reproduce the datasets used in the studies, or to adapt them to newly sequenced genomes. In this paper, we show that carefully selected small probes can be efficiently used to construct datasets. Once a set of probes is identified – and published –, datasets for whole genome comparisons can be produced, and reproduced, with elementary algorithms; decisions about what is considered an occurrence of a probe in a genome can be criticized and reevaluated; and the structure of a newly sequenced genome can be obtained rapidly, without the need of gene annotations or intensive computations.

1. Introduction

The study of genome rearrangements started at the beginning of the last century when evidence of inversions of large segments of DNA were actually observed by Dobzhansky and Sturtevant in the chromosomes of *Drosophila pseudoobscura* [8]. Their technique, which is best described as *visual hybridization* of paired homologous rearranged chromosomes, yielded the first dataset that could be used to infer phylogenetic relationship between species using "gene" order. In that study, the word "gene" referred to sections of chromosomes and were identified by a combination of numbers and letters.

Since then, numerous techniques have been developed to compare the structure of genomes of different species. Biological experiments, such as chromosome painting [16], or hybridization with probes [11], are costly and lengthy procedures that are no longer necessary with sequenced genomes.

For well-annotated genomes, the straightforward approach of detecting whether a given species has a certain gene works only for the most elementary DNA molecules, such as animal mitochondrial genomes [4]. In bacterial genomes, for example, gene fusions lead either to the elimination of valuable information, or to the aberrant fusion of distinct gene families [13].

A way to circumvent this problem is to work directly with raw sequences, bypassing the annotation step: whole genomes are compared against each other, and the genomes of each species are cut into large blocks of "conserved synteny" [5]. This usually requires large computational resources, and if new species are added to the study, the computation must be started over again.

The main problems with these various techniques are thus the technical and financial difficulties of reproducing independently the datasets, and of including new sequenced genomes in existing dataset, or even "revised" genomes (this is the case for genome assembly projects which represent an ongoing process, and where assembly errors can easily be interpreted as large scale rearrangements [3]). It would thus be extremely valuable to have a simple and efficient method to generate datasets for the study of whole genome rearrangements.

In this paper, we propose a technique of *virtual hybridization* based on sets of small probes – up to a few hundred nucleotides –, whose presence(s), absence, order and orientation can be quickly and accurately determined in a given genome. We give two explicit sets of probes, one for the mammalian chromosome X, and one for the chloroplast genomes.

2. Virtual Hybridization

Approximate string matching is defined as identifying, in a text, substrings that are *similar* to a given string p. In biological applications, the text is typically a genomic sequence, and similarity is defined by scoring possible alignments between s and p. Numerous algorithms and scoring schemes are available to identify approximate occurrences of short sequences in genomic sequences, the best known being the BLAST [1] heuristic and variations of the Smith-Waterman algorithm [15].

In the following, *probes* refer to short sequences of nucleotides, and *virtual hybridization* refers to the detection of occurrences of these probes in a genomic sequence. We detect an occurrence of a probe p in a genomic sequence if there exists an alignment between a substring p' of p and a substring s of the sequence that with at least I % identity and such that the length of p' is at least L % of the length of p, with default values $I = 80$ and $L = 80$.

Given a chromosome C, and a set P of probes, the result of a virtual hybridization experiment is a signed sequence $p_1 p_2 \ldots p_n$ which gives the order and orientation of the occurrences of probes of P in chromosome C. A probe can have more than one occurrence, or be absent from a given chromosome.

2.1. *Probe Selection*

The construction of a set of probes can be done in several different ways, the easiest being the use of already identified sets of markers common to different species. This approach is used in Section 3 in order to construct sets of probes for the mammalian chromosomes X. A alternate approach is described in Section 4 in which we present a software tool that can assist the selection procedure.

The two approaches to probe selection are first based on a multiple alignment of a small set of genomes, called *reference genomes*, in which probes are selected. The selected probes can then be hybridized with genomes different from the reference genomes. For example, in the chromosome X study, the reference genomes are the human, mouse and rat assemblies used in [5]. The set of probes was then used to analyze rearrangements in the dog and Rhesus monkey chromosomes X.

Any method of probe selection implies a series of choices that can be discussed and revised. However, once a set of probes is fixed, the information obtained in the comparison of genomes is easily and completely reproducible. The sets of probes discussed in this paper, and software to generate datasets, are available at cgl.bioinfo.uqam.ca/vhybridization.

2.2. *Probe Usefulness*

Genome rearrangement studies are all ultimately based on datasets that are signed sequences of markers. These markers can be genes, introns, exons, domains, probes or larger segments of DNA. An *occurrence* of a marker in a genome is specified by its start and end points, and its orientation (+ or -). We assume that the markers are non-overlapping in each genome in the study. The *dataset* D of the study is thus a set of signed sequences corresponding to the order and orientation of occurrences of the markers in various chromosomes.

Definition 2.1. A set P of probes is *useful* with respect to a given study if the dataset D of the study can be reconstructed using virtual hybridization.

We say that a probe – or its reverse complement – *detects* a marker m in a chromosome if: 1) It has exactly one occurrence within each occurrence of m, and the orientation of both occurrences are equal. 2) It has no occurrences outside of occurrences of m.

In order to prove that a set of probes can reconstruct the dataset of a study, it suffices to show that each marker of the study is detected by at least one probe. Given a set of n different probes that detect a set of n different markers, if $C = (m_1, m_2, \ldots\ldots, m_k)$ is a sequence in dataset D that describes a chromosome, then the virtual hybridization of the set of n probes on this chromosome will yield the sequence C. In Figure 1, for example, the set of probes {a, e, h} can be used to reconstruct the sequence $(m_1, m_2, m_3, -m_1, m_3)$, while capturing more rearrangements.

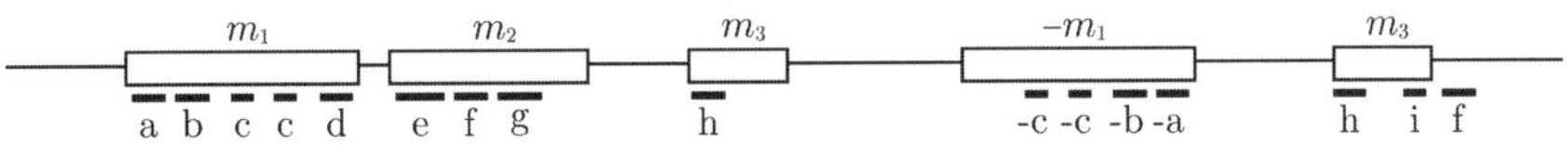

Figure 1. An example of the relations between small probes, in black, and larger markers, in white. The subset of probes {a, e, h} can be used to reconstruct the order of the markers.

For the chromosome X study, we constructed three different sets of probes, all of which can be used to reconstruct the order of synteny blocks of [5]. However, since genome assembly are often revised, we could apply the virtual hybridization procedure to the most recent assemblies of the three reference genomes, even if the probes were constructed using the older assemblies.

For chloroplast genomes, we made sure that each annotated gene of the chloroplast of *Arabidopsis thaliana* was detected by at least one probe. As a result, we can reconstruct the datasets of studies such as [7], that use the set of annotated genes common to chloroplasts.

Adapting datasets to revised genomes, or reconstructing existing datasets is a first step. The real challenge is to be able to identify rearrangements in newly sequenced genomes, or genomes that are different from the reference genomes. We will discuss some aspects of this problem in the next section.

3. From Chromosome X Anchors to Sets of Probes

In order to develop sets of probes to investigate rearrangements in the mammalian chromosomes X, we used the set of 12866 three-way anchors identified in the comparison of the human, mouse and rat chromosomes X [5]. We selected the corresponding sequences in the human chromosome X (Apr. 2003 assembly), and retained only anchors that were longer than 75 nucleotides. This initial set of probes was hybridized against the most recent assemblies of the human (Mar. 2006), mouse (Feb. 2006) and rat (Nov. 2004) chromosomes X. With a threshold of 80% identity over 80% of the length of the probes, the initial set of anchors was reduced to 1593, after duplicate and missing hits were removed. This set of probes is called P-1593 in the following experiments.

The 1593 probes define three signed permutations that exhibit 100 conserved segments, meaning that, in all three reference chromosomes, the order and orientation of these segments are conserved. In each of these conserved segments, we chose the probe that had maximal percentage of identity with the mouse genome. The resulting set of probes is called P-100. The average length of the probes in P-100 is 277, ranging from 76 to 1548 nucleotides.

Finally, we repeated the above selection process with a threshold of 70% identity over 70% of the length of the probes, yielding 6858 probes common to the three reference genomes, that regrouped into 334 conserved segments. Again, we chose the probes that had maximal percentage of identity with the mouse genome, yielding the set of probes P-334.

We first investigated how these sets of probes captured the rearrangements of the three reference chromosomes compared to the 16 synteny blocks defined in [5]. Table 1 shows that even the set P-100 captures much more rearrangements than the 16 blocks. Note that the distances are equal for the sets P-100 and P-1593, which is a consequence of how the set P-100 was constructed.

The distances obtained in Table 1 for the three sets of probes are similar to distances that take into account both macro and micro rearrangements [5]. Lowering the threshold to 70% identity over 70% of the length predictably increases the inversion distance, since the permutations obtained by hybridization with the P-334 set have more than three times the number of breakpoints of the permutations obtained by hybridization with the P-100 set.

Using the three sets of probes, we next hybridized the dog (Jul. 2005) and Rhesus

Table 1. Inversion distances between reference chromosomes according to different sets of probes.

Pair of species	16 synteny blocks	P-1593	P-100	P-334
Human and mouse	10	33	33	115
Human and rat	10	59	59	166
Mouse and rat	10	45	45	134

monkey (Jan. 2006) chromosomes X, with the same thresholds that were used in the construction of the probes. In each experiment, about a third of the probes were not found either in the dog or the Rhesus chromosomes X. Since these are still draft assemblies, we did not investigate further the missing probes. Table 2 gives the inversion distance between pairs of genomes with respect to each of the three sets of probes.

Table 2. Inversion distances between pairs of species according to different sets of probes.

Pair of species	P-1593	P-100	P-334
Human and Rhesus	3	2	4
Human and dog	14	5	18
Rhesus and dog	13	3	14

Interestingly, for each experiment, detected rearrangements were all non-overlapping inversions. It was also possible to assign each inversion to a specific lineage. Table 2 raises some questions on the size and construction of the set of probes. Clearly, the method of selection of the P-100 set has a considerable impact in assessing the rearrangements of the dog compared to the primates. For such comparisons, the set P-1593 seems more appropriate, since some of the conserved segments between the human and rodents appear to have been broken in the dog lineage.

4. Ab-initio Probes for Chloroplast Genomes

A second project was to obtain a set of probes for chloroplast chromosomes. Given the relatively small size of these sequences, we used a semi-automated approach that relies on visual inspection. We first identified a set of *candidate probes* using global alignments of the non-duplicated regions of the references chloroplast chromosomes of Table 3.

Table 3. Reference chloroplast chromosomes.

Species	Accession	Sequence (gi)
Arabidopsis thaliana	NC_000932	7525012
Calycanthus floridus	NC_004993	32480822
Pinus thunbergii	NC_001631	7524593
Triticum aestivum	NC_002762	14017551
Adiantum capillus	NC_004766	30352011
Psilotum nudum	NC_003386	18860289
Huperzia lucidula	NC_006861	60117151
Chaetosphaeridium globosum	NC_004115	22711893

A global alignment was obtained with MultiPipMaker [14]. The sequence of *Arabidopsis thaliana* was chosen as base sequence for the multiple alignment, which explains that most of the probes belong to the *Arabidopsis thaliana* chloroplast genome.

The resulting alignment was parsed using a visualization software called PipViewer. This software tool provides a representation of the multiple alignment with a color gradient, from red to green, standing respectively for low to good score. We developed PipViewer to quickly display large portions of a multiple alignment, and to select and mark blocks of contiguous nucleotides in the base sequence. When a block s is selected, PipViewer computes the virtual hybridization scores of s on the remaining sequences.

A good score is a non-ambiguous answer to the question "Does the probe hybridize at this place in the considered species ?". The first two columns of Table 4 show an example of a candidate probe of length 186 that hybridizes well with all species except *Pinus thunbergii*. The last two columns show an example of a rejected candidate probe of length 286: percentages of identity between 55 % and 70 % are considered ambiguous and yield to the rejection of the candidate probe.

Table 4. Examples of accepted and rejected candidate probes.

	Accepted candidate ($l = 186$)		Rejected candidate ($l = 286$)	
Genome	% Identity	% Probe length	% Identity	% Probe length
Arabidopsis thaliana	100.0	100.0	100.0	100.0
Calycanthus floridus	93.0	99.5	80.9	100.0
Pinus thunbergii	92.1	54.3	68.4	43.5
Triticum aestivum	93.5	100.0	79.4	100.0
Adiantum capillus	81.7	100.0	65.8	99.2
Psilotum nudum	82.7	99.5	67.6	100.0
Huperzia lucidula	91.9	100.0	71.8	100.0
Chaetosphaeridium globosum	88.2	100.0	75.8	100.0

Additional probes were added to this initial set to cover annotated genes of the reference chromosomes that were not detected by the initial set of candidates. The resulting set of candidate probes had 212 elements.

The second phase of the selection procedure was to eliminate overlapping candidates. We used the containment clustering algorithm implemented in ICAass [12] to detect total or partial containment between probes. Members of each cluster were hybridized on the eight reference chromosomes, and the most specific probe was selected. The resulting set of probes has currently 160 elements, ranging from 65 bp to 288 bp, with average length 144 bp. Table 5 gives the number of occurrences of probes in each of the 8 reference chromosomes. Note that a probe can have more than one occurrence, thus the total number of occurrences can be greater than 160.

4.1. *Investigating Rearrangements in Chloroplast Inverted Repeats*

In most chloroplast chromosomes, the presence of a large inverted repeat, with variable gene content, is a challenge to current models of genome rearrangements. One of our main

Table 5. Hits of the 160 probes on the reference chromosomes.

Genome	Single hits	Double hits (×2)	Triple hits (×3)	Total
Arabidopsis thaliana	110	34	0	178
Calycanthus floridus	115	31	0	177
Pinus thunbergii	102	6	0	114
Triticum aestivum	96	29	2	160
Adiantum capillus	40	14	0	68
Psilotum nudum	74	19	0	112
Huperzia lucidula	87	16	0	119
Chaetosphaeridium globosum	52	13	0	78

goal in developing the virtual hybridization technique was to create a common dataset to study these types of rearrangements.

Chloroplast chromosomes are usually depicted as circular molecules divided in 4 regions (Fig. 2): a long single copy (LSC), a short single copy (SSC), and two repeated regions (IRa and IRb). For example, in the *Arabidopsis thaliana* chloroplast chromosome, the two repeated regions have 100 % identity over 26264 bp long. However, there is ample evidence [2] that chloroplast chromosome molecules exist in many other configurations, such as the right part of Figure 2.

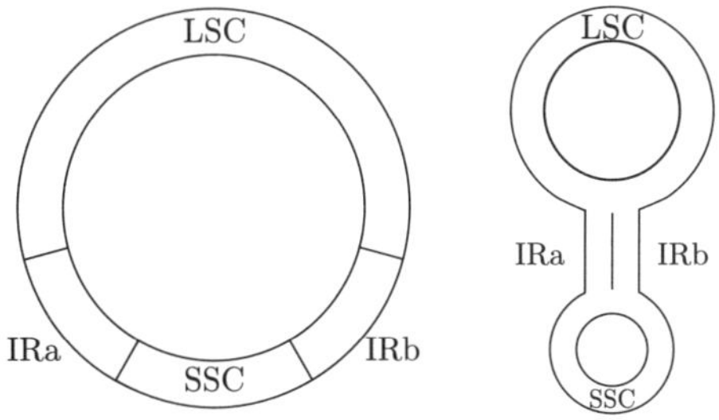

Figure 2. Chloroplast chromosomes are often depicted as round molecules divided in 4 regions: LSC, SSC, IRa, and IRb. Regions IRa and IRb are the exact inverted Watson-Crick complement of each other, thus the gene content of a chloroplast chromosome can be analyzed using the configuration on the right hand side.

Among the 160 probes, 23 of them cover the SSC region and parts of the neighboring IR region. This subset is particularly suitable to study rearrangements that occur in the IRa-SSC and SSC-IRb junctions. Table 6 gives the addresses of these 23 probes, together with a one letter code that will allow us to represent the order of these probes.

Figure 3 gives the linear order of the 23 probes in seven chloroplast chromosomes, illustrating the complex dynamics of rearrangements around the SSC region. These rearrangements cannot be explained by the classic models of inversions, tandem duplications and losses in linear sequences.

However, using a representation similar to the right-hand side of Figure 2, the relative order of the 23 probes can be compared with stem-loop diagrams (Fig. 4) that show the "Ebb and flow of the chloroplast inverted repeat" [9] in a very clear way.

Table 6. Addresses of the 23 probes that span the SSC region.

Code	Sequence (gi)	Address	Code	Sequence (gi)	Address
A	14017551	98822-98897	B	7525012	123386-123510
C	7525012	123051-123169	D	22711893	99559-99740
E	30352011	125951-126108	F	7525012	111571-111716
G	7525012	112035-112194	H	7524593	106257-106331
I	7525012	114271-114351	J	32480822	115274-115356
K	14017551	106037-106135	L	7525012	116136-116278
M	7525012	117387-117565	N	7525012	117964-118080
P	7525012	119384-119489	Q	7525012	120115-120388
R	7525012	120856-120989	S	7525012	121481-121619
T	7525012	122013-122151	U	7525012	122648-122796
V	30352011	126930-127041	W	7525012	127115-127231
X	18860289	107775-107877			

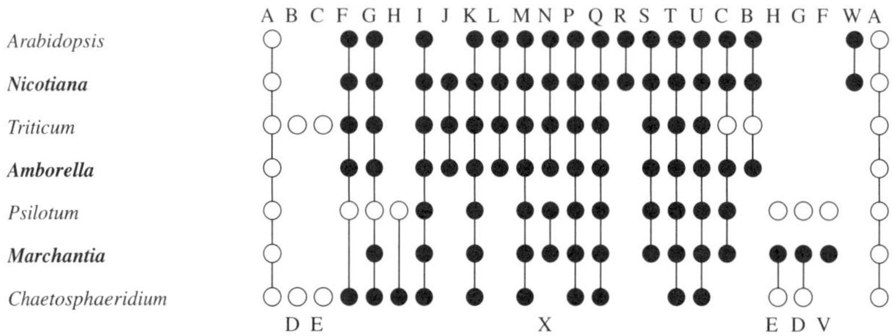

Figure 3. Linear order of 23 probes on 7 chloroplast chromosomes, showing numerous gains and losses. The SSC region of each chromosome is represented by black dots, white dots represent probes that belong to the inverted repeat region. Names in bold indicate species that are not in the reference chromosomes.

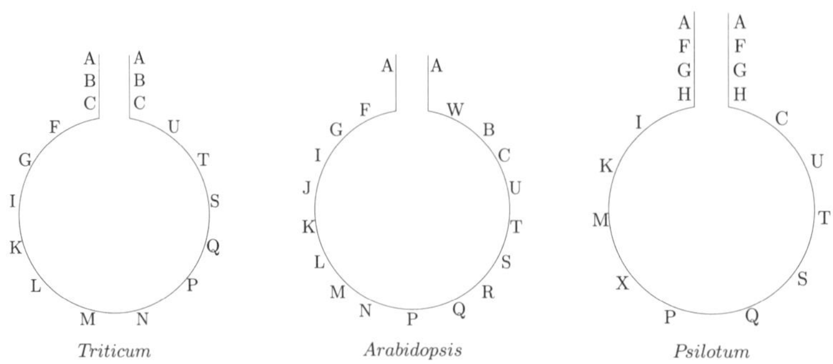

Figure 4. Respective order of probes of the SSC region and part of the inverted repeat of *Triticum*, *Arabidopsis*, and *Psilotum*. Probes B and C slip from the inverted repeat of *Triticum* to the right of SSC region of *Arabidopsis*, while probes F and G slip from the inverted repeat of *Psilotum* to the left of SSC region of *Arabidopsis*.

5. Conclusion

This paper presented a new approach to the construction of datasets used in genome rearrangement studies. We began to develop this approach when it became clear that it was extremely difficult to share or to reproduce the data used in published papers. Many decisions must be made when producing permutations or sequences that compare gene orders in different species. Our goal was to be able to give, in a compact way, all the tools necessary to reproduce our experiments.

Chloroplast chromosomes are small, and we could do probe selection and validation using elementary tools. The corresponding set of probes seems to be able to capture most of the rearrangements occurring in chloroplasts.

Chromosomes X, on the other hand, are huge molecules. Consequently, rearrangements occur at very different scales. A small set of probes, such as P-100, can be used to detect large scale rearrangements in species that are close to the reference genomes. However, we saw that such a set of probes becomes insufficient to analyze rearrangements in farther species such as the dog, and that the set P-1593 was more adequate. The influence of the phylogenetic spectrum spanned by both the reference genomes used to select probes, and the analyzed genomes, seems then to be an issue that should be addressed, in particular the question of when a new set of probes needs to be constructed for the current set of genomes.

In this work, we did not consider bacterial genomes. However, the nature of evolutionary events that affect them – duplications, lateral transfer and gene losses in particular – induces many non trivial gene families. The analysis of bacterial gene orders is thus challenging (see [6]), and involves sophisticated algorithms. It would be interesting to use the principle of virtual hybridization, instead of all-against-all comparisons of protein sequences, with such genomes.

References

1. Altschul, S.F., Madden, T.L., Schaffer, A.A., Zhang, J., Zhang, Z., Miller, W., Lipman, D.J.: Gapped BLAST and PSI-BLAST: a new generation of protein database search programs. *Nucleic Acids Research* 25:3389–3402, 1997.
2. Bendich, A.: Circular Chloroplast Chromosomes: The Grand Illusion. *The Plant Cell*, Vol 16, 1661–1666, 2004.
3. Bérard, S., Bergeron, A., Chauve, C.: Conserved structures in evolution scenarios. *Comparative Genomics RECOMB 2004 Workshop, LNCS/LNBI*, 3388:1–15, 2005.
4. Boore, J.L.: Animal mitochondrial genomes. *Nucleic Acids Research*, 27(8):1767–1780, 1999.
5. Bourque, G., Pevzner, P.A., Tesler, G.: Reconstructing the genomic architecture of ancestral mammals: Lessons from human, mouse, and rat genomes. *Genome Research*, 14(4):507–516, 2004.
6. Blin, G, Chateau, A., Chauve, C., Gingras, Y. Inferring positional homologs with common intervals of sequences. *Comparative Genomics RECOMB 2006 Workshop, LNCS/LNBI*, 4205:24–38, 2006.
7. Cui, L., Yue, F., dePamphilis, C., Moret, B.M.E., and Tang, J.: Inferring ancestral chloroplast genomes with inverted repeat. *Proceedings of the 2006 International Conference on Bioinformatics and Computational Biology (Biocomp'06)*, Las Vegas, 75–81, 2006.
8. Dobzhansky, T., Sturtevant, A.T.: Inversions in the Chromosomes of *Drosophila pseudoobscura*. *Genetics*, 23:28–64, 1938.
9. Goulding, S.E., Olmstead, R.G., Morden, C.W., Wolfe, K.H.: Ebb and flow of the chloroplast inverted repeat. *Molecular and General Genetics*,252:195–206, 1996.
10. Matsuoa, M., Itob, Y., Yamauchib, R., Obokataa, J.: The Rice Nuclear Genome Continuously

Integrates, Shuffles, and Eliminates the Chloroplast Genome to Cause ChloroplastNuclear DNA Flux. *The Plant Cell*, 17:665–675, 2005.
11. Olmstead, R.G., Palmer, J.D.: A chloroplast DNA phylogeny of the Solanaceae: subfamilial relationships and character evolution. *Annals of the Missouri Botanical Garden*, 79:346–360, 1992.
12. Parsons, J.D.: Improved Tools for DNA Comparison and Clustering. *Computer Applications in the Biosciences*, 11:603–613, 1995.
13. Pasek, S., Bergeron, A., Risler, J.-L., Louis, A., Ollivier, E., Raffinot, M.: Identification of genomic features using microsyntenies of domains: domain teams. *Genome Research*, 15(6):867–74, 2005.
14. Schwartz, S., Elnitski, L., Li, M., Weirauch, M., Riemer, C., Smit, A., Green, E.D., Hardison, R.C., Miller, W.: MultiPipMaker and supporting tools: alignments and analysis of multiple genomic DNA sequences. *Nucleic Acids Research*, 31(13):3518–3524, 2003.
15. Smith, T.F., Waterman, M.S.: Identification of Common Molecular Subsequences. *Journal of Molecular Biology*, 147:195–197, 1981.
16. Speicher, M.R., Ballard, S.G., Ward, D.C.: Karyotyping human chromosomes by combinatorial multi-fluor FISH. *Nature Genetics*, 12:368–376, 1996.
17. Wolf, P.G., Karol, K.G., Mandoli, D.F., Kuehl, J., Arumuganathan, K., Ellis, M.W., Mishler, B.D., Kelch, D.G., Olmstead, R.G., Boore, J.L.: The first complete chloroplast genome sequence of a lycophyte, Huperzia lucidula (Lycopodiaceae). *Gene*, 350(2):117–28, 2005.

TWO PLUS TWO DOES NOT EQUAL THREE: STATISTICAL TESTS FOR MULTIPLE GENOME COMPARISON

NARAYANAN RAGHUPATHY[1*], ROSE HOBERMAN[2*] AND DANNIE DURAND[1,2]

[1] *Dept. of Biological Sciences, Carnegie Mellon University, Pittsburgh, PA*

[2] *Computer Science Department, Carnegie Mellon University, Pittsburgh, PA*

E-mail: {rnarayan, roseh, durand}@cmu.edu

Gene clusters that span three or more chromosomal regions are of increasing importance, yet statistical tests to validate such clusters are in their infancy. Current approaches either conduct several pairwise comparisons, or consider only the number of genes that occur in all the regions. In this paper, we provide statistical tests for clusters spanning exactly three regions based on genome models of typical comparative genomics problems, including analysis of conserved linkage within multiple species and identification of large-scale duplications. Our tests are the first to combine evidence from genes shared among all three regions and genes shared between pairs of regions. We show that our tests of clusters spanning three regions are more sensitive than existing approaches and can thus be used to identify more diverged homologous regions.

1. Introduction

An essential task in comparative genomics is to identify chromosomal regions that descended from a single ancestral region, either through speciation or duplication. Conserved homologous regions can be used to find evidence of functional selection or shared regulatory regions, and to analyze the history of large-scale duplications and rearrangements. In distantly related genomes, homologous genes are used as markers for identifying homologous regions. Gene content and order, although initially conserved, will diverge through local rearrangements, gene loss, and duplications[5]. Thus, distantly related homologous regions appear as *gene clusters*, distinct chromosomal regions that share a number of homologous gene pairs, where neither gene order nor gene content is perfectly preserved.

In order to distinguish regions that arose from the same ancestral region from unrelated regions that share homologous gene pairs, it is necessary to show that local similarities in gene content could not have occurred by chance. There is an emerging body of work on statistical tests for this purpose[2,3,4,9,10,14,15,18]. However, this work focuses almost exclusively on tests for comparisons of two regions. With the rapid rate of whole genome sequencing, analysis of gene clusters that span three or more chromosomal regions is of increasing interest.

When comparing two regions, the number of shared homologs (x, shown in Fig. 1(a)) is typically used as the measure of similarity. However, this approach cannot be directly

*these authors have contributed equally to this work

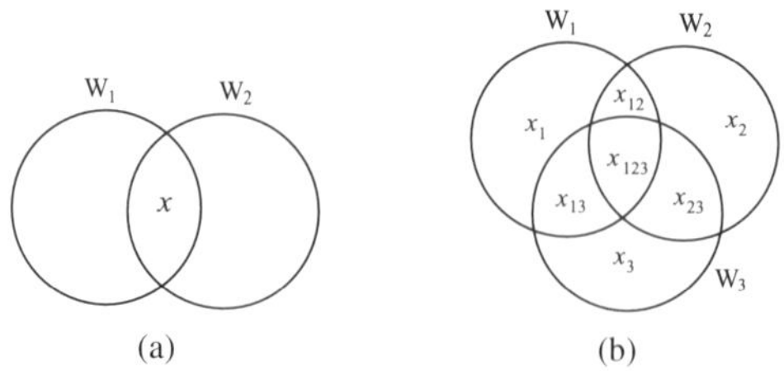

Figure 1. Venn diagram representation of shared homologs in windows sampled from distinct chromosomal regions. (a) Pairwise comparison of windows, W_1 and W_2, which share x homologous genes. (b) Three-way comparison of W_1, W_2, and W_3, in which x_{123} homologs appear in all three windows. The variables x_{ij} represent the number of genes that appear in only W_i and W_j, and x_i represents the number of genes that appear in only a single window W_i.

extended for tests of clusters spanning more than two regions. When comparing three regions (W_1, W_2, and W_3), there are many more quantities to consider (Fig. 1(b)): the number of homologs observed in all three regions (x_{123}), the number of homologs observed in each pair of regions (x_{12}, x_{13} and x_{23}), and the number of genes observed only in a single window (x_1, x_2, and x_3). Evidence for homology comes not only from the set of homologs that appear in *all* the regions being compared (x_{123}), but also from the number of homologs that appear in only a subset of the regions ($x_{ij}'s$). How best to combine evidence from different subsets of regions remains an unsolved problem.

In this paper, we develop the first attempt to address this issue, for the problem of clusters spanning exactly three regions. Given a set of three windows sampled from three genomes, each containing r consecutive genes, we wish to determine whether the windows share more homologous genes than expected by chance. (If duplications are under consideration, two of the windows will be sampled from distinct regions of a single genome.) This problem, while restricted to three regions, exihits the basic challenges that arise in the more general problem.

Statistical tests for gene clusters in multiple regions may be useful either because the researcher is studying more than two genomic regions or because comparison with additional genomes may increase confidence that a pair of regions arose from a single ancestral region. To identify regions duplicated in a whole genome duplication (WGD), in particular, comparisons with related genomes may be necessary. Although evidence of WGD can sometimes be found by comparing a genome with itself and looking for pairwise clusters, in many cases duplicated regions may not be identifiable by direct comparison due to *reciprocal gene loss*: Following a WGD, in many cases there is no immediate selective advantage for retaining a gene in duplicate, so one copy of most duplicates is lost. As a result, the gene content of duplicated regions is often disjoint.

A solution to this problem is comparison with the genome of a closely related species that diverged shortly before the whole genome duplication (a *pre-duplication* species). If two regions in the *post-duplication* species both have significant similarity to a single region in the pre-duplication species, they are likely to be homologous even if they share few or no homologous genes. This strategy provides more statistical power to detect duplicated regions and has been successfully employed to analyze duplications in fish[6], plants[8,16,17] and several yeast species[7,11].

The most common strategy for testing significance of multiple regions is to conduct multiple pairwise comparisons (reviewed by Simillion *et. al.*[12]). If region W_1 is significantly similar to W_2, and W_2 is significantly similar to region W_3, then homology between all three regions is inferred, even if W_1 and W_3 share few genes. This approach allows the use of existing statistical methods, which are designed for comparing two regions. However, this strategy is conservative as it will only identify a three-way cluster if at least two of the three pairwise comparisons are independently significant. Furthermore, it does not explicitly recognize the additional significance of genes that occur in all three regions.

In a second approach, once a significantly similar pair of regions is identified, the genes in these regions are merged to approximate their common ancestral region[12]. Then additional pairwise comparisons are conducted with this inferred ancestral segment as the search query. This approach still allows the use of pairwise statistical tests, but is more powerful than the above approach, since the second step considers the genes that occur in W_1 as well as those that occur in W_2 when searching for a third homologous region. However, it still requires that at least one pair of regions is independently significant. Moreover, when comparing with a third region, W_3, it does not consider the additional significance of genes that appear in W_1 *and* W_2, compared to genes that appear in only one of the regions.

The previous two approaches use sequential pairwise comparisons. Another model has been proposed that allows for *simultaneous* comparison of multiple regions[3]. However this model only considers x_{123}, the number of genes that are conserved in all regions. This approach is also conservative as it does not consider genes that occur in only a subset of the regions (the $x_{ij}'s$). Thus, current approaches account for either the genes that occur in all three regions, or those that occur in pairs of regions, but not both.

In this paper, we develop the first statistical tests that consider both the quantities x_{123} *and* x_{ij} simultaneously. We obtain expressions for the probability—under the null hypothesis of random gene order—that the number of shared genes is at least as large as the number observed. These expressions are derived for genome models that are appropriate for two common types of comparative genomics problems: (1) analyses of conserved linkage of genes in three regions from three genomes, and (2) identification of segments duplicated by a whole genome duplication, via comparison with the genome of a related, pre-duplication species. We show through simulations that our tests for comparing three regions are more sensitive than existing approaches, and have the potential to detect more diverged homologous regions.

2. Statistical Tests for Three Regions

The significance of a cluster depends not only on properties of the windows (Fig. 1), but also on the properties of the genomes (Fig. 2). The relevant properties of the genomes are the total number of genes in each genome and the *gene content overlap*— the fraction of genes shared among the three genomes. Depending on which biological questions are being investigated, the processes of gene loss differ, and an appropriate model of gene con-

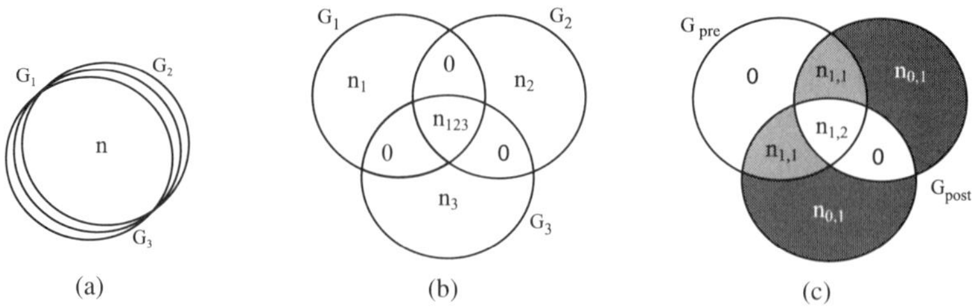

Figure 2. Gene content overlap models. The set of genes in each genome is represented as a circle. (a) Identical gene content model: all genes are shared between all three genomes. (b) Shared gene content model: n_{123} genes are shared between all three genomes. The remaining genes are singletons. (c) Pre/post duplication model: G_{pre} is the union of two ancestral, duplicated genomes embedded within it. $n_{1,2}$ genes appear twice in G_{post} (once in each embedded genome) and once in G_{pre}. These are the genes that are retained in duplicate. $n_{1,1}$ genes appear once in G_{pre} and once in G_{post}. These are the genes that were preferentially lost. $n_{0,1}$ genes appear once in G_{post} but do not appear in G_{pre}. These are the genes retained in singleton in G_{post} but lost in G_{pre}.

tent overlap will also differ. Here, we develop statistical tests for three different models of gene content overlap. The first two models are designed for comparisons of three genomes, while the third is for detection of duplicated regions by comparison with a pre-duplication genome. For each model we give analytical expressions for three statistical tests, and compute cluster probabilities for typical parameter values using Mathematica. We investigate the impact of different gene content overlap models and alternative test statistics on cluster significance, and compare the sensitivity of our tests with that of existing approaches.

2.1. *Identical Gene Content Model*

We model a genome G_i as an ordered set of N_i genes, $G_i = 1, 2, \ldots N_i$. We ignore chromosome breaks and physical distance between genes, and assume genes do not overlap. In this, the simplest model, each genome contains n identical genes, *i.e.*, $n = N_1 = N_2 = N_3$ (Fig. 2(a)). Each gene in genome G_i has exactly one homolog each in G_j and G_k.

In order to determine the significance of gene clusters, we require test statistics that capture the essential properties of the clusters of interest. In the pairwise case, given a pair of chromosomal regions containing x observed homologs, significance is typically demonstrated by showing that $P(X \geq x)$ is small under the null hypothesis, where X is a random variable representing the number of homologs shared between the two regions. This probability can be computed using a combinatorial approach, counting the number of ways the two windows can be filled with genes, such that they share at least x genes, and normalizing by the number of ways of filling the windows without restrictions.

We illustrate this approach for the simpler case of a pairwise cluster, then present analytical expressions for the probabilities of three-region clusters under the null hypothesis. Given two windows, W_1 and W_2 of size r_1 and r_2, sampled from two genomes containing n identical genes, the number of ways the windows can share *exactly* x genes is $\binom{n}{x}\binom{n-x}{r_1-x}\binom{n-r_1}{r_2-x}$. The first binomial is the number of ways of choosing the x shared genes, and the remaining two binomials give the number of ways of choosing two sets of genes to fill the remainder of each window, such that the sets are disjoint. We normalize by the total number of ways of choosing genes to fill two windows of size r_1 and r_2 is $\binom{n}{r_1}\binom{n}{r_2}$. Thus, the probability that these windows share *exactly* x genes is[3]

$$P_2(X=x) = \frac{\binom{n}{x}\binom{n-x}{r_1-x}\binom{n-r_1}{r_2-x}}{\binom{n}{r_1}\binom{n}{r_2}} = \frac{\binom{n}{x,r_1-x,r_2-x}}{\binom{n}{r_1}\binom{n}{r_2}}, \tag{1}$$

where we define[a]

$$\binom{n}{i_1, i_2, ..., i_k} \equiv \binom{n}{i_1}\prod_{j=1}^{k-1}\binom{n-\sum_{l=1}^{j} i_l}{i_{j+1}} = \frac{n!}{i_1!i_2!\ldots(n-i_1-i_2\ldots-i_k)!}.$$

Thus, the probability that two windows share *at least* x genes is

$$P_2(X \geq x) = \sum_{h=x}^{r} P_2(X=h). \tag{2}$$

We use an analogous approach and notation for computing the probabilities for comparisons of three regions. In addition, we define $\vec{x} = (x_{123}, x_{12}, x_{13}, x_{23})$ and use $\vec{X} = \vec{x}$ as shorthand for $X_{123} = x_{123}$, $X_{12} = x_{12}$, $X_{13} = x_{13}$, and $X_{23} = x_{23}$. As above, we first derive an expression for the probability of observing exactly $\vec{x}$ genes, then sum over this expression to find the probability of observing at least as many shared genes.

In the above pairwise comparison, we counted the number of ways to form three different sets: the x shared genes, the $r_1 - x$ genes unique to W_1, and the $r_2 - x$ genes unique to W_2. Computing the probability of three windows containing *exactly* the observed number of shared genes is a direct extension of the two-window problem, except there are seven sets to be selected (Fig. 1(b)) instead of three sets:

$$P_3(\vec{X}=\vec{x}) = \frac{1}{\binom{n}{r_1}\binom{n}{r_2}\binom{n}{r_3}} \cdot \binom{n}{x_{123},\ x_{12},\ x_{13},\ x_{23},\ x_1,\ x_2,\ x_3}. \tag{3}$$

The probability of observing *at least* $\vec{x}$ shared genes is obtained by summing over all possible values of X_{123} and X_{ij},

$$P_3(\vec{X} \geq \vec{x}) = \sum_{v_{123}=x_{123}}^{u_{123}} \sum_{v_{12}=x_{12}}^{u_{12}} \sum_{v_{13}=x_{13}}^{u_{13}} \sum_{v_{23}=x_{23}}^{u_{23}} P_3(\vec{X} = \vec{v}), \tag{4}$$

where $u_{123} = \min(r_1, r_2, r_3)$, $u_{12} = \min(r_1, r_2) - v_{123}$, $u_{13} = \min(r_1 - v_{12}, r_3) - v_{123}$, $u_{23} = \min(r_2 - v_{12}, r_3 - v_{13}) - v_{123}$, and $\vec{v} = (v_{123}, v_{12}, v_{13}, v_{23})$. In the worst case, evaluating this expression takes $O(r^4)$ time. In practice, the computation time is substantially reduced, because the summand decreases exponentially as x_{123} and the $x_{ij}'s$ increase. Only the smallest values will contribute to the final probability, and most of the terms can be disregarded.

It might seem natural to use the probability of observing the *exact* number of shared homologs directly to test cluster significance. However, such an approach is risky. As shown in Fig. 3(a), for small values of x_{ij}, $P(\vec{X}=\vec{x})$ underestimates $P(\vec{X} \geq \vec{x})$ by several

[a]Note that this is a non-standard use of the multinomial notation since we do not require that $n = i_1 + i_2 + \ldots i_k$.

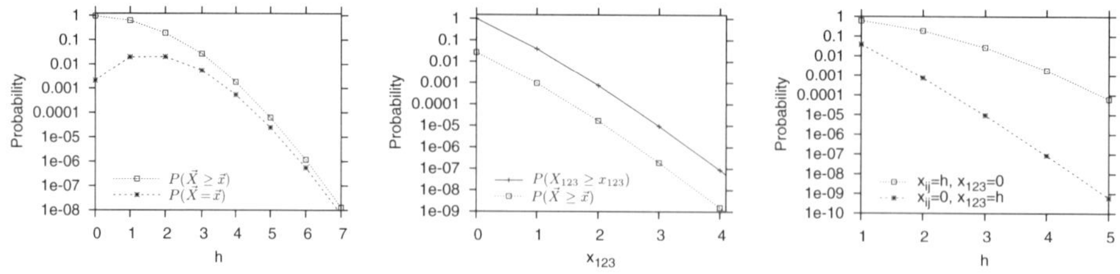

(a) (b) (c)

Figure 3. (a) A comparison of $P(\vec{X} \geq \vec{x})$ with $P(\vec{X} = \vec{x})$ for $n = 5000$, $r = 100$, $x_{123} = 0$, and $x_{12} = x_{13} = x_{23} = h$, as h ranges from zero to seven. (b) A comparison of $P(X_{123} \geq x_{123})$ with $P(\vec{X} \geq \vec{x})$ for $n = 5000$, $r = 100$, $x_{12} = x_{13} = x_{23} = 3$, as x_{123} ranges from zero to four.(c) A comparison of $P(\vec{X} \geq (h, 0, 0, 0))$ and $P(\vec{X} \geq (0, h, h, h))$, showing the impact of x_{123} and $x_{ij}'s$ on cluster significance, when $n = 5000$, $r = 100$.

orders of magnitude. For example, given the parameters in Fig. 3(a), when the three regions share *no* genes ($x_{123} = x_{ij} = 0$), the exact test reports a probability significantly less than one! This test will lead to false positives. However, as x_{ij} increases, the probabilities converge. This suggests that, for sufficiently large values of x_{ij}, the exact probability may be used as a fast approximation.

In order to assess the additional sensitivity gained by incorporating genes that are shared between only two of three regions into the statistical test, we compare $P(\vec{X} \geq \vec{x})$ with $P(X_{123} \geq x_{123})$, the probability of observing at least x_{123} homologs shared between all three windows. To ensure that all three windows share *exactly* x_{123} genes with no restrictions on the $x_{ij}'s$, it is necessary to select x_{12}, x_{13} and x_{23} so that they have no homologs in common. Otherwise, X_{123} would be greater than rather than equal to x_{123}. This can be achieved using the following expression for the number of windows that share *exactly* x_{123} genes:

$$q(X_{123} = x_{123}) = \sum_{x_{12}=0}^{r_1 - x_{123}} \binom{r_1}{x_{123}, x_{12}} \binom{n - r_1}{r_2 - x_{123} - x_{12}} \binom{n - x_{123} - x_{12}}{r_3 - x_{123}}, \quad (5)$$

where the second term ensures that W_1 and W_2 share exactly x_{12} genes, and the third term ensures that exactly x_{123} genes are shared in all three windows. We then obtain the probability of observing *at least* x_{123} genes in common by summing over q as follows:

$$P(X_{123} \geq x_{123}) = \binom{n}{r_2}^{-1} \binom{n}{r_3}^{-1} \sum_{k=x_{123}}^{u_{123}} q(X_{123} = k). \quad (6)$$

We analyzed the impact of considering the $x_{ij}'s$, by comparing Eq. 6 with Eq. 4 (Fig. 3(b)). $P(X_{123} \geq x_{123})$ is consistently two orders of magnitude greater than $P(\vec{X} \geq \vec{x})$. This is because a test based only on x_{123} fails to capture evidence of homology from genes that occur in only a subset of the windows (*i.e.*, the $x_{ij}'s$), and will severely underestimate cluster significance. For example, given a significance threshold of $\alpha = .01$ and the parameters used in Fig. 3(b), a cluster with $x_{12} = x_{13} = x_{23} = 3$ and $x_{123} = 1$ would not be considered significant using a test based on x_{123} alone, even though such a cluster is unlikely to arise by chance.

To further understand the relative importance of x_{123} and x_{ij}, we analyzed how much more a gene shared by all three windows contributes to significance than a gene shared by only two windows. Consider a cluster in which h genes are shared by all three windows (*i.e.,* $x_{123} = h, x_{ij} = 0$), compared to a cluster where there are h distinct genes shared between each *pair* of windows (*i.e.,* $x_{123} = 0, x_{ij} = h$). Notice that in both cases, each pair of windows shares h genes. However, in the first case each region only contains h shared genes, whereas in the second case each region shares $2h$ genes with the other regions. Although the total number of shared genes is larger in the second scenario, Fig. 3(c) shows that the first scenario is always much more significant. Even a small increase in x_{123} results in a large increase in significance—much more so than an increase of an equivalent number of homologous matches between pairs of regions.

2.2. *Shared Gene Content Model*

In contrast to the assumptions of the identical gene content model, in most cases, a genome will have *singleton* genes that do not have a detectable homolog in related genomes. How does this difference affect cluster significance? In the shared gene content model, we assume the genomes share a common set of $n_{123} \leq N_i$ homologs (Fig. 2(b)). In addition, each genome G_i contains $n_i = N_i - n_{123}$ singleton genes. Homology between gene pairs that have no homolog in the third genome is disregarded, with such genes being treated as singletons. This models the situation that would result if homologs were identified according to the triangle method used in COGs [13].

To compute the probability of observing exactly $\vec{x}$ shared genes, we must count the number of ways of choosing the $\vec{x}$ shared genes, as well as the genes that are unique to each window (x_1, x_2, and x_3). As in the case of identical gene content, the shared genes must be selected from the n_{123} genes common to the three genomes. However, the x_i genes that are unique to each window W_i can be selected either from the remaining common genes, or from the singletons of that genome (n_i). In the former case, care must be taken to ensure that a gene is only assigned to one window. As a result, two additional summations are required, since the number of ways to choose the x_3 genes unique to W_3 depends on how many genes from the n_{123} common genes were used to fill W_1 and W_2. The probability is:

$$P_S(\vec{X}=\vec{x}) = \binom{N_1}{r_1}^{-1}\binom{N_2}{r_2}^{-1}\binom{N_3}{r_3}^{-1}\binom{n_{123}}{x_{123},\, x_{12},\, x_{13},\, x_{23}} \sum_{i=0}^{x_1}\sum_{j=0}^{x_2}\binom{n_{123}-s}{i,j}\binom{n_1}{x_1-i}\binom{n_2}{x_2-j}\binom{N_3-s-i-j}{x_3}, \tag{7}$$

where $s = x_{123} + x_{12} + x_{13} + x_{23}$ is the total number of shared genes. $P_S(\vec{X} \geq \vec{x})$, the probability of observing *at least* as many shared genes under this model, can be computed from Eq. 7 by summing over $P_S(\vec{X}=\vec{x})$, similar to Eq. 4.

We use this expression to study how cluster significance depends on the extent of gene content overlap among the genomes. As the proportion of singleton genes in the genomes increases from 0.3 to 0.9, the probability of observing a cluster drops from 0.01 to 10^{-5}

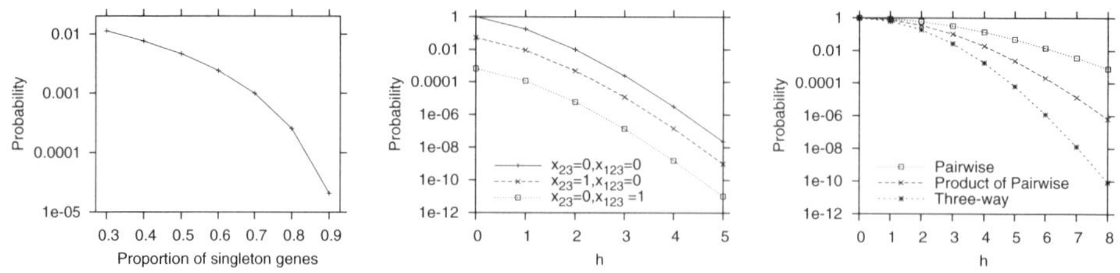

(a) (b) (c)

Figure 4. (a) The effect of singleton genes on cluster significance. The x-axis shows the proportion of singletons in each genome ($1 - n_{123}/N$). The y-axis shows the probability $P_S(\vec{X} \geq (1,1,1,1))$, when $N = N_1 = N_2 = N_3 = 5000$, and $r = 100$. (b) The effect of reciprocal loss on cluster significance in comparing pre- and post-duplication genomes, when $n_{1,2} = 450, n_{1,1} = 3600, n_{0,1} = 500, r = 50$, and $x_{12} = x_{13} = h$, as h ranges from 0 to 5. (c) Comparing pairwise probabilities, the product of two pairwise probabilities, and three-way probabilities, when $N = 5000$, $r = 100$, $x_{123} = 0$, and $x_{12} = x_{13} = x_{23} = h$.

(Fig. 4(a)). This is because as fewer homologs are shared between the genomes, it is more surprising to find them clustered together. This shows the importance of considering the extent of gene content overlap among the genomes when evaluating cluster significance.

2.3. *Pre/Post Duplication Model*

We propose a third genome overlap model specifically for analyzing duplications. Let G_{post} be a genome that has undergone a WGD and G_{pre} be a genome that diverged prior to the WGD (Fig. 2(c)). Let $n_{i,j}$ be the number of genes that appear i times in G_{pre} and j times in G_{post}, where $i \leq 1, j \leq 2$. This model only recognizes paralogs that arose through WGD, ignoring lineage specific duplications. Thus, it assumes that each gene in G_{post} has at most one paralog and that genes in G_{pre} have no paralogs; *i.e.*, $n_{2,0} = n_{2,1} = n_{2,2} = 0$. Furthermore, this model assumes that every gene that appears twice in the post-duplication genome also has a homolog in the pre-duplication genome; *i.e.*, $n_{0,2} = 0$. This assumption is based on the rationale that genes retained in duplicate are functionally important and, hence, are retained in G_{pre} as well. This assumption is supported by empirical observation. For example, in post-WGD yeast species over 95% of genes retained in duplicate are also present in each pre-WGD yeast genome[1]. Similarly, in this model every gene in G_{pre} has at least one homolog in G_{post} ($n_{1,0} = 0$). We use the convention that W_1 is the window sampled from G_{pre}, and W_2 and W_3 are sampled from G_{post}.

To compute the probability of observing *exactly* $\vec{x}$ shared homologs under the null hypothesis, we make the additional assumption that at most one copy of a duplicated gene appears in a given window. Given this condition,

$$P_D(\vec{X} = \vec{x}) = \frac{\binom{n_{1,2}}{x_{123}, x_{23}} \binom{N_{pre} - x_{123} - x_{23}}{x_{12}, x_{13}} \binom{N_{pre} - s}{x_1} \binom{N_{post} - n_{1,2} - s - x_1}{x_2, x_3}}{\binom{N_{pre}}{r_1} \sum_{i=0}^{\min(r_2, r_3)} \binom{n_{1,2}}{i} \binom{N_{pre} + n_{0,1} - i}{r_2 - i} \binom{N_{pre} + n_{0,1} - r_2}{r_3 - i}},$$

where $N_{pre} = n_{1,2} + n_{1,1}$ and $N_{post} = 2n_{1,2} + n_{1,1} + n_{0,1}$. $P_D(\vec{X} \geq \vec{x})$, the probability of observing *at least* $\vec{x}$ shared homologs under the null hypothesis, is then obtained as before by summing over $P_D(\vec{X} = \vec{x})$.

We calculated $P_D(\vec{X} \geq \vec{x})$ with parameter values based on a recent study of pre- and post-duplication yeast species[1,11]. In our simulations, $N_{post} = 5000$ and $n_{1,2} = 450$, consistent with the observation that only 16% of genes in *S. cerevisiae* are duplicate genes that arose during the WGD. Since the number of genes that occur twice in G_{post} is small, even small values of x_{123} and x_{23} will have a large impact on cluster significance. Fig 4(b) compares the significance of clusters for three reciprocal gene loss scenarios: when no genes are shared between the two regions selected from the post-duplication genome ($x_{123} = 0, x_{23} = 0$), when a single gene is shared ($x_{123} = 0, x_{23} = 1$), and when a single gene is shared among all three regions ($x_{123} = 1, x_{23} = 0$). The shape of the three curves is similar, but the probabilities drop by an order of magnitude from one to the next. Even the addition of a single gene retained in duplicate has a large impact on cluster significance! This is particularly noteworthy because current methods compare the pre-duplication region independently with each of the post-duplication regions, and thus ignore the values of x_{23} and x_{123}[6,7,8,11,16,17]. Our results show that these current methods could fail to detect clearly significant clusters, thus resulting in a substantial decrease in sensitivity.

3. Discussion

We have presented three different models of gene content overlap and proposed novel statistical tests for evaluating the significance of gene clusters spanning three regions. Our tests are the first to combine evidence from genes shared among all three regions and genes shared only between pairs of regions.

How do our three-way tests compare to current approaches reviewed in Sec. 1? Unlike tests that consider only x_{123}, our tests also consider $x_{ij}'s$, and thus can detect significant clusters even when x_{123} is small (Fig. 3(b)). Our tests also have advantages over current approaches based on pairwise statistical tests alone. These approaches construct multi-region clusters by merging pairwise clusters. However, this method does not explicitly consider the number of genes shared among all three regions. Our results (Fig. 3(c)) show that even a few genes conserved in all three regions dramatically increases the statistical significance of gene clusters. This effect is particularly strong when the shared gene content of the genomes is small (Fig. 4(a)). Thus, unlike pairwise tests, our approach can detect related regions where each pair of regions share only a few genes (*i.e.*, $x_{ij}'s$ are small), but where a few genes are also shared among all the regions (*i.e.*, x_{123} is non-zero but small).

Even when $x_{123} = 0$, we gain sensitivity over pairwise approaches. This is because the pairwise approach requires two of the three pairwise tests to be independently significant, whereas our approach considers the three regions jointly. Figure 4(c) illustrates this difference, for a scenario in which $n = 5000$ and $r = 100$. In this case, given a significance threshold of $\alpha = 0.01$, for a *pair* of regions to be significantly similar ($P_2(X \geq x)$), they must share at least seven genes. Thus, to find a three-way cluster with the pairwise approach, W_1 must share seven genes each with W_2 and W_3. In contrast, using our test $P_3(\bar{X} \geq \bar{x})$, a cluster is significant when each pair of regions shares only four genes, even when none of these genes appear in all three regions.

Since the comparison of two windows W_1 and W_2 is independent of the comparison of W_1 and W_3, one could try using the *product of two pairwise probabilities* as an approximation of the joint probability of all three windows. This approximation, though closer to the three-way probabilities, still underestimates the multi-region significance (Fig. 4(c)). This is because the product of pairwise probabilities fails to consider the genes shared between the third pair of windows (W_2 and W_3), and also does not give more weight to the genes that are shared among all the windows. Thus, we argue that pairwise tests are not always sufficient and multi-region tests will be able to identify more distantly related homologous regions. Here, we have presented initial results in this direction, yet many important problems remain. A more general test would take all paralogs into account. In addition, to investigate hypotheses of multiple WGDs within the same lineage, tests for more than three regions sampled from the same genome are required.

Acknowledgments

D.D. was supported by NIH grant 1 K22 HG 02451-01 and a David and Lucille Packard Foundation fellowship. R.H. was supported in part by a Barbara Lazarus Women@IT Fellowship, funded in part by the Alfred P. Sloan Foundation. We thank A. McLeod, A. Goldman, J. Joseph, M. Stolzer, and N. Song for help with manuscript preparation.

References

1. K. P. Byrne and K. Wolfe. The Yeast Gene Order Browser: combining curated homology and syntenic context reveals gene fate in polyploid species. *Genome Res*, 15(10):1456–1461, 2005.
2. P. Calabrese, S. Chakravarty, and T. Vision. Fast identification and statistical evaluation of segmental homologies in comparative maps. *Bioinformatics*, 19(suppl 1):i74–80, 2003.
3. D. Durand and D. Sankoff. Tests for gene clustering. *J Comput Biol*, 10(3–4):453–482, 2003.
4. R. Hoberman, D. Sankoff, and D. Durand. The statistical analysis of spatially clustered genes under the maximum gap criterion. *J Comput Biol*, 12(8):1081–1100, 2005.
5. L. Hurst, C. Pál, and M. Lercher. The evolutionary dynamics of eukaryotic gene order. *Nat Rev Genet*, 5(4):299–310, 2004.
6. O. Jaillon et al. Genome duplication in the teleost fish *Tetraodon nigroviridis* reveals the early vertebrate proto-karyotype. *Nature*, 431(7011):946–957, 2004.
7. M. Kellis, B. Birren, and E. Lander. Proof and evolutionary analysis of ancient genome duplication in the yeast *Saccharomyces cerevisiae*. *Nature*, 428(6983):617–624, 2004.
8. H. Ku, T. Vision, J. Liu, and S. Tanksley. Comparing sequenced segments of the tomato and *Arabidopsis* genomes: large-scale duplication followed by selective gene loss creates a network of synteny. *PNAS*, 97(16):9121–9126, 2000.
9. A. McLysaght, K. Hokamp, and K. Wolfe. Extensive genomic duplication during early chordate evolution. *Nat Genet*, 31(2):200–204, 2002.
10. N. Raghupathy and D. Durand. Individual gene cluster statistics in noisy maps. In *RECOMB 2005 Workshop on Comparative Genomics*, volume 3678 of *LNBI*, pages 106–120. Springer-Verlag, 2005.
11. D. Scannell, K. Byrne, J. Gordon, S. Wong, and K. Wolfe. Multiple rounds of speciation associated with reciprocal gene loss in polyploid yeasts. *Nature*, 440(7082):341–345, 2006.
12. C. Simillion, K. Vandepoele, and Y. Van de Peer. Recent developments in computational approaches for uncovering genomic homology. *Bioessays*, 26(11):1225–35, 2004.

13. R. Tatusov, E. Koonin, and D. Lipman. A genomic perspective on protein families. *Science*, 278(5338):631–637, 1997.
14. Z. Trachtulec and J. Forejt. Synteny of orthologous genes conserved in mammals, snake, fly, nematode, and fission yeast. *Mamm Genome*, 3(12):227–231, 2001.
15. K. Vandepoele, Y. Saeys, C. Simillion, J. Raes, and Y. Van De Peer. The automatic detection of homologous regions (ADHoRe) and its application to microcolinearity between *Arabidopsis* and rice. *Genome Res*, 12(11):1792–801, 2002.
16. K. Vandepoele, C. Simillion, and Y. Van de Peer. Detecting the undetectable: uncovering duplicated segments in *Arabidopsis* by comparison with rice. *Trends Genet*, 18(12):604–6, 2002.
17. K. Vandepoele, C. Simillion, and Y. Van de Peer. Evidence that rice and other cereals are ancient aneuploids. *Plant Cell*, 15(9):2192–2202, 2003.
18. J. Venter et al. The sequence of the human genome. *Science*, 291(5507):1304–1351, 2001.

THE DISTANCE BETWEEN RANDOMLY CONSTRUCTED GENOMES

WEI XU

Department of Mathematics and Statistics
University of Ottawa, Canada K1N 6N5
email: wxu060@uottawa.ca

In this paper, we study the exact probability distribution of the number of cycles c in the breakpoint graph of two random genomes with n genes or markers and χ_1 and χ_2 linear chromosomes, respectively. The genomic distance d between the two genomes is $d = n - c$. In the limit we find that the expectation of d is $n - \frac{2\chi_1\chi_2}{2\chi_1+2\chi_2-1} - \frac{1}{2}\ln\frac{n+\min(\chi_1,\chi_2)}{\chi_1+\chi_2}$.

1. Introduction

The study of genome rearrangements has developed a sophisticated technology for inferring a minimizing sequence of operations necessary to transform one genome into another, where the genomes are represented by signed permutations on $1, \cdots, n$ and the operations are modeled on the biological processes of inversion, reciprocal translocation, chromosome fusion and fission, transposition of chromosomal segments, excision and reintegration of circular chromosomal segments, among others. Once these inferences are made, however, there is a need for some way to statistically validate both the inferences and the assumptions of the evolutionary model.

Our approach has been to see to what extent there is an signal remaining in the comparative structure of the two genomes, or whether evolution has largely scrambled the order of each one with respect to the other, in terms of the evolutionary model assumed. This has led to the study of completely scrambled, i.e., randomized, genomes as a null baseline for the detection of a evolutionary signal. Insofar as a pair of genomes retain some evidence of evolutionary relationship, this should be detectible by contrast to randomized genomes. In previous papers, we have worked out the statistical properties of random genomes consisting of one or more circular chromosomes, [1] and those of two random genomes containing the same number c of linear chromosomes.[2]. The latter paper concentrated on showing that the number of circular chromosomes inevitably associated with random linear chromosomes is very small with realistic numbers of chromosomes. It only included a rough estimation of the statistical properties of the linear chromosomes.

The present paper introduces a new way of representing the comparison of linear genomes, requiring only a single source/sink vertex in the breakpoint graph of the two genomes, instead of the numerous "chromosomal caps" used in other treatments. This facilitates a more rigorous treatment of the case of linear chromosomes, including the more realistic situation where the number of linear chromosomes may be different (χ_1 and χ_2) in the two genomes being compared.

2. Genome Rearrangement with Linear Chromosomes

In our framework, each genome consists of n markers (genes, chromosomal segments, etc.), divided among a number of disjoint chromosomes. We fix the number of linearly ordered chromosomes, but in our construction of random genomes we will permit some additional, circularly ordered, chromosomes as well. In graph-theoretical terms, we usually represent each marker by two distinct vertices, marking the beginning and end of the marker, respectively. We call all of these *inner vertices*. For each linear chromosome, two extra vertices, named *caps* are added to represent the ends of the chromosome. In comparing two genomes containing different numbers χ_1 and χ_2 of linear chromosomes, we equalize their numbers at $\chi = \max(\chi_1, \chi_2)$ by adding an appropriate number of null chromosomes, each of which consists only of two caps, to one of the genomes.

2.1. *The Breakpoint Graph*

When two genomes, say a red one and a black one, containing the same n markers, are compared, we use red edges to connect the the nearest vertices of two adjacent markers according to their order in the red genome – this may be the end of one mark and the beginning of the other, or two ends, or two beginnings, depending on the orientation or "strandedness" of the markers on the chromosome. The first and last inner vertices are connected to caps. Each cap may only be connected to one inner vertex. We also connect the two caps of any null chromosome in the red genome by a red edge. Similarly, we use black edges to connect the vertices and caps in the black genomes. There are thus $2n$ inner vertices, 2χ caps, $n + \chi$ red edges and $n + \chi$ black edges in the graph.

Since each vertex is connected to one red and one black edge (one adjacency in each genome), a 2-regular graph is formed. A 2-regular graph always can be decomposed into number of cycles c, and in our bicoloured graph, the edge colours alternate around each cycle. Yancopoulos, Attie and Friedberg[3] showed that the edit distance d is related to the number of cycles c by

$$d = n + \chi - \max c, \tag{1}$$

when block interchanges (each counting as two operations) are allowed besides inversions and reciprocal translocations. The number of cycles depends on which red chromosome and which black chromosome are incident to the same cap, a choice which is left free in the graph definition. The maximal number of cycles in equation (1) refers to the optimal choice of this cap assignment. We refer to this particular graph as the breakpoint graph of the two genomes.

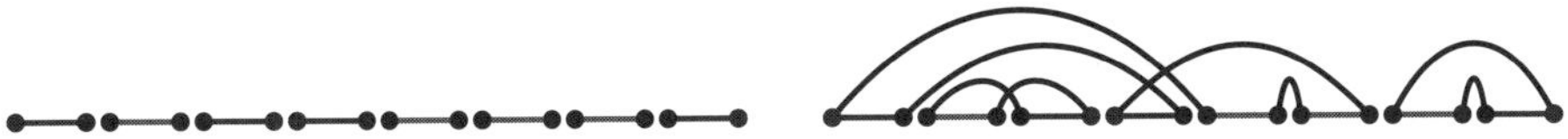

(a) (b)

Figure 1. The construction of a random breakpoint graph. We start with the red genome, represented by a set of cap edges (in blue) and a set of inner edges (in red), and add the black edges randomly, one by one, until every vertex is connected by one black edge. In (b) there are 3 cycles. Caps denoted by blue dots and inner vertices by black ones.

2.2. *Random Genomes*

Were we to construct genomes by successively adding markers or caps in random order, it would be very difficult to say anything precise about the breakpoint graph, because the linearity condition on chromosomes induces great complexity to the events whose probabilities we wish to calculate. Instead, we introduce the randomness directly in the construction of the breakpoint graph, leading to simple expressions for probabilities of the sizes and numbers of cycles. This simplicity comes at a cost, however, since the construction of a random genome at the level of the breakpoint graph does not exclude some circular chromosomes. As we shall mention later, there is good reason to believe that this feature does not affect our results on the limits of expectations.

To obtain two genomes randomized with respect to each other, it suffices to fix the gene order in one of them, say the red genome, and to introduce randomness into the black genome only. Because we are interested in calculations pertaining to the breakpoint graph, we simply postulate that at each step a black edge may be added to connect any two inner vertices that are not already incident to black edges. We do not at this stage really connect caps to inner vertices using black edges, because these edges are implicitly determined by the cycle optimization procedure applied to the rest of the graph. Thus we start with $2n + 2\chi$ vertices (inner vertices and caps), with red edges connected. We distinguish between two kinds of edges: 2χ *cap edges* incident to a cap and $n - \chi$ *inner edges* not incident to a caps. To construct the random breakpoint graph, we connect two inner vertices at random by a black edge until every vertex is incident to a black edge. Note that in randomly adding black edges we are not guaranteed to end up with linear chromosomes, since there is the possibility that the black genome so constructed will contain one or more circular chromosomes, with no caps. As χ becomes large, the number of such circles and the number of markers in them, will be small. Nevertheless this possibility is not part of the original problem involving two random genomes with linearly ordered chromosomes. Fortunately, partial mathematical results indicate that in the limit, the possible presence of circular chromosomes does not affect the probability structure of the breakpoint graph[4].

2.3. *Cap Optimization*

In the procedure of cap optimization, the breakpoint graph is decomposed into cycles and 2χ paths (whose two ends are caps or inner vertices incident to only one cap edge). The ψ_{HO} *homogenous paths* terminate with caps via two red edges (type 1) or with two inner vertices (type 2), with an equal number of the two types, and the ψ_{HE} *heterogenous paths* end with one cap and one inner vertex. The optimization principle developed by Hannenhalli and Pevzner[5] and Tesler[6], comes down to, in the reformulation by Yancopoulos *et al.*[3], to the addition of two black edges joining one homogenous path of type one to another homogeneous path of type 2 to form a cycle and the addition of a single black edge to each heterogeneous path to form a cycle. It can be seen that the maximized cap cycle number ψ is

$$\max\psi = \chi + \frac{1}{2}\psi_{HE} \tag{2}$$

2.4. *The* Flower *Representation*

To facilitate the construction of the random breakpoint graph, including the cycle optimization, we abandon the regular graph representation and introduce a modified model as follows.

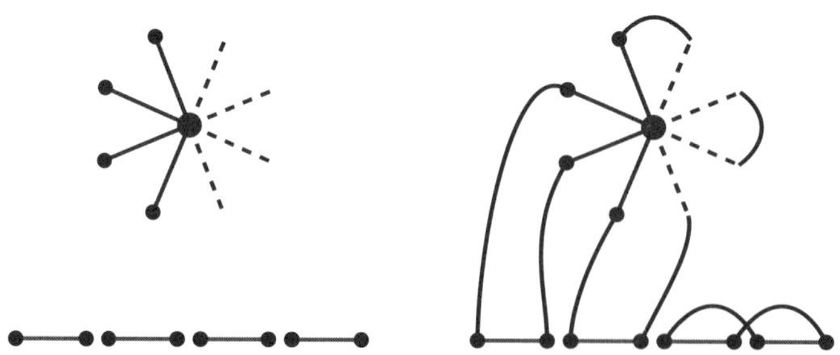

(a) (b)

Figure 2. The illustration of the modified model. At the initial state (a), all the caps have been merged into one source/sink vertex C. The dashed black edges are reserved for the 2χ black cap edges to be added later. At the end (b), all the cap edges should be connected via inner edges, except for some that are composed of two cap edges or a single edge with C at both ends. The rest of the inner edges form the inner cycles. In the figure, two homogenous paths, two heterogenous paths and one inner cycle are depicted.

We replace all the caps by a single source/sink vertex C. Then we may portray the cap edges as distributed around C as in Figure 2, while the inner edges are unaffected. In Figure 2(a), there are 2χ red cap edges and the same number of dashed edges incident to C indicating where the black cap edges will eventually connect. Some same-coloured pairs of these cap edges may represent null chromosomes. The construction proceeds by adding black edges one by one at random as detailed in the next section, and terminates when a complete structure as in Figure 2(b) is achieved.

The cycles are of two sorts, those in the flower structure, named *cap cycles* and the rest, *inner cycles*. In the next section, recurrence equations will derived for both kinds. Note that each "petal" of the flower, connected to the source/sink vertex, represents a path, either a homogenous or heterogeneous. The cap cycles are not explicitly depicted in the graph. Their total number is determined by the capping optimization formula.

3. The Recurrence Equations

3.1. *The Number of Heterogenous Paths ψ_{HE}*

From the cap optimization principle, the number of cap cycles should be equal to $\chi+\frac{1}{2}\psi_{HE}$.

During the construction, at each step it suffices to keep track only of the number of extended red cap edges, where this includes paths with C connected to a red edge at one end and a red edge at the other, the number of extended black cap edges, where this includes either dashed edges of paths with C connected to a black edge at one end and a red edge at the other.

We start from a general situation where there are r extended red cap edges and s extended black cap edges. The problem is denoted as (r, s).

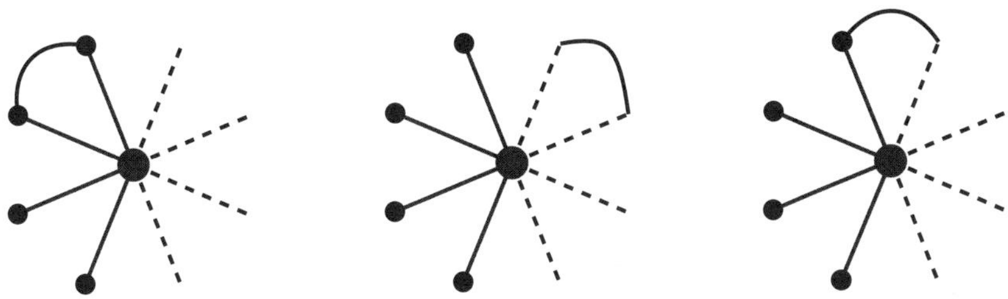

(a) (b) (c)

Figure 3. The three possible ways of completing a cap path. Homogenous paths are shown in (a) and (b) and a heterogenous path in (c).

At each step, one black edge is added, connecting two extended red cap edges, two dashed or extended black cap edges or one extended red cap edge and one dashed or extended black cap edge. Once a path forms, the total number of extended paths (i.e., the edges that remain to be connected) decreases by 2. The three possible ways of adding the black edge lead to the smaller problems $(r-2, s)$, $(r, s-2)$, $(r-1, s-1)$, respectively. The numbers of ways of doing each are $\binom{r}{2}$, $\binom{s}{2}$ and rs, respectively. Only in the last situation is a heterogenous path completed. Denote $\mathbf{n}(\psi_{HE}, r, s)$ as the total number of ways to get a breakpoint graph with ψ_{HE} heterogenous paths for the (r, s) problem. Since each problem with size (r, s) can be constructed from three smaller problems of sizes $(r-2, s)$, $(r, s-2)$ and $(r-1, s-1)$, respectively, we have the recurrence :

$$\mathbf{n}(\psi_{HE}, r, s) = \binom{r}{2}\mathbf{n}(\psi_{HE}, r-2, s) + \binom{s}{2}\mathbf{n}(\psi_{HE}, r, s-2) + rs\,\mathbf{n}(\psi_{HE}-1, r-1, s-1) \tag{3}$$

Denote by $\bar{\psi}_{HE}(r, s)$ the average number of heterogenous paths in the breakpoint graph for (r, s), defined as:

$$\begin{aligned}\bar{\psi}_{HE}(r, s) &= \frac{\sum_{\psi_{HE}=0}^{min(r,s)} \mathbf{n}(\psi_{HE}, r, s)\,\psi_{HE}}{\sum_{\psi_{HE}=0}^{min(r,s)} \mathbf{n}(\psi_{HE}, r, s)} \\ &= \frac{\sum_{\psi_{HE}=0}^{min(r,s)} \mathbf{n}(\psi_{HE}, r, s)\,\psi_{HE}}{\prod_{i=0}^{\frac{r+s}{2}} \binom{r+s-2i}{2}}\end{aligned}$$

where $\sum_{\psi_{HE}=0}^{min(r,s)} \mathbf{n}(\psi_{HE}, r, s) = \prod_{i=0}^{\frac{r+s}{2}} \binom{r+s-2i}{2}$ is the total number of ways to construct the breakpoint graph.

By summing over equation (3), we get the recurrence equation for the average number of heterogenous paths.

$$\bar{\psi}_{HE}(r, s) = \frac{\binom{r}{2}\bar{\psi}_{HE}(r-2, s) + \binom{s}{2}\bar{\psi}_{HE}(r, s-2) + rs\bar{\psi}_{HE}(r-1, s-1) + rs}{\binom{r+s}{2}} \tag{4}$$

Equation (4) has a probabilistic interpretation, since (r, s) can be decomposed into $(r-2, s)$, $(r, s-2)$ and $(r-1, s-1)$ with probabilities $\frac{r(r-1)}{(r+s)(r+s-1)}$, $\frac{s(s-1)}{(r+s)(r+s-1)}$ and $\frac{2rs}{(r+s)(r+s-1)}$, respectively.

3.2. *The Number of Inner Cycles*

The number of the inner cycles depends on the number of inner edges not used by the paths. Suppose we start with $2m$ extended cap edges (the extended red cap edges and the dashed or extended black edges) and l inner edges, which it will be convenient to denote (m, l).[a] In the random construction of the breakpoint graph, each addition of one black edge can lead to four different situations:

(1) two cap edges are connected – there are $\binom{2m}{2}$ ways of doing this – and the size of the problem becomes $(m-1, l)$
(2) one cap edge and one inner edge are connected – there are $4ml$ ways of doing this – and the size of the problem becomes $(m, l-1)$

[a] Rather than $(2m, l)$.

(3) two different inner edges are connected – there are $2l(l-1)$ ways of doing this – and the size of the problem becomes $(m, l-1)$

(4) the two ends of the same inner edges are connected – there are l ways of doing this. The size of the problem becomes $(m, l-1)$ and the number of inner cycles increases by one.

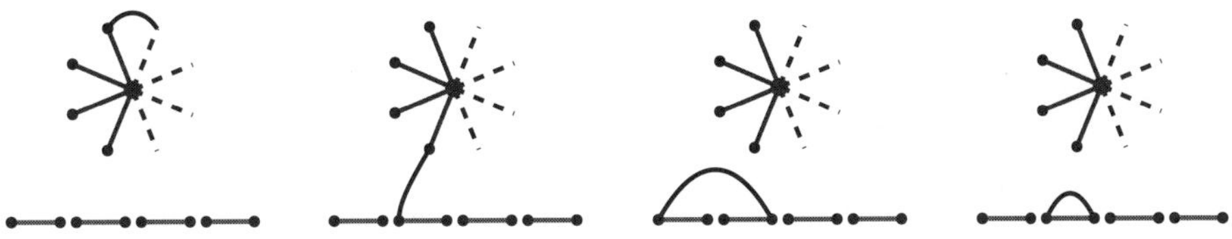

Figure 4. The four possible ways to build a black edge in counting the inner cycles. Two cap edges are connected (a); one cap edge and one inner edge are connected (b); two different inner edges are connected (c); the two ends of the same inner edge are connected (d). And only in the last case, one inner cycle is formed.

Denote by $\mathbf{n}(\kappa, m, l)$ the number of ways to get a breakpoint graph with κ inner cycles for a (m, l) problem. Similarly define $\bar{\kappa}(m, l)$ as the average number of inner cycles for the problem (m, l)

$$\bar{\kappa}(m,l) = \frac{\sum_{\kappa=0}^{l} \mathbf{n}(\kappa,m,l)\,\kappa}{\sum_{\kappa=0}^{l} \mathbf{n}(\kappa,m,l)} = \frac{\sum_{\kappa=0}^{l} \mathbf{n}(\kappa,m,l)\,\kappa}{\prod_{i=0}^{m+l} \binom{2m+2l-2i}{2}}$$

We then get the corresponding recurrences

$$\mathbf{n}(\kappa,m,l) = \binom{2m}{2}\mathbf{n}(\kappa,m-1,l) + [4ml + 2l(l-1)]\,\mathbf{n}(\kappa,m,l-1) + l\,\mathbf{n}(\kappa-1,m,l-1) \tag{5}$$

$$\bar{\kappa}(m,l) = \frac{\binom{2m}{2}\bar{\kappa}(m-1,l) + [4ml + 2l(l-1)]\,\bar{\kappa}(m,l-1) + l\,\bar{\kappa}(m,l-1) + l}{\binom{2m+2l}{2}} \tag{6}$$

Equation (6) also has a probabilistic interpretation, associating the probabilities $\frac{2m(2m-1)}{(2m+2l)(2m+2l-1)}$, $\frac{8ml}{(2m+2l)(2m+2l-1)}$, $\frac{4l(l-1)}{(2m+2l)(2m+2l-1)}$ and $\frac{2l}{(2m+2l)(2m+2l-1)}$ with the four possible smaller problems.

4. The Solution to the Problems

4.1. *The Cap Cycles*

The recurrence equations (4) and (6) enable rapid calculation of $\bar{\psi}_{HE}$ and $\bar{\kappa}$, but there is no easy way to convert them into a closed form solution.

We can, however, deduce these quantities through another combinatoric approach. The total number of ways to form any kind of flower structure is $\prod_{i=0}^{\frac{r+s}{2}} \binom{r+s-2i}{2}$. The ways to form a result with $2\psi_{HE}$ heterogenous paths (which should be always even) is

$$\begin{aligned}\mathbf{n}(2\psi_{HE}, r, s) &= \binom{\frac{r+s}{2}}{2\psi_{HE}} \frac{(r)!}{(r-2\psi_{HE})!} \frac{(s)!}{(s-2\psi_{HE})!} \binom{\frac{r+s}{2} - 2\psi_{HE}}{\frac{r}{2} - \psi_{HE}} \\ &\quad \prod_{i=0}^{\frac{r+s}{2}} \binom{r-\psi_{HE}-2i}{2} \prod_{i=0}^{\frac{r+s}{2}} \binom{s-\psi_{HE}-2i}{2} \\ &= \frac{(r)!(s)!(r+s)!}{(\frac{r+s}{2})!} \frac{4^{\psi_{HE}}}{(2\psi_{HE})!(\frac{r}{2}-\psi_{HE})!(\frac{s}{2}-\psi_{HE})!} \end{aligned} \tag{7}$$

Averaging over $\mathbf{n}(2\psi_{HE}, r, s)$ and rewriting $r = 2\chi_1$ and $r = 2\chi_2$, we get

$$\bar{\psi}_{HE}(\chi_1, \chi_2) = \frac{4\chi_1\chi_2}{2\chi_1 + 2\chi_2 - 1} \tag{8}$$

So the average number of cap cycles is

$$\psi = \max(\chi_1, \chi_2) + \frac{2\chi_1\chi_2}{2\chi_1 + 2\chi_2 - 1} \tag{9}$$

When $\chi_1 = \chi_2 = \chi$, it becomes $\chi + \frac{2\chi^2}{4\chi-1}$, approaching 1.5χ as χ becomes large, confirming a result which we have previously derived in another way.[2]

4.2. *The Inner Cycles*

In the flower structure where the numbers of chromosomes are equal, suppose we traverse all the edges, starting with a black cap edge, and each time we visit C, we choose an outgoing edge of colour different from the incoming edge. This will order the edges as in Figure 5(a). The last edge will be a red cap edge and C will be the last vertex. We then add the edges in the inner cycles to the right of the flower structure edges.

We define the position x of an edge, as the number of edges to the left of, and including, that edge. We assume there are χ linear chromosomes in each genome. So the smallest value possible for x is 1 and the largest one is $n - \chi$. The 2χ cap edges occupy random positions in the sequence. The constraints on the model are that the last cap edge should have C on its right, the ith cap edge can only distributed from $x_{i-1}+1$ to $n+\chi-(2\chi-i) = n-\chi+i$. Only the inner edges to the right of the variable $x_{2\chi}$, the position of the last cap edge, are in inner cycles. Once we know the distribution of $x_{2\chi}$, we then use the formula[1] for the expected number of cycles in circular genomes to calculate the number of inner cycles.

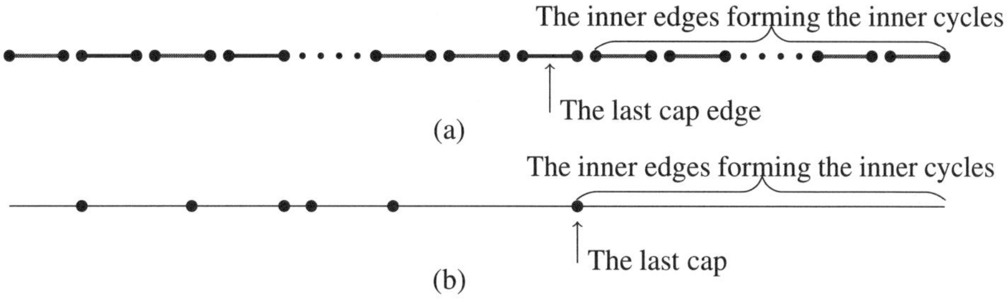

Figure 5. The exact model (a) for counting the inner cycle number and the approximate model (b). Model (a) is discrete. The cap in the last cap edge should be in the right side in order to correspond to the flower structure. Model (b) is a continuous approximation of model (a) when the number of inner edges are large enough and has no constraint on the last cap edge.

When n becomes large, we may define a continuous approximation to this construction. The x_i become the order statistics of χ uniformly distributed points on $(0, n + \chi)$. Using the distribution for the position of the x_χ, we find the expected number of inner cycles is

$$c(m = 2\chi, l = n - \chi) = \frac{1}{2} \ln \frac{\chi + n}{2\chi} + B \tag{10}$$

where B is some constant.[2]

Note that equation (10) is the asymptotic solution of equation (6), with $m = 2\chi, l = n - \chi$. B can be found from an initial condition: when $n = \chi$, then there are no inner edges and hence no inner cycles. So $\frac{1}{2} \ln \frac{\chi+\chi}{2\chi} + B = 0$, i.e., $B = 0$. In numerical comparison as well, the equation $c = \frac{1}{2} \ln \frac{\chi+n}{2\chi}$ confirms the recurrence equation (6).

4.3. *Two Genomes Having Different Numbers of Linear Chromosomes*

Suppose the two genomes being compared have χ_1 and χ_2 linear chromosomes, respectively. We have already found the formula for the cap cycles which is $\max(\chi_1, \chi_2) + \frac{2\chi_1\chi_2}{2\chi_1+2\chi_2-1}$. For the number of inner cycles, the approximate model only deals with the case where $\chi_1 = \chi_2 = \chi$. But that solution is also the asymptotic solution for the recurrence equation (6), which depends only on m and l. We can thus substitute the values for m and l in the case of unequal number of linear chromosomes. Note that in the case of equality, $m = 2\chi$ and $l = n - \chi$ and in the unequal case $m = \chi_1 + \chi_2$ and $l = n - \max(\chi_1, \chi_2)$.

$$\begin{aligned} c &= \frac{1}{2} \ln \frac{\chi + n}{2\chi} = \frac{1}{2} \ln \frac{2\chi + n - \chi}{2\chi} = \frac{1}{2} \ln \frac{m + l}{m} \\ &= \frac{1}{2} \ln \frac{\chi_1 + \chi_2 + n - \max(\chi_1, \chi_2)}{\chi_1 + \chi_2} \\ &= \frac{1}{2} \ln \frac{n + \min(\chi_1, \chi_2)}{\chi_1 + \chi_2} \end{aligned} \tag{11}$$

Hence in the limit the total number of cycles is

$$c = \max(\chi_1, \chi_2) + \frac{2\chi_1\chi_2}{2\chi_1 + 2\chi_2 - 1} + \frac{1}{2}\ln\frac{n + \min(\chi_1, \chi_2)}{\chi_1 + \chi_2} \tag{12}$$

And the genomic distance d is

$$d = n - \frac{2\chi_1\chi_2}{2\chi_1 + 2\chi_2 - 1} - \frac{1}{2}\ln\frac{n + \min(\chi_1, \chi_2)}{\chi_1 + \chi_2}. \tag{13}$$

5. Conclusion

The mathematical essence of the question with two genomes with linear chromosomes, is the number of the cycles in the 2-regular breakpoint graph whose vertices consist of a set of labeled vertices and another set of interchangeable caps vertices. We have shown that collapsing all the caps to a single source/sink facilitates the optimal capping problem as well as the calculation of cycle expectations.

The final result equation (13) can be applied to the comparison of two genomes with the same or different number of linear chromosomes plus any number of circular chromosomes. This is true under the condition that inversions, translocations and block interchanges are the mechanism of genomic rearrangement, where the latter count as if they were each two operations.[3]

References

1. Sankoff, D. and Haque, L. 2006. The distribution of genomic distance between random genomes. Journal of Computational Biology 13, 1005–1012.
2. Xu, W., Zheng, C. and Sankoff, D. 2006. Paths and cycles in breakpoint graphs of random multichromosomal genomes. Proceedings of RECOMB Satellite Conference on Comparative Genomics 2006, G. Bourque and N. El-Mabrouk, eds., Lecture Notes in Computer Science 4205. Heidelberg: Springer, 51–62.
3. Yancopoulos, S., Attie, O. and Friedberg, R. 2005. Efficient sorting of genomic permutations by translocation, inversion and block interchange. Bioinformatics 21, 3340 – 3346
4. Kim, J.H. and Wormald, N.C. 2001. Random matchings which induce Hamilton cycles, and Hamiltonian decompositions of random regular graphs, Journal of Combinatorial Theory, Series B 81, 20–44.
5. Hannenhalli, S. and Pevzner, P.A. 1995. Transforming men into mice (polynomial algorithm for genomic distance problem. Proceedings of the IEEE 36th Annual Symposium on Foundations of Computer Science. 581–92.
6. Tesler, G. 2002. Efficient algorithms for multichromosomal genome rearrangements. Journal of Computer and System Sciences 65, 587–609.

COMPUTING THE BREAKPOINT DISTANCE BETWEEN PARTIALLY ORDERED GENOMES

ZHENG FU AND TAO JIANG

Computer Science Department, University of California, Riverside

The total order of the genes or markers on a chromosome is crucial for most comparative genomics studies. However, the current gene mapping efforts might only suffice to provide a partial order of the genes on a chromosome. Several different genes or markers might be mapped at the same position due to the low resolution of gene mapping or missing data. Moreover, conflicting datasets might give rise to the ambiguity of gene order. In this paper, we consider the reversal distance and breakpoint distance problems for partially ordered genomes. We first prove that these problems are NP-hard, and then give an efficient heuristic algorithm to compute the breakpoint distance between partially ordered genomes. The algorithm is based on an efficient approximation algorithm for a natural generalization of the well-known feedback vertex set problem, and has been tested on both simulated and real biological datasets. The experimental results demonstrate that our algorithm is quite effective for estimating the breakpoint distance between partially ordered genomes and for inferring the gene (total) order.

1. Introduction

The total order of the genes or markers on a chromosome is very important for most comparative genomics studies. The *breakpoint distance* [12,8] and *reversal distance* [11,5] are commonly used as the evolutionary distances between genomes, and they work on the premise that the total order of the genes on each chromosome has been identified. However, except for a few model genomes, most genomes have not been completely sequenced yet. For these partially sequenced/assembled genomes, only partial gene maps are available, which might have a low resolution, missing genes/markers, or conflicting ordering information among each other. Combining these partial gene maps together might only suffice to provide a partial order of genes and markers. Hence, Zheng, Lenert and Sankoff [13,14] recently proposed a new general representation of a genome in terms of genes where each chromosome is a directed acyclic graph (DAG) rather than a permutation. Any linearization of the DAGs represents a possible total order of the genome. They generalized the sorting by reversal problem to assess the distance between two partially ordered genomes. The idea is to resolve the partial orders into two total orders (*i.e.* two linearizations of the DAGs corresponding to the two genomes) with the minimum reversal distance. In the same paper, a depth-first branch-and-bound search algorithm for computing the reversal distance is presented, which runs in exponential time in the worst case.

In this paper, we study efficient computation of the reversal distance and breakpoint distance problems between two partially ordered genomes. We show that these two problems are NP-hard. We also present an efficient heuristic algorithm to compute the breakpoint dis-

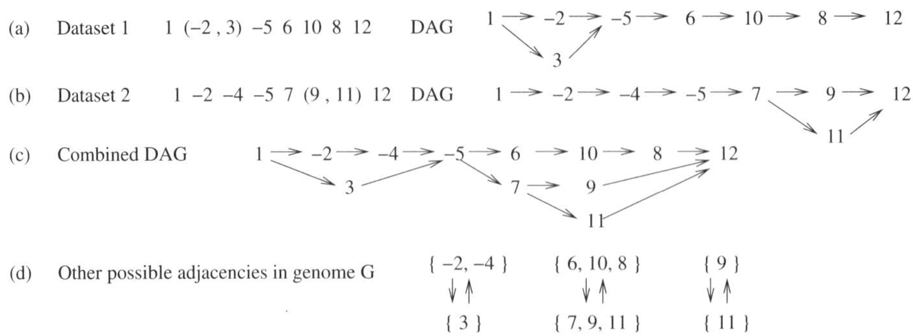

Figure 1. An example of DAG representation for a partially ordered genome.

tance between two partially ordered genomes, called *BDPOG*. The algorithm also reports a pair of total orders for the input genomes realizing the breakpoint distance. It runs in $O(n^3)$ time and uses an efficient approximation algorithm for a natural generalization of the well-known feedback vertex set problem as a subroutine. The BDPOG algorithm has been tested on both simulated and real biological datasets. The experimental results demonstrate that it is quite effective for estimating the breakpoint distance between partially ordered genomes and inferring the total gene orders.

The rest of the paper is organized as follows. We first introduce some preliminary facts and definitions in Section 2. Section 3 presents the NP-hardness results. Section 4 describes the algorithm BDPOG. Section 5 presents the experimental results on both simulated and real genome datasets. Finally, some concluding remarks are given in Section 6.

2. Preliminaries

Genes or markers are usually represented by signed (+ or -) symbols from a alphabet A, where the signs represent the strand of the genes. A totally ordered genome could be modeled as an ordered string of genes. However, the existing gene mapping efforts might only suffice to partially order the set of genes on a chromosome. If the order of some genes (*e.g.* $a_1, a_2, \cdots, a_n$) cannot be decided in a gene map, we will use $(a_1, a_2, \cdots, a_n)$ to represent the uncertainty of the ordering among them. For example, in the gene map $1\ (-2,\ 3)\ -5\ 6\ 10\ 8\ 12$, the ordering of all the genes has been decided except between genes 2 and 3.

Two or more gene maps constructed from different kinds of data or using different methodologies can be combined to form a more complicated partial order. As Zheng, Lenert, and Sankoff proposed in recent studies [13,14], directed acyclic graphs (DAGs) rather than linear permutations could be used to represent partially ordered genomes. In each DAG, all genes are represented by vertices, while the ordering relation between the genes is represented by arcs (see Figure 1).

Let Π and Γ be partially ordered genomes of size n, and the DAG representations for Π and Γ denoted as DAG(Π) and DAG(Γ). A linearization of DAG(Π) represents a possible ordering of genome Π. Let $L(\Pi)$ be the set of all possible linearizations of the DAG(Π).

Then we define the reversal distance between Π and Γ as

$$d_r(\Pi, \Gamma) = \min_{\pi \in L(\Pi), \gamma \in L(\Gamma)} d_r(\pi, \gamma) \quad (1)$$

Similarly, the breakpoint distance is

$$d_b(\Pi, \Gamma) = \min_{\pi \in L(\Pi), \gamma \in L(\Gamma)} d_b(\pi, \gamma) \quad (2)$$

We define the problem of computing $d_r(\Pi, \Gamma)$ as the *partial-order reversal distance (PRD)* problem and the problem of computing $d_b(\Pi, \Gamma)$ as the *partial-order breakpoint distance (PBD)* problem.

Clearly, all possible pairwise adjacency relationships in all possible linearizations of a DAG can be represented by the arcs of a DAG plus two arcs of opposite directions between all pairs of vertices which are not ordered by the DAG (see Figure 1d). We say that a pair of genes forms a *possible adjacency* in genome Π if they are possibly adjacent in any linearization of DAG(Π). We say that a pair of genes a and b is a *possible common adjacency* and write $a \cdot b$ if they are a possible adjacency in both genomes Π and Γ. And a is the *left end* of $a \cdot b$ and b is the *right end* of $a \cdot b$. Let $\mathcal{S}$ be the set of all possible common adjacencies. Define an order relation "$\leadsto$" between a pair of possible common adjacencies $a \cdot b$ and $c \cdot d$ in $\mathcal{S}$. We write $a \cdot b \leadsto_\Pi c \cdot d$ if one of the following four conditions is satisfied: (i) c or d is reachable from a or b in DAG(Π) (*i.e.* at least one of the genes in the second possible common adjacency is reachable from at least one of the genes in the first possible common adjacency in DAG(Π)); (ii) $a = c$ and $b \neq d$, or $b = d$ and $a \neq c$ (*i.e.* two different possible common adjacencies share the same left end or the same right end); (iii) $b = c$ (*i.e.* the right end of the first possible common adjacency is the same as the left end of the second possible common adjacency); (iv) let $a = u_{m-1}$, $b = u_m$, $c = u_1$ and $d = u_2$, and there exist possible common adjacencies $u_1 \cdot u_2$, $u_2 \cdot u_3$, $\cdots$, $u_{m-1} \cdot u_m$, $3 \leq m$ and another path in DAG(Π) from u_1 to u_m other than $u_1 \to u_2 \to \cdots \to u_{m-1} \to u_m$.

Based on the order relation "$\leadsto_\Pi$", we define a directed graph $\mathcal{G}_\Pi$, called the *adjacency-order graph*. The construction of $\mathcal{G}_\Pi$ is described as follows (see figure 2):

- Every possible adjacency in $\mathcal{S}$ is represented by a vertex.
- For every two possible common adjacencies, if $a \cdot b \leadsto_\Pi c \cdot d$, add an arc from the vertex $a \cdot b$ to the vertex $c \cdot d$.

A directed cycle in an adjacency-order graph usually represents a conflict among the possible common adjacencies in this cycle. And based on the construction of the adjacency-order graph, we have the following theorem.

Theorem 2.1. *All the possible common adjacencies in an acyclic adjacency-order graph $\mathcal{G}_\Pi$ could always co-exist in some linearization of DAG*(Π).

Proof. Omitted due to page limit. Please see the full version. □

The graph $\mathcal{G}_\Gamma$ can be constructed in the same way except that the arcs represent the relation $\leadsto_\Gamma$ instead of $\leadsto_\Pi$. Note that since the adjacency-order graphs $\mathcal{G}_\Pi$ and $\mathcal{G}_\Gamma$ are

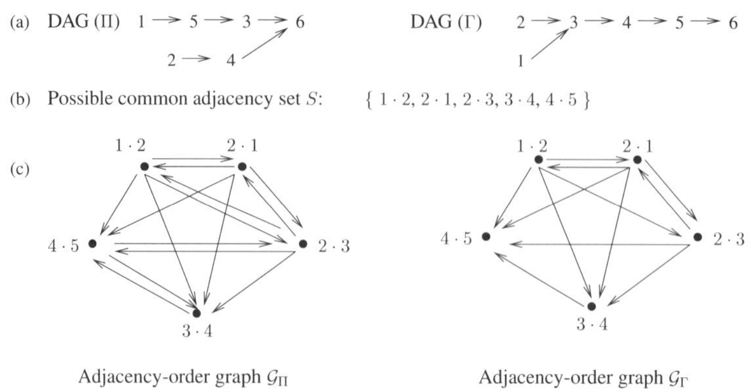

Figure 2. An example of the construction of adjacency-order graphs. In $\mathcal{G}_\Pi$, the arcs inserted by condition (i): $(1 \cdot 2, 2 \cdot 3)$, $(1 \cdot 2, 3 \cdot 4)$, $(1 \cdot 2, 4 \cdot 5)$, $(2 \cdot 1, 2 \cdot 3)$, $(2 \cdot 1, 3 \cdot 4)$, $(2 \cdot 1, 4 \cdot 5)$, $(2 \cdot 3, 3 \cdot 4)$, $(2 \cdot 3, 4 \cdot 5)$, $(4 \cdot 5, 2 \cdot 3)$, $(4 \cdot 5, 3 \cdot 4)$; the arcs inserted by condition (ii): $(2 \cdot 1, 2 \cdot 3)$, $(2 \cdot 3, 2 \cdot 1)$; the arcs inserted by condition (iii): $(1 \cdot 2, 2 \cdot 1)$, $(1 \cdot 2, 2 \cdot 3)$, $(2 \cdot 1, 1 \cdot 2)$, $(2 \cdot 3, 3 \cdot 4)$, $(3 \cdot 4, 4 \cdot 5)$; and the arc inserted by condition (iv): $(2 \cdot 3, 1 \cdot 2)$. Note that some arcs might satisfy several different conditions.

constructed by possible common adjacencies, they should share a same vertex set but may have different arc sets.

3. Computational Complexity of the PRD and PBD Problems

In this section, we show that both PRD and PBD problems are NP-hard, using different reductions.

Theorem 3.1. *The PRD problem is NP-hard.*

Proof. The proof is based on a careful analysis of the structure of the breakpoint graph for two partially ordered genomes [13,14], the Hannenhalli-Pevzner formula [5] for the reversal distance between two totally ordered genomes, and a reduction from the NP-hard problem MAX-ACD [2]. The details are omitted due to page limit. Please see the full version. □

By using a different reduction, we can prove the NP-hardness of the the breakpoint distance between two partially ordered genomes.

Theorem 3.2. *The PBD problem is NP-hard.*

Proof. We prove that the decision version of the PBD problem is NP-hard by a reduction from the decision version of *minimum feedback vertex set problem.*

Minimum Feedback Vertex Set Problem (MFVS)

INSTANCE: A directed graph $G(V, A)$ and a positive integer k.

QUESTION: Is there a subset $X \subseteq V$ with $|X| \leq k$ such that deleting all the vertices from X and their incident arcs will leave G acyclic?

Let directed graph $G(V, A)$ and positive integer k make up an arbitrary instance of the MFVS problem. The reduction to the breakpoint distance problem between partial ordered permutations (Π and Γ) works as follows: (a) For every vertex v_i in G, make two genes v_i^1

and v_i^2. (b) Add another $n+1$ genes $\{x_1, x_2 \cdots, x_{n+1}\}$, where $n = |V|$. (c) Construct a totally ordered genome $\Gamma = x_1\ v_1^1\ v_1^2\ x_2\ v_2^1\ v_2^2\ \cdots\ x_n\ v_n^1\ v_n^2\ x_{n+1}$. (d) Construct a partially ordered genome $\Pi = x_{n+1}\ (p_1, p_2, \cdots, p_m)\ x_1\ \cdots\ x_n$, where $m = |E|$ and each p_i, $i \in [1, m]$, represents an ordered pair of vertices. If there is an arc directed from vertex v_u to vertex v_w in G, we will have a pair $p_i = v_u^1\ v_w^2$, which means that in the genome Π gene v_u^1 is ordered before gene v_w^2. Finally, the order between p_i and p_j, $i \neq j$, is unknown. Figure 3 gives a simple example for this reduction.

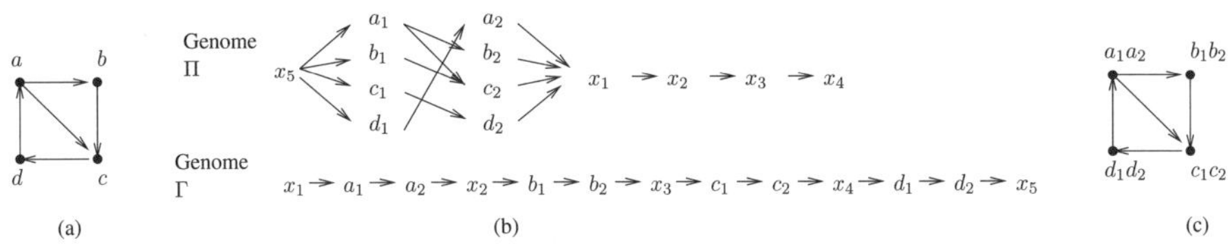

Figure 3. An example of the reduction from the minimum feedback vertex set problem to the breakpoint distance problem. (a) Directed graph $G(V, A)$. (b) Genome Π and Γ, where Γ is a totally ordered genome. (c) Adjacency-order graph of Π, $\mathcal{G}_\Pi$, which is isomorphic to $G(V, A)$.

This reduction guarantees that for Π and Γ, the set of all possible common adjacencies $\mathcal{S} = \{v_1^1 \cdot v_1^2, v_2^1 \cdot v_2^2, \cdots, v_n^1 \cdot v_n^2\}$. Adjacency-order graph $\mathcal{G}_\Pi$ is isomorphic to $G(V, A)$, while adjacency-order graph $\mathcal{G}_\Gamma$ is acyclic since Γ is totally ordered. Based on the special construction of Π and Γ, the cardinality of the minimum feedback vertex set of $\mathcal{G}_\Pi$, or graph $G(V, A)$, is exactly $d_b(\Pi, \Gamma) - 2n - 2$. Therefore, the feedback vertex set problem of $G(V, A)$ and k could be resolved by computing the $d_b(\Pi, \Gamma)$. The result of Theorem 3.2 hence follows. □

4. An Efficient Heuristic Algorithm for Computing the Breakpoint Distance

Let Π and Γ be two partially ordered genomes with possible common adjacency set $\mathcal{S}$. Computing the breakpoint distance $d_b(\Pi, \Gamma)$ is actually the problem of finding two linearizations of Π and Γ containing the maximum number of possible common adjacencies. In other words, we want to delete the smallest number of possible common adjacencies from $\mathcal{S}$ while leaving the rest of possible common adjacencies conflict free (*i.e.* they could co-exist in some linearizations). One way to delete order conflicts among possible common adjacencies is using the adjacency-order graph.

By Theorem 2.1, if the adjacency-order graph is acyclic, all the possible common adjacency vertices could be linearized by topological sort and partially ordered genomes could be totally ordered based on such a topological sort. Hence, deleting the smallest number of vertices to make both adjacency-order graphs (*i.e.* $\mathcal{G}_\Pi$ and $\mathcal{G}_\Gamma$) acyclic simultaneously could approximate the $d_b(\Pi, \Gamma)$. Formally,

Definition 4.1. *Minimum Double Feedback Vertex Set (MDFVS) problem*
Given two directed graphs with the same vertex set and different arc sets, find the minimum-cardinality subset of the vertices whose deletion leaves both graphs acyclic simultaneously.

The output vertex set is called a *minimum double feedback vertex set.*

4.1. *An Efficient Approximation Algorithm for the Minimum Double Feedback Vertex Set Problem*

Recall that the minimum feedback vertex set (MFVS) problem deals with a single graph, *i.e.*, the goal is to find the subset of vertices with the minimum cardinality whose deletion will leave the (single) input graph acyclic. We know that for the minimum feedback vertex set problem, the best-known approximation algorithm [4,10] in directed graphs achieves a performance ratio of $O(\log n \log\log n)$, where n is the number of vertices of the digraph, although the algorithm requires to the solution of a linear program. Another useful approximation algorithm [3] (denoted *APPROX-MFVS*) achieves a performance ratio bounded by the length, in terms of the number of vertices, of a longest simple cycle in the input digraph. Based on the strong relationship between the MFVS problem and the MDFVS problem, we could prove the following theorem.

Theorem 4.1. *There exists a polynomial 2λ-approximation algorithm for the* MDFVS *problem, where λ is the maximum length, in terms of the number of vertices, of a longest simple cycle in any of the two input graphs.*

Proof. In the MDFVS problem, we are given two directed graphs, say G_1 and G_2, which have the same vertex set and different arc sets. Utilizing the approximation algorithm APPROX-MFVS for the MFVS problem as a subroutine, we can easily design an approximation algorithm, denoted *APPROX-MDFVS* (see Figure 4), for the MDFVS problem as follows. Run APPROX-MFVS on G_1 and G_2 separately to get the feedback vertex sets FVS(G_1) and FVS(G_2), respectively. Denote the union of FVS(G_1) and FVS(G_2) as DFVS(G_1, G_2). DFVS(G_1, G_2) is certainly a double feedback vertex set, although not necessarily minimal. In fact, it might contain some vertices whose deletion will not affect the property of DFVS. Hence, the algorithm in its last step greedily removes vertices from DFVS(G_1, G_2) as much as possible, as long as the remaining vertices still form a DFVS. Let OPT_1 and OPT_2 be the optimal values of MFVS on G_1 and G_2 respectively. Let OPT be the optimal value of MDFVS on G_1 and G_2. It is obvious that $\text{OPT}_1, \text{OPT}_2 \leq$ OPT. Since $|\text{FVS}(G_1)| \leq \lambda_1 \text{OPT}_1$, where λ_1 is the length of a longest simple cycle in G_1, and $|\text{FVS}(G_2)| \leq \lambda_2 \text{OPT}_2$, where λ_2 is the length of a longest simple cycle in G_2, we get $\text{DFVS}(G_1, G_2) \leq 2\lambda\text{OPT}$, where $\lambda = max\{\lambda_1, \lambda_2\}$. Since the algorithm APPROX-MFVS can be implemented in $O(n^3)$ worst-case running time, the algorithm APPROX-MDFVS also runs in $O(n^3)$ time. □

4.2. *The Final Heuristic Algorithm for Breakpoint Distance*

Following the above discussion, we present an efficient heuristic algorithm, denoted as BDPOG, to calculate $d_b(\Pi, \Gamma)$ in four steps, given DAG(Π) and DAG(Γ):

(1) Add two vertices (*e.g.* v_0 and v_{n+1}) to the two input DAGs. In each DAG, add arcs from v_0 to all the vertices with in-degree 0, and add arcs from all the vertices with

```
Algorithm APPROX-MDFVS(G1 (V, A1),G2(V, A2) )
/* G1 and G2 are two directed graphs with the same vertex set and different arc sets.*/
1. FVS(G1) ← APPROX-MFVS(G1)
2. FVS(G2) ← APPROX-MFVS(G2)
3. DFVS ←FVS(G1) ⋃FVS(G2)
5. for each w ∈ DFVS
6.    if G1(V\ DFVS ∪{w}) and G2(V\ DFVS ∪{w}) are both acyclic
7.    then DFVS ← DFVS \{w}
8. Output DFVS
```

Figure 4. The approximation algorithm for MDFVS.

out-degree 0 to v_{n+1}.

(2) Derive the possible common adjacency set $\mathcal{S}$ from the DAGs and construct the adjacency-order graphs $\mathcal{G}_\Pi$ and $\mathcal{G}_\Gamma$.

(3) Find a double feedback vertex set for $\mathcal{G}_\Pi$ and $\mathcal{G}_\Gamma$, denoted as DFVS$(\mathcal{G}_\Pi, \mathcal{G}_\Gamma)$, by applying the APPROX-MDFVS algorithm.

(4) Output $n + 1 - |\mathcal{S}| + |\text{DFVS}(\mathcal{G}_\Pi, \mathcal{G}_\Gamma)|$ as $d_b(\Pi, \Gamma)$ and the corresponding total orders of Π and Γ.

It is obvious that the performance of the BDPOG algorithm directly depends on the performance of the APPROX-MDFVS algorithm. The construction of the adjacency-order graphs in step 2 takes $O(n^3)$ time, where n is the total number of genes, since it involves a transitive closure construction. Since the APPROX-MDFVS algorithm runs in $O(n^3)$ time, the overall running time of the BDPOG algorithm is $O(n^3)$.

5. Experimental Results

In order to test the performance of the BDPOG algorithm, we have applied it to both simulated data and real biological data. We will also use an example from the Comparative Grass Genomics database (http://www.gramene.org) to illustrate the application of our method on real data.

5.1. *Simulated Data*

We use simulated data to assess the performance of our algorithm on computing the breakpoint distance between two partially ordered genomes. The simulated data is generated as follows. Start from a genome G with n distinct symbols whose signs are generated randomly. Perform r reversals on the genome G to obtain another genome H. The boundaries of these reversals are uniformly distributed within the range of the genome. The maps of these two simulated genomes are generated according to two parameters: the *group rate* p corresponding to the probability of a gene being placed at the same position as the next gene, and the *missing rate* q that determines how many genes are missing from the map. Each gene is subjected independently to these two events. Note that every gene has to exist in at least one map of each genome. Then we combine all the map datasets for each genome into a DAG. Clearly, these two DAGs represents two partially ordered genomes g

and h generated from genomes G and H. The quadruple (n, r, p, q) specifies the parameters for generating two partially ordered genomes as test data.

We run BDPOG on 20 random instances for each combination of parameters. The average breakpoint distance between partially ordered genomes g and h, computed by BDPOG, is compared with the average breakpoint distance between totally ordered genomes G and H. The results are shown in Figure 5. As we can see from the figure, our heuristic algorithm is quite reliable in computing the breakpoint distance between two partially ordered genomes. On average, the distance computed by BDPOG algorithm is very close to the real breakpoint distance between the totally ordered genomes. The difference between two breakpoint distances generally increases as two genomes become more related, or the uncertainty of gene orders increases, *e.g.*, increasing (p, q) from $(0.2, 0.1)$ to $(0.4, 0.2)$.

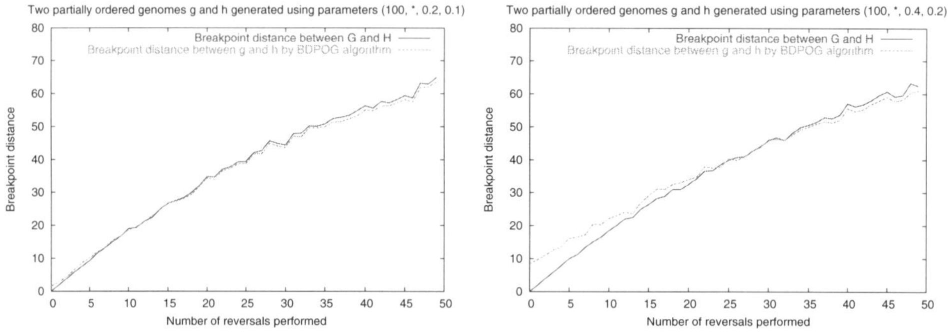

Figure 5. Performance of our heuristic algorithm BDPOG on simulated data.

5.2. *Real Data*

We use the X chromosomes of human (*Homo sapiens*, UCSC hg18, March 2006), mouse (*Mus musculus*, UCSC mm8, March 2006), and rat (*Rattus norvegicus*, UCSC rn4, November 2004) genomes in our real data test. In these three datasets downloaded from the UCSC Genome Browser [6] website (http://genome.ucsc.edu), all the genes are totally ordered. We perform the test between each pair of genomes, where we extract all the gene orthologs between the two compared genomes. Then we generate two partially ordered genomes for the compared genomes using the method described in the previous section, although we need to specify the group rate p and missing rate q here. By using our heuristic algorithm, the breakpoint distance between the simulated partially ordered genomes and the possible total order for each genome can be determined. We run our heuristic algorithm BDPOG on ten random instances, and compare the average estimated breakpoint distance and the gene orders with the real ones. The results are shown in Table 1. For example, if we generate the partially ordered chromosomes for the X chromosomes of human and mouse by using parameters $p = 0.2$ and $q = 0.1$, we get 44.15 as the average estimated breakpoint distance. In the total gene orders output by our algorithm, an average of 384.05 gene adjacencies among 388 are kept for the human X chromosome and an average of 382.9 gene adjacencies among 388 are kept for the mouse X chromosome. Note that, the average estimated breakpoint distance 44.15 is smaller than the real breakpoint distance between human and mouse, *i.e.*, 45. A possible reason is that a small amount of uncertainties in gene order might actually decrease the number of reversals between two genomes. Overall, the results demonstrate that our algorithm performs very well on estimating breakpoint distance and recovering the gene orders for partially ordered genomes.

Table 1. Comparison of the estimated breakpoint distances and the gene orders with the real ones. $^{\varsigma}$ The number of the common gene adjacencies exist in both the real genome G and the total order of the partially ordered genome g obtained by BDPOG.

G/H	#orthologs	$d_b(G, H)$	g and h ($p = 0.2, q = 0.1$)			g and h ($p = 0.4, q = 0.2$)		
			estimated $d_b(g, h)$	common adjs in g and G^{ς}	common adjs in h and H	estimated $d_b(g, h)$	common adjs in g and G	common adjs in h and H
human/mouse	389	45	44.15	384.05	382.9	63.75	380.3	379.4
human/rat	132	22	21.3	129.9	131	27.2	128.5	126.4
mouse/rat	126	17	15.65	124.3	124.05	20	123.45	121.95

To further illustrate the application of our method on real data, we use an example from the Comparative Grass Genomics database (http://www.gramene.org). We examine two closely related genomes, maize and sorghum. We used the "IBM2 neighbors 2004" and the "IBM neighbors maps" [9] for chromosome 1 of maize, and compared it with the "Paterson 2003" [1] and the "Klein 2004" [7] maps for the chromosome labeled C and $LG{-}01$, respectively, of sorghum. All markers of maize indicated as having a homolog in one of the datasets of sorghum are extracted, and vice versa. We extracted 21 markers in total. The two DAGs constructed from the maize datasets and sorghum datasets and the total order of the DAGs output by our algorithm are shown in Figure 6.

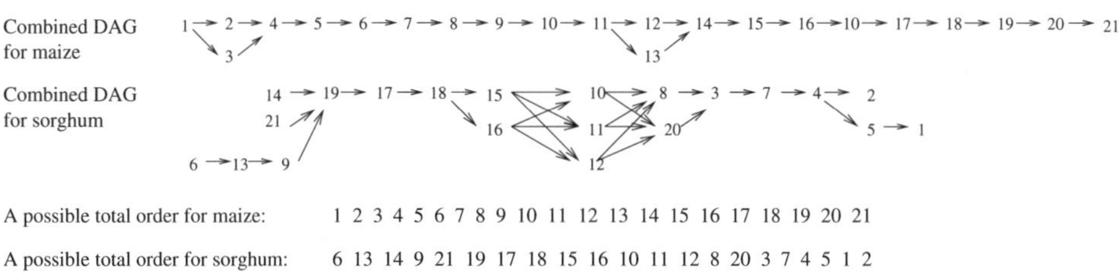

Figure 6. A comparison of maize and sorghum chromosomes using partially ordered data from the Gramene database.

6. Conclusion

In this paper, we have presented some complexity and algorithmic results for the problem of comparing two partially ordered genomes. In particular, we proposed an efficient heuristic algorithm to estimate the breakpoint distance between two partially ordered genomes and infer the corresponding linearizations achieving the distance. In our construction, we defined a useful tool, called the adjacency-order graph, and introduced a new optimization problem (MDFVS), for which we designed an efficient approximation algorithm. Our preliminary experiments on simulated and real data have demonstrated that our algorithm performs very well on estimating breakpoint distance and recovering the gene orders for partially ordered genomes. Considering the breakpoint distance is just the first step. In the future, we plan to look into other distances between partially ordered genomes, *e.g.*, the reversal distance, and try to design more efficient algorithms.

7. Acknowledgement

This project is supported in part by NSF grant CCR-0309902, National Key Project for Basic Research (973) grant 2002CB512801, NSFC grant 60528001, and a Changjiang Visiting Professorship at Tsinghua University.

References

1. J.E. Bowers *et al.*, "A high-density genetic recombination map of sequence-tagged sites for sorghum, as a framework for comparative structural and evolutionary genomics of tropical grains and grasses," *Genetics*, 165:367-386, 2003.
2. A. Caprara, "Sorting by reversal is difficult," *Proc. 1St RECOMB*, pp.75-83, 1997.
3. C. Demetrescu and I. Finocchi, "Combinatorial Algorithms for Feedback Problems," *Information Processing Letters*, 86(3):129-136, 2003.
4. G. Even *et al.*, "Approximating minimum feedback sets and multi-cuts in directed graphs," *Proc. 4th Int. Conf. on Integer Prog. and Combinatorial Optimization, Lecture Notes in Comput. Sci.*, 920, Springer-Verlag, 14-28, 1995.
5. S. Hannenhalli and P. Pevzner, "Transforming cabbage into turnip (Polynomial algorithm for sorting signed permutations by reversals)," *Proc. of 27th Annual ACM Symposium on the Theory of Computing*, pp.178-187, 1995.
6. D. Karolchik *et al.*, "The UCSC Genome Browser Database," *Nucleic Acids Res.*, 31(1): 51-54, 2003.
7. M.A. Menz *et al.*, "A high-density genetic map of Sorghum bicolor (L.) Moench based on 2926 AFLP, RFLP and SSR markers," *Plant molecular biology* 48:483-499, 2002
8. J. Nadeau and B. Taylor, "Lengths of chromosomal segments conserved since divergence of man and mouse," *Proc. Natl. Acad. Sci.*, 81:814-818, 1984.
9. M.L. Polacco and Jr Coe E., "IBM neighbors: a consensus GeneticMap," 2002
10. P.D. Seymour, "Packing directed circuits fractionally," *Combinatorica*, 15:281-288, 1995.
11. D. Sankoff, "Mechanisms of genome evolution: models and inference," *Bull. Int. Stat. Institut.*, 47:461-475, 1989.
12. G. Watterson *et al.*, "The chromosome inversion problem," *J. Theor. Biol.*, 99:1-7, 1982.
13. C. Zheng and D. Sankoff, "Genome rearrangements with partially ordered chromosomes," *COCOON*, 2005.
14. C. Zheng, A. Lenert, and D. Sankoff, "Reversal distance for partially ordered genomes," *Bioinformatics*, 21(Suppl.1):i502-i508, 2005.

INFERRING GENE REGULATORY NETWORKS BY MACHINE LEARNING METHODS

JOCHEN SUPPER, HOLGER FRÖHLICH, CHRISTIAN SPIETH, ANDREAS DRÄGER, ANDREAS ZELL*

Centre for Bioinformatics Tübingen (ZBIT), Sand 1, 72076 Tübingen
jochen.supper@uni-tuebingen.de

The ability to measure the transcriptional response after a stimulus has drawn much attention to the underlying gene regulatory networks. Several machine learning related methods, such as Bayesian networks and decision trees, have been proposed to deal with this difficult problem, but rarely a systematic comparison between different algorithms has been performed. In this work, we critically evaluate the application of multiple linear regression, SVMs, decision trees and Bayesian networks to reconstruct the budding yeast cell cycle network. The performance of these methods is assessed by comparing the topology of the reconstructed models to a validation network. This validation network is defined *a priori* and each interaction is specified by at least one publication. We also investigate the quality of the network reconstruction if a varying amount of gene regulatory dependencies is provided *a priori*.

1. Introduction

Transcriptional data sets provide valuable insight to cellular processes under various conditions. These data sets can be analyzed by cluster analysis, thereby providing undirected gene relations. In order to model gene regulatory networks (GRN) directed relations between genes have to be considered. Most GRNs can be represented as interconnections between genes, each indicating that one gene influences the expression of another gene.

Today, modeling of GRNs is guided by a rich flow of experimental data. The stream is still widened by an increasing pool of measurement techniques. Despite of all this information, detailed knowledge regarding network models is still almost exclusively collected by biologists. They collect and integrate data, expand and refine their models and finally validate them. For our modeling efforts, we will concentrate on the regulatory information that can be extracted solely from transcriptional response data. The restriction to transcriptional response data provides us with a large number of measured genes along with a small sampling rate. This, of course, leads to a high level of ambiguity for every GRN reconstruction method.

Several approaches for GRN reverse engineering have emerged during the last years. These approaches include analytical methods such as Boolean networks[12], (non)-linear

*This work was supported by the National Genome Research Network (NGFN II) of the Federal Ministry of Education and Research in Germany under contract number 0313323.

networks[19] and differential equations[3] but also machine learning methods such as decision trees[16] and Bayesian networks[9].

To reconstruct a GRN, a set of transcriptional response measurements has to be available. Given this data, one of the above mentioned reconstruction methods may be employed to untangle the underlying topological structure of the interaction network. One problem thereby is that it is very hard to validate the performance of the proposed approaches. This makes it difficult to compare methods and even more to judge, if certain approaches are helpful at all. In previous publications GRN models have been validated mostly by co-citation[16] or on artificial data[18]. Despite these efforts no general validation method has emerged.

In this work, we present one network that has been investigated thoroughly and where the interactions are known in many cases. This network is a subset containing 20 genes involved in the budding yeast cell cycle defined by Chen *et al.*[2] for which Spellman *et al.*[17] and Cho *et al.*[4] publicly provide time-series measurement data. This enables us to build a validation network, for which the interactions can be specified. Additionally, it allows us to systematically investigate, how prior knowledge on parts of the networks changes the validity of results obtained by an automatic GRN reconstruction. Thereby, we concentrate on machine learning methods, such as Bayesian networks[8], multiple linear regression, CART decision trees[1] and SVMs[5]. For the last three we closely follow the framework proposed by Soinov *et al.*[16]. They used a so-called wrapper approach[11] in combination with decision trees to learn the minimal subsets of genes, which best predict the up/down regulation of a considered gene. By comparing these different approaches we build upon the work of Husmeier *et al.*[9], who performed a sensitivity and specificity analysis of GRN reconstruction for Bayesian networks.

2. Materials and Methods

2.1. *Data*

2.1.1. *Budding Yeast Cell Cycle*

The biological model used for this research is the budding yeast cell cycle, which has been thoroughly investigated over many years. Cho *et al.*[4] and Spellman *et al.*[17] contributed to these investigations by publicly providing a large transcriptional data set. They measured the progression of the cell cycle with different synchronization techniques. Altogether this results in 73 time point measurements. These measurements were performed on microarrays[6], each consisting of 6178 data points, from which we select a subset containing 20 genes. This is done according to Chen *et al.*[2], who did an extensive literature search to set up a system of differential equations to define the topology of the GRN. In addition to the interactions provided by the differential equations we searched TRANSFAC[20], Entrez Gene[13] and the *Saccharomyces* Genome Database (SGD[10]) for known dependencies between a pair of genes. The entire network contains 56 interactions and is depicted in Figure 1. It will serve as our validation network for the studies performed in this paper. Although this validation network might contain some false interactions or others, which were not active at the time the measurements were taken, we can nevertheless rank our reconstructions with regard to their closeness to this network. That means, the closeness of an inferred GRN to the validation network should not be understood in an absolute, but in a relative sense.

2.1.2. *Preprocessing and Availability of the Data*

The data set is normalized by the average $\log_2$ ratio, which implicitly describes a non-linear relationship between the genes. We also performed pre-experiments without normalization and with normalization through a sigmoidal function, but found the results to be inferior.

The data set as well as the described models are all available from public data sources. An SBML version of the topological validation network is available on our homepage[a].

2.2. *Machine Learning Methods for GRN Reconstruction*

Our starting point is a gene-expression matrix $\mathbf{X} \in \mathbb{R}^{20\times 73}$, where each row represents a gene and each column represents a sample taken at a specific time step. That means, an element X_{ij} of $\mathbf{X}$ indicates the expression level of gene i in sample j. We consider Bayesian networks (BN), multiple linear regression (MLR), CART decision trees (CART) and Support Vector Machines (SVMs) for GRN reconstruction from this data.

2.2.1. *Bayesian Networks*

For learning the GRN with a BN we discretized the data in the following way: For each gene i we distinguish only the two states "expressed above average" and "expressed below average". That means we transform each entry X_{ij} to an entry Y_{ij} , which is defined as:

$$Y_{ij} := \begin{cases} 1 \; X_{ij} \geq \bar{X}_i, \text{ where } \bar{X}_i \text{ is the average} \\ \quad \text{expression level of gene } i \\ 0 \text{ otherwise} \end{cases}$$

This is done in accordance to Soinov *et al.*[16]. We then learn the structure of a dynamic BN using a MCMC search in the structure space as proposed by Husmeier *et al.*[9]. Thereby we use the MATLAB™ code provided on his homepage[b]. After training the dynamic BN we construct a GRN by only considering those dependencies, for which the expected posterior probability is above average.

2.2.2. *Multiple Linear Regression*

The GRN reconstruction by means of MLR resembles the framework by Soinov *et al.*[16]: For each gene i we identify two prediction problems:

---

[a]www-ra.informatik.uni-tuebingen.de/mitarb/supper/ml/

[b]http://www.bioss.sari.ac.uk/~dirk

- The prediction of the expression of gene i in sample j from all other genes in sample j.
- The prediction of the expression of gene i in sample j from all other genes in sample $j-1$.

In both cases we search for a minimal combination of genes that allows to predict the expression of gene i reliably. This is achieved by considering only those genes, for which the Pearson correlation of the expression level with gene i is at least 60 %. These genes are subsequently selected to train a MLR model that predicts the expression level of gene i. If the 10-fold cross-validated mean correlation of the model output with the true expression level of gene i is above 60 %, then the MLR model is considered as reliable and the selected genes are considered as probable regulators for gene i in the GRN reconstruction. The bottom line is that the MLR reconstruction can be viewed as a correlation network with directed edges.

2.2.3. *Decision Trees*

In case of the GRN inference by means of CART we formulate the two prediction tasks from the last paragraph as classification rather than regression tasks. This allows to follow the framework by Soinov *et al.* directly. More specifically, we now have three prediction problems instead of the two stated above:

- the prediction of the state Y_{ij} of gene i in sample j from the expression levels of all other genes in sample j.
- the prediction of the state Y_{ij} of gene i in sample j from the expression levels of all other genes in sample $j-1$.
- the prediction of the *change* of state Y_{ij} of gene i in sample j from the state changes of all other genes in sample j.

The change of state Y_{ij} is either "equal", "regulated up" or "regulated down". That means in the first two cases we have a binary and in the last one a three-class classification problem. We use the CART implementation provided in the MATLAB™ 7.0 statistics toolbox with pruning turned on and the Gini diversity index as node split criterion. This way selecting a good combination of genes, which allow forecasting the state of gene i reliably, is embedded into the learning of the decision tree. Similar to above, we set an accuracy threshold of 75% beyond which we consider the predictions made by the CART model as acceptable.

2.2.4. *Support Vector Machines*

SVMs have attracted a high interest within the bioinformatics community during the last years due to their good prediction performance for various tasks. They rely on principles from statistical learning theory[15]. The idea is to construct an optimal hyperplane between two classes +1 and -1 such that the margin, i.e. the distance of the hyperplane to the point closest to it, is maximized. To allow for nonlinear classification, so-called kernel functions are employed, which can be thought of as special similarity measures. They implicitly map the original data into some high dimensional feature space, in which the optimal hyperplane can be found.

In our case we consider linear kernels $k(\mathbf{x}, \mathbf{x}') = \langle \mathbf{x}, \mathbf{x}' \rangle$ as well as polynomial kernels of degree 2 $k(\mathbf{x}, \mathbf{x}') = \langle \mathbf{x}, \mathbf{x}' \rangle^2$, where $\mathbf{x}$ and $\mathbf{x}'$ are the expression levels of all genes except of gene i in sample j. The polynomial kernel implicitly computes all pairwise products between expression levels of two genes. This way not only linear, but also nonlinear dependencies between gene expressions can be captured.

In addition to a kernel function, a soft margin parameter C has to be fixed. In our case we choose C from the grid $2^{-2} ... 2^{14}$ by means of 5-fold cross-validation. To determine for each gene i, which genes are suited best to predict its state, we employ the RFE[7]. This algorithm successively eliminates that gene, which influences the size of the margin least. The termination of this procedure is determined by an additional 10-fold cross-validation.

Of course, a direct comparison of the different methods for GRN reconstruction introduced is not unproblematic, because each algorithm depends on certain parameter settings and different data formats are in use. Nevertheless, we think that a comparative study, even if we should not forget about its limitations, might be useful to gain some insights.

2.3. *Network Validation*

2.3.1. *Statistical Stability of The Solution*

For validating all of the above approaches we are interested in those parts of the true network, which are reconstructed in a *statistically stable* way by each single method. That means, we are interested in those inferred gene regulatory dependencies, which are not sensitive to the respective training data, but to the underlying biological process. For this purpose we use 10-fold cross-validation: We randomly split the measurement data into 10 parts, train our model on 9 parts and then test the model on the remaining part. This procedure is iterated until each part is left out exactly once for testing. At the end we only use those connections consistently inferred during the 10-fold cross-validation. Hence, the resulting network can be seen as a consensus model, integrating results from different data splits.

2.3.2. *Validating the Topology*

We validate the network topologies obtained from the different consensus models for each algorithm by calculating the following statistics:

(1) the fraction of correctly identified edges in the validation network (*recovered* connections)
(2) the fraction of correctly constructed edges in relationship to all constructed edges (*direct* connections)
(3) the fraction of constructed edges connecting genes with topological distance 2 in relationship to all constructed edges (*indirect* connections): If in the validation

network we have $a \to b \to c$ and we reconstruct $a \to c$, then the included edge directly models an indirect regulatory influence.

(4) the same as 3. with distances > 2 (*spurious* connections).

(5) the graph edit distance (*GED*) between the constructed and the correct GRN: The graph edit distance describes the minimal number of edit operations (edge insertion/edge removal) that transform one graph into another one. In our case we use the algorithm by Kelly *et al.*[14] to calculate this distance.

The first statistic can be viewed as a sensitivity measure for the GRN reconstruction algorithm, whereas 2. - 4. describe the specificity. Discriminating between different types of inferred edges here seems beneficial, because it allows a better insight into the quality of the reconstruction. The graph edit distance, on the other hand, is a combined statistic capturing both, the number of correctly recovered dependencies and the number of inferred edges, which do not exist in the validation network. We think that the discrimination between sensitivity and specificity of the GRN reconstrucion is necessary, because, there is a trade-off between the fraction of correctly identified relations in the validation network and the fraction of all inferred connections, which are correct. A maximization of the first goal could be achieved trivially by connecting every gene in the network, which would be rather naive. In contrast, a pure maximization of the second goal would lead to an edge free graph, obviously containing no false connection. A good GRN reconstruction should therefore find a fair balance between a high number of inferred edges and a low number of spurious connections.

3. Results

3.1. *Comparison of Different Methods*

We validated the machine learning methods introduced in the last section as described above. The results in Table 1 show a relatively low statistical stability for all methods. The Bayesian network reconstruction (Fig. 1) leads to the lowest number of inferred edges (only 7) and hence to a very low sensitivity (only 1 connection of the true network was recovered). At the same time the fraction of indirect and spurious dependencies among these 7 was relatively high. Also the graph edit distance was the highest among all methods.

MLR, CART, linear and polynomial SVMs all recovered a substantially higher number of relations of the validation network. Thereby the recovery rate of the validation network was clearly highest for the linear SVM reconstruction and second highest for the MLR reconstruction. At the same time the fraction of direct connections in the inferred GRN was highest for the MLR reconstruction. The fraction of indirect connections was highest in the CART model. The linear SVM had the highest fraction of spurious edges. However, at the same time the graph edit distance was lowest for this model and second lowest for the polynomial SVM model.

All in all we observe that the dynamic BN is outperformed by the other methods. Among these we favor the linear SVM model, since it has the lowest graph edit distance, indicating a fair trade-off between the number of recovered edges from the true validation network and spurious or false inferred dependencies.

Table 1. Validation of the GRN reconstruction with different methods. All statistics in % (see subsection 2.3 for explanation).

	BN	MLR	CART	lin. SVM	poly. SVM
recov.	1.79	7.14	3.57	10.71	5.36
direct	14.29	25.0	18.18	16.67	15.79
indir.	42.86	43.75	54.55	38.89	31.58
spur.	42.86	31.25	27.27	44.44	36.84
GED	35	32	32	**30**	31

3.2. *Effect of Prior Knowledge*

In a second study we concentrated on the linear SVM reconstruction method and investigated the influence of incorporating prior knowledge of certain relations in the GRN. For this purpose we modified the procedure described in the last section such that for a gene k, which is known to influence gene i, the influence on the margin is explicitly set to ∞. This way the RFE algorithm is forced to rank such a gene highest. Furthermore, known relations are drawn in the GRN even if the classification accuracies for gene i are below the prescribed threshold of 75%.

In Figure 2a we depict the influence of prior knowledge on the sensitivity and specificity statistics, if 10, 20, ..., 50% randomly selected relations of the validation network are known. The results are averaged over 10 trials. As one can see the number of recovered edges increases in a piecewise linear fashion with the increase of the prior knowledge. The increase from no prior knowledge to 10% known edges is higher than e.g. from 10% to 20%. While at the beginning we have a gain of almost 15%, thereafter the gain decreases to around 10% only and is hence at the same level as the number of edges additionally provided by prior knowledge. The fraction of direct edges increases parallel to the fraction of recovered edges. The fraction of indirect and spurious connections decreases in a roughly linear fashion with the fraction of known relations. While at the beginning the largest fraction of all constructed edges is spurious, with 20% knowledge it is roughly at same level and with 30% below the fraction of direct relations.

As expected, the total number of relations in the inferred GRN, which also includes the connections drawn by prior knowledge, increases with the fraction of known edges (Fig. 2b). In contrast, the number of newly inferred edges decreases with the increase of prior knowledge, which seems surprising at the first glance. However, this phenomenon might be due to a lower number of missing edges with an increasing number of already known ones. In the graphs from Figure 2a we see the same piecewise linear behavior as in Figure 2b. Again, the increase from 0 to 10% prior knowledge has a higher impact than e.g. from 20 to 30%.

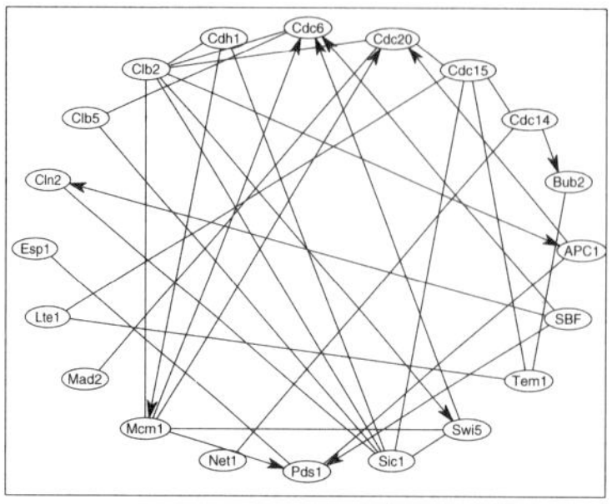

(a) The validation network: An arrow $a \longrightarrow b$ indicates that b is regulated by a. Edges with no arrows indicate a mutual influence $a \longrightarrow b$ and $b \longrightarrow a$.

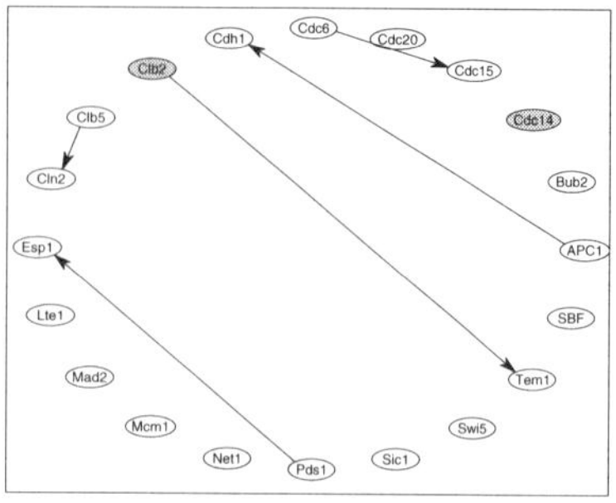

(b) The GRN reconstructed by the Bayesian network. Gray nodes indicate self-regulation.

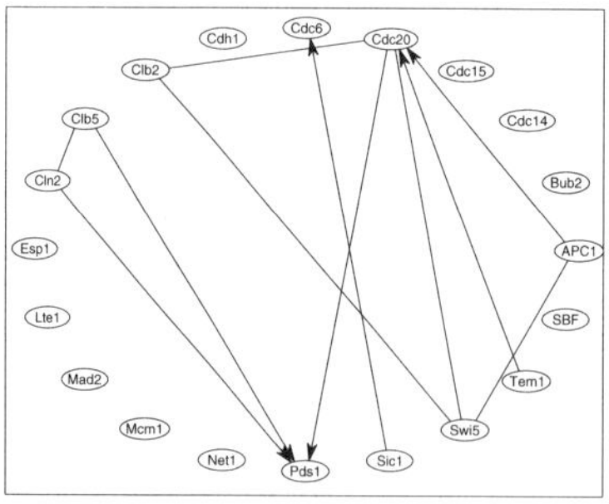

(c) The GRN reconstructed by multiple linear regression.

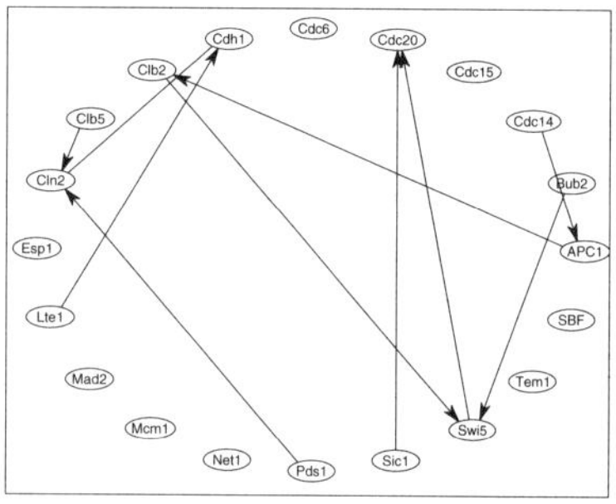

(d) The GRN reconstructed by the CART method.

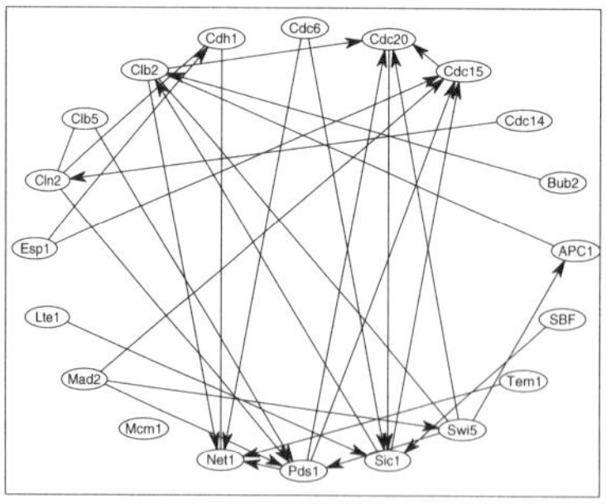

(e) The GRN reconstructed by the linear SVM method.

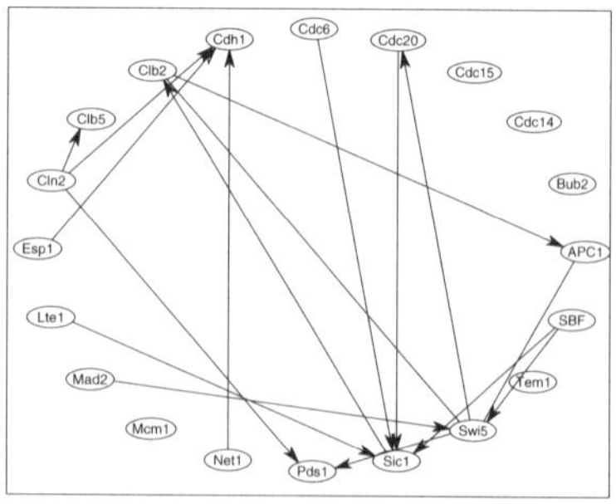

(f) The GRN reconstructed by the polynomial SVM method.

Figure 1. Literature network and reconstructions by different methods.

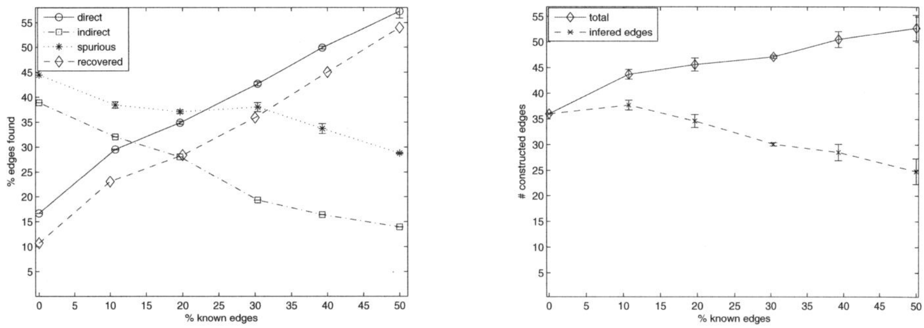

(a) Effect of incorporating prior knowledge on randomly selected edges: sensitivity and specificity statistics. Average values over 10 trials ($\pm$ std. dev.).

(b) Effect of incorporating prior knowledge on randomly selected edges: total number and number of newly inferred edges. Average values over 10 trials ($\pm$ std. dev.).

Figure 2. Effect of incorporating prior knowledge on randomly selected edges.

4. Conclusion

In this paper we systematically compared different machine learning methods for GRN reconstruction. We considered Bayesian networks, multiple linear regression, CART decision trees, linear and nonlinear SVMs. We developed a framework for evaluating the inference methods with regard to their statistical stability by using 10-fold cross-validation. A well investigated biological data set from the literature served as our basis. This enabled us construct a validation network against which we could compare our results by means of sensitivity and specificity analysis. We found linear SVMs to produce slightly superior reconstructions compared to the other methods, especially to dynamic Bayesian networks. Thereby the inference scheme closely followed the method proposed by Soinov *et al.*[16] for decision trees. However, it has to be remarked that our results depend on the specific parameter and data format settings for the individual algorithms.

We additionally investigated the influence of prior knowledge on the quality of the learned network topology. We found that adding known connections has a relatively large positive impact, when no prior knowledge existed before, whereas if further increasing the prior knowledge the gain becomes smaller.

We think that the main benefit of this work lies in a first trial to compare different machine learning methods for GRN reconstruction in a systematic manner, which to our best knowledge has rarely been done before. In our future work we will extend our research on different data sets from the one used here and evaluate the results in a similar manner.

References

1. L. Breiman, J. Friedman, R. Olshen, and C. Stone. *Classification and Regression Trees*. Wadsworth and Brooks, 1984.
2. Katherine C Chen, Laurence Calzone, Attila Csikasz-Nagy, Frederick R Cross, Bela Novak, and John J Tyson. Integrative analysis of cell cycle control in budding yeast. *Mol Biol Cell*, 15(8):3841–3862, Aug 2004.
3. Kuang-Chi Chen, Tse-Yi Wang, Huei-Hun Tseng, Chi-Ying F Huang, and Cheng-Yan Kao. A stochastic differential equation model for quantifying transcriptional regulatory network in Saccharomyces cerevisiae. *Bioinformatics*, 21(12):2883–90, Jun 2005.
4. R. J. Cho, M. J. Campbell, E. A. Winzeler, L. Steinmetz, A. Conway, L. Wodicka, T. G. Wolfsberg, A. E. Gabrielian, D. Landsman, D. J. Lockhart, and R. W. Davis. A genome-wide transcriptional analysis of the mitotic cell cycle. *Mol Cell*, 2(1):65–73, Jul 1998.
5. C. Cortes and V. Vapnik. Support vector networks. *Machine Learning*, 20:273 – 297, 1995.
6. M. B. Eisen and P. O. Brown. DNA arrays for analysis of gene expression. *Methods Enzymol*, 303:179–205, 1999.
7. I. Guyon, J. Weston, S. Barnhill, and V. Vapnik. Gene Selection for Cancer Classification using Support Vector Machines. *Machine Learning*, 46:389 – 422, 2002.
8. D. Heckerman. A tutorial on learning with bayesian networks. *Data Mining and Knowledge Discovery*, 1:79 – 119, 1997.
9. Dirk Husmeier. Sensitivity and specificity of inferring genetic regulatory interactions from microarray experiments with dynamic Bayesian networks. *Bioinformatics*, 19(17):2271–2282, Nov 2003.
10. Laurie Issel-Tarver, Karen R Christie, Kara Dolinski, Rey Andrada, Rama Balakrishnan, Catherine A Ball, Gail Binkley, Stan Dong, Selina S Dwight, Dianna G Fisk, Midori Harris, Mark Schroeder, Anand Sethuraman, Kane Tse, Shuai Weng, David Botstein, and J. Michael Cherry. Saccharomyces Genome Database. *Methods Enzymol*, 350:329–346, 2002.
11. R. Kohavi and G. John. Wrappers for Feature Subset Selection. *Artificial Intelligence*, 97(12):273 – 324, 1997.
12. S. Liang, S. Fuhrman, and R. Somogyi. Reveal, a general reverse engineering algorithm for inference of genetic network architectures. *Pac Symp Biocomput*, 1:18–29, 1998.
13. Donna Maglott, Jim Ostell, Kim D Pruitt, and Tatiana Tatusova. Entrez Gene: gene-centered information at NCBI. *Nucleic Acids Res*, 33(Database issue):D54–D58, Jan 2005.
14. A. Robes-Kelly and E. Hancock. Edit distance from graph spectra. In *Proc. 9th IEEE Int. Conf. Comp. Vis.*, volume 1, pages 234 – 241, 2003.
15. B. Schölkopf and A. J. Smola. *Learning with Kernels*. MIT Press, Cambridge, MA, 2002.
16. Lev A Soinov, Maria A Krestyaninova, and Alvis Brazma. Towards reconstruction of gene networks from expression data by supervised learning. *Genome Biol*, 4(1):R6, 2003.
17. P. T. Spellman, G. Sherlock, M. Q. Zhang, V. R. Iyer, K. Anders, M. B. Eisen, P. O. Brown, D. Botstein, and B. Futcher. Comprehensive identification of cell cycle-regulated genes of the yeast Saccharomyces cerevisiae by microarray hybridization. *Mol Biol Cell*, 9(12):3273–3297, Dec 1998.
18. Jochen Supper, Christian Spieth, and Andreas Zell. Reverse engineering non-linear gene regulatory networks based on the bacteriophage lambda ci circuit. In *IEEE Symposium on Computational Intelligence in Bioinformatics and Computational Biology (CIBCB)*, 2005.
19. D. C. Weaver, C. T. Workman, and G. D. Stormo. Modeling regulatory networks with weight matrices. *Pac Symp Biocomput*, 1:112–23, 1999.
20. Edgar Wingender. TRANSFAC, TRANSPATH and CYTOMER as starting points for an ontology of regulatory networks. *In Silico Biol*, 4(1):55–61, 2004.

A NOVEL CLUSTERING METHOD FOR ANALYSIS OF BIOLOGICAL NETWORKS USING MAXIMAL COMPONENTS OF GRAPHS

MORIHIRO HAYASHIDA AND TATSUYA AKUTSU

Bioinformatics Center, Institute for Chemical Research, Kyoto University
Uji-city, Kyoto 611-0011, Japan
E-mail: {morihiro, takutsu}@kuicr.kyoto-u.ac.jp

HIROSHI NAGAMOCHI

Department of Applied Mathematics and Physics, Graduate School of Informatics, Kyoto University
Kyoto-city, Kyoto 606-8501, Japan
E-mail: nag@amp.i.kyoto-u.ac.jp

This paper proposes a novel clustering method for analyzing biological networks. In this method, each biological network is treated as an undirected graph and edges are weighted based on similarities of nodes. Then, maximal components, which are defined based on edge connectivity, are computed and the nodes are partitioned into clusters by selecting disjoint maximal components. The proposed method was applied to clustering of protein sequences and was compared with conventional clustering methods. The obtained clusters were evaluated using P-values for GO (GeneOntology) terms. The average P-values for the proposed method were better than those for other methods.

1. Introduction

Clustering is one of fundamental techniques in bioinformatics. Indeed, many clustering methods have been developed and/or applied for analyzing various kinds of biological data. Among them, such hierarchical clustering methods as the single-linkage, complete-linkage and average-linkage methods have been widely used [3,9]. However, these clustering methods are based on similarities between two elements or two clusters, and relations with other elements or clusters are not so much taken into account.

Relations between biological entities are often represented as networks or (almost equivalently) graphs. For example, nodes are proteins in a protein-protein interaction network, and two nodes are connected by an edge if the corresponding proteins interact with each other. For another example, nodes are again proteins in a sequence similarity network of proteins, and two nodes are connected by an edge if the corresponding protein sequences are similar to each other. Moreover, in this case, similarity scores are assigned as weights of edges. Since these networks are considered to have much information, clustering based on network structures might be useful. Of course, conventional clustering methods can be applied to clustering of nodes in these networks [3,9]. But, information on network structure is not so much taken into account by these methods. For an extreme example, suppose that the network is a complete graph and all edges have the same weight. Then, all the nodes should be put into one cluster and sub-clusters should not be created. However, conventional clustering methods create many sub-clusters. Therefore, clustering methods that utilize structural information on a network should be developed. Though clustering methods utilizing structural information have been developed [7,12], many of these are heuristic and/or recursive and thus it is unclear which properties are satisfied for the final clusters.

On the other hand, in graph theory and graph algorithms, the Gomory-Hu tree is well-known [5] where it is defined for an undirected network with weighted edges. This tree essentially contains all information on minimum cuts for all pairs of nodes. It is known that a Gomory-Hu tree can be computed efficiently using a maximum flow algorithm. Furthermore, maximal components can be efficiently computed from a Gomory-Hu tree [11], where a maximal component is a set of nodes with high connectivity (the precise definition is given in Sec. 2). It is known that a set of maximal components constitutes a laminar structure, which is essentially a hierarchical structure.

Based on the above facts, we develop a novel clustering method for an undirected network. In this method, nodes are partitioned into clusters by selecting disjoint maximal components. The method works in $O(n^2 m \log(n^2/m))$ time, where n and m are the numbers of nodes and edges, respectively. The Gomory-Hu tree was already applied to analysis of protein folding pathways [8,10,13]. However, to our knowledge, it was not applied to analysis of large scale protein sequence networks. Moreover, as to be shown in Sec. 3, our method employs additional ideas to effectively utilize the Gomory-Hu tree.

In this paper, we apply the proposed method to clustering of protein sequences and compare with the single-linkage, complete-linkage and average-linkage methods. We evaluate the computed clusters using *P*-values for GO (GeneOntology) terms. The results suggest the effectiveness of the proposed method.

The organization of the paper is as follows. In Sec. 2, we briefly review maximal components of undirected graphs and conventional clustering methods. In Sec. 3, we present our clustering method based on maximal components. In Sec. 4, we show the results on computational experiment. Finally, we conclude with future work.

2. Preliminaries

In this section, we review edge-connectivity and maximal components [11]. We also review three conventional hierarchical clustering methods: single-linkage, average-linkage and complete-linkage clustering methods.

2.1. *Edge-connectivity*

Let $G = (V, E)$ be an undirected edge-weighted graph with a vertex set V and an edge set E, where each edge e is weighted by a nonnegative real $c_G(e) \in \Re^+$. We define the edge-connectivity $\lambda_G(u, v)$ between two nodes u and v as follows:

$$\lambda_G(u, v) = \min_{\{X \subseteq V | u \in X, v \in V - X\}} \sum_{p \in X, q \in V - X} c_G(p, q). \quad (1)$$

A subset X of V is called (u, v)-cut if $u \in X$ and $v \in V - X$, or $u \in V - X$ and

$v \in X$. Among them, a (u, v)-cut X which gives a minimum $\lambda_G(u, v)$ is called a minimum (u, v)-cut.

2.2. *Maximal Components*

Definition 1. A subset X of V is called a *maximal component* if it satisfies the following conditions,

$$\forall u, v \in X \qquad \lambda_G(u, v) \geq l, \tag{2}$$

$$\forall u \in X, \forall v \in V - X \ \lambda_G(u, v) < l, \tag{3}$$

where $l = \min_{u,v \in X} \lambda_G(u, v)$. Such a subset X is also called an *l-edge-connected component.*

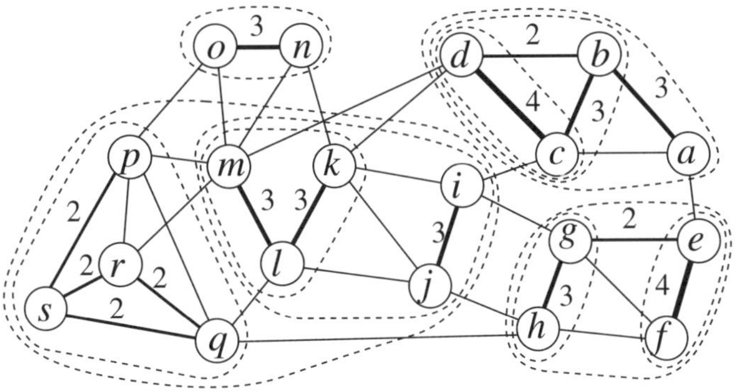

Figure 1. Illustration for maximal components of a graph $G = (V, E)$ with $V = \{a, b, c, d, e, f, g, h, i, j, k, l, m, n, o, p, q, r, s\}$ and E, where each number denotes the weight of the edge, and edges without numbers are weighted by 1. Each set of nodes surrounded with a dashed line is a maximal component of G. For example, the set $X = \{a, b, c, d\}$ is a maximal component because $\lambda_G(u, v) \geq 5$ for any $u, v \in X$, $\lambda_G(u, v) < 5$ for any $u \in X$ and $v \in V - X$, and $\min_{u,v \in X} \lambda_G(u, v) = 5$. It is also called a 5-edge-connected component.

Figure 1 shows an example of maximal components. Definition 1 means that the internal nodes of a maximal component are connected with each other more strongly than with any other external nodes. Moreover, nodes of internal maximal components are connected with each other more strongly than (and equally to) those of external maximal components which include the internal maximal components.

Definition 2. A family $\chi \subseteq 2^V$ is called *laminar* if $X \cap Y = \emptyset$, $X \subseteq Y$, or $Y \subset X$ for any sets $X, Y \in \chi$.

A laminar family χ is represented by a rooted tree $\boldsymbol{\tau} = (\boldsymbol{\nu}, \boldsymbol{\epsilon})$. The node set $\boldsymbol{\nu}$ is defined by $\boldsymbol{\nu} = \chi \cup \{V\}$, where V corresponds to the root of $\boldsymbol{\tau}$. Let t_X denote a node corresponding to a set $X \in \boldsymbol{\nu}$. For two nodes t_X and t_Y in $\boldsymbol{\tau}$, t_X is a child of t_Y if and only if $X \subset Y$ holds and χ contains no set Z with $X \subset Z \subset Y$.

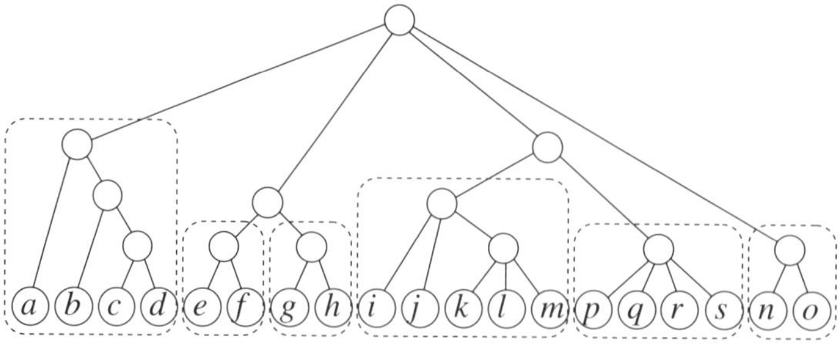

Figure 2. The rooted tree representation of maximal components $\chi(G)$ of the graph G in Fig. 1. The six sets of nodes surrounded by dashed lines are the resulting clusters provided by the procedure SelectLaminar.

Theorem 1. *Let $\chi(G)$ denote the set of all maximal components of G. Then, $\chi(G)$ is a laminar family.*

Proof. We assume that there exist three nodes x, y and z so that $x \in X - Y$, $y \in Y - X$, and $z \in X \cap Y$ for two maximal components $X, Y \in \chi(G)$, where X is an l-edge-connected component and Y is an h-edge connected component. We can assume without loss of generality that $l \geq h$. From $x, z \in X$ and Eq. 2 of the definition of maximal components for X, we have $\lambda_G(x, z) \geq l \geq h$. On the other hand, from $x \notin Y, z \in Y$ and Inequality (3) for Y, we have $\lambda_G(x, z) < h$. It contradicts our assumption. □

2.3. *Linkage Methods*

We briefly review three linkage clustering methods: single linkage (or nearest neighbor method), complete linkage (or farthest neighbor method), and average linkage. Each method starts with a set of clusters, where each cluster consists of a single distinct node. Then, two clusters having the minimum distance are merged into one cluster. This procedure is repeated until there is only one cluster. The distance $D(X, Y)$ between two clusters X and Y is defined in a different way depending on a clustering method:

$$D(X,Y) = \begin{cases} \min\limits_{x \in X, y \in Y} d(x,y) & \text{(for single linkage)} \\ \max\limits_{x \in X, y \in Y} d(x,y) & \text{(for complete linkage)} \\ \dfrac{1}{|X||Y|} \sum\limits_{x \in X, y \in Y} d(x,y) & \text{(for average linkage)} \end{cases}, \quad (4)$$

where $d(x, y)$ denotes the distance between two nodes x and y.

It should be noted that the distance between two nodes should be small if the similarity between these nodes is high, whereas the weight of the edge between these two nodes should be large. Since we are going to use the similarity score (which is high for similar nodes), we use modified versions of these clustering algorithms. In the modified versions, the clusters with the maximum score are merged, instead of the clusters with the minimum distance. Moreover, 'min' and 'max' in Eq. 4 are exchanged.

3. Selection of Disjoint Clusters from Hierarchical Structure

The set of all maximal components $\chi(G)$ of a graph G provides a hierarchical structure which can be represented as a rooted tree $\tau(G)$ because the set $\chi(G)$ is a laminar family. This structure gives a kind of hierarchical clustering. However, what we need is a set of disjoint clusters because we are interested in classification of protein sequences. That is, input nodes should be partitioned into disjoint clusters. Thus, we propose a method to find disjoint clusters from $\chi(G)$. In our method, a set of maximal components $\chi(G)$ of the graph G is first computed using a Gomory-Hu tree. And then, disjoint maximal components are selected in a bottom-up manner, based on the tree structure $\tau(G)$. The detailed procedure is given below.

Procedure SelectLaminar
  Input : a laminar family χ
  Output : a set of clusters $\chi_c \subseteq \chi$
Begin
  $\tau :=$ the rooted tree made from χ
  $\chi_c := \emptyset$
  repeat
    $X_p :=$ a parent node of not marked deepest leaves of τ
    repeat
      $X_s := X_p$
      $X_p :=$ the parent node of X_s
    until X_p has a child X_t except X_s such that $|X_t| \geq 2$
    Add all the child nodes of X_p to χ_c
    Mark all the descendant leaves of X_p in τ
  until all the leaves of τ are marked
  return χ_c
End

It should be noted that $|X_t|$ denotes the number of nodes in G that are contained in X_t. This procedure outputs a subset $\chi_c = \{X_1, \ldots, X_m\}$ from the laminar family $\chi(G)$ of all maximal components of a graph G such that $X_i \cap X_j = \emptyset$ holds for any two sets $X_i \neq X_j \in \chi_c$, and $\bigcup_{i=1}^{m} X_i = V$ holds. Figure 2 shows an example. This procedure provides the clusters according to the hierarchical structure.

4. Experimental Results

4.1. *Data and Implementation*

In order to evaluate the proposed clustering method, we applied clustering methods to classification of protein sequences based on the pairwise similarity. We used 5888 protein sequences (The file name is "orf_trans.20040827.fasta") from Saccharomyces Genome Database (SGD) [6]. This file contains the translations of all systematically named ORFs except dubious ORFs and pseudo-genes. We calculated the similarities between all pairs of the proteins using a BLAST search [1] with an E-value threshold of 0.1. An edge between two nodes exists only when the E-value between the proteins is less than or equal to 0.1. All isolated nodes (i.e., nodes with degree 0) are removed. As a result, 32484 pairwise similarities and 4533 nodes were detected.

As an edge-weight, we used the integer part of $-3000\log_{10} h$ for the E-value h of $10^{-\frac{1000}{3}} < h \leq 0.1$, and 10^6 for $0 \leq h \leq 10^{-\frac{1000}{3}}$. This mapping was injective for all the E-values of the data. It should be noted that the similarity between proteins is large when the E-value is small, and comparison operations of floating point numbers can cause incorrect results.

We solved maximum flow problems with HIPR (version 3.5) [15] which is an implementation of the algorithm developed by Goldberg and Tarjan [4], and constructed a Gomory-Hu tree [5] for an edge-weighted graph G to obtain all the maximal components of G from the tree.

4.2. *Results*

To evaluate the performance of our clustering method, we used GO-TermFinder (version 0.7) [2]. To find the most suitable GO term for a specified list (cluster) of genes, this software calculates a P-value using the hypergeometric distribution as follows:

$$P = 1 - \sum_{i=0}^{k-1} \frac{\binom{M}{i}\binom{N-M}{n-i}}{\binom{N}{n}} = \sum_{i=k}^{n} \frac{\binom{M}{i}\binom{N-M}{n-i}}{\binom{N}{n}} \tag{5}$$

where N is the total number of genes, M is the total number of genes annotated by the specific GO term, n is the number of genes in the cluster, and k is the number of genes annotated by the specific GO term in the cluster. P-value means the probability of seeing k or more genes with an annotation by a GO term among n genes in the list, given that M in the population of N have that annotation. For example, $P = 1$ holds if none of the genes in the specified list are annotated by the GO term. On the other hand, if all the genes are annotated, $P = \frac{M(M-1)\cdots(M-n+1)}{N(N-1)\cdots(N-n+1)}$ is very small because M is usually much smaller than N.

In order to avoid that a lot of false positive GO terms are chosen, GO-TermFinder can use corrected P-values. We employed these corrected P-values to evaluate clustering results.

We used three types of ontologies on biological processes, cellular components, and molecular functions (Their file names are "process.ontology.2005-08-01","component.ontology.2005-08-01", and "function.ontology.2005-08-01"). We obtained these files also from SGD [6].

We compared the proposed method with other clustering methods using single linkage, complete linkage, and average linkage. These clustering methods usually produce a hierarchical clustering. In order to obtain non-hierarchical clustering results, we applied our proposed procedure in the previous section, SelectLaminar, to their results.

Table 1 shows the averages of logarithms of corrected P-values over all 4533 proteins. Among these proteins, there were some proteins which could not be annotated by GO-TermFinder. Therefore, we regarded a corrected P-value as 1 for such proteins, and calculated the averages. We see from the table that our clustering method using maximal components outperformed other methods. For every ontology, the average of our method was lower than that of others. It means that our method classified protein sequences into protein functions better than others.

Table 1. Results for three ontologies on biological processes, cellular components, and molecular functions, by four clustering methods using maximal components, single linkage, complete linkage, and average linkage. Left column: the average of logarithm of corrected P-values. Right: the number of annotated proteins.

Method	Process		Component		Function	
Maximal component	-8.9462	2618	-5.9189	2641	-10.657	2624
Single linkage	-5.2346	2947	-4.5076	2970	-4.7721	2903
Complete linkage	-3.0674	3258	-2.3149	3391	-3.8539	3050
Average linkage	-3.2556	3692	-2.4423	3761	-4.1007	3508

Figures 3, 4 and 5 show logarithms of corrected P-values on 800 lowest proteins for the ontologies on biological processes, cellular components, and molecular functions, respectively. For every ontology, corrected P-values of our method were lower than others. The distributions of complete linkage and average linkage had similar behavior. For the ontologies on biological processes and cellular components, corrected P-values of single linkage were close to those of our method. In particular, our method provided good results for molecular functions.

Table 2 shows GO terms with lowest 8 corrected P-value for the ontology of biological processes in resulting clusters of clustering methods using maximal components, single linkage, complete linkage, and average linkage. In both complete linkage and average linkage, the same GO term (GO:0006319 Ty element transposition) was annotated to the first and second lowest clusters. It means that a cluster having the GO term was divided into two or more clusters by the methods.

As for CPU time, the proposed method is reasonably fast. Though the worst case time complexity of the proposed method is $O(n^2 m \log(n^2/m))$, it is expected to work faster in practice. Indeed, the proposed method took 6.3 sec. for clustering of a graph with 4533 nodes on a Linux PC with Xeon 3.6GHz CPU and 4GB memory. Though the single-linkage clustering took only 0.024 sec., our proposed method produced better results.

5. Conclusion

We developed a clustering method using maximal components, where a maximal component can be characterized as a subgraph having maximal edge connectivity. We compared

Table 2. GO terms with lowest 8 corrected *P*-value for the ontology of biological processes in resulting clusters of clustering methods using maximal components, single linkage, complete linkage, and average linkage.

Rank	Maximal component	
1	GO:0006319 (Ty element transposition)	2.7522e-190
2	GO:0006468 (protein amino acid phosphorylation)	2.4181e-113
3	GO:0008643 (carbohydrate transport)	1.2509e-43
4	GO:0006865 (amino acid transport)	1.0052e-37
5	GO:0006511 (ubiquitin-dependent protein catabolism)	9.4800e-24
6	GO:0006810 (transport)	1.2224e-21
7	GO:0006081 (aldehyde metabolism)	8.1134e-21
8	GO:0016567 (protein ubiquitination)	1.1405e-19
Rank	**Single linkage**	
1	GO:0006319 (Ty element transposition)	3.0396e-176
2	GO:0006081 (aldehyde metabolism)	4.0950e-19
3	GO:0006530 (asparagine catabolism)	3.5363e-16
4	GO:0006166 (purine ribonucleoside salvage)	5.0151e-15
5	GO:0045039 (mitochondrial inner membrane protein import)	3.1055e-14
6	GO:0046839 (phospholipid dephosphorylation)	7.8293e-14
7	GO:0005992 (trehalose biosynthesis)	3.4570e-13
8	GO:0006913 (nucleocytoplasmic transport)	4.4109e-13
Rank	**Complete linkage**	
1	GO:0006319 (Ty element transposition)	3.7058e-81
2	GO:0006319 (Ty element transposition)	9.1783e-55
3	GO:0008645 (hexose transport)	1.1156e-20
4	GO:0006319 (Ty element transposition)	4.1098e-18
5	GO:0000209 (protein polyubiquitination)	5.7229e-17
6	GO:0006530 (asparagine catabolism)	3.5363e-16
7	GO:0006081 (aldehyde metabolism)	2.1634e-15
8	GO:0006166 (purine ribonucleoside salvage)	5.0151e-15
Rank	**Average linkage**	
1	GO:0006319 (Ty element transposition)	4.4023e-79
2	GO:0006319 (Ty element transposition)	9.1783e-55
3	GO:0008645 (hexose transport)	1.1156e-20
4	GO:0006081 (aldehyde metabolism)	4.0950e-19
5	GO:0006319 (Ty element transposition)	4.1098e-18
6	GO:0006530 (asparagine catabolism)	3.5363e-16
7	GO:0006166 (purine ribonucleoside salvage)	5.0151e-15
8	GO:0045039 (mitochondrial inner membrane protein import)	3.1055e-14

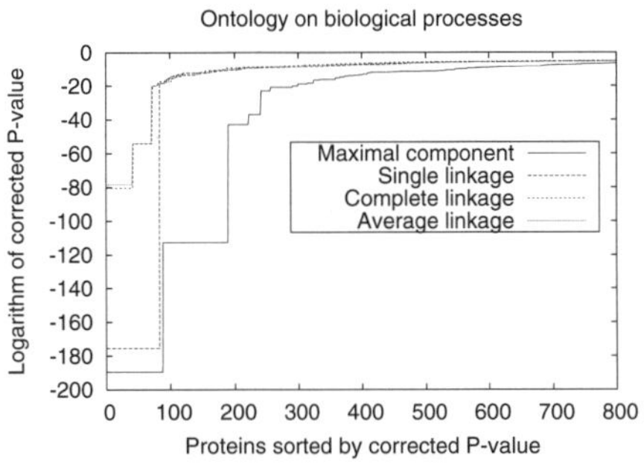

Figure 3. Logarithms of corrected P-values on 800 lowest proteins for ontology on biological processes.

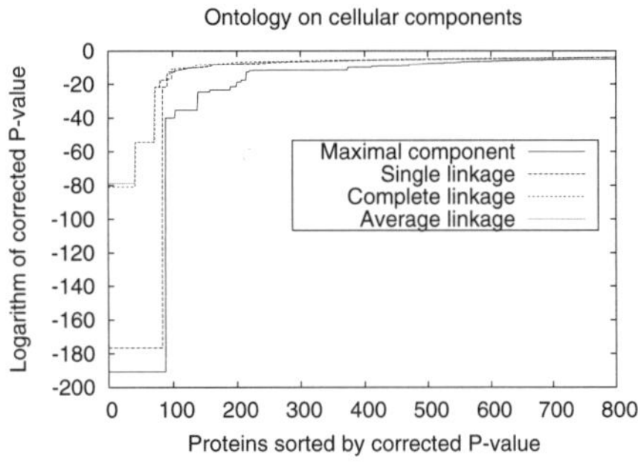

Figure 4. Logarithms of corrected P-values on 800 lowest proteins for ontology on cellular components.

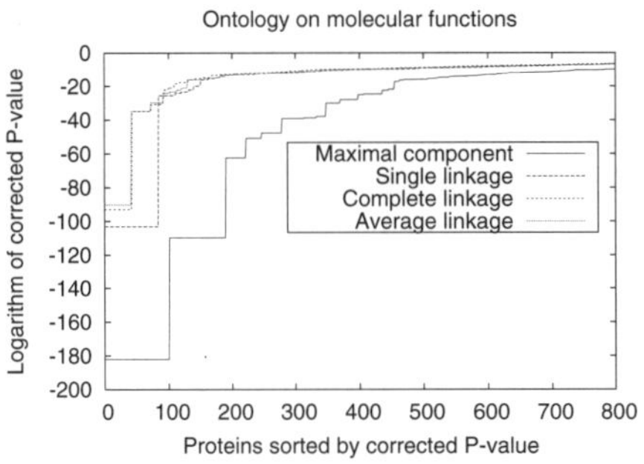

Figure 5. Logarithms of corrected P-values on 800 lowest proteins for ontology on molecular functions.

the proposed method with the single linkage, complete linkage, and average linkage clustering methods using protein sequence data. Our proposed method outperformed these three methods in terms of the corrected P-values provided by GO-TermFinder, and classified protein sequences into protein functions better than the three methods.

Although we did not compare clustering methods other than the linkage methods with our method in this study, many clustering methods have been proposed. For example, the k-means method [14] is well known as a non-hierarchical clustering method. However, it cannot be directly applied to edge-weighted graphs because it is difficult to define the center of a cluster and the distance between the center and any node in the graph.

There are several future works. We used log of E-values as edge-weights. However, this weighting method is not necessarily the best. Thus, finding better weighting method is important future work. We developed a simple method in order to select disjoint clusters from a set of maximal components. However, better results may be obtained by using a more elaborated method. Thus, improvement of selection of disjoint clusters should be done. We have applied the proposed clustering method to clustering of protein sequences. However, our method is not limited to analysis of protein sequences. For example, clustering of gene expression data is one of extensively studied problems. Therefore, application to analysis of gene expression data is also important future work.

Acknowledgments

This work is supported in part by Grants-in-Aid #1630092 and "Systems Genomics" from the Ministry of Education, Science, Sports, and Culture of Japan.

References

1. S.F. Altschul, T.L. Madden, A.A. Schäffer, J. Zhang, Z. Zhang, W. Miller and D.J. Lipman, Gapped BLAST and PSI-BLAST: a new generation of protein database search programs, *Nucleic Acids Res.*, 25:3389-3402, 1997.
2. E.I. Boyle, S. Weng, J. Gllub, H. Jin, D. Botstein, J.M. Cherry and G. Sherlock, GO::TermFinder–open source software for accessing Gene Ontology information and finding significantly enriched Gene Ontology terms associated with a list of genes, *Bioinformatics*, 20:3710-3715, 2004.
3. A.J. Enright and C.A. Ouzounis, GeneRAGE: a robust algorithm for sequence clustering and domain detection, *Bioinformatics*, 16:451–457, 2004.
4. A. Goldberg and R. Tarjan, A new approach to the maximum flow problem, *Journal of the Association for Computing Machinery*, 35:921-940, 1988.
5. R.E. Gomory and T.C. Hu, Multi-terminal network flows, *SIAM Journal of Applied Mathematics*, 9:551-570, 1961.
6. E.L. Hong, R. Balakrishnan, K.R. Christie, M.C. Costanzo, S.S. Dwight, S.R. Engel, D.G. Fisk, J.E. Hirschman, M.S. Livestone, R. Nash, J. Park, R. Oughtred, M. Skrzypek, B. Starr, C.L. Theesfeld, R. Andrada, G. Binkley, Q. Dong, C. Lane, B. Hitz, S. Miyasato, M. Schroeder, A. Sethuraman, S. Weng, K. Dolinski, D. Botstein and J.M. Cherry, Saccharomyces Genome Database, ftp://ftp.yeastgenome.org/yeast/
7. H. Kawaji, Y. Takenaga and H. Matsuda, Graph-based clustering for finding distant relationships in a large set of proteins, *Bioinformatics*, 20:243–252, 2004.
8. J.M. Kleinberg, Efficient algorithms for protein sequence design and the analysis of certain evolutionary fitness landscapes, *Proc. 3rd Int. Conf. Computational Molecular Biology (RECOMB)*, 226–237, 1999.
9. E.V. Koonin, R.L. Tatusov and K.E. Rudd, Sequence similarity analysis of Escherichia coli proteins: Functional and evolutionary implications, *Proc. Natl. Acad. Sci. USA*, 92:11921–11925, 1995.
10. S.V. Krivov, and M. Karplus, Hidden complexity of free energy surfaces for peptide (protein) folding, *Proc. Natl. Acad. Sci. USA*, 101:14766–14770, 2004.
11. H. Nagamochi, Graph algorithms for network connectivity problems, *Journal of the Operating Research Society of Japan*, 47:199–223, 2004.
12. G. Yona, N. Linial and M. Linial, ProtoMap: automatic classification of protein sequences, a hierarchy of protein families, and local maps of the protein space, *PROTEINS: Structure, Function, and Genetics*, 37:360–378, 1999.
13. M.J. Zaki, V. Nadimpally, D. Bardhan and C. Bystroff, Predicting protein folding pathways, *Bioinformatics*, 20:i386–i393, 2004.
14. J.B. MacQueen, Some methods for classification and analysis of multivariate observations, *Proceedings of the Fifth Berkeley Symposium on Mathematical Statistics and Probability*, 1:281–297, 1967.
15. http://www.avglab.com/andrew/soft.html

GENE REGULATORY NETWORK INFERENCE VIA REGRESSION BASED TOPOLOGICAL REFINEMENT

JOCHEN SUPPER*, HOLGER FRÖHLICH*, ANDREAS ZELL

Centre for Bioinformatics Tübingen (ZBIT), Sand 1, 72076 Tübingen
jochen.supper@uni-tuebingen.de

Inferring the structure of gene regulatory networks from gene expression data has attracted a growing interest during the last years. Several machine learning related methods, such as Bayesian networks, have been proposed to deal with this challenging problem. However, in many cases, network reconstructions purely based on gene expression data not lead to satisfactory results when comparing the obtained topology against a validation network. Therefore, in this paper we propose an "inverse" approach: Starting from *a priori* specified network topologies, we identify those parts of the network which are relevant for the gene expression data at hand. For this purpose, we employ linear ridge regression to predict the expression level of a given gene from its relevant regulators with high reliability. Calculated statistical significances of the resulting network topologies reveal that slight modifications of the pruned regulatory network enable an additional substantial improvement.

1. Introduction

Transcriptional datasets provide valuable insights into specific cellular processes under various conditions. To control these processes the cell utilizes regulatory mechanisms, whereas for each specific process only a small fraction of the complete regulatory network is affected. Therefore, a gene regulatory network (GRN) is a large graph covering regulatory mechanisms for various stimuli. Yet for a specific observation only a small fraction of the GRN can be inferred and linked to the respective transcriptional data. Thus, the fixed topological structure of a GRN can be detached from the dynamic structure comprising a subset of the fixed topological structure associated with a quantitative observation.

Today, modeling of GRNs is guided by a rich flow of experimental data. The stream is still widened by an increasing pool of measurement techniques including mRNA microarray technology[17], chromatin immunoprecipitation (ChIP)[1], quantitative RT-PCR[7] and microarray-based immunoassays[13]. Despite of all this information, detailed knowledge regarding the topology of network models is still almost exclusively collected by biologists. They collect and integrate data, expand and refine their models and finally validate them. For our modeling efforts, we will use this qualitative knowledge provided by the biological observations to compile *a priori* topologies of the GRN. Within such topologies, we search for a subset of connections which is in good accordance to the transcriptional data and therefore prune the topological network using a reverse engineering approach.

*These authors contributed equally to this work.

Several of such reverse engineering approaches for GRNs have emerged during the last years. They include analytical methods such as Boolean networks[12], (non)-linear networks[23,21], S-Systems[20] and differential equations[3], but also machine learning methods such as decision trees[18] and Bayesian networks[10]. Most of these approaches have been applied solely to transcriptional data, thereby neglecting the *a priori* information available. Thus, they implicitly assume that the topological network can be connected arbitrarily, which leads to a huge number of potential regulatory mechanisms. However, some inference methods previously published consider *a priori* information. For instance Someren *et al.* compiled a validation network based on a co-citation approach[22], others have restricted the regulators to transcription factors[15], and, as in this work, the putative *cis*-regulatory elements have been employed as initial network structure[16].

Despite the use of *a priori* information, a set of transcriptional response measurements has to be available in order to reconstruct GRNs. Given this data, one of the above mentioned reconstruction methods can be employed to untangle the underlying topological structure of the interaction network. One of the principle problems thereby is the ambiguity inherent in reconstructing the GRN from given expression data, which is due to the small sampling rate along with the high noise level of the measurements. Even if the total number of microarray measurements is growing at a tremendous rate, the number of measurements utilizable for a specific observation is still limited. The bottom line is, that typical GRN inference methods produce topologies, which contain a certain fraction of false or unverifiable interactions. This fraction of spurious edges has to be traded off against the fraction of correctly identified regulatory dependencies. Husmeier *et al.*[10] systematically investigate this trade-off for a Bayesian network reconstruction of an artificially created network.

This motivated us to take an "inverse" approach to the usual GRN reconstruction methods: based on a given network topology from literature or by putative *cis*-regulatory elements (which may also contain wrong interactions), our approach is first, to identify the part of this network, which is relevant for the experimental data and second, to eventually modify this "pruned" network modestly to better fit the data. The basic assumption behind our approach is, that today, in many cases, a good starting network topology, which subsequently has to be refined in order to fit the experimental measurements, can be obtained from public databases or by the rich biological literature. In this work, the refinement is achieved through a machine learning based approach using linear ridge regression[9], which resembles the framework proposed by Soinov *et al.*[18]. In favor of our approach, we are able to show on publicly available Yeast genome datasets, that the prediction accuracy of gene expression levels is significantly higher in our fitted topology than in the original or a random network.

The remainder of this paper is organized as follows: in the next section, we introduce the datasets used and explain our approach in detail. In Section 3 we present and discuss the results obtained on our investigated datasets. Finally, in Section 4 the conclusions are drawn.

2. Materials and Methods

2.1. *Data*

2.1.1. *Budding Yeast Cell Cycle*

The biological model used for this research is the budding yeast cell cycle. This model displays a small subset of the whole network, where it is known that the chosen genes play an important role in the respective processes. The dataset was taken from Spellman *et al.*[19] and Cho *et al.*[4] , who measured the cell cycle of the budding yeast. The budding yeast cell cycle is known at a high level of detail. Spellman *et al.* and Cho *et al.* measured the progression of the cell cycle with different synchronization techniques. Overall they provided four records of time series measurements, which can be used for modeling purposes. The sampling number of these datasets ranges from 14 to 24, the samples are taken at equidistant time-steps for each series. However the time-steps are not equal for the different series, they range from 7 to 30 minutes. In this study we use the α factor, cdc 15 and cdc 28 arrest, as well as the elutriation time series. This results in 73 time point measurements altogether. These measurements have been taken on microarrays[5], each consisting of 6178 data genes. We imputed missing values in the gene expression measurements by the SVD method described[6].

2.1.2. *Chen Dataset*

Of the above described 6178 genes we chose a subset according to Chen *et al.*[2] , who used a set of differential equations to define the topology of the GRN. In addition to the interactions provided by the differential equations we searched TRANSFAC[24], Entrez Gene[14] and the *Saccharomyces* Genome Database (SGD[11]) for known dependencies between a pair of genes.

The entire network contains 56 interactions and is depicted in Figure 1. It will serve as our first *a priori* network for the reconstruction.

2.1.3. *Cis-Regulatory Elements*

The major control in transcriptional gene regulation is mediated by transcription factors (TFs) that bind to the promotor region of a gene. We used these connections between TFs and genes to construct the second validation network. In this network, the inferred genes are still restricted to the genes from the Chen *et al.* dataset, whereas the TFs are not. To establish these TF-gene connections, we extracted the *cis*-regulatory elements from the SCPD (The Promoter Database of *Saccharomyces cerevisiae*) database. This database was developed by Zhang *et al.*[25] and contains experimentally mapped TF binding sites as primary data entries and predefined putative regulatory elements using matrix and consensus methods. To extract the binding sites, we restricted the search to the 500 bp upstream sequence and searched for the consensus patterns contained.

The entire network contains 145 interactions and is depicted in Table 1.

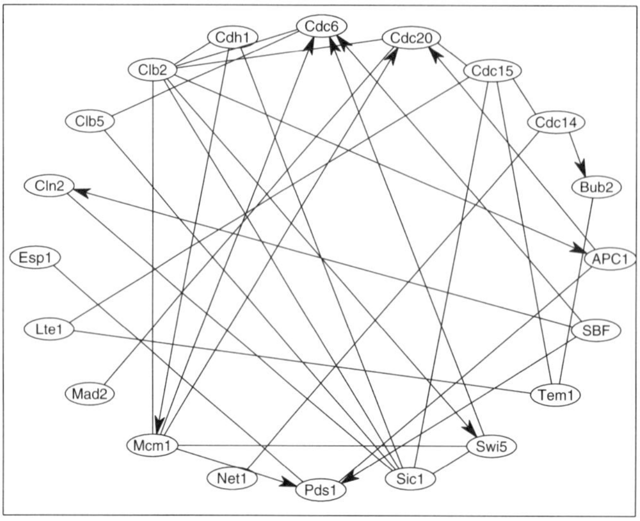

Figure 1. Literature network for the Chen data: An arrow $a \to b$ indicates that b is regulated by a. Edges with no arrows indicate a mutual influence $a \to b$ and $b \to a$.

Table 1. Network composed by *cis*-regulatory elements.

Gene		**Regulators**
Cdc14	←	GCN4, TBP, INO1, PHO4, ABF1, Swi5, STE12
Cdc15	←	GCR1, BAS2, ABF1, GCN4, ADR1, MSE, YOX1, YHP1
Cdc20	←	YHP1, GCR1, RAP1, MSE, GCN4, STE12, YOX1, Swi5
Cdc6	←	GCR1, GCN4, MCB, Mcm1, Swi5, BAS2, TBP
Cdh1	←	MSE, GCR1, ADR1, GCN4, UASPHR, REB1, MSE
Clb2	←	AP-1, GCN4, GCR1, BAS2, GCN4, ADR1, TBP
Clb5	←	ACE2, Swi5, Mcm1, GCR1, GCN4, YOX1, YHP1, MSE, BAS2, AP-1
Cln2	←	STE12, YOX1, MSE, TBP, GCN4, BAS2, YHP1
Esp1	←	GCN4, PHO4, ABF1, YHP1, GCR1, BAS2, TBP, YOX1, RAP1, TBP
Lte1	←	GCN4, BAS2, ACE2, Swi5, TBP, ADR1, YOX1, SCB, YHP1, MCB, MATalpha2
Mad2	←	GCN4, TBP, Swi5
Mcm1	←	BAS2, GCN4, YHP1, YOX1, Swi5
Net1	←	YHP1, GCN4, YOX1, ABF1, REB1, STE12
Pds1	←	CPF1, PHO4, GCN4, TBP, YHP1, GCR1, MATalpha2, MSE, MCB, BAS2, YOX1
Sic1	←	GCN4, ABF1, BAS2, ADR1, Swi5, MCB, TBP, GCN4, MATalpha2, SCB
Swi5	←	GCN4, BAS2, Mcm1, GCR1, YHP1, YOX1, ADR1, PHO4, MATalpha2, ATF
Tem1	←	GCN4, ADR1, Swi5, GCR1, AP-1, YHP1, SCB, RAP1, Mcm1, YOX1, TBP
APC1	←	ADR1, GCR1, MIG1, MSE, YHP1, YOX1, Swi5, GCN4

2.2. *Regression Based Network Refinement*

Given one of the above described literature or *cis*-regulatory networks our goal is to fit their topology to our datasets. This is done in a framework resembling that proposed by Soinov *et al.*[18]: For each gene g we know a set of possible regulators $R_g = \{r_1, ..., r_{n_g}\}$. From these regulators we would like to select the subset $\tilde{R}_g$ of R_g, which allows the highest prediction performance of the expression level of g. Thereby, predicting the expression level of g can be either done within one time step t, or from time step $t-1$ to t. We consider both prediction tasks and in the end merge the obtained regulator subsets $\tilde{R}_g^1$ and $\tilde{R}_g^2$ to the final subset $\tilde{R}_g$. We use linear ridge regression as the prediction machinery with ridge constant $\tau = 10^{-5}$.

To prune the "irrelevant" regulators for g, we adopt the RFE algorithm originally proposed for SVM feature selection[8]: In linear ridge regression one estimates a hyperplane $f : \mathbb{R}^{n_g} \to \mathbb{R},\ f(\mathbf{x}) = \langle \mathbf{w}, \mathbf{x} \rangle + b$, where $\mathbf{w}$ is the normal vector of the hyperplane and b a bias term. The components of the normal vector $\mathbf{w}$ can be understood as weights for the individual features in $\mathbf{x}$. Therefore, we can prune the regulator, for which the component in $\mathbf{w}$ is the smallest. Then the regression function is re-estimated and the whole procedure iterated until all regulators are removed. Overall, we receive a ranking of all regulators depending on their time of removal from the regulator set R_g. The optimal number of regulators in our case is determined via 5-fold cross-validation, where we use the mean squared correlation between predicted and true gene expression values to measure the prediction performance.

After pruning the original set of regulators we allow to add one extra regulator, which was not in R_g before, if this further improves the 5-fold cross-validated mean squared correlation. This slight modification of the pruned network takes into account that there might exist interactions, which are not covered by the $apriori$ provided network structure and otherwise could not be detected. Nevertheless, the selected subset $\tilde{R}_g^i$ $(i = 1, 2)$ of regulators, which typically consists only of a few genes, is completely rejected, if the mean 5-fold cross-validated correlation between the true and the predicted expression levels of g is below 60%. This ensures that only interactions with a high statistical confidence are inserted into the network.

As a last step the final regulator subset $\tilde{R}_g = \tilde{R}_g^1 \cup \tilde{R}_g^2$ is evaluated with respect to the two prediction tasks described in the beginning. This is done via 5-fold cross-validation, measuring the mean squared correlation between the predicted and the true expression level of g.

2.3. *Comparison Scheme*

The bottom line from the above described approach is that for all genes in the inferred network we obtain an estimate, how well their expression level can be predicted from their putative regulators. This can be viewed as a measure for the reliability of the network. Which allows us to compare different network topologies with respect to this measure and to compute statistical significances. More specifically, we are interested, how well a network computed with the method described in the last subsection performs relative to the following reference topologies:

- the original literature and *cis*-regulatory network

- a random network with the same number of connections as the literature network
- a fully connected network

Comparing the prediction performances obtained from these network topologies to our refined network allows the computation of p-values, via Wilcoxon's signed rank test. In the following section we describe the results we achieved this way for the two datasets investigated in this paper.

3. Results

3.1. *Chen Dataset*

Our method from subsection 2.2 yielded 6 genes with a nonempty regulator set. It is depicted in Figure 2a). For comparison reasons we also computed the network, which was obtained by just pruning the literature network and not introducing any further interactions. It had only 5 genes with a nonempty regulator set and is shown in Figure 2b). It is worth mentioning that both networks can differ in more than one regulator per gene: If in the modification step an additional regulator is added to the pruned regulator set, the prediction performance of the prediction model can increase such that it exceeds the minimum prescribed threshold of 60% mean squared correlation between predicted and true gene expression values. Hence, the whole regulator set is added to the pruned network at once in this case.

In Table 2 we compare the resulting average 5-fold cross-validated mean squared correlations for these genes with those obtained from the literature, the random and the fully connected network (see last section). As shown, the pruned network with additional modifications yielded a high significant improvement compared to the random and the literature network. In contrast, the pruned network without additional modifications had a much worse p-value compared to the random network and was not significantly better than the original literature network. A direct comparison of the pruned networks with and without additional modifications reveals a high statistical significant difference between both (p-value = 0.0059).

Table 2. 5-fold cross-validated correlations (r) with true expression levels for genes with nonempty regulator sets (Chen dataset). For the inferred networks (pruned+modification, pruned only) the p-values are calculated as described in subsection 2.3.

Topology	$r\pm$std. err.	p-val. lit.	p-val. rand.	p-val. full
pruned+mod.	$\mathbf{72.07 \pm 1.94}$	$\mathbf{9.76 \cdot 10^{-4}}$	$\mathbf{0.0024}$	0.4697
pruned only	65.74 ± 3.77	0.1953	0.0645	0.2324
literature	42.52 ± 2.76			
random	39.58 ± 2.55			
full	51.87 ± 2.83			

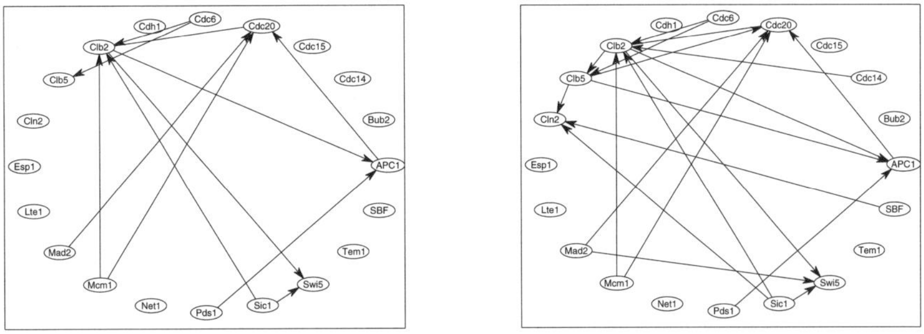

(a) with pruning only (b) with pruning and modification

Figure 2. Inferred networks for the Chen data.

Table 3. Network inferred by pruning the *cis*-regulatory elements.

Gene		**Regulators**
Cdc14	←	Swi5, ABF1, STE12, PHO4
Cdc15	←	—
Cdc20	←	GCN4, Swi5, GCR1, MSE, YOX1, YHP1
Cdc6	←	—
Cdh1	←	UASPHR
Clb2	←	—
Clb5	←	GCN4, GCR1, BAS2, MSE, YOX1, Mcm1, AP-1, ACE2
Cln2	←	GCN4, TBP, STE12, BAS2, MSE, YOX1, YHP1
Esp1	←	—
Lte1	←	—
Mad2	←	—
Mcm1	←	—
Net1	←	—
Pds1	←	PHO4, GCR1, BAS2, YOX1, YHP1, MCB
Sic1	←	—
Swi5	←	GCN4, BAS2, ADR1, YOX1, YHP1
Tem1	←	TBP, Swi5, SCB
APC1	←	ADR1, Swi5, GCR1

3.2. *Cis-Regulatory Elements*

Our method from subsection 2.2 on this dataset yielded 11 genes with a nonempty regulator set (Table 4). Again, we also computed the network, which was obtained by just pruning the initial network network without introducing any further interactions. It had only 9 genes with a nonempty regulator set and is shown in Table 3. Like in the last subsection, in Table 4 we compared the resulting average 5-fold cross-validated mean squared correlations for these genes with those obtained from the *cis*-regulatory, the random and the fully connected network. As seen, the pruned network with additional modifications yielded a high significant improvement compared to the random, the *cis*-regulatory network and the full network. The pruned network without additional modifications in all cases had much

Table 4. Network inferred by pruning the *cis*-regulatory elements network and some allowed subsequent modifications. Regulators introduced additionally to the pruned network are marked bold. Regulators, which were not in the *a priori* network are written italic as well.

Gene		Regulators
Cdc14	←	**TBP**, PHO4, ABF1, Swi5, STE12, ***RAP1***
Cdc15	←	—
Cdc20	←	GCN4, ***APC1***, Swi5, GCR1, MSE, YOX1, YHP1
Cdc6	←	—
Cdh1	←	UASPHR, ***Sic1***
Clb2	←	**GCR1, BAS2, ADR1**, ***REB1***
Clb5	←	GCN4, GCR1, BAS2, MSE, YOX1, ***Cdc6***, Mcm1, AP-1, ACE2
Cln2	←	GCN4, TBP, STE12, BAS2, MSE, YOX1, YHP1, ***AP-1***
Esp1	←	—
Lte1	←	—
Mad2	←	—
Mcm1	←	—
Net1	←	—
Pds1	←	PHO4, GCR1, BAS2, YOX1, YHP1, MCB, ***Tem1***
Sic1	←	**GCN4, TBP, ABF1, Swi5, BAS2, *Cdc6*, SCB, MATalpha2**
Swi5	←	GCN4, BAS2, ADR1, YOX1, YHP1, ***ACE2***
Tem1	←	TBP, Swi5, **GCR1, AP-1**, SCB, ***Sic1***
APC1	←	ADR1, Swi5, GCR1, ***INO1*, MSE, MIG1**

worse p-values, especially compared to the fully connected network. A direct comparison of the pruned networks with and without additional modifications reveals a high statistical significant difference between both (p-value = 0.0036).

4. Conclusion

We introduced a method to refine a GRN topology obtained from the literature or from public databases such that it fits a given gene expression dataset. Thereby our criterion was the estimated generalization performance achieved by a linear ridge regression model, which was trained to predict the expression level of each gene in the network from the expression levels of its regulators. An algorithm was developed to find a minimal regulator subset for each gene, which allows the highest prediction rate. Thereby slight modifications of the pruned literature network were allowed in order to take into account the possible incompleteness or defectiveness of the original network topology.

We performed evaluations of our method on publicly available datasets from Yeast genome and compared our approach against the original *a priori* network, a random network with the same number of interactions as the *a priori* network and a fully connected network. We were able to show that our inferred networks on both datasets could significantly improve on the original, the random, and in case of the second dataset, also on the fully connected dataset. Furthermore, an interesting finding was that allowing slight modifications of the pruned *a priori* network in all cases lead to much better p-values than without allowing these changes.

Altogether we think that a main contribution of this work was first, the introduction of a network refinement method from a given starting topology, and second the possibility to compute statistical significances for the inferred network, which to our knowledge has not been possible so far.

Table 5. 5-fold cross-validated correlations (r) with true expression levels for genes with nonempty regulator set (*cis*-regulatory elements). For the inferred networks (pruned+modification, pruned only) the p-values are calculated as described in subsection 2.3.

Topology	$r\pm$std. err.	p-val. *cis*-reg.	p-val. rand.	p-val. full
pruned+mod.	$\mathbf{73.85 \pm 2.71}$	$\mathbf{5.96 \cdot 10^{-5}}$	$\mathbf{4.61 \cdot 10^{-5}}$	$\mathbf{2.14 \cdot 10^{-4}}$
pruned only	68.3 ± 4.09	0.0181	$3.86 \cdot 10^{-4}$	0.0312
cis-reg.	52.04 ± 3.32			
random	45.13 ± 2.84			
full	50.87 ± 2.81			

References

1. Michael J Buck and Jason D Lieb. ChIP-chip: considerations for the design, analysis, and application of genome-wide chromatin immunoprecipitation experiments. *Genomics*, 83(3):349–60, Mar 2004.
2. Katherine C Chen, Laurence Calzone, Attila Csikasz-Nagy, Frederick R Cross, Bela Novak, and John J Tyson. Integrative analysis of cell cycle control in budding yeast. *Mol Biol Cell*, 15(8):3841–3862, Aug 2004.
3. Kuang-Chi Chen, Tse-Yi Wang, Huei-Hun Tseng, Chi-Ying F Huang, and Cheng-Yan Kao. A stochastic differential equation model for quantifying transcriptional regulatory network in Saccharomyces cerevisiae. *Bioinformatics*, 21(12):2883–90, Jun 2005.
4. R. J. Cho, M. J. Campbell, E. A. Winzeler, L. Steinmetz, A. Conway, L. Wodicka, T. G. Wolfsberg, A. E. Gabrielian, D. Landsman, D. J. Lockhart, and R. W. Davis. A genome-wide transcriptional analysis of the mitotic cell cycle. *Mol Cell*, 2(1):65–73, Jul 1998.
5. M. B. Eisen and P. O. Brown. DNA arrays for analysis of gene expression. *Methods Enzymol*, 303:179–205, 1999.
6. S. Friedland, A. Niknejad, and L. Chihara. A simultaneous reconstruction of missing data in dna microarrays. *Linear Algebra Appl.*, 2005. to appear.
7. David G Ginzinger. Gene quantification using real-time quantitative PCR: an emerging technology hits the mainstream. *Exp Hematol*, 30(6):503–12, Jun 2002.
8. I. Guyon, J. Weston, S. Barnhill, and V. Vapnik. Gene Selection for Cancer Classification using Support Vector Machines. *Machine Learning*, 46:389 – 422, 2002.
9. A. Hoerl and R. Kennard. Ridge regression: Biased estimation for nonorthogonal problems. *Technometrics*, 12(1):55 – 67, 1970.
10. Dirk Husmeier. Sensitivity and specificity of inferring genetic regulatory interactions from microarray experiments with dynamic Bayesian networks. *Bioinformatics*, 19(17):2271–2282, Nov 2003.
11. Laurie Issel-Tarver, Karen R Christie, Kara Dolinski, Rey Andrada, Rama Balakrishnan, Catherine A Ball, Gail Binkley, Stan Dong, Selina S Dwight, Dianna G Fisk, Midori Harris, Mark Schroeder, Anand Sethuraman, Kane Tse, Shuai Weng, David Botstein, and J. Michael Cherry. Saccharomyces Genome Database. *Methods Enzymol*, 350:329–346, 2002.

12. S. Liang, S. Fuhrman, and R. Somogyi. Reveal, a general reverse engineering algorithm for inference of genetic network architectures. *Pac Symp Biocomput*, 1:18–29, 1998.
13. Gavin MacBeath. Protein microarrays and proteomics. *Nat Genet*, 32 Suppl:526–32, Dec 2002.
14. Donna Maglott, Jim Ostell, Kim D Pruitt, and Tatiana Tatusova. Entrez Gene: gene-centered information at NCBI. *Nucleic Acids Res*, 33(Database issue):D54–D58, Jan 2005.
15. D. Pe'er, A. Tanay, and A. Regev. Minreg: A scalable algorithm for learning parsimonious regulatory networks in yeast and mammals. *J. Machine Learning Research*, 7:167 – 189, 2006.
16. Jianhua Ruan and Weixiong Zhang. A bi-dimensional regression tree approach to the modeling of gene expression regulation. *Bioinformatics*, Nov 2005.
17. A. Schulze and J. Downward. Navigating gene expression using microarrays–a technology review. *Nat Cell Biol*, 3(8):E190–5, Aug 2001.
18. Lev A Soinov, Maria A Krestyaninova, and Alvis Brazma. Towards reconstruction of gene networks from expression data by supervised learning. *Genome Biol*, 4(1):R6, 2003.
19. P. T. Spellman, G. Sherlock, M. Q. Zhang, V. R. Iyer, K. Anders, M. B. Eisen, P. O. Brown, D. Botstein, and B. Futcher. Comprehensive identification of cell cycle-regulated genes of the yeast Saccharomyces cerevisiae by microarray hybridization. *Mol Biol Cell*, 9(12):3273–3297, Dec 1998.
20. Christian Spieth, Felix Streichert, Nora Speer, and Andreas Zell. A memetic inference method for gene regulatory networks based on s-systems. In *Proceedings of the IEEE Congress on Evolutionary Computation (CEC 2004)*, pages 152–157, 2004.
21. J. Supper, C. Spieth, and A. Zell. Reverse engineering non-linear gene regulatory networks based on the bacteriophage lambda ci circuit. In *IEEE Symposium on Computational Intelligence in Bioinformatics and Computational Biology (CIBCB)*, 2005.
22. E. P. van Someren, B. L T Vaes, W. T. Steegenga, A. M. Sijbers, K. J. Dechering, and M. J T Reinders. Least Absolute Regression Network Analysis of the murine osteoblast differentiation network. *Bioinformatics*, Dec 2005.
23. D. C. Weaver, C. T. Workman, and G. D. Stormo. Modeling regulatory networks with weight matrices. *Pac Symp Biocomput*, 1:112–23, 1999.
24. Edgar Wingender. TRANSFAC, TRANSPATH and CYTOMER as starting points for an ontology of regulatory networks. *In Silico Biol*, 4(1):55–61, 2004.
25. J. Zhu and M. Q. Zhang. SCPD: a promoter database of the yeast Saccharomyces cerevisiae. *Bioinformatics*, 15(7-8):607–611, 1999.

ALGORITHM ENGINEERING FOR COLOR-CODING TO FACILITATE SIGNALING PATHWAY DETECTION

FALK HÜFFNER, SEBASTIAN WERNICKE, AND THOMAS ZICHNER

Institut für Informatik, Friedrich-Schiller-Universität Jena
Ernst-Abbe-Platz 2, D-07743 Jena, Germany
E-mail: {hueffner,wernicke,tzi}@minet.uni-jena.de

To identify linear signaling pathways, Scott et al. [RECOMB, 2005] recently proposed to extract paths with high interaction probabilities from protein interaction networks. They used an algorithmic technique known as color-coding to solve this NP-hard problem; their implementation is capable of finding biologically meaningful pathways of length up to 10 proteins within hours. In this work, we give various novel algorithmic improvements for color-coding, both from a worst-case perspective as well as under practical considerations. Experiments on the interaction networks of yeast and fruit fly as well as a testbed of structurally comparable random networks demonstrate a speedup of the algorithm by orders of magnitude. This allows more complex and larger structures to be identified in reasonable time; finding paths of length up to 13 proteins can even be done in seconds and thus allows for an interactive exploration and evaluation of pathway candidates.

1. Introduction

Motivation. Accompanying the availability of genome-scale protein–protein interaction data, various approaches have been proposed to datamine the corresponding networks[a] for biologically meaningful substructures such as dense groups of interacting proteins[3,6,9] and loop structures[2]. A special role—with respect to both biological meaning as well as algorithmic tractability—is played by the most simple structures, that is, *linear* pathways. These are easy to understand and analyze and, as demonstrated by Ideker et al.[8] for the yeast galactose metabolism, they can serve as a seed structure for experimental investigation of more complex mechanisms. Initiated by Steffen et al.[12], the automated discovery of linear pathways in protein interaction networks is hence a promising undertaking.

Unfortunately, finding linear pathways—that is, paths in a graph where each vertex occurs at most once—is an NP-hard problem[5]. However, a randomized algorithmic technique called "color-coding"[1] is known to solve this problem efficiently for small path lengths. (The details of color-coding are explained in Section 2). Consequently, Scott et al.[10] recently proposed to employ color-coding to datamine protein interaction networks for signaling pathways. They were able to demonstrate that the color-coding approach is indeed capable of identifying biologically meaningful pathways. Their implementation is limited, however, to path lengths of around 10 vertices and, moreover, it requires some hours of

[a]We use the term "network" for fields outside mathematics and computer science and "graph" for discussing algorithmic aspects.

runtime for these path lengths. In this work, we give various novel improvements for color-coding, both from a worst-case perspective as well as under practical considerations. This allows us to find pathways consisting of more than 20 vertices in some hours and the task of finding pathways of length 10 can be accomplished in a few seconds.

Structure of this work. The color-coding technique is explained in Section 2. Our algorithmic improvements and data structures are discussed in Section 3. The improved color-coding algorithm has been implemented in C++. Section 4 discusses experimental results that were obtained by using our implementation on the *S. cerevisiae* (yeast) interaction network of Scott et al.[10], the *D. melanogaster* (fruit fly) interaction network of Giot et al.[7], and random networks that are structurally similar to protein interaction networks. Our experiments demonstrate that the algorithmic improvements proposed in this work facilitate the detection of larger, more complex pathway candidates and opens the possibility for interactive exploration of smaller structures.

The source code of our color-coding implementation can be downloaded as free software from `http://theinf1.informatik.uni-jena.de/colorcoding/`.

2. The Color-Coding Technique

We model protein interaction networks as undirected graphs where each vertex is a protein and each edge is weighted by the negative logarithm of the interaction probability for the two proteins it connects. Following Scott et al.[10], we formalize the problem of pathway candidate detection to a NP-hard problem called MINIMUM-WEIGHT PATH.

> MINIMUM-WEIGHT PATH
> **Input:** An undirected edge-weighted graph $G = (V, E)$ with $n := |V|$ and $m := |E|$ and an integer k.
> **Task:** Find a length-k path in G that minimizes the sum over its edge weights.

Color-Coding. Alon et al.[1] proposed a technique called *color-coding* to solve MINIMUM-WEIGHT PATH. The idea is to randomly color the vertices in the input graph with k colors[b] and then search for *colorful* paths, that is, paths where no color occurs twice. Given a fixed coloring of vertices, finding the minimum-weight colorful path is accomplished by dynamic programming: Assume that for some $i < k$ we have computed a value $W(v, S)$ for every vertex $v \in V$ and cardinality-i subset S of vertex colors; this value denotes the minimum weight of a path that uses every color in S exactly once and ends in v. Clearly, this path is simple because no color is used more than once. We can now use this to compute the values $W(v, S)$ for all cardinality-$(i + 1)$ subsets S and vertices $v \in V$ because a colorful length-$(i+1)$ path that ends in a vertex $v \in V$ can be composed of a colorful length-i path that does not use the color of v and ends in a neighbor of v. More precisely, we let

$$W(v, S) = \min_{e=\{u,v\}\in E} \Big(W(u, S \setminus \{\text{color}(v)\}) + w(e) \Big). \tag{1}$$

[b] Section 3.1 shows that using only k colors is suboptimal from a runtime perspective.

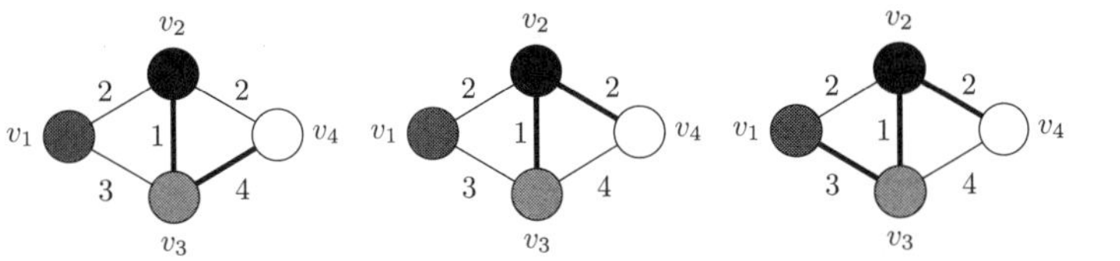

Figure 1. Example for solving MINIMUM-WEIGHT PATH using the color-coding technique. Using Equation (1) a new table entry (right) is calculated using two already known entries (left and middle).

as exemplified in Figure 1.

It is easy to verify that the dynamic programming takes $O(2^k m)$ time.[c] Whenever the minimum-weight length-k path in the input graph is colored with k colors (i.e., every vertex has a different color), then it is found. The problem, of course, is that the coloring of the input graph is random and hence many coloring *trials* have to be performed to ensure that the minimum-weight path is found with a high probability. While the result is only optimal with a (user-specifiable) probability, setting the error probability ε to say, 0.1%, is likely to be acceptable in practice (even more so because only the logarithm of the error probability goes into the overall runtime and hence, very low error probabilities are efficient to achieve).

A particularly appealing aspect of the color-coding method is that it can be easily adapted to many practically relevant variations of the problem formulation: For example, the set of vertices where a path can start and end can be restricted (such as to force it to start in a membrane protein and end in a transcription factor[10]). Also, recently it has been demonstrated that pathway queries to a network, that is, the task of finding a pathway in a network that is as similar as possible to a query pathway, can be handled with color-coding[11].

Unless otherwise noted, we use the following variant of MINIMUM-WEIGHT PATH that matches the experiments by Scott et al.[10]: With an error probability of $\varepsilon = 0.1\%$, we seek 100 minimum-weight paths which must differ from each other in at least 30% of the vertices (to ensure that they are not only small modifications of the global minimum-weight path).

3. Improving the Efficiency of Color-Coding

This section presents several algorithmic improvements for color-coding that lead to large savings in time and memory consumption. Whereas the improvement in Section 3.2 is of heuristic nature, the improvements in Sections 3.1 and 3.3 make color-coding also more efficient in a worst-case scenario. Note that these improvements are generally applicable to color-coding and not restricted to the protein interaction network scenario.

---

[c] Literature usually states the weaker bound $O(2^k km)$ that is obtained when representing the sets S explicitly instead of using a table.

3.1. *Speedup by Increasing the Number of Colors*

Clearly, we need at least k colors when trying to find a length-k path using the color-coding technique. Increasing the number of used colors beyond this leads to a tradeoff: Fewer trials have to be performed to ensure the same error bound, yet each trial takes longer. More specifically, assume that in order to detect a path of length k we are using the color-coding technique with $k + x$ colors for some positive integer x. Then the probability P_c of a path in the input graph being colorful becomes

$$P_c = \frac{\binom{k+x}{k} \cdot k!}{(k+x)^k} = \frac{(k+x)!}{x!(k+x)^k} = \prod_{i=1}^{k} \frac{i+x}{k+x} \tag{2}$$

because there are $(k + x)^k$ ways to color k vertices with $k+x$ colors and $\binom{k+x}{k} \cdot k!$ of these use mutually different colors. The overall runtime $t_{\mathcal{A}}$ of the algorithm to ensure an error probability of at most ε is a product of two factors, namely the runtime of a single trial and the number of trials $t(\varepsilon)$ to perform. As discussed in Section 2, the worst-case runtime for each trial is $O(2^{k+x} \cdot m)$ and we obtain

$$t_{\mathcal{A}} \le t(\varepsilon) \cdot O(2^{k+x} \cdot m) = \left\lceil \frac{\ln \varepsilon}{\ln(1 - P_c)} \right\rceil \cdot O(2^{k+x} \cdot m). \tag{3}$$

We should choose x such that the right-hand side of (3) is minimized. All works we are aware of use $x = 0$ for the analysis, which yields $t_{\mathcal{A}} = O(|\ln \varepsilon| \cdot e^k \cdot 2^k m) = O(|\ln \varepsilon| \cdot 5.44^k m)$. While this choice can be argued for with respect to memory requirements for a trial (after all, these are a major bottleneck for dynamic programming algorithms), it is not optimal concerning $t_{\mathcal{A}}$:

Theorem 3.1. *The worst-case runtime of color-coding for* MINIMUM-WEIGHT PATH *with* $1.3k$ *colors and error probability* ε *is* $O(|\ln \varepsilon| \cdot 4.32^k m)$.

Proof. To estimate the factorials in Equation (2), we use the double inequality

$$\sqrt{2\pi} n^{n+1/2} \cdot \exp(-n + 1/(12n+1)) < n! < \sqrt{2\pi} n^{n+1/2} \cdot \exp(-n + 1/(12n))$$

derived from Stirling's approximation. This yields

$$\begin{aligned} P_c &\ge \frac{\sqrt{2\pi}(k+x)^{k+x+1/2} \cdot \exp\left(-k - x + \frac{1}{12k+12x+1}\right)}{\sqrt{2\pi} x^{x+1/2} \cdot \exp\left(-x + \frac{1}{12x}\right)} \cdot (k+x)^{-k} \\ &= \left(\frac{k}{x} + 1\right)^{x+1/2} \cdot \exp\left(-k - \frac{1}{12x} + \frac{1}{12k + 12x + 1}\right). \end{aligned}$$

Setting $x := 0.3k$ and using the inequality $\ln(1 - P_c) < -P_c$ (which is valid because the probability P_c satisfies $0 < P_c < 1$) we obtain

$$t_{\mathcal{A}} \le \left\lceil \frac{\ln \varepsilon}{\ln(1 - P_c)} \right\rceil \cdot O(2^{k+x} \cdot m) < \left(\frac{\ln \varepsilon}{-P_c} + 1\right) \cdot O(2^{k+x} \cdot m)$$

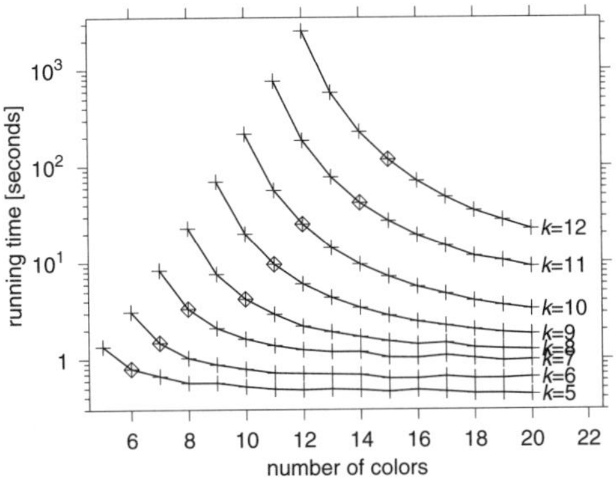

Figure 2. Runtimes for finding the 20 minimum-weight paths of different lengths in the yeast protein interaction network of Scott et al.[10]. No lower bound function (Section 3.2) was used. The highlighted point of each curve marks the optimal choice when assuming worst-case trial runtime.

where

$$\frac{1}{P_c} < 4.33^{-0.3k-1/2} \cdot \exp\left(k + \frac{1}{12x}\right) = O\left(\frac{e^k}{1.552^k}\right) = O(1.752^k)$$

which finally yields

$$t_{\mathcal{A}} \leq |\ln \varepsilon| \cdot (O(1.752^k) + 1) \cdot O(2^{1.3k} m) = O(|\ln \varepsilon| \cdot 4.32^k m)$$

as claimed by the theorem. □

Numerical evaluation suggests that setting x close to $0.3k$ (whether to round up or down should be determined numerically for a concrete k) as done in the theorem is actually an optimal choice from a runtime perspective.

For a practical implementation, while we could fix the number of colors at the worst-case optimum $k + x$, it is most likely beneficial to choose x even larger, because various algorithmic tweaks and the underlying graph structure can keep the runtime of a trial significantly below the worst-case estimate. This in turn causes the increase in runtime per trial by choosing a larger x to be even more overcompensated by a decrease in the total number of trials needed, as is demonstrated in Figure 2. In fact, for a small path length of 8–10 we can choose the number of colors to be the maximum our implementation allows (that is, 31), and get by with a very small number of trials (≈15–30). (Based on such observations, our implementation uses an adaptive approach to the number of colors, starting with the maximum of 31 and decreasing this in case a trial runs out of memory.)

3.2. *Speedup by Lower Bounds and Cache Preheating*

In a color-coding trial, every vertex carries entries for up to 2^{k+x} color sets, each of them representing a partial colorful path with a certain weight. Because each entry may get expanded to an exponentially large collection of new entries, pruning even a small fraction

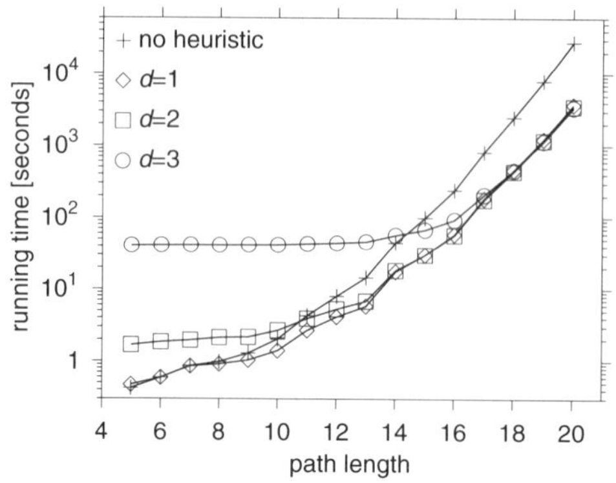

Figure 3. Runtime comparison with heuristic evaluation functions for different values of d (seeking the 20 lowest-weight paths in the yeast network of Scott et al.[10] that differ in at least 30% of participating vertices).

of them can lead to a significant speedup. The pruning strategy that we employ makes use of the fact that we are only looking for a fixed number of minimum-weight paths. As soon as we have found this number of candidates, we can always remove entries where the weight of the corresponding partial path is certain to exceed the weight of the worst known path in the current collection of paths when completed.

Consider an entry $W(u, S)$ corresponding to some partial path. To obtain a length-k path, we need to append another $k - |S|$ edges, and so a lower bound for the total weight of a length-k path expanded from this entry is $W(u, S) + (k - |S|)w_{\min}$, where $w_{\min}$ is the minimum weight of any edge in the graph. We improve upon this simple bound by dividing the path length left not into single edges, but short segments of d edges, calculating a lower bound separately for each segment, and summing up these bounds.

We prepare the lower bound calculation in a preprocessing phase by dynamic programming on the uncolored graph. Clearly, there is a trade-off between the time invested in the preprocessing (depending on d) and the time saved in the main algorithm. For the yeast network of Scott et al.[10], setting $d = 2$ seems to be a good choice with an additional second of preprocessing time. For $d = 3$, the preprocessing time increases to 38 seconds, an amount of time that is only recovered when searching for paths of length at least 19 (see Figure 3).

Using lower bounds is only effective once we have already found as many paths as we are looking for. Therefore, it is important to quickly find some low-weight paths early in the process. We achieve this acquisition of lower bounds by prepending a number of trials with a thinned-out graph, that is, for some $0 < t < 1$, we consider a graph that contains only the $t|E|$ lightest edges of the input graph. (Especially in database applications, similar techniques are known as "preheating the cache.") Trials for a certain value of t are repeated with different random colorings until the lower bound does not improve any more. By default, t is increased in steps of $1/10$; should we run out of memory, this step size is halved. This allows to successfully complete trials in the thinned out graphs, making trials feasible on the original graphs by providing them with powerful bounds for pruning.

Table 1. Basic properties of the network instances YEAST (Scott el al.[10]) and DROSOPHILA (Giot et al.[7]). The *clustering coefficient* is the probability that $\{u, v\} \in E$ for $u, v, x \in V$ with $\{u, x\} \in E$ and $\{x, v\} \in E$.

	vertices	edges	clustering coefficient	average degree	maximum degree
YEAST	4 389	14 319	0.067	6.5	237
DROSOPHILA	7 009	20 440	0.030	5.8	175

3.3. *Efficient Storage of Color Sets*

Since one is not only interested in the weight of a solution, but in the vertices (that is, proteins) that it consists of, it is common to not only store the weight of a partial colorful path in Equation (1) but also a concrete sequence of vertices that realizes this weight. This accounts for the bulk of the memory requirement of a color-coding implementation because $k\lceil\log|V|\rceil$ bits per stored path are required. We propose to save memory here by noting that it suffices to store only the order in which the colors appear on a path: after completing a color set at some vertex u, the path can be recovered by running a shortest path algorithm (e.g., Dijkstra's algorithm[4]) for the source vertex u while allowing it to only travel edges that match the color order. This reduces the memory cost per entry to $k\lceil\log k\rceil$ bits, which, for our application, amounts to a saving factor of about 2–4. Because of the resulting increase in computer cache effectiveness, this usually also leads to a speedup except when either short path lengths are used (where memory is not an issue anyway) or when many solution paths are found and have to be reconstructed.

As to the data structure for the color sets at each vertex, they are managed as a Patricia tree, that is, a compact representation of a radix tree[4] where any node which is an only child is merged with its parent. A color set is represented as a bit string of fixed length. The Patricia tree allows for very quick insertions and iterations with a moderate memory overhead of, e.g., 12 bytes per color set on a 32-bit system.

4. Experimental Results

Method and Results. We have implemented the color-coding technique with the improvements described in the last section. The source code of the program is available from `http://theinf1.informatik.uni-jena.de/colorcoding/`; it is written in the C++ programming language and consists of approximately 1200 lines of code. The testing machine is an AMD Athlon 64 3400+ with 2.4 GHz, 512 KB cache, and 1 GB main memory running under the Debian GNU/Linux 3.1 operating system. The program was compiled with the GNU g++ 4.2 compiler using the options "-O3 -march=athlon".

The real-world network instances used for speed measurements were the *Saccharomyces cerevisiae* interaction network used by Scott el al.[10] and the *Drosophila melanogaster* interaction network described by Giot et al.[7]. Some properties of these networks, which we will refer to as YEAST and DROSOPHILA, are summarized in Table 1.

To explore the sensitivity of the runtime to various graph parameters (namely, the number of vertices, the clustering coefficient, the degree distribution, and the distribution of edge weights), the implementation was also run on a testbed of random graph instances

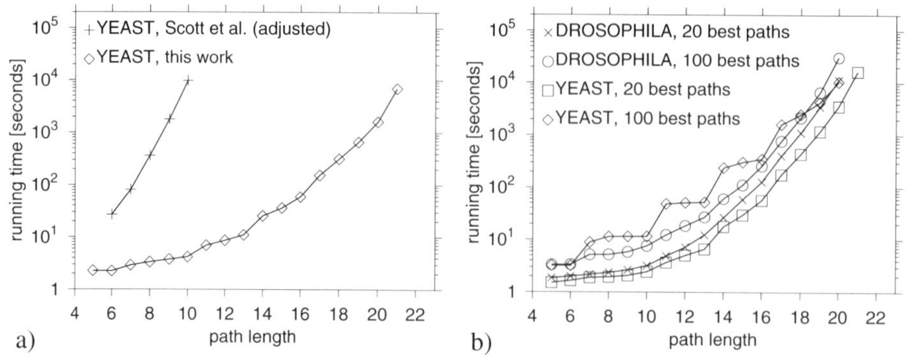

Figure 4. **(a)** Runtimes for YEAST as reported by Scott et al. [10] (adjusted for speed difference of the testing machines) and measured with our implementation. In both cases, paths must start at a membrane protein, and end at a transcription factor. Memory requirements were, e. g., 3 MB for $k = 10$ and 242 MB for $k = 21$.
(b) Comparison of the runtimes of our implementation when applied to YEAST and DROSOPHILA for various path lengths, seeking after either 20 or 100 minimum-weight paths that mutually differ in at least 30% of their vertices. There were no restrictions as to the sets of start and end vertices.

that were generated with the algorithm described by Volz[13]. The results of all experiments and details as to the experimental setting are given in Figures 4 and 5.

Note that Scott et al.[10] obtained their runtimes on a dual 3.0 GHz Intel Xeon processor with 4 GB main memory. To make their runtimes comparable with ours, Figure 4 does not report their original times here, but divides them by 1.2 (which is a very conservative estimate in favor of Scott et al.[10] that most likely overestimates the speed of our machine).

Discussion. Compared to the (machine-speed adjusted) runtimes from Scott et al.[10], our implementation is faster by a factor of 10 to 2 000 on YEAST (see Figure 4a). Scott et al. discuss findings for paths up to a length of 10 which they were able to find in about three hours. These can be found within seconds by our implementation, allowing for interactive queries and displays. The range of feasible path lengths is more than doubled.

Figure 4b shows that the runtimes for both YEAST and DROSOPHILA are roughly equal. The only exception is the search for the best 100 paths within YEAST which not only takes unexpectedly long but also displays step-like structures. Most likely, these two phenomena can be attributed to the fact that certain path lengths allow for much fewer well-scoring paths than others in YEAST, causing the lower-bound heuristic to be less effective. Figure 4b also demonstrates that a major factor in the runtime is actually the number of paths that is sought after. This is because a larger number of paths worsens the lower bound of the heuristic which cannot cut off as many partial solutions and maintaining the list of paths and checking the "at least 30% of vertices must differ" criterion becomes more involved.

Figures 5a, 5b, and 5d show that the runtime of the color-coding algorithm appears to be somewhat insensitive to the size of the graph (increasing linearly with increasing graph size) as well as the clustering coefficient and the distribution of edge weights. The somewhat unexpectedly high runtimes for graphs with less than 500 vertices in Figure 5a are explained by the fact that the number of length-10 and length-15 paths in these networks

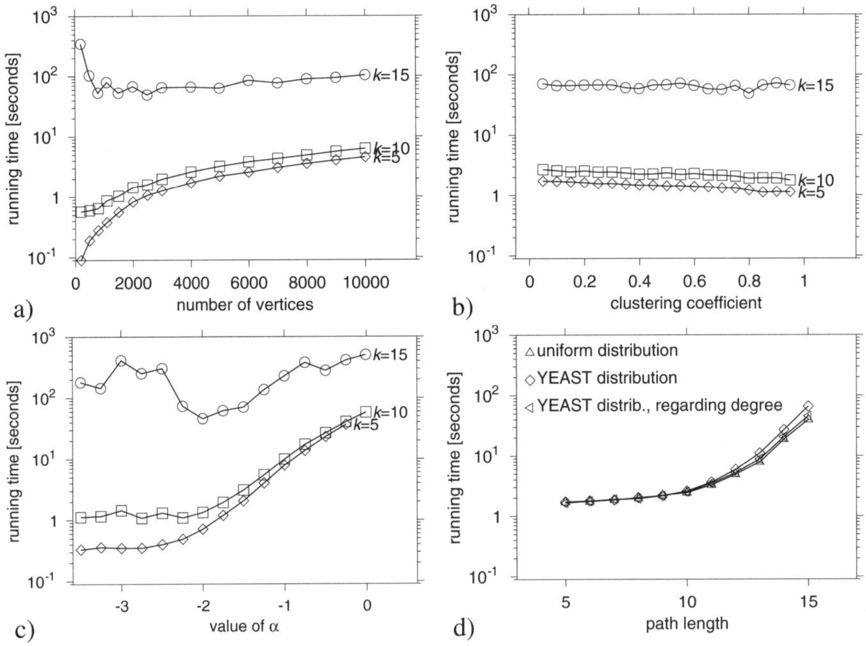

Figure 5. Runtime for our color-coding implementation on random networks, seeking after 20 minimum-weight paths. Unless a parameter is the variable of a measurement, the following default values are used (we have empirically found them to result in networks that are quite similar to YEAST): 4 000 vertices; degree distribution is a power law with exponential cutoff, that is, the fraction p_k of vertices with degree k satisfies $p_k \sim k^{\alpha} \cdot e^{-k/1.3} \cdot e^{-45/k}$; the default value for α is -1.6; edge weights are distributed as in YEAST; the clustering coefficient is 0.1. The data shown reports the average runtime over five runs each. **(a)** Dependency on the number of vertices. **(b)** Dependency on the clustering coefficient. **(c)** Dependency on the parameter α of the power law distribution. **(d)** Dependency on the distribution of edge weights for three different distributions: A uniform $[0, 1]$-distribution, the distribution of YEAST, and the distribution of YEAST under consideration of vertex degree.

is very low, causing the heuristic lower bounds to be rather ineffective (this also explains why the effect is worse for $k = 15$ than it is for $k = 10$).

Figure 5c shows that the algorithm is generally faster when the vertex degrees are unevenly distributed. This comes as no surprise because for low-degree vertices, fewer color sets have to be maintained in general and the heuristic lower bounds are often better. For $k = 15$, two points in the curve require further explanation: First, the drop-off in running time for $\alpha < -3$ is explained by the random graph "disintegrating" into small components. Second, the increased runtime for $-3 \leq \alpha \leq -2$ can most is most likely due to a decrease in the total number of length-15 paths as compared to larger values of α.

5. Conclusion

We have given various algorithmic improvements that enable the color-coding technique as a tool both for fast exploration of small pathway candidates as well as for finding larger structures than previously possible. The protein interaction networks of yeast and fruit fly are the so-far most extensively analyzed and understood; as more high-quality data becomes available in the future, it would be interesting to further investigate the practical application of color-coding beyond the detection of linear pathways. Most of the improvements we have given are also useful for detecting other structures that can also be handled with color-coding such as cycles and trees. Finally, we plan further research into the applicability of color-coding to querying pathways in a network (as recently done by Shlomi et al.[11]) and are working on extending our color-coding implementation to a user-friendly tool for pathway candidate detection.

Acknowledgments

Falk Hüffner was supported by the Deutsche Forschungsgemeinschaft, Emmy Noether research group PIAF (fixed-parameter algorithms), NI 369/4, Sebastian Wernicke was supported by the Deutsche Telekom Stiftung, and Thomas Zichner was supported by the Deutsche Forschungsgemeinschaft project PEAL (Parameterized Complexity and Exact Algorithms), NI 369/1. The authors are grateful to Jacob Scott (Cambridge, MA) for providing them with the yeast interaction network discussed in Ref. 10 and to Hannes Moser and Rolf Niedermeier (Jena) for discussions and comments.

References

1. N. Alon, R. Yuster, and U. Zwick. Color-coding. *J. ACM*, 42(4):844–856, 1995.
2. J. S. Bader, A. Chaudhuri, J. M. Rothberg, and J. Chant. Gaining confidence in high-throughput protein interaction networks. *Nature Biotech.*, 22(1):78–85, 2004.
3. T. Can, O. Çamoğlu, and A. K. Singh. Analysis of protein–protein interaction networks using random walks. In *Proc. 5th ACM SIGKDD Workshop on Data Mining in Bioinformatics (BIOKDD '05)*, 2005.
4. T. H. Cormen, C. E. Leiserson, R. L. Rivest, and C. Stein. *Introduction to Algorithms*. MIT Press, 2001.
5. M. R. Garey and D. S. Johnson. *Computers and Intractability: A Guide to the Theory of NP-Completeness*. Freeman, 1979.
6. G. K. Gerber, Z.-B. Joseph, T. I. Lee, et al. Computational discovery of gene modules and regulatory networks. *Nature Biotech.*, 21(11):1337–1342, 2003.
7. L. Giot, J. S. Bader, C. Brouwer, et al. A protein interaction map of Drosophila melanogaster. *Science*, 302(5651):1727–1736, 2003.
8. T. Ideker, V. Thorsson, J. A. Ranish, et al. Integrated genomic and proteomic analyses of a systematically perturbed metabolic network. *Science*, 292(5518):929–934, 2001.
9. T. Ito, T. Chiba, R. Ozawa, et al. A comprehensive two-hybrid analysis to explore the yeast protein interactome. *PNAS*, 98(8):4569–4574, 2001.
10. J. Scott, T. Ideker, R. M. Karp, and R. Sharan. Efficient algorithms for detecting signaling pathways in protein interaction networks. *J. Comp. Biol.*, 13(2):133–144, 2006. Preliminary version appeared in *Proc. RECOMB'05*.
11. T. Shlomi, D. Segal, E. Ruppin, and R. Sharan. QPath: a method for querying pathways in a protein–protein interaction network. *BMC Bioinformatics*, 7:199, 2006.
12. M. Steffen, A. Petti, J. Aach, P. D'haeseleer, and G. Church. Automated modelling of signal transduction networks. *BMC Bioinformatics*, 3:34, 2002.
13. E. Volz. Random networks with tunable degree distribution and clustering. *Phys. Rev. E*, 70:056115, 2004.

DE NOVO PEPTIDE SEQUENCING FOR MASS SPECTRA BASED ON MULTI-CHARGE STRONG TAGS

KANG NING, KET FAH CHONG, HON WAI LEONG

Department of Computer Science,
National University of Singapore,
3 Science Drive 2, Singapore 117543

This paper presents an improved algorithm for de novo sequencing of multi-charge mass spectra. Recent work based on the analysis of multi-charge mass spectra showed that taking advantage of multi-charge information can lead to higher accuracy (sensitivity and specificity) in peptide sequencing. A simple de novo algorithm, called GBST (Greedy algorithm with Best Strong Tag) was proposed and was shown to produce good results for spectra with charge > 2. In this paper, we analyze some of the shortcomings of GBST. We then present a new algorithm GST-SPC, by extending the GBST algorithm in two directions. First, we use a larger set of multi-charge strong tags and show that this improves the theoretical upper bound on performance. Second, we give an algorithm that computes a peptide sequence that is optimal with respect to shared peaks count from among all sequences that are derived from multi-charge strong tags. Experimental results demonstrate the improvement of GST-SPC over GBST.

1. Introduction

De novo peptide sequencing of tandem mass (MS/MS) is a challenging problem in proteomics and high-throughput generation of MS/MS spectra with modern proteomics technology is compounding the problem. As the volume of MS/MS spectra grows, the accompanying algorithmic technology for automatically interpreting these spectra has to keep pace. An increasingly urgent problem is the interpretation of *multi-charge* spectra – MS/MS spectra with charge 3, 4, and 5 are available from the publicly accessible GPM (Global Proteome Machine) dataset [5]; and those with charge 3 are available from the ISB (Institute for Systems Biology) dataset [10]. It is foreseen that increasingly there will be more multi-charge spectra produced and so the problem of accurate interpretation of these spectra will become more important with time.

Many existing algorithms for peptide sequencing have focused largely on interpreting spectra of charge 1, even when dealing with multi-charge spectra, and only a few algorithms [4, 8, 11] take account for higher charge ions. Recent work by the authors [4] using this approach has shown that the sensitivity accuracy of Lutefisk [15] and PepNovo [8] (both of which consider only ion-types of charge 1 and 2) are very low (less that 25%) when applied to higher charge spectra from the GPM dataset. Their experimental study also showed that there is significant potential improvement in the performance if multiple charges are taken into consideration during the sequencing process. A simple de novo algorithm, called *GBST* (Greedy algorithm with Best Strong Tags) was also presented that uses multiple charges to achieve good results for multiple charge spectra. The GBST algorithm consists of two phases: in the first phase, a set of "best" strong tags are computed based on strong evidence in the spectrum (charge 1, *b*-ions and *y*-ions, and no neutral loss, and direct connectivity); in the second phase, the GBST algorithm then link the set of "best" strong tags by taking into account more ion-types (charges) and greater connectivity. A standard algorithm was then used to generate the set of paths that corresponds to the top *k* predicted peptide sequences.

In this paper, we present an improved algorithm called *GST-SPC* that improves on GBST algorithm. In the first phase, the GST-SPC algorithm computes a *larger* set of strong tags – the set of all *maximal* multi-charge strong tags. We show that this improves the theoretical upper bound on the sensitivity. In the second phase, the GST-SPC algorithm computes a peptide sequence that is *optimal* with respect to *shared peaks count* (*SPC*) from among all sequences that are derived from strong tags. Our evaluation shows that the GST-SPC algorithm improves on GBST, especially on multi-charge spectra.

2. Review of Related Work and Problem Formulation

We first give a quick review of related work on de novo peptide sequencing for MS/MS. De novo algorithms [1, 3, 4, 6, 8, 11, 15] are used to predict sequences or partial sequences for novel peptides or for peptides that are not in the protein database. Most de novo sequencing algorithms [3, 6, 8, 15] uses a *spectrum graph* approach to reduce the search space of possible solutions. Given a mass spectrum, the spectrum graph [6] is a graph where each vertex corresponds to some ion type interpretation of a peak in the spectrum. Edges represent amino acids which can interpret the mass difference between two vertices. Each vertex in this spectrum graph is then scored using *Dancik scoring* based on its supporting peaks in the spectrum (see [6] for details). Given such a scoring the predicted peptide represents the optimal weighted path from the *source vertex* (of mass 0) to the *end vertex* (of mass *M*).

PepNovo [8] uses a spectrum graph approach similar to [6], but uses an improved scoring function based on a probability network of different factors which affect the peptide fragmentation and how they conditionally affect each other (represented by edges from one vertex to another). The algorithm PEAKS [11] does not explicitly construct a spectrum graph but builds up an optimal solution by finding the best pair of prefix and suffix masses for peptides of small masses until the mass of the actual peptide is reached. A fast dynamic programming algorithm is used.

Problem Formulation of Multi-Charge Peptide Sequencing: Our formulation of multi-charge peptide sequencing follows that in [4]. We summarize it here to facilitate our discussion of the GBST algorithm (see [4] for detailed discussion).

Consider a multi-charge MS/MS spectrum *S* of charge α for a peptide $\rho = (a_1a_2...a_n)$ where a_j is the jth amino acid in the sequence. The *parent mass M*, of the peptide is given by $m(\rho) = M = \sum_{j=1}^{n} m(a_j)$. A peptide fragment $\rho_k = (a_1a_2...a_k)$ $(k \leq n)$ has fragment mass $m(\rho_k) = \sum_{j=1}^{k} m(a_j)$. The peaks in the spectrum *S* come from peptide fragmentation. Each peak *p* can be characterized by its ion-type given by $(z, t, h) \in (\Delta_z \times \Delta_t \times \Delta_h)$, where z is the charge of the ion, t is the basic ion-type, and h is the neutral loss incurred by the ion. In this paper, we use $\Delta = (\Delta_z \times \Delta_t \times \Delta_h)$, where $\Delta_z = \{1,2,\ldots,\alpha\}$, $\Delta_t = \{a,b,y\}$, and $\Delta_h = \{\phi, -H_2O, -NH_3\}$. The (z, t, h)-*ion* of the peptide fragment ρ_k will produced an observed peak p in the spectrum S, that has a mass-to-charge ratio of $mz(p)$, that can be computed by the following formula [4]:

$$m(\rho_k) = mz(\rho) \cdot z + (\delta(t) + \delta(h)) + (z - 1) \qquad (1)$$

where $\delta(z)$ and $\delta(z)$ are the mass difference for the respective ion-type and neutral-loss.

The *theoretical spectrum of charge* α [4] for ρ is defined by $TS_\alpha^\alpha(\rho) = \{p \mid p$ is observed peak for (z, t, h)-*ion* of peptide fragment ρ_k, for all $(z, t, h) \in \Delta$ and $k=0,1,\ldots,n\}$. It represents the set of *all possible* observed peaks that may be present in any experimental spectrum for the peptide ρ.

In peptide sequencing, we are given an experimental spectrum $S = \{p_1, p_2, \ldots, p_n\}$, where each peak p_k is described by two parameters: $mz(p_k)$, the observed mass-to-charge ratio and $intensity(p_k)$, its intensity. The problem is to determine the peptide ρ that produced S. In practice, of course, only a small fraction of these peaks in $TS_\alpha^\alpha(\rho)$ are present in S and there are also noise peaks as well.

Extended Spectrum and Spectrum Graph: To account for the different ion-types considered by different algorithms, the notions of *extended spectrum* and *extended spectrum graph* were introduced S_β^α were introduced in [4], where α denotes the maximum charge for S, and β denotes the maximum charge considered by the algorithm ($\beta = 2$ for PepNovo [8] and Lutefisk [15]). In the *extended spectrum* S_β^α, for each peak $p_j \in S$ and ion-type $(z, t, h) \in (\{1,2,\ldots,\beta\} \times \Delta_t \times \Delta_h)$, we generate a pseudo-peak denoted by $(p_j, (z, t, h))$ with a corresponding *assumed* fragment mass given by (1). Then, the *extended spectrum graph* of connectivity d is a graph $G_d(S_\beta^\alpha)$ in which each vertex represents a pseudo-peak $(p_j, (z, t, h))$ in the extended spectrum S_β^α, namely to the (z, t, h)-*ions* for the peak p_j. Two special vertices are added – the *start vertex* v_0 to represent mass 0 and the *end vertex* v_M for the parent M. For each vertex v, we define *prefix residue mass* of v, denoted by $PRM(v)$, to be the *prefix mass* of the interpreted peptide fragment mass for vertex v. It is defined as $PRM(v) = m(v)$ if v is a prefix ion type, and $PRM(v) = M - m(v)$ if v is a suffix ion type, where M is the parent mass. There is a directed edge (u, v) from vertex u to vertex v if we can find a directed path of at most d amino acid with total mass equal to the $(PRM(v) - PRM(u))$. (The standard spectrum graph use $d = 1$.) Note that the number of possible paths to be searched is $O(20^d)$, which increases exponentially with d. In this paper, we use d=2, unless otherwise stated. The extended spectrum is a generalization because when β=1, all peaks are assumed to be of charge 1, and so $S_1^\alpha = S$ – namely, there is no extension. In the extended spectrum S_2^α, only ions of charge 1 or 2 are considered (even for spectra with charge $\alpha > 2$). Algorithms such as PepNovo [8] and Lutefisk [15] uses some subsets of S_2^α and $G_2(S_2^\alpha)$.

Upper Bound on Sensitivity: Given any spectrum graph G defined on an experimental spectrum S from a *known* peptide ρ, the notion of *theoretical upper bound on sensitivity* was defined in [4] as follows: Given G, we can compute the path in G that *maximizes* the number, p^*, of amino acids from the (*known*) peptide ρ. Then, $U(G) = p^*/|\rho|$ is an upper bound on the *sensitivity* for *any sequencing algorithm* based on the spectrum graph approach using the graph G. Then $U(G_d(S_\beta^\alpha))$ is the *theoretical upper bound on sensitivity* for the extended spectrum graph $G_d(S_\beta^\alpha)$, namely using the extended spectrum S_β^α with all ion types in Δ and a connectivity of d. PepNovo and Lutefisk which considers charge of up to 2 (and connectivity of up to 2) are bounded by $U(G_2(S_2^\alpha))$ and there is a sizeable gap between $U(G_2(S_2^5))$ and $U(G_2(S_5^5))$.

3. Evaluation of Greedy Strong Tag Algorithm for Multi-Charge Spectra

The GBST algorithm [4] was a simple algorithm that takes into account higher charge ions. It performs well on multi-charge spectra compared to other de novo algorithms. However, we show in this section that there is still a big gap in performance with respect to the theoretical upper bound $U(G_2(S_\alpha^\alpha))$.

The GBST Algorithm: The GBST algorithm first computes a set, *BST*, of "best" (or reliable) strong tags. To find strong tags, they use ion-types that appear most frequently, namely, charge 1, *b-ions* and *y-ions* with no neutral loss. The restricted set is given by $\Delta^R = (\Delta_z^R \times \Delta_t^R \times \Delta_h^R)$, where $\Delta_z^R = \{1\}$, $\Delta_t^R = \{b, y\}$, and $\Delta_h^R = \{\phi\}$. They also define $G_1(S_1^\alpha, \Delta^R)$, the spectrum graph $G_1(S_1^\alpha)$ where the ion types considered are restricted to those in Δ^R. Then, a *strong tag* T of ion-type $(z, t, h) \in \Delta^R$ is a maximal path $\langle v_0, v_1, v_2, \ldots, v_r \rangle$ in the graph $G_1(S_1^\alpha, \Delta^R)$, where every vertex $v_i \in T$ is of a (z, t, h)-*ion*. In each "component" of this graph, GBST compute a "best" strong tag with respect to some scoring function [4]. Then, the set *BST* is the set comprising the best strong tag for each component in the spectrum graph $G_1(S_1^\alpha, \Delta^R)$.

After the set of best tag, *BST*, is computed, the GBST algorithm then proceeds to find the best sequence that result from paths obtained by "extending" the tags from *BST* using all possible ion-types. It search for paths in the graph $G_2(BST)$ defined as follows: the vertices are the strong tags in *BST*, and we have a directed edge from the tail vertex u of a strong tag T_1 to the head vertex v of another strong tag T_2 if there is a directed edge (u,v) in the graph $G_2(S_\alpha^\alpha)$. We note two major difference between $G_2(BST)$ and the extended spectrum graph $G_2(S_\alpha^\alpha)$ – firstly, the number of vertices in $G_2(BST)$ is smaller; and secondly, the number of edges is also much smaller since only strong tags are linked in a head-to-tail manner. However, all ion types are considered in the graph $G_2(BST)$.

Upper Bounds on Sensitivity for GBST: Since the GBST algorithm uses a restricted set of ion-types Δ^R in its search for best strong tags, we let $U(R) = U(G_1(S_1^\alpha, \Delta^R))$ be the upper bound on sensitivity *with ion-type restriction*. For the second phase, we define $U(BST) = U(G_2(BST))$, the upper bound on sensitivity *with best strong tag restriction*.

Datasets Used: To evaluate the performance of GBST vis-à-vis the upper bounds, we used spectra that are annotated with their corresponding peptides – the GPM-Amethyst dataset [5] (Q-star data with good resolution[1]) and the ISB dataset [10] (Ion-Trap data with low resolution). For each dataset, we selected subsets of spectra with annotated

---

[1] Though these GPM spectra are high resolution spectra, they have been pre-processed using deconvolution [16], and so charge state determination using mono-isotopes is not possible.

peptides validated with an X-correlation score (Xcorr) greater than 2.0. The selected GPM dataset we use contains 2328 spectra, with 756, 874, 454, 205, and 37 each of charge 1,2,3,4, and 5, respectively, with an average of 46.5 peaks per spectrum.
The selected ISB dataset contains 995 spectra, with 16, 489, and 490 of charge 1,2, and 3, respectively, with an average of 144.9 peaks per spectrum.

The Evaluation Results: We have computed these upper bounds on sensitivity for both the GPM and the ISB datasets and the results are shown in Figure 1, together with the actual sensitivity obtained by the GBST algorithm. The results in Figure 1 show that for GPM datasets, *U*(*BST*) is near to *U*(*R*), but the GBST results have sensitivities about 10% less than *U*(*BST*). This indicates that the GBST has not been able to fully utilize the power of *BST*. For the ISB datasets, even *U*(*BST*) is far from *U*(*R*). Therefore, it is natural the GBST algorithm can not perform well on ISB datasets.

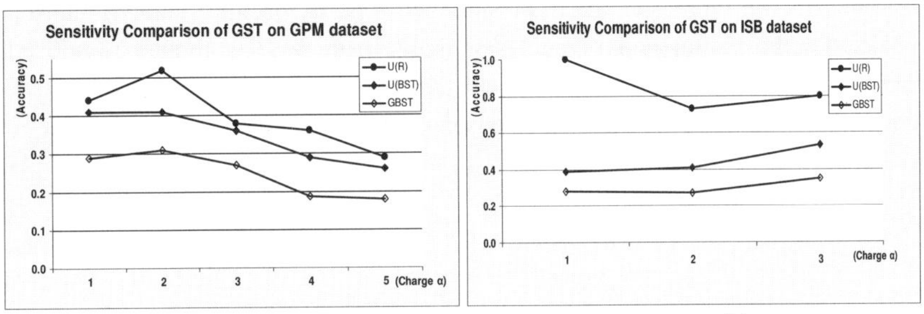

(a) (b)

Figure 1. The comparison of sensitivity results of *GBST* with theoretical upper bounds. *U*(*R*) and *U*(*BST*) on (a) GPM dataset, and (b) ISB datasets.

4. An Improved Algorithm – GST-SPC

In this paper, we present an improved algorithm, called GST-SPC, for de novo sequencing of multi-charge spectra, which improves the GBST algorithm in two ways: (a) by selecting a larger set of multi-charge strong tags, and second, and (b) by improving the sequencing algorithms for a given set of multi-charge strong tags.

(a) Using a Larger Set of Strong Tags: A straight-forward improvement of GBST [4] is to expand the set of strong tag under consideration. We do this as follows: (i) when searching for strong tags, we include multi-charge ions (using S_{α}^{α} instead of just S_{1}^{α}), and (ii) instead of choosing *only one* "best" strong tag from each component of the graph $G_1(S_1^{\alpha}, \Delta^R)$, we allow a set of all *multi-charge strong tags* in each component of the graph $G_1(S_{\alpha}^{\alpha}, \Delta^R)$ to be chosen. Namely, a *multi-charge strong tags* of ion-type $(z^*, t, h) \in \Delta^R$ is a maximal path $\langle v_0, v_1, v_2, \ldots, v_r \rangle$ in $G_1(S_{\alpha}^{\alpha}, \{\Delta^R\})$, where every vertex v_i is of a (z^*, t, h)-ion, in which t and h should be the same for all vertices, but z^* can be different number from $\{1, \ldots \alpha\}$. We let *MST* denote this set. The algorithm for computing the *MST* is the almost identical to that for tag generation (a depth-first search) with slight modification to store the *MST*. Running the GBST algorithm with the MST (instead of the BST) improves the results slightly (the details not shown here).

Theoretically, the size of the *MST* may be exponential. However, in practice, our experiments show that the MST does not exhibit exponential growth from BST. For GPM datasets (average of about 46 peaks) the increase in the average number of strong tags is from 10 to about 50. For ISB datasets (average of 145 peaks) the increase is from 15 to about 90. The average length of strong tags in MST is 4.65 amino acids for GPM datasets, and 2.26 for ISB datasets.

We define $U(MST) = U(G_2(MST))$ the *theoretical upper bound on sensitivity with respect to the set MST* of multi-charge strong tags. The increase from *U*(*BST*) to *U*(*MST*) is shown in Figure 2. From Figure 2, it is easy to see that the introduction of MST has pushed up the theoretical upper bounds for both datasets. For GPM dataset, the best sequencing results obtainable from MST is about 5% higher in accuracies than BST. We also note that *U*(*MST*) is very close to the *U*(*R*), the theoretical upper bounds with Δ^R. For ISB datasets, the increase is more pronounced – partly because the ISB datasets have more peaks. The best sequencing results obtainable from *MST* is about 10%~60% higher in accuracies than *BST*, and with in 20% to the theoretical upper bounds. This shows a great potential for sequencing algorithms based on *MST*.

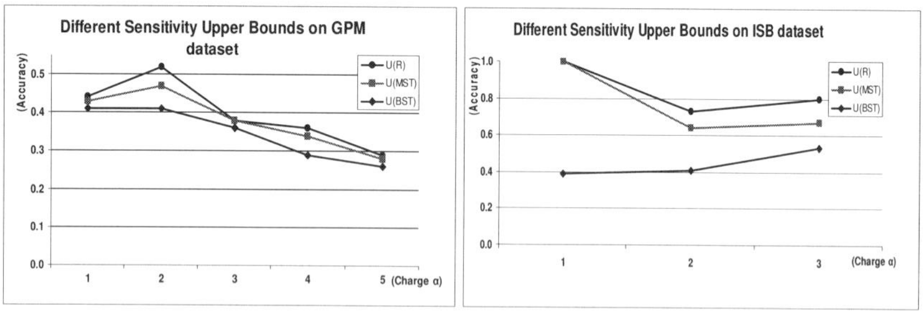

(a) (b)

Figure 2. Comparing the theoretical upper bounds on sensitivity for MST and BST. Results are based on (a) GPM dataset, and (b) ISB datasets.

(b) Optimal Shared Peaks Count: While the GBST algorithm modified to use MST (in place of BST) is slightly better, there is still a gap in performance. This motivates us to formulate the problem of *maximizing the shared peaks counts* with respect to the computed set of multi-charge strong tags. The shared peak count (SPC) is a commonly used and fairly objective criterion for determining the "quality" of de novo peptide sequencing. We also show that we can solve this problem *optimally in polynomial time*.

Suppose that we are given the set, say MST, of strong tags. Define a *multi-charge strong tag path Q* to be a path from v_0 to v_M given by $Q = (q_0\ T_1\ q_1\ T_2\ q_2\ T_3\ q_3\ \ldots\ q_{k-1}\ T_k\ q_k)$ where each T_j is a strong tag in MST and each q_j is a path of at most *two* amino acids, or mass difference that "links" the preceding tag to the succeeding tag in the usual head-to-tail fashion. A strong tag path *Q* gives rise to a peptide sequence *P*(*Q*) obtained by interpreting the "gaps" in the path *Q*. A example of P(Q) is "[50]CGV[100]PK". Given the peptide sequence *P*(*Q*), we can compute the shared peaks count of *P*(*Q*). Then our problem can be stated as the following: Among all the possible strong tag paths, we want to find an *optimal multi-charge strong tag path* Q^* that *maximize the shared peak count* in the peptide sequence $P(Q^*)$.

Our solution to this problem is to form the graph $G_2(MST)$ defined in the same ways as the graph $G_2(BST)$. We first pre-compute the shared peaks count for each tag in *MST*. For each edge (u, v) connecting two tags T_u and T_v, we compute the path Q of length with at most two amino acids that locally maximizes that shared peak count of Q against experimental spectrum. Then we can compute the path with maximum shared peaks count in the graph $G_2(MST)$, which is a DAG. Additional processing has to be done if either end vertex is not connected to the first (or last) vertex in the path, or the sparse areas are not connectable – we connect this via mass difference. It is easy to see that this algorithm optimizes the shared peaks count among all peptide sequences obtained by extending the multi-charge strong tags in MST via connectivity 2. Next, we present an algorithm that produces provably better result.

Improving the Spared Peaks Counts using *H(MST)*: We can further improve the shared peaks count if we increase the maximum connectivity *d*. However, this will cause the running time to grow exponentially due to the number of paths to be searched. We propose a graph $H(MST)$, a superset of $G_2(MST)$ which is simple to define, and yet not too computationally expensive. In $H(MST)$, we have an edge from the tail vertex u of T_u to the head vertex v of T_v if the mass difference $(PRM(v)–PRM(u))$ is in the range [57.02, 186.08] Da, where 57.02 and 186.08 are the minimum and maximum mass of an amino acid, respectively. In addition, we can also pre-compute the path from u to v that locally maximizes the shared peak count. For this sub-problem, we have fast procedure that does this efficiently. The length of the computed path from u to v varies depending on the mass difference. The rest of the algorithm is to interpret edges in $H(MST)$.

Algorithm GST-SPC: Finally, our GST-SPC algorithm uses the multi-charge strong tag set *MST* and the graph $H(MST)$ to compute a peptide with optimal *shared peaks count.*

5. Performance Evaluation of Algorithm GST-SPC

We have compared the performance of our algorithms with two other algorithms with freely available implementation, Lutefisk [15] and PepNovo [8]. For specific spectrum and algorithm, the sequencing results with best scores are compared. To compare performance of GST-SPC with the GBST [4], Lutefisk [15], and PepNovo [8], we use the following accuracy measures:

Sensitivity = # correct / $|\rho|$ Tag-Sensitivity = # tag-correct / $|\rho|$

Specificity = # correct / $|P|$ Tag-Specificity = # tag-correct / $|P|$

where #correct is the "*number of correctly sequenced amino acids*" and #tag-correct is "*the sum of lengths of correctly sequenced tags* (*of length* > *1*)". The number of correctly sequence amino acids is computed (approximated) as the longest common subsequence (lcs) of the correct peptide sequence ρ and the sequencing result P. The *sensitivity* indicates the quality of the sequence with respect to the correct peptide sequence and a high sensitivity means that the algorithm recovers a large portion of the correct peptide.

The *tag-sensitivity* accuracy takes into consideration of the continuity of the correctly sequenced amino acids. For a fairer comparison with algorithms like PepNovo that only outputs the highest scoring tags (subsequences) we also use specificity and tag-specificity measures, which measures how much of the results are correct.

The comparison of the different algorithms based on the four accuracy measures is summarized in Figure 3 (for the GPM datasets) and Figure 4 (for the ISB datasets). Overall, the results obtained by our GST-SPC algorithm using the shared peaks count scoring functions are promising. On the GPM datasets, the GST-SPC outperforms the other algorithms. For example, it has higher sensitivity than Lutefisk (by 10% for charge ≥ 2) and PepNovo (by about 10%) in sensitivity and tag-sensitivity. It has comparable specificity and tag-specificity PepNovo for charge 1 and 2. It is constantly better than GBST and Lutefisk (for charge > 1) on all accuracy measures.

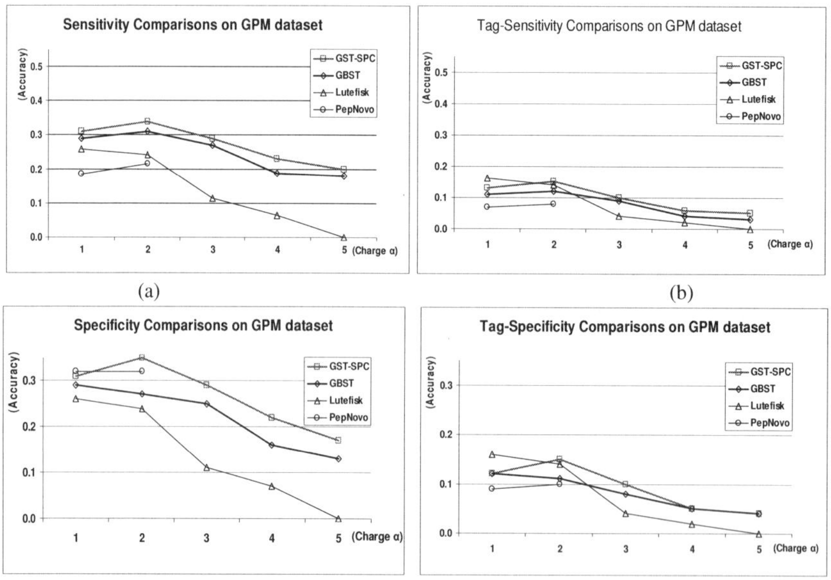

(c) (d)

Figure 3. Comparison of different algorithms on GPM dataset – based on (a) sensitivity, (b) tag-sensitivity, (c) specificity and (d) tag-specificity. PepNovo only has results for charge 1 and 2.

For the ISB dataset, the results shows that the ranking as follows: (PepNovo, GST-SPC, GBST, Lutefisk) for all the accuracy measures. The ISB datasets contains many noises and PepNovo has a sophisticated scoring function that may account for its best performance, especially on datasets with charge 1. For spectra with charge 2, the difference in performance is not as high. However, since PepNovo do not (as yet) handle spectra with charge greater than 2, there was no way to compare results for charge 3. That comparison would be interesting given the apparent trend exhibited in the results.

We also compare the algorithm with respected to the number of *completely correct* identified peptide sequences. Our results (not shown here due to space limitations) show that the GST-SPC algorithm out-performs Lutefisk, but is slightly worse than PepNovo. We have also listed (in Table 1) a few sample "good" interpretations of the GST-SPC algorithm, on which Lutefisk does not provide good results. It is interesting to note that GST-SPC algorithm can identify more correct amino acids – illustrating the power of using multi-charge strong tags.

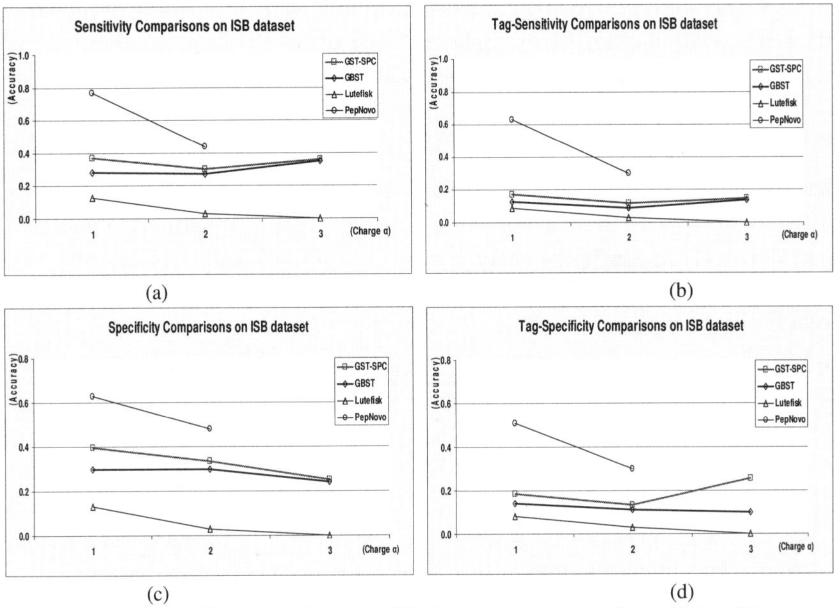

Figure 4. Comparison of different algorithms on ISB dataset – based on (a) sensitivity, (b) tag-sensitivity, (c) specificity and (d) tag-specificity. PepNovo only has results for charge 1 and 2.

Table 1: The sequencing results of Lutefisk, PepNovo and GST-SPC algorithm on some spectra. The accurate subsequences are labeled in bold and red. "-" means there is no result.

M/Z	Z	Real	Lutefisk	PepNovo	GST-SPC
1219.8	2	VAQLEQVYIR	[170.1]E***LEKVYLR***	GL***QLEQVYLR***	AVE***IEQVYIR***
1397.9	2	ELEEIVQPIISK	[242.1]***EEL***AVG[LP]***LSK***	***EELVKPLLSK***	***EIEEI***A[101.0]QH***ISK***
1644.9	2	PAAPAAPAPAEKTPVKK	[AP]***AAPA***[HS]AP[198.1]***PA***AA[CS]	***AAPA***DFEAMTNLPK	***APAAPAPA***[56.1]APAMTKVPK
1838.8	3	SSYSLSGWYENIYIR	[172.1]L[303.2][243.1][NP][MT]***LYLR***	-	***SS***IYI[27.3]IIEPCE***IYIR***
2000.2	4	PAAPAAPAPAEKTPVKKKAR	[323.1]R***PA***[AP]EKTN[LP]K[199.1]R	-	***APAAPA***MWNYNHKPYIR
1936.1	4	SIRVTQKSYKVSTSGPR	[199.1][PW][259.1]L[250.1]***KVSTSGPR***	-	VVIS***VTQK***[63.8]W***KVSTSGPR***
2101.1	4	KIETRDGKLVSESSDVLPK	[243.1]***LV***R[TY]YT***SESS***AE[PV]R	-	IKQHTHECY***SESSDVIPK***
2359.0	5	CDKDLDTLSGYAMCLPNLTR	-	-	AF***CDYA***[417.2]RNQKIRCP***TR***

6. Conclusion

In this paper, we propose a novel algorithm, GST-SPC for de novo sequencing of multi-charge MS/MS spectra. Our algorithm is based on the idea of using multi-charge strong tags to assist in reducing the size of the problem space to be searched. For a fixed set of strong tags, the GST-SPC algorithm optimizes the shared peaks count among all possible augmentations of the tags to form peptide sequences. The experimental results on ISB and GPM datasets show that GST-SPC is better than the GBST algorithm and Lutefisk. Against PepNovo, it performs better on GPM datasets and is worse on the ISB datasets. We have also showed theoretical upper bound results for our algorithms.

However, it is interesting to note that none of these algorithms is close to the theoretical upper bound of the sensitivity (based on Δ^R restriction) shown in Figure 2. This indicates that there is hope that there can be an algorithm based on *MST* that outperforms all of these algorithms. Since there is still room for algorithms based such ideas to improve, the idea itself is very promising. Other research directions are also possible – we are currently looking at more flexible method to connect strong tags rather than the head-to-tail manner, for example; and statistical significance (rather than SPC) of the strong tags and peptide sequencing results are also important for us to investigate.

Acknowledgments

The authors would like to thank Pavel Pavzner and Ari Frank of UCSD for insightful discussion and help with the PepNovo program. The work was partially supported by the National University of Singapore under grant R252-000-199-112.

References

1. Bandeira, N., Tang, H., Bafna, V. and Pevzner, P. Shotgun Sequencing by tandem mass spectra assembly. *Analytical Chemistry*, 76:7221-7233, 2004.
2. Birkinshaw, K. Deconvolution of mass spectra measured with a non-uniform detector array to give accurate ion abundances. *J Mass Spectrom.*, 38:206-210, 2001.
3. Chen, T., Kao, M.-Y., Tepel, M., Rush, J. and Church, G. M. A Dynamic Programming Approach to De Novo Peptide Sequencing via Tandem Mass Spectrometry. *Journal of Computational Biology*, 8:325-337, 2001.
4. Chong, K. F., Ning, K. and Leong, H. W. De Novo Peptide Sequencing For Multiply Charged Mass Spectra. *To appear APBC2006*, 2006.
5. Craig, R., Cortens, J. P. and Beavis, R. C. Open source system for analyzing, validating, and storing protein identification data. *J Proteome Res.*, 3:1234-1242, 2004.
6. Dancik, V., Addona, T., Clauser, K., Vath, J. and Pevzner, P. De novo protein sequencing via tandem mass-spectrometry. *J. Comp. Biol.*, 6:327-341, 1999.
7. Eng, J. K., McCormack, A. L. and John R. Yates, I. An approach to correlate tandem mass spectral data of peptides with amino acid sequences in a protein database. *JASMS*, 5:976-989, 1994.
8. Frank, A. and Pevzner, P. PepNovo: De Novo Peptide Sequencing via Probabilistic Network Modeling. *Anal. Chem.*, 77:964 -973, 2005.
9. Han, Y., Ma, B. and Zhang, K. SPIDER: Software for Protein Identification from Sequence Tags with De Novo Sequencing Error. *2004 IEEE Computational Systems Bioinformatics Conference (CSB'04)*, 2004.
10. Keller, A., Purvine, S., Nesvizhskii, A. I., Stolyar, S., Goodlett, D. R. and Kolker, E. Experimental protein mixture for validating tandem mass spectral analysis. *OMICS*, 6:207-212, 2002.
11. Ma, B., Zhang, K., Hendrie, C., Liang, C., Li, M., Doherty-Kirby, A. and Lajoie, G. PEAKS: Powerful Software for Peptide De Novo Sequencing by MS/MS. *Rapid Communications in Mass Spectrometry*, 17:2337-2342, 2003.
12. Mann, M. and Wilm, M. Error-tolerant identification of peptides in sequence databases by peptide sequence tags. *Analytical Chemistry*, 66:4390-4399, 1994.
13. Perkins, D. N., Pappin, D. J. C., Creasy, D. M. and Cottrell, J. S. Probability-based protein identification by searching sequence databases using mass spectrometry data. *Electrophoresis*, 20:3551-3567, 1999.
14. Tabb, D., Saraf, A. and Yates, J. r. GutenTag: high-throughput sequence tagging via an empirically derived fragmentation model. *Anal Chem.*, 75:6415-21, 2003.
15. Taylor, J. A. and Johnson, R. S. Implementation and uses of automated de novo peptide sequencing by tandem mass spectrometry. *Anal Chem.*, 73:2594-2604, 2001.
16. Zheng, H., Ojha, P. C., McClean, S., Black, N. D., Hughes, J. G. and Shaw, C. Heuristic charge assignment for deconvolution of electrospray ionization mass spectra. *Rapid Commun Mass Spectrom.*, 17:429-436, 2003.

COMPLEXITIES AND ALGORITHMS FOR GLYCAN STRUCTURE SEQUENCING USING TANDEM MASS SPECTROMETRY*

BAOZHEN SHAN, BIN MA AND KAIZHONG ZHANG

Department of Computer Science
University of Western Ontario
London, Ontario, Canada N6A 5B7
E-mail: bxshan, bma, kzhang@csd.uwo.ca

GILLES LAJOIE

Department of Biochemistry
University of Western Ontario
London, Ontario, Canada N6A 5B7
E-mail: glajoie@uwo.ca

Determining glycan structures is vital to comprehend cell-matrix, cell-cell, and even intracellular biological events. Glycan structure sequencing, which is to determine the primary structure of a glycan using MS/MS spectrometry, remains one of the most important tasks in proteomics. Analogous to the peptide *de novo* sequencing, the glycan *de novo* sequencing is to determine the structure without the aid of a known glycan database. We show in this paper that glycan *de novo* sequencing is NP-hard. We then provide a heuristic algorithm and develop a software program to solve the problem in practical cases. Experiments on real MS/MS data of glycopeptides demonstrate that our heuristic algorithm gives satisfactory results on practical data.

1. Introduction

The carbohydrates of glycoproteins and glycolipids are commonly referred as glycans. The glycan moieties cover a range of diverse biological functions. The rapid progress in proteomics has generated an increased interest in the full characterization of glycans [1].

A glycan is assembled from simple sugars by removal of water during the linkage of simple sugars. In terms of mass, there are eight usual types of simple sugars. There are two types of glycans, O-linked and N-linked glycans. The primary structure of a glycan is characterized by its two dimensional linkages (Figure 1a), which can be represented by labelled trees with node labels representing the type of simple sugars (Figure 1b).

Glycan structure sequencing is to determine primary structures of glycans using tandem mass spectrometry [6]. In an MS/MS experiment, a glycan will fragment into different

*This research was undertaken, in part, thanks to funding from NSERC, PREA, the Canada Research Chairs program, and NSF of China 60553001. BM's work was partially done when he visited the Center for Advanced Study at Tsinghua University.

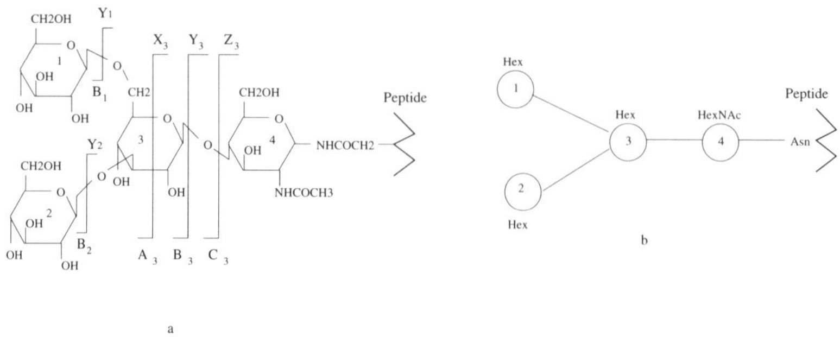

Figure 1. (a) Four simple sugars are linked together to form a glycan on a peptide. The long verticle lines indicate some possible fragmentations of the glycan in an MS/MS experiment. A_i, B_i, C_i, X_i, Y_i, and Z_i are the names of the resulted fragment ions. (b) The abstract tree representation of the glycan in (a).

fragment ions at any of the several labeled locations. Modified Domon and Costello's [7] nomenclature for glycan fragmentation is used in this paper (Figure 1a). There are three types of fragmentation, generating six types of ions A, B, C, X, Y and Z. A, B, and C types of ions correspond to subtrees. Whereas, X, Y, and Z types of ions correspond to the precursor (the whole structure including the peptide) minus subtrees. Usually, Y and B ions dominate in the spectrum [18]. So, for simplicity, we consider only Y and B ions to demonstrate our models.

The MS/MS spectrum of a glycan consists of many peaks, each of which is presumably generated by many copies of one fragment ion. The position of the peak indicates the mass to charge ratio of the corresponding fragment ion, and the intensity of the peak indicates the relative abundance of the fragment ion. Consequently, different glycans usually produce different MS/MS spectra. It is thus possible, and now a common practice, to use the spectrum of a glycan to determine its sequence structure.

One method to interpret the MS/MS spectrum involves searching the glycan database for a glycan, of which the theoretical spectrum matches the observed spectrum well. However, the determination of novel glycan structures requires *de novo* sequencing, which computes the primary structure directly from the spectrum without the help of a glycan database. Because the glycan database is rather incomplete, currently *de novo* sequencing is more important in these two approaches.

Tandem mass spectrometry has perviously been used to sequence peptides, where the linear amino acid sequence of a peptide is derived from the MS/MS spectrum (see [3,13,2] for a few examples). Glycan *de novo* sequencing in some sense is more difficult as it tries to derive a tree structure instead of a linear sequence. There have been attempts to solve glycan *de novo* sequencing problem [4,5,8,9,10,14,15,16]. Gaucher *et al.* [9] reported a glycan topology analysis tool, STAT, which generates all possible glycan primary structures. Obviously, the number of possible glycan structures grow exponentially. Consequently, STAT is only feasible for small glycans containing up to ten monosaccharides.

Similar to spectrum graph approach of peptide *de novo* sequencing [2], relationship between peaks in the spectrum is used to reduce the computation in glycan *de novo* sequencing. Mizuno et al. [14] first employed the relationship for compositional analysis. Ethier et al. [8] further built relationship tree for glycan *de novo* sequencing. Relationship tree approach has some difficulties to deal with. First, it requires very high quality of spectrum. Second, the restriction of certain structural topology makes it inappropriate for general glycan structures.

Shan *et al.* [15] reported a heuristic algorithm for glycan *de novo* sequencing from the MS/MS spectra of glycopeptides. The algorithm first generates many acceptable small subtrees, which are then joined together in a repetitive process to obtain larger and larger suboptimal subtrees until reaching the desired mass. At each size of the subtree, only limited number of subtrees are kept for later use. Experiments on real MS/MS data showed that the heuristic algorithm can determine glycan structures.

Tang *et al.* [16] reported a dynamic programming approach to determine glycan structures by recording the k best solutions at each iteration step (by default $k = 200$). The algorithm prefers linear structure to branching structure. Another algorithm called Cartoonist [10] has been created to interpret spectra of N-glycans. For each peak, Cartoonist computes all plausible structures with scores indicating the probability of correctness.

The classical methods for the characterization of glycoproteins by mass spectrometry were to cleave glycans with enzymes and then analyze the structures of the released glycans. Therefore, most of reported algorithms focus on interpreting MS/MS spectra of released glycans [4,5,8,9,10,14,16]. Recently, biochemists began to analyze glycopeptides derived from trypsin digestion of glycoproteins directly [18,12]. Therefore, there is a need to provide computational tools to assist the analyses. The algorithm in [15] was designed for both released glycan and glycopeptide. However, it only worked for the N-linked glycopeptides. The other type, O-linked glycopeptides often contains more than one glycan moieties.

Before this research, no software has been reported for interpreting spectrum of glycopeptide with multiple glycan moieties. In addition, the computational complexity of glycan *de novo* sequencing has remained unknown. The paper aims to solve these problems. More specifically, the contributions of this paper are the following:

(1) A polynomial time algorithm is provided under a simple model of glycan *de novo* sequencing.
(2) A more realistic model of glycan *de novo* sequencing is proved to be NP-hard.
(3) A new heuristic algorithm for the glycan *de novo* sequencing is introduced.
(4) A software program, GlycoMaster, is developed. Experiments on real data demonstrate our method works very well in practice. The software is available at http://bif.csd.uwo.ca/glycomaster.

2. Modelling Glycan *De Novo* Sequencing Problem

In the MS/MS spectrum of a glycopeptide, peaks of Y and B ions are located at separate regions because Y fragments contain peptide part which usually has relatively large mass. Let Σ be the alphabet of simple sugars. A glycan tree T is an unordered rooted tree with bounded degree whose nodes are labelled by letters from Σ. The degree of glycan trees is bounded by four because each sugar has at most five linkages. The root of T is linked to a peptide $P = a_1 \ldots a_k$ where a_i is from an amino acid alphabet Σ_a.

We assume that there is a post order numbering of the nodes in T. We use t_i to represent the ith node in T and $T[i]$ to represent the subtree rooted at t_i. When there is no confusion, we also use t_i to represent the sugar located at node t_i. Let $|T'|$ denote the size of a tree T'. Because of the post order numbering, for each $1 \leq i \leq |T|$, there is an $\hat{i} = i - |T[i]| + 1$ such that $t_{\hat{i}}, t_{\hat{i}+1}, \ldots, t_i$ represent all the nodes in $T[i]$.

For a sugar $g \in \Sigma$, we use $\|g\|$ to denote its mass. For an amino acid $a \in \Sigma_a$, we use $\|a\|$ to denote its mass. Let $\|T\| = \sum_{i=1}^{|T|} \|t_i\|$ and $\|P\| = \sum_{i=1}^{|P|} \|a_i\|$, then the actual mass, M, of the precursor ion of tree T linked with peptide P is $\|T\| + \|P\| + 18$ because of an extra H_2O in the peptide. For each subtree $T[i]$, let B_i represent the B-ion associated with $T[i]$ and Y_i represent the Y-ion associated with T linked with P and subtree $T[i]$ removed. Let $b_i = \sum_{k=\hat{i}}^{i} \|t_k\|$ and $y_i = M - b_i$, the actual mass of B_i is $b_i + 1$ (because of a proton added) and the mass of Y_i is y_i. We use $Y_{i_1,\ldots,i_k}$, where $T[i_1], \ldots, T[i_k]$ are non-overlapping subtrees, to denote the Y-ion associated with T linked with P and subtrees $T[i_1], \ldots, T[i_k]$ removed. Let $y_{i_1,\ldots,i_k} = M - (\sum_{l=1}^{k} b_{i_l})$, then the mass of $Y_{i_1,\ldots,i_k}$ is $y_{i_1,\ldots,i_k}$.

For simplicity, in this section, we only consider B-ions and Y-ions of the form of B_i and Y_i.

Let $\mathcal{M} = \{(m_i, h_i)\}$ be a spectrum of a glycan, where m_i is the mass and h_i is the intensity of the peak. For each mass value m, according to the intensity of the peaks around m, a score function $g(m)$ can be defined. Let T be a glycan tree, then the score of T, $S(T)$ is defined to be the summation of $g(m)$ for all the mass values m of the fragment ions of T. And the glycan structure *de novo* sequencing problem is then to find a tree structure T such that the mass of T is equal to a given value M', and $S(T)$ is maximized.

Notice that several different fragment ions of T may give the same mass value m. In such a case, whether the $g(m)$ score is counted several times or only once changes the definition of $S(T)$. Consequently, the difficulties of the glycan structure *de novo* sequencing problem under these two definitions are very different. The following two sections discuss these two definitions.

2.1. *A Simple Model*

Given a mass spectrum $\mathcal{M}$, there are different ways to define the score function $g(m)$. In this paper we simply assume $g(m)$ is given. Let $M' = M - \|P\| - 18$, where M is the precursor mass and P is the peptide. Let T be a glycan tree. The score of T is defined as follows: $S(T) = \sum_{i=1}^{|T|} g(b_i + 1) + g(y_i)$. Furthermore, because $y_i = M - b_i$, we denote $f(m) = g(m+1) + g(M - m)$. Then the score of a tree T becomes

$$S(T) = \sum_{i=1}^{|T|} f(b_i). \tag{1}$$

It turns out that under such definition, we can in polynomial time calculat the optimal tree T such that $\|T\| = M'$ and $S(T)$ is maximized. Two dynamic programming algorithms can be designed based on Lemma 2.1 and 2.2, respectively. The proofs of the lemmas are omitted here and will be provided in the full version of the paper. It is not hard to prove that the time complexities of the two algorithms are $O(M^4)$ and $O(M^2)$, respectively.

Lemma 2.1. *Let $S(m)$ be the maximum score of a glycan tree with mass m, then*

$$S(m) = \max_{\substack{g \in \Sigma \\ 0 \le m_1 \le m_2 \le m_3 \le m_4 \\ m_4 = m - \|g\| - m_1 - m_2 - m_3}} f(m) + S(m_1) + S(m_2) + S(m_3) + S(m_4)$$

Lemma 2.2. *Let $S(m)$ be the maximum score of a glycan tree with mass m and $S_2(m)$ be the maximum score of a glycan forest with at most two glycan trees and mass m, then*

$$\begin{aligned} S(m) &= \max_{g \in \Sigma; 0 \le m_1 \le m - \|g\| - m_1} f(m) + S_2(m_1) + S_2(m - \|g\| - m_1) \\ S_2(m) &= \max_{0 \le m_1 \le m - m_1} S(m_1) + S(m - m_1) \end{aligned}$$

The simple model in this section works in practice if the spectrum has good quality and the structure is relatively simple. The problem with this model is that if there are two different subtrees $T[i]$ and $T[j]$ with the same mass, $m(T[i]) = m(T[j])$, then in the score for T, $S(T)$, this mass will be used twice. Since these two ions generate the same peak in the spectrum, we should only use the peak once.

2.2. *A More Realistic Model*

In order to avoid repeatedly using peaks, the score function defined in (1) needs to be modified as follows.

Let $\Gamma(T) = \{b_i, y_i \mid 1 \le i \le |T|\}$. Define $S(T) = \sum_{m \in \Gamma(T)} g(m)$. Because $\Gamma(T)$ is a set, the definition here only uses each mass value once, even if there are multiple ions giving the same value. Further, the existence of the peptide separates Y and B ions in the MS/MS spectrum of a glycopeptide. If we let $\Delta(T) = \{b_i \mid 1 \le i \le |T|\}$. Then $S(T)$ can be rewritten as

$$S(T) = \sum_{m \in \Delta(T)} f(m). \tag{2}$$

With this model, the approach of Lemma 2.2 no longer works. In fact, the complexity of computing the optimal solution under this model becomes NP-hard.

3. NP-hardness of Glycan *De Novo* Sequencing

We will reduce Exact Cover by 3-Sets [11] to glycan structure sequencing problem.

Exact Cover by 3-Sets

INSTANCE: A finite set $E = \{e_1, e_2, \ldots, e_n\}$ where $n = 3q$ and a collection S of 3-element subsets of E.

QUESTION: Does S contain an exact cover for E, that is, a subcollection $S' \subseteq S$ such that every element of E occurs in exactly one member of S'?

Lemma 3.1. *There are n positive integers $z_1 < z_2 \ldots < z_n$ such that*

(1) $z_i + z_j = z_{i'} + z_{j'}$ implies $\{i, j\} = \{i', j'\}$.
(2) $z_i + z_j + z_k = z_{i'} + z_{j'} + z_{k'}$ implies $\{i, j, k\} = \{i', j', k'\}$.
(3) $z_i \leq poly(n)$.

Proof. We determine $z_1, \ldots, z_n$ inductively. The first three integers are $z_1 = 1$, $z_2 = 2$, and $z_3 = 3$. Suppose that we already determined $z_1, \ldots, z_{k-1}$ satisfying conditions 1 and 2, we now prove that z_k can be found in the range of $z_{k-1} + 1$ and $z_{k-1} + n^5$.

Consider the following two equations where $1 \leq i, j, l, i', j' \leq k - 1$.

- $x + z_{i'} = z_i + z_j$
- $x + z_{i'} + z_{j'} = z_i + z_j + z_l$

There are in total less than $\frac{n^3}{2} + \frac{n^5}{5!}$ equations above, each has only one solution of x. Therefore, in the range of $z_{k-1}+1$ and $z_{k-1}+n^5$, there is an integer that is not the solution of any of the equations. The integer can be trivially found in polynomial time and is used as z_k.

From the construction, we have $z_1 < \ldots < z_n = O(n^6)$. □

Lemma 3.2. *There are n positive integers $z_1, \ldots, z_n$ satisfying the conditions in Lemma 3.1 and*

(4) if $i \neq j$, $|z_i - z_j| \geq n^6 + 2$.
(5) if $\{i, j\} \neq \{i', j'\}$, $|z_i + z_j - z_{i'} - z_{j'}| \geq n^6 + 2$.
(6) if $\{i, j, k\} \neq \{i', j', k'\}$, $|z_i + z_j + z_k - z_{i'} - z_{j'} - z_{k'}| \geq n^6 + 2$.

Proof. Mutiply each integer determined in Lemma 3.1 by $n^6 + 2$. This will not violate any conditions in Lemma 3.1. □

Let Σ contain only one letter g, and $m(g) = 1$.

Theorem 3.1. *The glycan structure sequencing problem is NP-hard under the realistic tree scoring model and an arbitrary mass scoring scheme.*

Proof. Due to page limit, we only give the reduction and the idea of the proof. More details of the proof will be provided in the full version of the paper.

Given an instance of the exact cover by 3-sets with $E = \{e_1, e_2, \ldots, e_n\}$ and $S = \{s_1, \ldots, s_q\}$ where $n = 3q$ and $s_l = \{e_i, e_j, e_k\}$.

Our idea is to design n subtrees, $T_1, \ldots, T_n$, each corresponding to an e_i. By carefully designing the spectrum, i.e., assigning values of $f(m)$ at different m, we can ensure that when there is an exact cover, then the optimal solution will be a tree like Figure 2, where

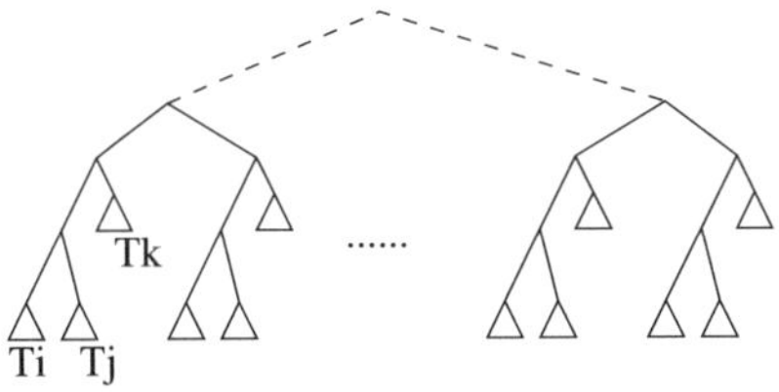

Figure 2. Optimal tree of our construction in NP-hardness proof.

each three-subtree group corresponds to a 3-set. And the score of the tree is equal to n. However, if there is no exact cover, then the score of the tree is smaller than n.

Let $z_1, \ldots, z_n$ be as in Lemma 3.2 and $N = n \times \max z_i$. Let $M = 4 \times N$. Let $x_i = M + z_i$. Then $x_1, \ldots, x_n$ also satisfy Lemma 3.2.

Let e_i correspond to x_i. Let $s_l = \{e_i, e_j, e_k\}$ correspond to $z_l = x_i + x_j + x_k + 2$. Let $X = \{x_i\}$. Let $Y = \{y_k \mid y_k = x_i + x_j + 1\}$. Let $Z = \{z_l\}$.

Let $D_1 = \{|x_i - x_j| - 1\}$, $D_2 = \{|x_i + x_j - x_{i'} - x_{j'}| - 1\}$, and $D_3 = \{|x_i + x_j + x_k - x_{i'} - x_{j'} - x_{k'}| - 1\}$. Let $D = D_1 \cup D_2 \cup D_3$, then $|D| < n^6$ and for any $d \in D$, $n^6 < d < N$.

In the spectrum, let

$$f(m) = \begin{cases} -1, & m \in D \\ 0, & 1 \le m \le M - N \text{ and } m \notin D \\ 1, & m \in X \\ -1, & M - N < m \le M + N \text{ and } m \notin X \\ 0, & m \in Y \\ -1, & 2M < m \le 2M + N \text{ and } m \notin Y \\ 0, & m \in Z \\ -1, & 3M < m \le 3M + N \text{ and } m \notin Z \\ 0, & 3tM < m \le 3tM + N,\ 2 \le t \le n/3 \\ -1, & \text{else} \end{cases}$$

Let the total mass be $\sum x_i + n - 1 = nM + \sum z_i + n - 1 < nM + N$.

Note that the score function $f()$ can be computed in $poly(n)$ time. Because of the construction of $f()$, the following property can be proved.

Claim 3.1 There is a subtree T_i for each x_i, such that $m(T_i) = x_i$ and $f(m(T')) = 0$ for each subtree T' of T_i.

If there is an exact cover $S' \subset S$, for every $s_i = \{e_i, e_j, e_k\} \in S'$, we can link T_i, T_j, and T_k together as in Figure 2. This will give us a total score n because all $f(m) = 1$ for $m \in X$ are included, while no $f(m) = -1$ is used.

On the other hand, if there is a tree with score n, then all $f(m)$ for $m \in X$ are included and none of $f(m) = -1$ is used. This can be used to prove that in the optimal tree, there is one and only one subtree with mass x_i. and for each such subtree, its parent or grand parent is of mass $x_i + x_j + x_k + 2$ where $\{e_i, e_j, e_k\}$ is in S. This means that we have an exact cover. □

4. An Algorithm for Glycan *De Novo* Sequencing

In previous sections, we showed that Glycan *De Novo* sequencing problem has polynomial algorithm to find optimal solution with the simple model. However, with more realistic model it is NP-hard. In this section, we extend the polynomial time algorithm of the simple model to a heuristic algorithm of the realistic model.

4.1. *A Heuristic Algorithm*

A glycan forest is represented as $f = < t_1, \ldots, t_n >$ where t_i is a glycan tree and n is called the degree of f. We use t_0 denotes an empty tree. We use $F(m)$ to represent a set of glycan forests with mass m.

$$F(m) = \{f \mid f \text{ is a forest such that } m(f) = m\}$$

We use $g \oplus f$ to represent the glycan tree rooted at g and each tree of f is a child of g. Let $f \ominus < t_i, t_j, t_k, t_l >$ represent the glycan forest resulting from removing t_i, t_j, t_k, t_l from f.

Given a $g \in \Sigma$ and a glycan forest f, we use $g \otimes f$ to represent a set of glycan forests generated by g and $f = < t_1, \ldots, t_n >$.

$$g \otimes f = \{< g \oplus < t_i, t_j, t_k, t_l >, f \ominus < t_i, t_j, t_k, t_l > \; > \mid 0 \leq i \leq j \leq k \leq l \leq n\}$$

In our algorithm, the score of a forest is computed by scoring a set of masses of fragments which include Y-fragments, B-fragments and internal fragments (B-fragment losing one or more B-fragments). Fragments with same mass are used only once, because these fragments generate the same peak in the spectrum.

We now give a high level description of our heuristic algorithm. For each mass m we maintain a fixed number of forests with mass m and high scores in $F(m)$. We can compute $F(m)$ in two steps

(1) Compute a set of candidate forests $F_c(m) = \cup_{g \in \Sigma} \cup_{f \in F(m - ||g||)} g \otimes f$
(2) Compute $F(m)$ by evaluating the score for each of the forests in $F_c(m)$ and remove those with low scores.

Let M' be the desired mass of the glycan forest. Once $F(M')$ is computed, the solution is chosen from forests in $F(M')$ with maximum score.

Algorithm:

Input: Σ (simple sugars), M' (mass of the glycan) and $\mathcal{M}$ (a set of mass peaks)
Output: candidates of sequence structures with scores
(1) **for** $m = 1$ **to** M'
(2) $\quad F_c(m) = \emptyset$
(3) $\quad$ **for** each $g \in \Sigma$

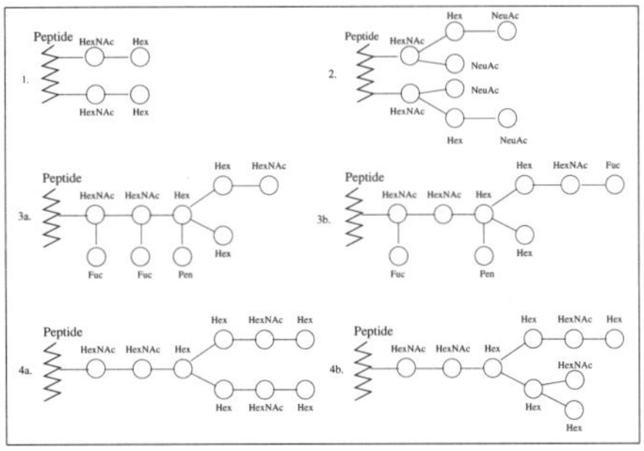

Figure 3. Sample structures computed by GlycoMaster.

(4) **for** each $f \in F(m - ||g||)$

(5) **for** each (i, j, k, l) s.t. $(t_i, t_j, t_k, t_l \in f)$

(6) $F_c(m) = F_c(m) \cup < g\oplus < t_i, t_j, t_k, t_l >, f\ominus < t_i, t_j, t_k, t_l > \; >$

(7) Scoring forests in $F_c(m)$ and put top $|F|$ in $F(m)$

(8) **return** $F(M')$

The time complexity is $O(d^4|\Sigma| \times |F| \times M')$, where d is the degree of forest.

4.2. *Experiments and Discussion*

Based on the above algorithm, a software program, GlycoMaster, has been developed for glycan *de novo* sequencing. We choose $|F| = 1000$ in the software implementation of the algorithm.

The software was tested using twenty MS/MS spectra of glycopeptides. The samples were derived from the cationic peanut peroxidase and rat bone osteopontin after tryptic digestion. The MS/MS spectra of the samples were obtained by using a Q-TOF2 in the positive ion ESI MS/MS mode with borosilicate nano tips. The correctness of the automated interpretation was evaluated by comparing with the structure determined by manual analysis [12,19]. Because most algorithms [4,5,8,9,14,10,16] were designed for MS/MS spectra of glycans only and therefore cannot handle glycopeptides data we obtained, here we only demonstrate that the algorithm in this paper improved our previous algorithm in [15].

The performance of our algorithm is shown in the following table.The compositions of all samples were correctly computed. The structures of nineteen out of twenty spectra are the same as deduced by manual interpretation. Two of them are O-linked glycopeptides with two glycan moieties linked to the peptide (1 and 2 in Figure 3).

Algorithm	Correct Structures	Wrong Structures	Partially Wrong
[15]	16/20	2/20	2/20
This work	19/20	0/20	1/20

Our algorithm also works for the MS/MS data of released glycans. As future work the algorithm will be tested with MS/MS data of released glycans. Currently, no public MS/MS data of glycans are available to us.

References

1. K. Aebersold and M. Mann. Mass spectrometry-based protemics. *Nature*, 422:198-207, 2003.
2. V. Bafna and N. Edwards. On de novo interpretation of tandem mass spectra for peptide identification. *RECOMB 2003*, 9-18, Berlin, Germany, 2003.
3. Chen, T., Kao, M-Y., Tepel, M., Rush J., and Church, G. 2001. A dynamic programming approach to *de novo* peptide sequencing via tandem mass spectrometry. *J. Comp. Biology* 8(3), 325-337.
4. C.A. Cooper, E. Gasteiger, and N.H. Packer. GlycoMod - A software tool for determining glycosylation compositions from mass spectrometric data. *Proteomics*, 1:340-349, 2000.
5. C.A. Cooper, H. Joshi, M. Harrison, M. Wilkins, and N Packer: GlycoSuiteDB: a curated relational database of glycoprotein glycan structures and their biological sources. *Nucleic Acids Res.*, 31:511-513, 2003.
6. A. Dell and H. R. Morris. Glycoprotein Structure Determination by Mass Spectrometry. *Science*, 291:2351-2356, 2001.
7. B. Domon and CE. Costello. A systematic nomenclature for carbohydrate fragmentations in FAB-MS/MS spectra of glycoconjugates. *Glycoconjugate J.*, 5:397-409, 1988.
8. M. Ethier, *et al.*. Application of the StrOligo algorithm for the automated structure assignment of complex N-linked glycans from tandem mass spectrometry.*Rapid Commun. Mass Spectrom.*, 17:2713-2720, 2003.
9. S.P. Gaucher, J. Morrow and J.A. Leary. A saccharide topology analysis tool used in combination with tandem mass spectrometry *Anal. Chem.*, 72:2231-2236, 2000.
10. D. Goldberg, M. Sutton-Smith, J. Paulson and Anne Dell. Automatic annotation of matrix-assisted laser desorption/ionization N-glycan spectra. *Proteomics*, 5:865-875, 2005.
11. Michael R. Garey and David S. Johnson. Computers and intractability: a guide to the theory of NP-completeness. pp221, W.H. Freeman and Company, San Franciso 1979.
12. K. Keykhosravani, A. Doherty-Kirby, C. Zhang, A. Goldberg, G.K. Hunter and G. Lajoie. Comprehensive identification of posttranslational modifications in rat bone osteopontin by mass spectrometry. *Biochemistry*, 44:6990-7003, 2005.
13. B. Ma et. al. PEAKS: Powerful Software for Peptide De Novo Sequencing by Tandem Mass Spectrometry. *Rapid Comm. in Mass Spectrom.* 17(20):2337-2342. 2003.
14. Y. Mizuno and T. Sasagawa. An automated interpretation of MALDI/TOF postsource decay spectra of oligosaccharides. 1. Automated Peak Assignment. *Anal. Chem.*, 71:4764-4771, 1999.
15. B. Shan, K. Zhang, B. Ma, C. Zhang and G. Lajorie. An Algorithm for Determining Glycan Structures from MS/MS Spectra. *Proceedings of ICBA 2004*, Florida, USA, 2004, in Advanced in bioinformatics and its applications, 414-425, World Scicntific, 2004.
16. H. Tang, Y. Mechref, and M.V. Novotny. Automated interpretation of MS/MS spectra of oligosaccharides. *Bioinformatics* 21:i431-i439, 2005.
17. M.E. Taylor and K. Drickamer. Introduction to Glycobiology. Oxford University Press, 2003.
18. J. Zala. Mass Spectrometry of Oligosaccharides. *Mass Spectrometry Reviews*, 23161-227, 2004.
19. C. Zhang, A. Doherty-Kirby and G. Lajoie. Investigation of Cationic Peanut Peroxidase Glycans by Electrospray Ionization Mass Spectrometry. *Phytochemistry*, 65:1575-88, 2004.

SEMI-SUPERVISED PATTERN LEARNING FOR EXTRACTING RELATIONS FROM BIOSCIENCE TEXTS

SHILIN DING[†], MINLIE HUANG[†], XIAOYAN ZHU[*]

State Key Laboratory of Intelligent Technology and Systems (LITS), Department of Computer Science and Technology, Tsinghua University, Beijing, 100084, China

dingsl@gmail.com, minlie.huang@hotmail.com, zxy-dcs@tsinghua.edu.cn

A variety of pattern-based methods have been exploited to extract biological relations from literatures. Many of them require significant domain-specific knowledge to build the patterns by hand, or a large amount of labeled data to learn the patterns automatically. In this paper, a semi-supervised model is presented to combine both unlabeled and labeled data for the pattern learning procedure. First, a large amount of unlabeled data is used to generate a raw pattern set. Then it is refined in the evaluating phase by incorporating the domain knowledge provided by a relatively small labeled data. Comparative results show that labeled data, when used in conjunction with the inexpensive unlabeled data, can considerably improve the learning accuracy.

1. Introduction

Knowledge extraction from bioscience texts has become an emerging field for both Information Extraction and Natural Language Processing communities. These tasks include recognizing biological named entities [10, 21], extracting relations between entities [4, 12, 19], and identifying biological events and scenarios [20]. The major challenges come from the fact that biomedical literatures contain abundant domain-specific knowledge, inconsistent terminologies, and complicated syntactic structures or expressions.

In this paper, the work is focused on extracting relations between biological entities, such as protein-protein interactions (PPI). Various methods and systems have been proposed. The most prevalent methods are rule-based or pattern-based. Such methods adopt hand-coded rules or automated patterns and then use pattern matching techniques to capture relations. Hand-coded patterns are widely used in the early stage of this research. For example, Ono [11] manually constructed lexical patterns to match linguistic structures of sentences for extracting protein-protein interactions. Such methods contribute high accuracy but low coverage. Moreover, the construction of patterns is time-consuming and requires much domain expertise.

Systems which can learn patterns automatically for general relation extraction include AUTOSLOG [14], CRYSTAL [17], SRV [6], RAPIER [1], ONBIRES [7, 8], and so forth. Most of them take annotated texts as input, and then learn patterns semi-

[†]The two authors have equal contributions

[*]Corresponding author: zxy-dcs@tsinghua.edu.cn Tel: 86-10-62796831 Fax: 86-10-62771138

automatically or automatically. But effective evaluation of these patterns remains a major unsolved problem. Moreover, most pattern-based applications require a well-annotated corpus for training [2, 5].

Since data annotation is expensive and time-consuming, the major problem in pattern-based methods is how to automatically learn patterns efficiently and effectively, with limited annotated data available. Unsupervised principle is preferable with the ability to exploit huge amount of unlabeled texts in biomedical domain. The crucial problem here is that patterns generated from unlabeled data may be erroneous or redundant. Therefore, pattern evaluation algorithm is indispensable. A systematic methodology based on ranking functions is widely used by most methods [3, 15, 18]. Such algorithms assign a score to each pattern according to the ranking functions and then keep the top n best patterns (n is a pre-specified threshold). In such algorithms, each pattern is evaluated independently. Thus, the redundancy among patterns is difficult to reduce.

To solve these problems, a semi-supervised [16, 22] model is proposed, combining both unlabeled and labeled data. A pattern generation algorithm is first implemented to mine relevant pattern structures from unlabeled data, where sentences are pairwise aligned by dynamic programming to extract identical parts as the pattern candidates. Since the generation algorithm does not require any annotation in the corpus, pattern evaluation algorithms with labeled information are then integrated to complete the learning procedure. Two types of pattern evaluation algorithms are investigated. The first is a ranking function based algorithm, which evaluates the effectiveness of every single pattern independently and delete the ones that have no contribution to the performance. The second is heuristic evaluation algorithm (*HEA*), which aims to search the optimal pattern set in a heuristic manner. Compared to the first method, the deletion of a pattern is determined by the current pattern set, not only itself. Comparative results show that our ranking function outperforms other prevalent ones and *HEA* exhibits advantages over ranking function based algorithms.

The paper is organized as follows. The first part of semi-supervised learning, pattern generation method with unlabeled data, is presented in Section 2. Then pattern evaluation method which relies on labeled data to curate the learning result is explained in Section 3. The experiments and conclusions are discussed in Section 4 and 5.

2. Pattern Generation

First of all, several definitions are presented here:

- A Sentence is a sequence of word-tag pairs: $STN = WTP_{1,2,\ldots,N}$, where each WTP_i is a word-tag pair (w_i,t_i). Here w_i is a word and t_i is the part-of-speech (POS) tag of w_i.
- Sentence Structure is defined as SS = {*prefix*, NE_1, *infix*, NE_2, *suffix*}. NE_1 and NE_2 are semantic classes of the named entities. The *prefix, infix, and suffix* are sequences of *WTPs* before NE_1, between NE_1 and NE_2, *and* behind NE_2, respectively.
- A pattern is defined as *PTN*={*pre-filler*, NE_1, *mid-filler*, NE_2, *post-filler*}. The *fillers* are sequences of *WTPs* before NE_1, between NE_1 and NE_2, and behind NE_2.

Examples for these definitions are shown in Table 1.

Table 1. Examples for sentences and patterns

	Examples
STN	Several/JJ recent/JJ studies/NNS have/VBP implicated/VBN P_00172/NN in/IN the/DT signaling/NN pathway/NN induced/VBN by/IN P_00006/NN ./.
WTP	induced/VB ; P_00172/NN
SS	{{*prefix*: {Several/JJ recent/JJ studies/NNS have/VBP implicated/VBN}} {NE_1: {PROTEIN/NN}} {*infix*: {in/IN the/DT signaling/NN pathway/NN induced/VBN by/IN}} {NE_2: {PROTEIN/NN}} {*suffix*: {NULL}}}
PTN	{{*pre-filler*: {NULL}} {NE_1: {PROTEIN/NN}} {*mid-filler*: {induced/VBN by/IN}} {NE_2: {PROTEIN/NN}} {*post-filler*: {NULL}}}

Sequence alignment algorithm is adopted to generate patterns by aligning pairwise sentences in training corpus. The identical parts in aligned sentences are extracted as pattern candidates. More formally, given two *SSs*: ($prefix^i$, NE_1^i, $infix^i$, NE_2^i, $suffix^i$) and ($prefix^j$, NE_1^j, $infix^j$, NE_2^j, $suffix^j$), the sequence alignment algorithm is carried out on three pairs – ($prefix^i$, $prefix^j$), ($infix^i$, $infix^j$), and ($suffix^i$, $suffix^j$) – to extract identical *WTPs* and form the three fillers of a *PTN*. The algorithm is shown in Figure 1.

Input: A sentence structure set $S = \{SS_1, SS_2, \cdots, SS_n\}$
Output: A set of patterns: P

1. **For** every pair in S: (SS_i, SS_j)□S (i ≠ j), do
2. **if** $SS_i.NE_1$!=$SS_j.NE_1$ **or** $SS_i.NE_2$!= $SS_j.NE_2$, **then** go to 1;
3. **else do**
4. $NE_1 = SS_i.NE_1$, $NE_2 = SSi_j.NE_2$
5. do alignment for SS_i.*prefix* and SS_j.*prefix*;
6. extract the identical *WTPs* to form the *pre-filler* of a candidate pattern *p*.
7. do the same operations in step 5 and 6 to form *mid-filler* and *post- filler*,.
8. **if** *p* already exists in *P*, **then** increase the count of *p* with 1.
9. **else** add *p* to *P* with a count of 1;
10. Output *P*

Figure 1. Pattern learning algorithm

This algorithm automatically learns patterns from sentences whose named entities have been identified by a dictionary-based method and requires no further annotation. It is almost unsupervised which is able to make better use of the enormous data available online, and release domain experts from the heavy burden of creating annotated corpora.

3. Pattern Evaluation

The pattern generation algorithm discussed in Section 2 does not require supervised information. It may produce erroneous patterns such as ("", PROTEIN, "shown to be",

PROTEIN, ""), which will match many false positive instances. Previous works usually depended on rule-based methods or manual selection to screen out the best patterns.

To automate the relation extraction system, we developed a pattern evaluation algorithm to assess patterns by a small annotated corpus. Here we discuss two types of evaluation algorithms: the first one utilizes ranking functions and the second one is a heuristic evaluation algorithm.

3.1. Ranking Function Based Algorithm

Ranking function based evaluation algorithms assess each pattern independently. They assign a score to each pattern by ranking functions, and then filter out those patterns with lower scores than a threshold. Previous pattern-based systems have adopted various ranking functions, which take into consideration the number of instances that are correctly or incorrectly matched by a pattern. Two ranking functions are surveyed here:

The first one is proposed by Cohen in their system RIPPER [3]:

$$Ripper(p) = \frac{p.positive - p.negative}{p.positive + p.negative} \tag{1}$$

where *p.negative* indicates the number of false instances matched by the pattern *p* and *p.positive* denotes the number of correct instances. In essence, this function only takes into consideration the ratio of *p.positive* to *p.negative* (*p/n* for short). The second function is proposed by Riloff [15], with two factors – *p/n* and *p.positive*:

$$Riloff(p) = \frac{p.positive}{p.positive + p.negative} * \log_2(p.positive) \tag{2}$$

The critical issue about these ranking functions presented above is that only two factors, *p/n* and *p.positive*, are considered. However, other factors should be considered, such as *p+n*. *Ripper* can not distinguish a pattern with 50 true positives and 50 negatives (50/50 for short) from (1/1) pattern, while the former pattern apparently contributes more to precision and recall. Although *Riloff* function works well for the two patterns by introducing the log(*p.positive*) term, it does not work for such 4 patterns: (1/4), (2/8), (3/12) and (4/16). These patterns, whose *p.positive* is larger, will have a higher rank by *Riloff* function. However, since p/n is very low, it is reasonable to determine that patterns with larger (*p+n*) are worse. *Riloff* function fails to handle the case. To involve more factors for pattern evaluation, we propose a novel ranking function as follows:

$$HD(p) = (\beta + \log_2 \frac{p.positive + 0.5}{p.negative + 0.5}) * \ln(p.positive + p.negative + 1) \tag{3}$$

where the parameter β is a threshold that controls *p/n*. If $p/n > 2^{-\beta}$, *HD* is an increasing function of (*p+n*), which means if several patterns have the same *p/n* that exceeds $2^{-\beta}$, a pattern with larger (*p+n*) has a higher rank. If $p/n < 2^{-\beta}$, the first term is negative, which means a pattern with larger (*p+n*) will have a lower rank. Thus different ranking strategies are used when different p/n are met. Experiments in Section 4.3 will illustrate how *HD* outperforms other functions, where the parameter β is set to 0.5 empirically.

3.2. *Heuristic Evaluation Algorithm (HEA)*

Ranking function based algorithms assess each pattern independently. It is not difficult to delete erroneous patterns by these algorithms. However, redundancy among patterns, which heavily impose computational burden on relation extraction tasks, can not be reduced effectively. For example, there are two patterns ("", PROTEIN, "bind to", PROTEIN, "") and ("", PROTEIN, "to bind to", PROTEIN, ""). Apparently, the second pattern is redundant since all instances it matches will also be captured by the first one. However, it cannot be filtered because its score is almost the same as the first ones. To remove erroneous and redundant patterns, we propose a heuristic evaluation algorithm (*HEA*), which aims to obtain the optimal pattern set in a heuristic manner.

Formally, given an evaluation corpus S and a pattern set P, we define an optimization function which maps the pattern set P to its performance on S:

$$\begin{aligned} M(S,\cdot) &: \Sigma \rightarrow R \\ P &\mapsto M(S,P) \end{aligned} \tag{4}$$

where Σ denotes the space of all possible pattern sets and R is the real number space. Starting from the initial pattern set P_0, we aim to obtain the optimal set P^* by maximizing $M(S,P)$ in a heuristic manner. The iterative procedure follows formula (5):

$$P_{k+1} = P_k - \arg_{p_i \in P_k} \nabla M(S, P_k) \tag{5}$$

where $\nabla M(S,P_k) = \frac{\partial M(S,P_k)}{\partial P_k} = \max_{p_i \in P_k} \{M(S,P_k - \{p_i\}) - M(S,P_k)\}$ is the gradient of $M(S,P_k)$ in k-th step.

The algorithm is shown in Figure 2. In practice, we store and index all possible matching results produced by the whole pattern set P_0 by preprocessing. Thus, for each iteration, evaluating the pattern set P_k is carried out by finding the results in the index (excluding all the patterns that are not in P_k), without a whole re-running of the program. This method makes the iterative procedure computationally feasible.

Input: an initial pattern set $P_0 = \{p_1, p_2, \cdots, p_n\}$, the training set S, the testing set T, an optimization function $M(S,P)$

Output: the optimal pattern set P^*

1. $k=0$, $P_k = P_0$
2. Calculate the gradient:
   $\nabla M(S,P_k) = \max_{p_i \in P_k} \{M(S,P_k - \{p_i\}) - M(S,P_k)\}$
3. Find the "worst" pattern to be deleted:
   $p_m = \arg_{p_i \in P_k} \nabla M(S,P_k)$
4. **If** $\nabla M(S,P_k) \geq \Delta_{th}$ **then do**
5. $\quad M(S,P_{k+1}) = M(S,P_k - \{p_m\})$
6. $\quad P_{k+1} = P_k - p_m$
7. $\quad$ Evaluate the performance of the pattern set P_{k+1} on the testing corpus T
8. $\quad k=k+1$, go to 2
9. **else** output $P^* = P_k$

Figure 2. Heuristic Evaluation Algorithm (*HEA*)

In this algorithm, an optimization function *M*(*S*,*P*) has to be determined. Note that the goal of *HEA* is to achieve the optimal pattern set out of an initial set. Thus, the direct target of F_1 score can be taken as an optimization function.

4. Experiment

Corpora used in the experiments are introduced in Section 4.1. Experiments of pattern generation on unlabeled data and pattern evaluation on labeled data are discussed in detail in Section 4.2 and 4.3. These sections are aimed to investigate the effectiveness of the semi-supervised learning model.

4.1. Data Preparation

The first corpus used for protein-protein interaction extraction is downloaded from http://www.biostat.wisc.edu/~craven/ie/ [13]. This corpus consists of 2,430 sentences gathered from Munich Information Centre for Protein Sequences (MIPS). This corpus is used for pattern generation.

The second is collected from the GENIA corpus [9] which consists of 2000 abstracts from MEDLINE. We have manually annotated the protein-protein interactions, and finally obtained a corpus with 4,221 PPIs in 2,561 sentences.

4.2. Semi-supervised Learning Model

In this section, we discuss the effectiveness of semi-supervised learning model by comparing the performance of refined patterns with that of original patterns.

First, 2,480 patterns are initially obtained from MIPS by the generation algorithm. Their performance is set as the baseline. Then the GENIA corpus with annotated relations is randomly partitioned to five parts for 5-fold cross-validation, one of the five

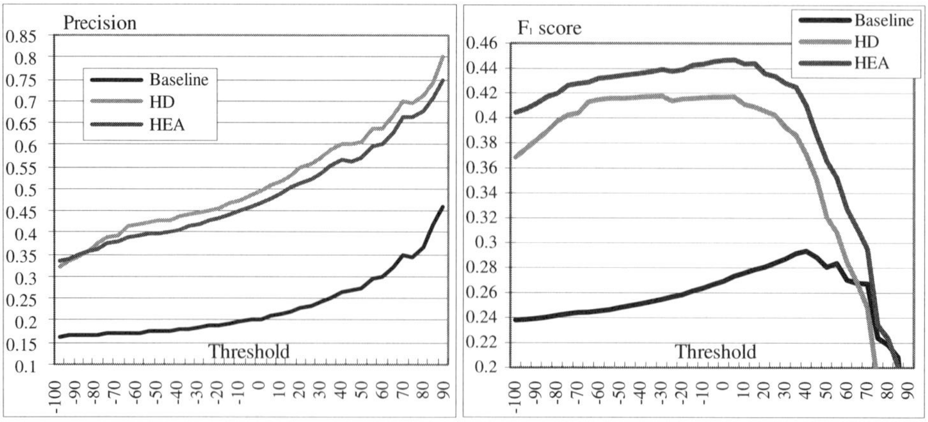

Figure 3. Performance of semi-supervised model

parts for testing and the remainder for pattern evaluation. For the two evaluation methods, the top 100 patterns are preserved to extract relations from the testing corpus.

Experiment results are shown in Figure 3 over different user-specified thresholds on the testing corpus. It shows that 1) the raw pattern set generated without labeled information is poor in accuracy, but has promising recall (about 45% to 50%). 2) Our proposed ranking function *HD* and *HEA* method both achieve significant improvements. The precision is improved by over 25% with little loss in recall, results in improvement of F_1 score by 16% to 19%. These results indicate that the pattern generation algorithm does extract useful patterns from unlabeled data, and pattern evaluation algorithm greatly improves the accuracy with labeled information.

4.3. Pattern Evaluation with Labeled Data

In this section, we discuss the difference among different evaluation algorithms, which is crucial in semi-supervised learning. The GENIA corpus is used in the same way as before for 5-fold cross-validation. The raw pattern set is also taken from the previous experiment. In this experiment, the pattern deletion order in ranking function based methods, including *Ripper*, *Riloff*, and *HD*, is determined by the corresponding functions. In other words, patterns with lower ranks (worse patterns) will be removed earlier. In *HEA*, which pattern to be deleted is determined dynamically as before. To provide a complete comparison, we delete all of the patterns in each algorithm, which means the parameter Δ_{th} in *HEA* could be set to a very small numerical value.

Table 2. Performance of optimal pattern sets determined by ranking function based algorithms and *HEA*

Method	Patterns	Precision	Recall	F_1 score	Impr. of F_1
Baseline (raw)	2480	19.0%	46.5%	27.0%	–
Ripper	1626	41.0%	40.8%	40.9%	+49.3%
Riloff	88	40.8%	44.5%	42.6%	+55.5%
HD	**92**	52.5%	38.8%	**44.5%**	**+62.4%**
HEA	**72**	43.5%	45.9%	**45.5%**	**+66.1%**

Table 2 shows the performance and cardinality of the optimal pattern sets achieved by different methods. The smallest pattern set and the best system performance are achieved by *HEA*, which means *HEA* can reduce redundancy maximally and guarantee the best system performance at the same time. Although the performance of *HD* and *HEA* is only slightly better than that of *Ripper* and *Riloff* function, further studies in Figure 4 will demonstrate why *HD* and *HEA* outperform the other two methods.

When the $2{,}480^{th}$~$1{,}600^{th}$ patterns determined by *Ripper* function are deleted, the performance is enhanced dramatically. However, when it starts to delete 50 more patterns, the performance degrades extremely. These patterns include ("", NE1, "inhibit", NE2), ("", NE1, "induce", NE2, "") with both large *p.positive* and large *p.negative* (79/142 and 308/220 in our experiment), but the p/n is not large enough (compared to the 3/1 or 7/3 patterns), which directly leads to low ranks. Therefore, *Ripper* function that involves only the p/n factor can not assess patterns properly.

Riloff function is also unable to evaluate patterns adequately. Firstly, the "worst" 900 patterns ranked by *Riloff* function are not the worst in fact, because deleting these patterns does not lead to remarkable improvements. However, deleting the 900th~100th patterns results in significant improvements. Hence these patterns should have much lower ranks. Secondly, although the best result (42.6% with 88 patterns) is very promising, the curve keeps rising until it reached the optimal point at a very narrow peak. Thus it is very difficult to determine the number of patterns to hold in practice (The system performance is very sensitive to the threshold).

In comparison, *HD* function exhibits advantages over traditional ranking functions. The *HD* curve shows that it removes the most undesirable patterns at position 2,460th ~1,700th, with 16 percentages improvement of F_1 score. And then the curve rises slowly, when deleting "medium" patterns, until it reaches the optimal point (44.5% with 92 patterns). After that point, deleting any pattern will cause a remarkable decline in the performance. This curve shows that *HD* ranks the patterns more precisely. And it also has a much broader "safe" area (from the 500th to 100th patterns). Thus the number of patterns to hold is much easier to set, compared to the narrow peak in *Riloff* curve.

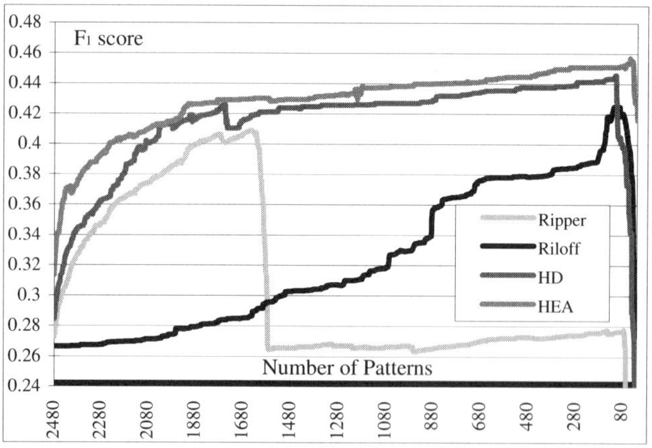

Figure 4. Comparison between ranking function based algorithm and *HEA*

In addition, the curve of *HEA* has almost the same trend as that of *HD*, which means *HEA* is also effective to remove incorrect and redundant patterns. The cardinality of the optimal pattern set obtained is smaller than that by *HD*, which means that *HEA* is more capable to reduce redundancy among patterns. From this figure, we can see that *HEA* outperforms other methods significantly.

The comparative experiments show that 1) algorithms based on traditional ranking functions fail to evaluate the contribution of the entire pattern set effectively because they take into account each pattern independently, while *HD* function is more reasonable by involving more factors; 2) *HEA* can remove erroneous and redundant patterns effectively and consequently achieves a highest F_1 score (45.5%) with the smallest pattern set (72 patterns). Thus, the semi-supervised learning procedure can be carried out effectively and achieves a state-of-art performance.

5. Conclusion

Pattern-based methods have been widely used for the task of relation extraction from bioscience texts. However, most of these methods either construct patterns manually, or require a well-annotated training corpus to learn patterns. In this paper, we have proposed a semi-supervised model to automatically learn patterns with unlabeled and labeled data. Little domain expertise is required and vast texts available in biomedical domain can be fully exploited. Moreover, two types of pattern evaluation algorithms based on labeled information are proposed to remove erroneous and redundant patterns. The first one is based on a novel ranking function *HD* which takes into account more factors than prevalent ranking functions. Experimental results show that *HD* function exhibits advantages over other ranking functions. The second one is a heuristic evaluation algorithm, which aims to obtain the optimal pattern set in iterative steps. This algorithm contributes improvement over ranking function based algorithms.

We also note that the major bottleneck of pattern-based IE systems is whether they have an effective NLP module to handle complex syntactic structures in the bioscience texts. Currently we have a shallow parsing module to enhance the results; however, the future work will still focus on developing more competitive NLP techniques.

Acknowledgments

The work was supported by Chinese Natural Science Foundation under grant No. 60572084 and 60321002.

References

1. M. Califf and R. Mooney. Bottom-Up Relational Learning of Pattern Matching Rules for Information Extraction.*Journal of Machine Learning Research*, 4, 2003, pp. 177–210.
2. J. Chiang and C. Yu. Literature extraction of protein functions using sentence pattern mining. *IEEE. Trans. On Knowledge and Data Engineering*, 17(8), 2005, pp. 1088-1098.
3. W.W. Cohen. Fast effective rule induction. In *Proceedings of International Conference on Machine Learning*, Lake Tahoe, CA, 1995.
4. M. Craven. Learning to extract relations from Medline. *AAAI-99 Workshop on Machine Learning for Information Extraction*, 1999.
5. N. Daraselia, A. Yuryev, S. Egorov, S. Novichkova, A. Nikitin, and I. Mazo. Extracting Human Protein Interactions from MEDLINE Using a Full-Sentence Parser. *Bioinformatics*, 20 (5), 2004, pp. 604- 611.
6. D. Freitag. Machine learning for information extraction in informal domains. *Machine Learning*, 39, 2000, pp. 169–202.

7. M.L. Huang, X.Y. Zhu, Y. Hao, D.G. Payan, K. Qu, and M. Li. Discovering patterns to extract protein-protein interactions from full texts. *Bioinformatics*, vol. 20, 2004, pp. 3604-3612.
8. M.L. Huang, X.Y. Zhu, S.L. Ding, H. Yu, and M Li. ONBIRES: ONtology-based BIological Relation Extraction System. In *Proceedings of the Fourth Asia Pacific Bioinformatics Conference*, Taiwan, China, 2006.
9. J.D. Kim, T. Ohta, Y. Teteisi, and J. Tsujii. GENIA corpus - a semantically annotated corpus for bio-textmining. *Bioinformatics*. 19(suppl. 1), 2003, i180-i182.
10. K.J. Lee, Y.S. Hwang, and H.C. Rim. Two-Phase Biomedical NE Recognition based on SVMs. In *Proceeding of the ACL 2003 Workshop on Natural Language Processing in Biomedicine*, Sappora, Japan, 2003, pp. 33-40.
11. T. Ono, H. Hishigaki, A. Tanigami, and T. Takagi. Automated extraction of information on protein-protein interactions from the biological literature. *Bioinformatics*, 17(2), 2001, pp. 155-161.
12. C. Plake, J. Hakenberg, and U. Leser. Optimizing Syntax Patterns for Discovering Protein-Protein Interactions. *ACM Symposium on Applied Computing*,195-201, 2005.
13. S. Ray, and M. Craven. Representing Sentence Structure in Hidden Markov Models for Information Extraction. In *Proceedings of the 17th International Joint Conference on Artificial Intelligence,* 2001, pp. 1273-1279.
14. E. Riloff. Automatically constructing a dictionary for information extraction tasks. In *Proceedings of the Eleventh National Conference on Artificial Intelligence*, Washington, D.C., 1993, pp. 811–816.
15. E. Riloff. Automatically Generating Extraction Patterns from Untagged Text. In *Proceedings of the Thirteenth National Conference on Artificial Intelligence (AAAI-96)*, 1996, pp. 1044-1049.
16. M. Seeger. *Learning with labeled and unlabeled data*, Technical Report, University of Edinburgh, 2001.
17. S. Soderland, D. Fisher, J. Aseltine, W. Lehnert. CRYSTAL: Inducing a Conceptual Dictionary. In *Proceedings of the Fourteenth International Joint Conference on Articial Intelligence* (IJCAI '95), 1995.
18. K. Sudo. *Unsupervised discovery of extraction patterns for information extraction.* Ph.D. Thesis, New York University, New York, NY. 2004.
19. J.M. Temkin, and M.R. Gilder. Extraction of protein interaction information from unstructured text using a context-free grammar. *Bioinformatics*, 19, 2003, pp. 2046 - 2053.
20. A. Yakushiji, Y. Tateisi, Y. Miyao, and J. Tsujii. Event extraction from biomedical papers using a full parser. In *Proceedings of Pacific Symposium on Bio-computing*, 2001, pp. 408-419.
21. G.D. Zhou, J. Zhang, J. Su, D. Shen, and C. Tan. Recognizing names in biomedical texts: a machine learning approach. *Bioinformatics*, 20 (7), 2004, pp.1178-1190.
22. X. Zhu. *Semi-Supervised Learning Literature Survey*, Technical Report No. 1530, Computer Sciences Department, University of Wisconsin, Madison, 2006.

FLOW MODEL OF THE PROTEIN-PROTEIN INTERACTION NETWORK FOR FINDING CREDIBLE INTERACTIONS

KINYA OKADA

Department of Computational Biology, Graduate School of Frontier Sciences, The University of Tokyo & Computational Biology Research Center, National Institute of Advanced Industrial Science and Technology, 5-1-5 Kashiwanoha, Kashiwa, Chiba 277-8561, Japan

KIYOSHI ASAI

Department of Computational Biology, Graduate School of Frontier Sciences, The University of Tokyo & Computational Biology Research Center, National Institute of Advanced Industrial Science and Technology, 5-1-5 Kashiwanoha, Kashiwa, Chiba 277-8561, Japan

MASANORI ARITA

Department of Computational Biology and PRESTO-JST, Graduate School of Frontier Sciences, The University of Tokyo & Institute of Advanced Biosciences, Keio University 5-1-5 Kashiwanoha, Kashiwa, Chiba 277-8561, Japan

Large-scale protein-protein interactions (PPIs) detected by yeast-two-hybrid (Y2H) systems are known to contain many false positives. The separation of credible interactions from background noise is still an unavoidable task. In the present study, we propose the relative reliability score for PPI as an intrinsic characteristic of global topology in the PPI networks. Our score is calculated as the dominant eigenvector of an adjacency matrix and represents the steady state of the network flow. By using this reliability score as a cut-off threshold from noisy Y2H PPI data, the credible interactions were extracted with better or comparable performance of previously proposed methods which were also based on the network topology. The result suggests that the application of the network-flow model to PPI data is useful for extracting credible interactions from noisy experimental data.

1 Background

1.1 Qualitative Assessment of Protein-protein Interactions

Well-known critique of yeast two-hybrid (Y2H) systems is their high false positive rates [1]. Separation of credible interactions from background noise is a necessary and crucial step in analyzing Y2H data. So far, three types of assessments have been proposed for this task: degree criterion, topological criterion, and intersection with other datasets.

Degree criterion uses node degree (number of connections) in the network to find false positives. Wet-lab scientists generally assume that proteins with too many interactions are sticky proteins or self-activators whose interactions are unimportant. On the other hand, bioinformatists may consider such highly connected nodes are biologically central, as the power-law hypothesis suggests that they are the scaffold of protein-protein interaction networks [2, 3]. Simplistic degree criterion is therefore not recommended for assessing observed interactions. Recent argument further complicates our understanding by showing that the power-law property of networks may results from biased data sampling [4].

To overcome such ambiguity, topological criterion uses local topology and statistics of adjacent interactions to extract credible interactions. Saito *et al.* presented a method to use network density to estimate credibility. Their "Interaction Generality" (IG) method is based on the idea that proteins which have many interacting partners that are independent (*i.e.* star-shaped nodes) are likely to be false positives [5]. Later they improved the method to incorporate further topological properties, such as the pattern of possible topological relationships of the common-neighbors of interacting protein pairs [6]. Its difficulty rests in the choice of appropriate criteria. There are many topological characteristics such as clustering coefficient, number of high (low) degree nodes, or distribution of node degrees and so on. Indeed, Bader *et al.* conclude that regression analysis using various topological parameters as explanatory variables could not determine conclusive factors for extracting interactions that appear in both Y2H and co-immuno precipitation analysis [7].

The hardness of detecting spurious interactions is due to the absence of true dataset. Currently, most reliable choice is the intersection with other datasets. Indeed, the intersection of multiple high-throughput datasets is enriched with confident interactions [8]. Unfortunately this operation leaves much fewer interactions for analysis [9], and is suitable only for identifying a very small, high-confidence set of interactions.

In summary, the current consensus is that edges of highly connected nodes are suspicious, but simple counting of node-degree is troublesome because the counting may overlook biologically central proteins (hubs) in the networks. Yet we do not have a decisive topological measure to distinguish between biological hubs and sticky proteins.

In this paper, we focus on the experimental characteristics of PPI data. In Y2H experiments, PPIs are detected using 'bait' proteins fused with a DNA binding domain, and 'prey' proteins fused with a transcription-activation domain of a transcription factor. Although physical interaction of biomolecules is considered equally coordinated in principle, experimentally observed interactions between a bait and a prey protein are not commutative: *i.e.* an interaction between a bait protein A and a prey protein B does not imply an interaction between B as a bait and A as a prey in most cases. Thus, a PPI network of Y2H systems represents a directed graph in which nodes and edges represent proteins and interactions, respectively. The reason for directionality is, at least in part, attributable to known biological mechanism. Some bait protein can activate transcription independently of an interaction with a prey protein (self-activator) [10]. Translated into terms of network analysis, this means that a bait protein of high out-degree and less in-degree is more likely to be spurious.

1.2 Network-flow Model of Interactions

By focusing on the previous observation for edge directionality, we propose a flow model of protein interactions. Our model assumes a hypothetical information flow throughout the network, where the flow coming from in-edges is evenly distributed among out-edges in a mass-balanced fashion (Fig. 1). It also assumes that the whole network is closed and in steady state: *i.e.* there is no influx or efflux from the outside world and the internal flow is stable. We consider the amount of steady-state flow as 'reliability score' for edges. Intuitively stated, a sticky protein has many out-edges, and their reliability score is low. Consequently, nodes to which such edges connect also obtain a little flow (or reliability score). Proteins with many in-edges, on the other hand, may collect more flow and therefore higher reliability score. In the steady state under the above-mentioned propagation scheme, nodes and edges can be ranked according to their flow amount. The model can be also considered a hybrid of the PageRank algorithm of the Google search engine [11] and the mass-balanced signaling network [12]. In the following, we will use the word 'score' to refer to flow amount in the network.

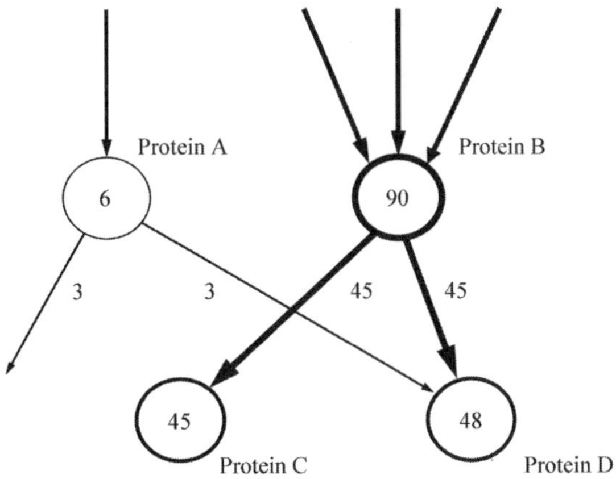

Fig 1. Protein *A* has one backward edge and two forward edges. The reliability score of protein *A* is 6; the reliability score of each forward edge is 6/2 = 3. Protein *D* has two backward edges. One is pointed at by protein *A* and its reliability score is 3, the other is pointed at by protein *B* and its reliability score is 45. The reliability score of protein *D* is therefore 3 + 45 = 48.

2 Algorithms

2.1 Simple Definition of Node Scores

We adopt the PageRank algorithm for determining the score through nodes and edges [11]. Let u be a node in PPI networks, $\mathbf{F}_u$ be the set of nodes with which u interacts as bait (forward edges), and $\mathbf{B}_u$ the set of nodes with which u interacts as prey (backward edges). Let c be a factor used for normalization (so that the total rank of all nodes is constant). Let us first define a score, **R**, which is a slightly simplified version of the final score used for assessment:

$$\mathrm{R}(u) = c \sum_{v \in \mathbf{B}_u} \frac{\mathrm{R}(v)}{|\mathbf{F}_v|}$$

where u and v are nodes in PPI networks. The equation indicates that the score of a node becomes less if its source nodes have small score and it points to many nodes. The score into a node is distributed evenly among its forward edges. After convergence, a consistent steady-state distribution is obtained (Fig. 2).

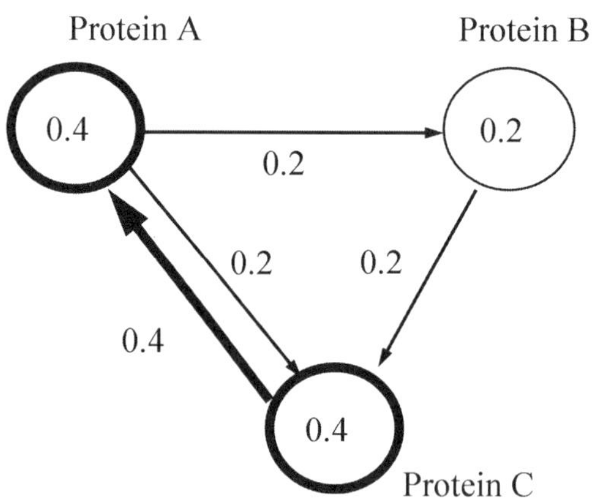

Fig 2. Consistent steady state of propagation of reliability score (or flow).

2.2 *Full Definition of Node Scores*

The definition of the score described above has an intuitive basis in stoichiometry. Each node represents a certain number of protein molecules, and they independently interact with their partners. Since no quantitative information for interactions is available, we assume that score is distributed evenly among forward edges. A compact network representation for this process is an adjacency matrix **A** where the rows and columns correspond to nodes as

$$\mathbf{A}_{uv} = \begin{cases} 1 & \text{if there is an edge from } u \text{ to } v \\ 0 & \text{otherwise} \end{cases}$$

We first model the amount of score for each node as inversely proportional to its out-degree, and define the normalized adjacency matrix **Q** as

$$\mathbf{Q}_{uv} = \begin{cases} 1/|\mathbf{F}_u| & \text{if there is an edge from } u \text{ to } v \\ 0 & \text{otherwise} \end{cases}$$

The transition-probability matrix **T** is defined as the transpose of **Q**. Given a column vector **R** over nodes as the steady-state distribution, we obtain

$$\mathbf{R} = \frac{1}{\lambda}\mathbf{TR}$$

That is, **R** is an eigenvector of **T** with eigenvalue λ. However, there is a problem with this simple function. Suppose that a group of nodes pointing to each other has no out-edges to other nodes. Then, their closed loop will only accumulate score and never distribute because there is no out-edges. Hence, no steady-state distribution exists (or zero flux, which is meaningless). To overcome this problem, a noise source is introduced. Let **E**(u) be some column vector over the nodes that corresponds to the uniform noise. Then we have **A'** = **A** + **E1** where **1** is the row vector consisting of all ones. Thus a matrix **Q'** is defined as

$$\mathbf{Q}'_{uv} = \frac{\mathbf{A}'_{uv}}{\sum_i \mathbf{A}'_{ui}}$$

and the transition-probability matrix **T'** is defined as the transpose of **Q'**. Finally, the reliability score for each node **R'** is defined as

$$\mathbf{R}' = \frac{1}{\lambda'}\mathbf{T}'\mathbf{R}'$$

In the original PageRank algorithm, the uniform noise is interpreted as moving to a different http address during net surfing. In the PPI network, it is interpreted as both experimental and biological noises. Since any interaction can be missed, or falsely detected with equal probability, we can safely introduce the uniform noise. In the present study, we set **E** uniform over all nodes with the value α =1.0e-5 after testing several parameter values. Changing the parameter (0.1< α <1.0e-8) had little effect on the resulting ROC scores (data not shown).

3 Results

3.1 Directional Dataset from Y2H Experiments

For the PPI network data, we used the Ito-core [13] and the Uetz dataset [14] for *Saccharomyces cerevisiae*. The intersection between the Ito and Uetz datasets was used as the true-positive samples, as in the method of Saito *et al.* to assess the credibility of interactions [5]. Both PPI networks were transformed to directed graphs, by drawing directed edges from bait protein to prey protein, ignoring multiple edges and self loops. The resulting non-redundant Ito-core dataset comprises of 789 interactions among 786 proteins, and the Uetz dataset, 1,407 interactions among 1,314 proteins. The size of common network was 114 interactions.

3.2 *Non-directional Dataset from Interaction Database*

The DIP-core, another reliable dataset made by filtering DIP-full interactions, is available at the DIP server (http://dip.doe-mbi.ucla.edu/) [9]. This network is undirected. The dataset was also used as the true-positive samples in the previous study for assessing PPI data [6]. From the dataset, redundant interactions and self loops were excluded to obtain the set of 5,785 interactions among 2,254 proteins. The dataset shared 231 interactions with the Ito-core dataset, and 394 interactions with the Uetz dataset. These edges were used as the DIP-intersection sample.

3.3 *Prediction by the Steady-flow Algorithm*

The steady flow of the PPI network was computed using the method in Section 2. The receiver operating characteristics (ROC) curve for each dataset was calculated by changing the cut-off score for extracting 114 true-positive samples from the Ito-core and Uetz dataset (Fig. 3). The area under ROC curves (ROC score) are shown in Table 1. In the same manner, DIP-intersection samples were predicted from the Ito-core and Uetz dataset (Table 2). We also reproduced the performance of the "interaction generality" (IG1) and its improved method (IG2) to assess interactions [5, 6], and listed their performance in both tables.

Table 1. ROC scores for the Ito-core and Uetz intersection samples.

dataset	This method	IG1	IG2
Ito-core	0.642	0.628	0.622
Uetz	0.724	0.648	0.623

Our method scored slightly better than the other two methods for the intersection between Ito-core and Uetz dataset, but its performance was only comparable for the DIP-intersection sample (Table 1, 2). This trend was unchanged for a wide range of noise factors ($0.1 < \alpha < 1.0e\text{-}8$), and was considered a robust conclusion. To confirm the contribution of PPI directionality on which our algorithm is based (and IG methods are not), we randomly flipped 'bait-prey' relationship in a protein pair for each dataset. The ROC score of the resulting random datasets were 0.514 and 0.523 for Ito-core and Uetz data, respectively (Fig. 3). The inefficiency of all three algorithms in the DIP-intersection sample is attributable to the independence of DIP-intersection samples to network topology. The difficulty of the prediction from network topology is further mentioned in Discussion.

Table 2. ROC scores for the DIP-intersection samples.

dataset	This method	IG1	IG2
Ito-core	0.602	0.608	0.636
Uetz	0.579	0.568	0.565

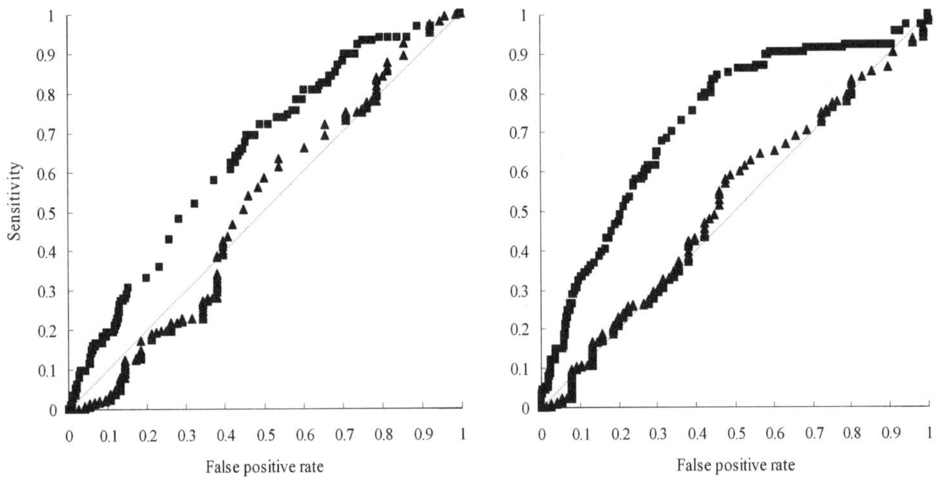

Fig 3. (LEFT) The ROC curve for the Ito-core dataset. Sensitivity and false positive rate are plotted according to the reliability score. Square plots represent Ito-core dataset, and triangle plots represent permuted dataset in which we randomly permuted 'bait-prey' relationship. The higher the plot, the better its performance. (RIGHT) The same ROC curves by using the Uetz dataset.

4 Discussion

4.1 Use of Global Network Structure

The PPI network is considered a superposition of biologically meaningful network and random noises. Especially, high degree nodes result from superposition of biological hubs and sticky proteins. To distinguish between them, our method computes the flow of `biological information' in PPI networks. The steady-state flow is conserved throughout network structure: the amount of flow at each node, *i.e.* the sum of flows from incoming edges, is evenly distributed among outgoing edges. Since the flow amount is interpreted as reliability of edges in our analysis, the key factor is the flow direction in the network structure.

The use of directionality in Y2H data is arguable, because biologically, all interactions must be commutative. What kind of benefit exists in utilizing experimental bias (directionality)? At least, it helps detect one eminent source of false positives, self-activators. Since they produce star-like interactions, our method lowers priorities of self-activators, and effectively removes false positives. Our method can lower reliability of hubs that have too many outgoing edges, and can increase reliability of nodes that are linked from important nodes.

The IG1 and IG2 methods both aim at extracting credible interactions using topology [5, 6]. The IG1 method basically computes a normalized number of common neighbors for each interacting protein pair, and disregards interactions with few common neighbors. The IG2 method additionally considers topological properties of interactions. Whereas the score of these methods is determined by local topology only, our score is computed as a steady-state, global distribution of flow in the PPI network. The improvement of the ROC score by our method indicates that the intersection of the Ito-core and Uetz datasets are correlated with network hubs in terms of network-flow (global information), rather than with hubs in terms of node degrees (local information). The effect of directionality was also confirmed using randomized PPI networks. While previous studies on biological hubs have been dealing with node degrees only [2, 16], our analysis showed the importance of more global properties of the network. Although the analysis is still in the preliminary stage using limited data, the application of this algorithm may reveal novel biological hubs in other networks.

4.2 Credibility of Intersection Data

The credibility of intersection data between different datasets is highly debatable. The percentile of the overlap in the interactions from the two Y2H systems used (reproducible samples) is small: only 14% of the interactions identified by the Ito-full dataset overlap with those present in the Uetz dataset, and only 8% of the interactions identified by the Uetz dataset overlap with those present in the Ito-full dataset. Likewise, the percentile of the intersectional samples with the DIP-core dataset (DIP-intersection samples) is also small: only 29% of the Ito-full dataset and only 28% of the Uetz dataset overlap with the DIP-core dataset. The same trend is observed in other high-throughput studies of protein interactions [1].

Even between manually curated databases such as the Database of Interacting Proteins (DIP) or the Munich Information Center for Protein Sequences (MIPS), a significant amount of data is not consistent: for example, 1,939 proteins out of 6,745 in the MIPS database do not show any interaction in other databases [17]. This forces us to use the limited datasets currently available. The same conclusion was reached in other validation studies [5, 6, 8]. As used in those studies, the reproducible samples and the DIP-intersection samples were used in the present study, but the percentile of the overlap between the DIP-intersection samples and the reproducible samples is small: only 33% of the intersectional DIP-intersection samples of the Ito-full dataset and only 19% of those of the Uetz dataset overlap with the reproducible samples.

Due to this statistic nature, it is not worthwhile to discuss individual protein interactions in our analysis. Rather, we focused on the elucidation of the large-scale organization of the network. The high correlation between highly reliable nodes and the credible datasets in our analysis suggests that our method detected at least as much information from the global network topology as did previous methods using local node degrees and local topology. The performance improvement using global topology was also seen in the work by J. Chen *et al.,* which was kindly suggested by one of the referees. The method used existence of 'alternative pathway' in the network (IRAP method) [18], together with the scores of IG1 as the bootstrap value for each edge. It is noteworthy that our method is a simple matrix operation and does not consider any biological parameters or bootstrap values.

Reliability measures from global topology, including ours, however, can never become biologically persuasive. Global, statistical information does not tell us about local interaction for prediction or verification. In this respect, our flow model should be combined with other local information such as protein functions or structures. The main contribution of the current work is therefore to show that the results of current reliability assessments are no better than a simple matrix algorithm, and that we need further research to effectively use large-scale, noisy interaction data.

Acknowledgments

We thank Professor Ito (University of Tokyo) for discussion on technical issues of Y2H systems. This work was supported by a Grant-in-Aid for Scientific Research on Priority Areas "Systems Genomics" and COE Program Grant "Genome Language" from the Ministry of Education, Culture, Sports, Science and Technology of Japan.

References

1. R. Mrowka, A. Patzak and H. Herzel. Is there a bias in proteome research? *Genome Res.* 11:1971-1973, 2001.
2. H. Jeong, S. P. Mason, A.-L. Barabasi, and Z. N. Oltvai. Lethality and centrality in protein networks. *Nature* 411:41-42, 2001.
3. E. Ravasz, A. L. Somera, D. A. Mongru, Z. N. Oltvai, and A.-L. Barabasi. Hierarchical organization of modularity in metabolic networks. *Science* 297:1551-1555, 2002.
4. M. P. Stumpf, C. Wiuf and R. M. May. Subnets of scale-free networks are not scale-free: Sampling properties of networks. *Proc. Natl. Acad. Sci. USA,* 102(12):4221—4224, 2005.
5. R. Saito, H. Suzuki, and Y. Hayashizaki. Interaction generality, a measurement to assess the reliability of a protein-protein interaction. *Nucleic Acids Res.* 30:1163-1168, 2002.
6. R. Saito, H. Suzuki, and Y. Hayashizaki. Construction of reliable protein–protein interaction networks with a new interaction generality measure. *Bioinformatics* 19:756-763, 2003.

7. J. S. Bader, A. Chaudhuri, J. M. Rothberg, and J. Chant. Gaining confidence in high-throughput protein interaction networks. *Nat. Biotechnol.* 22:78-85, 2003.
8. C. von Mering, R. Krause, B. Snel, M. Cornell, S. G. Oliver, S. Fields, and P. Bork. Comparative assessment of large-scale data sets of protein-protein interactions. *Nature* 417:399-403, 2002.
9. C. M. Deane, L. Salwinski, I. Xenarios, and D. Eisenberg. Protein interactions: two methods for assessment of the reliability of high throughput observations. *Mol. Cell Proteomics* 1:349-356, 2002.
10. M. E. Cusick, N. Klitgord, M. Vidal, and D. E. Hill. Interactome: gateway into systems biology. *Hum. Mol. Genet.* 14:R171-R181, 2005.
11. L. Page, S. Brin, R. Motwani, and T. Winograd. The pagerank citation ranking: bringing order to the web. *Technical report, Stanford Digital Library Technologies Project,* 1998.
12. J. A. Papin and B. O. Palsson. Topological analysis of mass-balanced signaling networks: a framework to obtain network properties including crosstalk. *J. Theor. Biol.* , 227(2): 283-297, 2004.
13. T. Ito, T. Chiba, R. Ozawa, M. Yoshida, M. Hattori, and Y. Sakaki. A comprehensive two-hybrid analysis to explore the yeast protein interactome. *Proc. Natl. Acad. Sci. USA* 98:4569-4574, 2001.
14. P. Uetz, L. Giot, G. Cagney, T. A. Mansfield, R. S. Judson, J. R. Knight, D. Lockshon, V. Narayan, M. Srinivasan, P. Pochart, A. Qureshi-Emili, Y. Li, B. Godwin, D. Conover, T. Kalbfleisch, G. Vijayadamodar, M. Yang, M. Johnston, S. Fields, and J. M. Rothberg. A comprehensive analysis of protein-protein interactions in Saccharomyces cerevisiae. *Nature* 403:623-627, 2000.
15. S. Oliver. Guilt-by-association goes global. *Nature* 403:601-603, 2000.
16. S. Wuchty. Interaction and domain networks of yeast. *Proteomics* 2:1715-1723, 2002.
17. S. H. Yook, Z. N. Oltvai, and A.-L. Barabasi. Functional and topological characterization of protein interaction networks. *Proteomics* 4:928-942, 2004.
18. J. Chen, W. Hsu, M. L. Lee, and S-K. Ng. Discovering reliable protein interactions from high-throughput experimental data using network topology. *Artif. Intell. Med.* 35(1-2):37-47, 2005.

ALL HITS ALL THE TIME:
PARAMETER FREE CALCULATION OF SEED SENSITIVITY

DENISE Y.F. MAK [1], GARY BENSON [2*]

[1] *Graduate Program in Bioinformatics, Boston University, Boston, MA 02215*

[2] *Dept. of Computer Science, Dept. of Biology, Boston University, Boston, MA 02215*

Standard search techniques for DNA repeats start by identifying *seeds*, that is, small matching words, that may inhabit larger repeats. Recent innovations in seed structure have led to the development of *spaced* seeds [8] and *indel* seeds [9] which are more sensitive than contiguous seeds (also known as k-mers, k-tuples, l-words, etc.). Evaluating seed *sensitivity* requires 1) specifying a homology model which describes types of alignments that can occur between two copies of a repeat, and 2) assigning probabilities to those alignments. *Optimal* seed selection is a resource intensive activity because essentially all alternative seeds must be tested [7]. Current methods require that the model and probability parameters be specified in advance. When the parameters change, the entire calculation has to be rerun. In this paper, we show how to *eliminate* the need for prior parameter specification. The ideas presented follow from a simple observation: given a homology model, the alignments *hit* by a particular seed remain the same regardless of the probability parameters. Only the weights assigned to those alignments change. Therefore, if we know all the hits, we can easily (and quickly) find optimal seeds. We describe a highly efficient preprocessing step, which is computed just *once* for each seed. In this calculation, strings which represent possible alignments are *unweighted* by any probability parameters. Then we show several increasingly efficient methods to find the optimal seed when given specific probability parameters. Indeed, we show how to determine exactly which seeds can *never* be optimal under any set of probability parameters. This leads to the startling observation that out of thousands of seeds, only a handful have any chance of being optimal. We then show how to find optimal seeds and the boundaries within probability space where they are optimal. We expect this method to greatly facilitate the study of seed space sensitivity, construction of multiple seed sets, and the use of alternative definitions of optimality.

1. Introduction

We are interested in solving the following problem. Given 1) a homology model (iid, Markov chain, hidden Markov model, etc.) which describes the types of alignments that occur between DNA repeats, 2) a maximum length for the alignments, and 3) a class of seeds (number of matches, number and type of wildcards), efficiently preprocess all the seeds in the class so that when given a set of probability parameters for the model, the optimal seed can be quickly identified.

As an example, assume the homology model is *match/mismatch iid*, where alignments of repeats are presumed to consist solely of matches and mismatches, a *representative string* for an alignment is a binary sequence of 1's (matches) and 0's (mismatches), and

*This research was supported in part by nsf grant dbi-0413462.

the probabilities for 1 and 0 at each position in the string are independent and identically distributed. Also assume that alignments have maximum length 100 and that the seed class has 11 matches and 7 match/mismatch wildcards (this is the PatternHunter seed class [8]). We seek to preprocess all the seeds in the class (there are 16 choose 7 or 11,440 of them and because mirror images have the same sensitivity, actually 5,720 different seeds) so that when given the parameters which fully describe the homology model we can quickly choose the optimal seed. In this case, if the parameters, specified as (probability of match, probability of mismatch, alignment length), are $(0.7, 0.3, 64)$, then we want to quickly identify the optimal seed, which is the PatternHunter seed (111*1**1*1**11*111) [8]. *But*, if we are then given another set of parameters which can be different in both probabilities and length, say $(0.75, 0.25, 50)$ we want to again quickly find the optimal seed, which is (111*1*1**11*1**111) [4]. Even more, we would like to know at what particular values (probability and length) optimality ends for one seed and begins for another.

Many existing methods [1, 5, 9] find the optimal seed for a single combination of homology model, parameter set, and seed class by enumerating *all patterns* for a given seed (*e.g.*, 128 for the PatternHunter seed) for *all seeds* in the class *e.g.*, 5,720 for the PatternHunter class) and testing these against the representative strings from the homology model. Best methods have time complexity in $O(LPwS)$ where L is the string length, P is the number of patterns per seed (an exponential in the number of wildcards), w is the length of a seed, and S is the number of seeds (the number of combinations of positions available in a seed for the wildcards). Importantly, when the parameter set changes, the entire calculation must be rerun. Choi *et. al* [4], in an extensive sampling of seed sensitivities, have taken this approach, rerunning a seed sensitivity algorithm for every combination of parameter values. Keich *et. al* [6] use a DP algorithm to find the optimal seed and Buhler *et. al* [3] sacrifice global optimality for locally optimal seeds.

Our seed preprocessing, which only needs to be calculated once, adds another factor to the time complexity which is dependent on the homology model (see section 3). But, it gives us three benefits. It allows repeated searches for the optimal seed with different parameters to proceed much more efficiently, it allows us to identify seeds that can *never* be optimal, and allows us to partition the parameter space into regions, each covered by a single optimal seed. (Note: a general method for parametric inference, which allows the partitioning of parameter space, has been described in [11].)

Our method has other advantages. It works with *sets* of seeds, and it allows for alternate definitions of optimality (see section 6). We expect this later feature will aide the discovery of highly sensitive *sets* of seeds.

The ideas presented in this paper follow from a simple observation: given a model, the alignments *hit* by a particular seed remain the same regardless of the probability parameters. Only the weights assigned to those alignments change. Therefore, if we know all the hits, we can easily (and quickly) find optimal seeds. In essence, our preprocessing identifies, for each seed, the specific set of representative strings it hits. Because the number of strings can be enormous (2^{100} for alignments of length 100 in the binary iid model) we cannot actually save the entire set of strings for each seed. Rather, we save *counts* for each probability class that the strings can occupy.

The remainder of the paper is organized as follows. In section 2 we give a formal definition of the problem. In section 3 we present the seed preprocessing method. In section 4 we show several increasingly efficient methods to find the optimal seed when given specific model parameters. In section 5 we show results from applying our parameter free calculation to several seed classes. Finally, in section 6 we discuss implications of this method and several uses of the information provided by preprocessing the seeds.

2. Problem Description

Formally, the problem we address is:
Seed All Hits Preprocessing

- **Given:** 1) a homology model (specified by probability variables), 2) a seed class (specified by seed width, number and type of wildcards), and 3) a maximum length for alignments called the target length t.
- **Compute:** A preprocessing of the seed class so that when values are assigned to the probability variables and a length L between 1 and t is selected, the optimal seed is returned.

For the remainder of this paper we will assume that the homology model is binary (match/mismatch) iid, *i.e.*, the Bernoulli model. The method extends to iid models with more parameters such as the ternary alphabet model (match/transition/mismatch) of Noé and Kucherov [10] and to Markov chains as well.

3. Method

We start with some definitions.

Definition 1: An alignment is described by a *representative string* over a binary alphabet: 1 indicates a match and 0 indicates a mismatch. A *probability equivalence class* contains representative strings that have the same probability when the probability parameters are specified.

In the Bernoulli model, strings belong to an equivalence class if they are the same length and contain the same number of 1's because all such strings have the same probability when $P(1)$ = the probability of 1, is defined. The number of equivalence classes is $t + 1$, where t is the target length.

Defintion 2: A *seed* is a string beginning and ending with a 1 and containing 1's and $*$'s, where 1 represents a match and $*$ is a wildcard that denotes either 1 (match) or 0 (mismatch). The length w of a seed is the total number of 1's and $*$'s. A *pattern* of a seed is any string of 1's and 0's which is obtained by replacing each wildcard in the seed by either a 1 or 0. A pattern *hits* a representative string if the pattern occurs in the string. Any pattern or set of patterns will hit a certain number of representative strings and these strings will fall into different equivalence classes. We indicate how many in each equivalence class by a vector of counts which we call the *PECC vector* (probability equivalence class counts vector) of length $t + 1$ with the classes ordered by the number of 1's from zero to t.

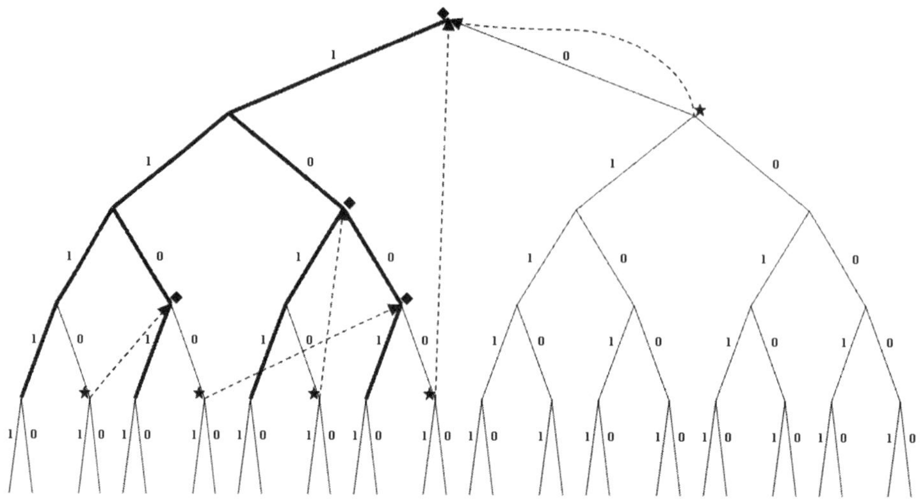

Figure 1. Aho-Corasick tree for seed 1**1 (bold) overlayed on all-strings tree. *Phantom* nodes are shown with ⋆, *fail-to* nodes are shown with ♦, and *phantom* nodes are linked to their *fail-to* nodes with - - →.

Our preprocessing algorithm derives from a recognition that seed hits can be summarized as hits within probability equivalence classes, that is, by PECC vectors (see figure 2). The result of the preprocessing for a single seed is a collection of PECC vectors, one for each representative string length L from 1 to t. Conceptually, the preprocessing, generates all possible representative strings for all lengths L from 1 to t, tests each to see if it contains a hit to the seed, and stores counts for the ones that do in the appropriate equivalence class. In actuality, strings are *not* generated. Instead, we collect and store PECC vectors which are used recursively in a dynamic programming algorithm.

Our data structure for collecting information is the Aho-Corasick tree (AC tree) for the seed patterns which is constructed as the initial step of the preprocessing. For the remainder of this section, we illustrate the preprocessing algorithm with a simple example using the seed 1**1, and a target length $t = 5$. In figure 1, we show the AC tree for 1**1 overlayed on the tree of all representative strings (*all-strings tree*). The following definitions refer to that figure.

Definition 3: For a node n, the number of edges in the path from the root to n is *depth(n)*. The *node string* for n, in the AC tree or all-strings tree, is the concatenation of edge labels from the root to n. The node string for a leaf (in either tree) is a *leaf string*. The leaf string for a leaf f in the AC tree is a pattern of the seed and is called pat_f. The number of 1's in a node string for node n is $ones(n)$. Each leaf, f, of the AC tree is the root of a subtree T_f in the tree of all-strings. A *fail node* in the AC tree is a non-leaf which has at least one *phantom* child, that is, a child node in the all-strings tree which does not occur in the AC tree. Each phantom p is the root of a subtree T_p in the all-strings tree. For each phantom p, its *fail-to* node ft_p is the node in the AC tree whose node string is the longest *proper* suffix of the node string for p.

Observations 3: Each phantom p has a fail-to node in the AC tree. The depth of a fail-to node ft_p is at most one less than the depth of p (alternately, at most the depth of the fail node parent of p).

Computing the PECC Vector

Base case: $L = w$. (For $L < w$, the PECC vector is all zeros.) The leaf strings of the AC tree are the only strings hit by the seed. The PECC vector for these strings can be determined when the tree is constructed.

Induction Step: $L = k + 1$. PECC vectors are already stored for $L = w, \ldots, k$. Consider the AC tree overlayed on the all-strings tree of length $k + 1$ (figure 1). We are interested in three types of AC tree nodes. For each node, we calculate a PECC vector which stores the leaf strings hit in the subtree (of the all-strings tree) rooted by the node.

- **Leaf.** Each leaf, f, of the AC tree is the root of a subtree T_f in the tree of all-strings. Every leaf string in T_f (from the root of the all-strings tree) is hit by the seed because it is an *extension* of pattern pat_f to length $k + 1$. Using regular expression notation, the leaf strings in T_f can be specified as

  $$pat_f\{0|1\}^{k+1-w}.$$

  The PECC vector for leaf f is the vector of counts for all the leaf strings in T_f. These strings fall into $k + 2 - w$ probability equivalence classes, where class i contains $ones(f) + i$ 1's with i in the range

  $$i = 0, \ldots, k + 1 - w.$$

  The number of strings for each i is determined by a combinations formula

  $$(k + 1 - w) \text{ choose } i.$$

  Notice that the subtrees rooted by all AC tree leaves are identical and can be treated identically, only the values $ones(f)$ differ. The combination formulas can be computed in advance and consulted when needed through a look-up table.
- **Phantom node.** Each phantom child p of a fail node n, is the root of a subtree T_p in the tree of all-strings. Unlike the subtrees T_f, not all leaf strings in T_p are hit by the seed. But, there is a simple way to determine which strings are hit. The information is already stored (as part of the dynamic programming) as a PECC vector in its fail-to node ft_p.
- **Fail-To node.** Each fail-to node ft must store a *series* of PECC vectors for subsequent iterations for look-up by phantom nodes that point to it. Each vector in the series represents a different depth of the all-strings tree. The PECC vectors for the root, which is also a fail-to node, are the output vectors for the seed, one for each length L from 1 to t.

For efficiency, the vector for each leaf, phantom, and fail-to node is pushed up the tree for addition into its nearest ancestor fail-to node. This assures that the additions are linear in the size of the tree. Determining where each vector must be added and the vector's depth relative to L can be done with a single traversal of the AC tree when the tree is constructed.

Time complexity

Theorem 1: *The time complexity for computing the PECC vectors for a single seed for all lengths between $L = 1$ and $L = t$ is $O(vtPx)$ where v is the number of probability equivalence classes, t is the target length, P is the number of patterns for the seed, and x is the number of ones in the seed. The time complexity for all seeds in a class is $O(vtPxS)$ where the class contains S seeds. This is a factor v more than current best calculations.*

Proof: Building the AC tree for all patterns of the seed, determining fail nodes and fail-to nodes is linear in the size of the AC tree which has size $O(Px)$. There are $t - w$ iterations through the tree. In each iteration, PECC vectors are updated for the leaves, fail nodes and fail-to nodes. The total number of these nodes and the total number of additions is linear in the size of the AC tree. Note that every addition requires adding two vectors of size at most v. □

Note: In the match/mismatch iid homology model, $v = t + 1$, so the time complexity is $O(t^2 Px)$.

Algorithm performance. Table 1 gives times for computing the PECC vectors for several seed classes.

4. Finding Optimal Seeds

Below we present three ways to find optimal seeds when given a probability value p and a homology region length L. We assume that the seeds in the class under consideration have already been preprocessed for their PECC vectors.

4.1. *Scan the PECC Vectors*

Using the PECC vector for length L from each seed, compute the total sensitivity of the seed and return the seed with maximum sensitivity. A PECC vector for length L contains

Table 1. Results for several seed classes for homologous regions of length 64. Note the very low number of dominant and optimal seeds in all classes. All computation was done on seeds after removal of mirrors. Time is for computing all PECC vectors in a single run for regions of length 50 to 64. Lengths under 50 were also computed in the same run, but PECC vectors were not saved. Calculations were made on a dual 1GHz PIII processor with 2GB RAM.

class		number of seeds				Time
1's	*'s	all	minus mirrors	dominant	optimal	(hr:min:sec)
9	6	1,716	868	7	4	0:03:40
10	6	3,003	1,519	6	4	0:06:27
11	7	11,440	5,720	12	5	0:47:00
12	6	8,008	4,032	10	3	0:19:06
12	7	19,448	9,752	36	10	1:24:35
13	6	12,376	6,216	13	7	0:32:25
13	7	31,824	15,912	20	5	2:28:32
14	6	18,564	9,324	20	7	0:49:51
14	7	50,388	25,236	22	3	4:09:25
15	6	27,132	13,608	24	4	1:17:58
15	7	77,520	38,760	23	5	6:29:26

$L + 1$ equivalences classes. The probability of a representative string which belongs to the ith equivalence class (specified by the number of 1's in the string, $i = 0, \ldots, L$) is $p^i q^{L-i}$ where p is the probability of 1 and $q = 1 - p$. Let the count in equivalence class i be C_i. Then the polynomial $C_0 \cdot p^0 q^L + C_1 \cdot p^1 q^{L-1} + \ldots + C_L \cdot p^L q^0$ gives the sensitivity of the seed. This method recomputes the sensitivity for each seed when p changes.

4.2. *Dominant Seeds*

Definition 4: For two seeds A and B with PECC vectors $\vec{V}_A$ and $\vec{V}_B$, if the difference vector $\vec{D} = \vec{V}_A - \vec{V}_B$ contains only zero values, then A and B are *equivalent*. If the difference vector contains no negative values and some values are positive, then A *dominates* B. If A is never dominated by another seed, then A is *dominant*.

If A dominates B, it means that in every probability equivalence class, A hits at least as many strings as B and in some classes hits more. In this situation, B can *never* be optimal and can be removed from the seed class. For this method, first determine the dominant seeds, then compute sensitivity only for their PECC vectors as in the previous section. Return the seed with maximum sensitivity. Determining which seeds are dominant occurs just once. When p changes, only the sensitivity of each dominant seed must be recomputed.

Amazingly, only a few seeds are dominant within the seed classes for the match/mismatch iid homology models we have investigated. For example, in the Pattern-Hunter seed class, with 5720 seeds (after removing equivalent mirrors), at length 64, only 12 seeds are dominant. Table 1 shows the number of dominant seeds in several seed classes.

4.3. *Partitioning the Parameter Space*

Definition 5: For two dominant seeds A and B with PECC vectors $\vec{V}_A$ and $\vec{V}_B$, if the difference vector $\vec{D} = \vec{V}_A - \vec{V}_B$ contains both positive and negative values, then A and B *flip*. A has more hits in some probability equivalence classes than B, and B has more hits in other classes. In this case, A and B will usually partition the probability space (figure 3) with A having better sensitivity in some regions of the space (where A *wins* relative to B) and B having better sensitivity in others (where B wins relative to A). A dominant seed A is *optimal* in some region of probability space only if A wins in that region relative to every other dominant seed.

Figure 2 shows the PECC vectors and difference vector for a pair of seeds that flip. To determine the winning regions for a seed pair A and B in the match/mismatch iid model, we solve for the roots of a polynomial equation with a single unknown variable p. Let the difference in counts in equivalence class i be D_i. Then our polynomial P_{AB}, containing $L+1$ terms has the form: $D_0 \cdot p^0 q^L + D_1 \cdot p^1 q^{L-1} + \ldots + D_L \cdot p^L q^0$ (see bottom of figure 2). Replacing q with $1 - p$, we solve the equation $P_{AB} = 0$ using the Mathematica function SOLVE [13]. There are L roots (complex and real) that exist but only real roots between 0 and 1 are of interest. Testing the PECC vector for A on either side of the roots reveals those regions where A wins. To find where a dominant seed A is optimal, we merge the regions where A wins relative to every other dominant seed. An example is shown in figure 3.

number of 1's	hit counts seed A	seed B	different in hit counts A - B
0	0	0	0
1	0	0	0
2	0	0	0
3	0	0	0
4	0	0	0
5	38	38	0
6	1520	1520	0
7	29640	29640	0
8	375332	375334	-2
9	3468729	3468766	-37
10	24928629	24928696	-67
11	144948172	144942776	5396
12	700456139	700364916	91223
13	2867679722	2866827015	852707
14	10087980802	10082362006	5618796
15	30819118260	30790682312	28435948
16	82428580640	82313087704	115492936
17	194186285079	193799799947	386485132
18	404765317464	403681622157	1083695307
19	748926920505	746353121681	2573798824
20	1232762203160	1227550871937	5211331223
21	1807611479754	1798588935398	9022544356
22	2362723608570	2349367736133	13355872437
23	2753640936852	2736789617398	16851319454
24	2861644355241	2843634715311	18009639930
25	2652331424044	2636186283832	16145140212
26	2194080691374	2182108728673	11971962701
27	1622400592905	1615198100214	7202492691
28	1075039055326	1071615007952	3424047374
29	640344850572	639104255291	1240595281
30	343883821799	343558130745	325691054
31	166774138475	166716560683	57577792
32	73001069494	72994947436	6122058
33	28759911342	28759587965	323377
34	10150595182	10150589139	6043
35	3190187285	3190187260	25
36	886163135	886163135	0
37	215553195	215553195	0
38	45379620	45379620	0
39	8145060	8145060	0
40	1221759	1221759	0
41	148995	148995	0
42	14190	14190	0
43	990	990	0
44	45	45	0
45	1	1	0

$$-2x^8(1-x)^{37} - 37x^9(1-x)^{36} - 67x^{10}(1-x)^{35} + 5396x^{11}(1-x)^{34} + 91223x^{12}(1-x)^{33} + 852707x^{13}(1-x)^{32} + 5618796x^{14}(1-x)^{31} + 28435948x^{15}(1-x)^{30} + 115492936x^{16}(1-x)^{29} + 386485132x^{17}(1-x)^{28} + 1083695307x^{18}(1-x)^{27} + 2573798824x^{19}(1-x)^{26} + 5211331223x^{20}(1-x)^{25} + 9022544356x^{21}(1-x)^{24} + 13355872437x^{22}(1-x)^{23} + 16851319454x^{23}(1-x)^{22} + 18009639930x^{24}(1-x)^{21} + 16145140212x^{25}(1-x)^{20} + 11971962701x^{26}(1-x)^{19} + 7202492691x^{27}(1-x)^{18} + 3424047374x^{28}(1-x)^{17} + 1240595281x^{29}(1-x)^{16} + 325691054x^{30}(1-x)^{15} + 57577792x^{31}(1-x)^{14} + 6122058x^{32}(1-x)^{13} + 323377x^{33}(1-x)^{12} + 6043x^{34}(1-x)^{11} + 25x^{35}(1-x)^{10}$$

Figure 2. Top) PECC vectors for seeds A: 111*1**1 and B: 111**1*1 (for the class with five 1's and three *'s and homologous region length 45) and the difference vector showing positive and negative values. Bottom) Polynomial constructed from the difference vector.

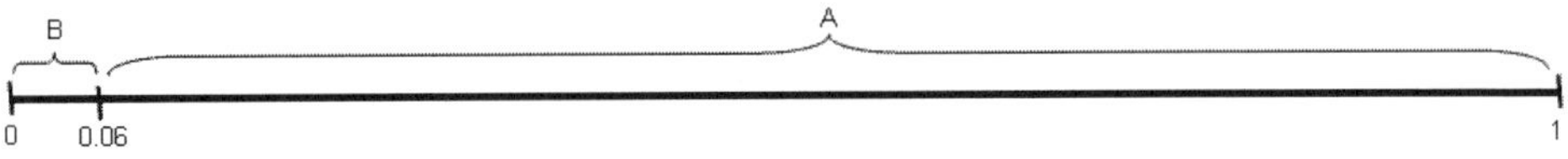

Figure 3. Partition of parameter space into regions won by seeds A and B from figure 2

For this method, the probability space is first partitioned, the optimal region is found for the specific value p, and the sensitivity of the optimal seed in that region is computed using the seed's PECC vector. Partitioning occurs only once. When p changes, the optimal region must again be determined and the sensitivity of a single seed computed.

5. Results

Table 1 shows the results of our preprocessing for dominant and optimal seeds on several seed classes. Note the extremely small number of dominant and optimal seeds in these classes. Table 2 shows the partition of probability space by the optimal seeds in each class. Note that some seeds are optimal in more than one region. Figure 4 illustrates the partition for a single seed class, the PatternHunter class, for homologous regions of length 64. More generally, we are interested in the partition of probability space for all seeds in the same weight class, not just a seed class. Figure 5 illustrates the partition of probability space by seeds drawn from all seed classes of weight 11, *i.e.*, those having 11 ones and any number of wildcards. Note that the contiguous seed is optimal only at the low end of the probability range. The number of wildcards in the optimal seed increases towards the high end of the range and ultimately decreases again at the highest levels.

6. Discussion

How optimal regions vary as homology region length varies. The issue of optimality in the limit with respect to length of the homologous regions has been addressed in [3, 5, 4]. Using our partition method, we can examine variations in optimal seeds and the regions of probability space they cover as the length of the homologous region varies. Figure 5 illustrates this variation for the weight 11 seed classes. Note that in this case, the set of optimal seeds remains relatively stable over a range of lengths, *e.g.*, seeds A, B, C, D,

Figure 4. Range for optimal seeds from the Pattern Hunter class at region length 64. A: 111*1**11*1*1**111, B: 111*1*1**11*1**111, C: (PH seed) 111*11**1*1**1*111, D: 11**111*1**1*111*1, E: 1111*1*11**1***111, F: 1111***1**11*1*111

Table 2. Optimal seeds and the range of probabilities where they are optimal for several seed classes at region length 64. Note that the same seed can be optimal in more than one region.

number of 1's	number of *'s	optimal seeds	probability range
9	6	A:111**1*1**11*11	0 - 0.1110266686
		B:111*1*1**1**111	0.1110266686 - 0.4327682188
		C:111***1*1*11*11	0.4327682188 - 0.9694790865
		B:111*1*1**1**111	0.9694790865 - 0.9991450536
		D:1111**11*1*1**1	0.9991450536 - 1
10	6	A:111**1*1*1*11*11	0 - 0.0231912575
		B:111*1**1*11**111	0.0231912575 - 0.0457879868
		C:111*1*11***11*11	0.0457879868 - 0.9436271851
		D:1111**1*1*11**11	0.9436271851 - 1
11	7	A:111*1**11*1*1**111	0 - 0.0524790924
		B:111*1*1**11*1**111	0.0524790924 - 0.0775105071
		C:111*11**1*1**1*111	0.0775105071 - 0.7304317142
		B:111*1*1**11*1**111	0.7304317142 - 0.9845899783
		D:11**111*1**1*111*1	0.9845899783 - 0.9997355115
		E:1111*1*11**1***111	0.9997355115 - 1
12	6	A:111*1*1*11*11**111	0 - 0.0125740804
		B:111*11**1*11*1*111	0.0125740804 - 0.9818956319
		C:11*111**1*111*1*11	0.9818956319 - 1
12	7	A:111*1**11*1*11**111	0 - 0.0269449089
		B:111*1*11*1**11**111	0.0269449089 - 0.0501533511
		C:1111**11**1*1*11*11	0.0501533511 - 0.1324581579
		D:1111*1**11**1*1*111	0.1324581579 - 0.1713344621
		E:1111**1*1*1**11*111	0.1713344621 - 0.2193278527
		F:1111*1*1**11**1*111	0.2193278527 - 0.5616667374
		G:1111*11**1*1*1**111	0.5616667374 - 0.8266957477
		H:1111*1*1**11*1**111	0.8266957477 - 0.9600165421
		I:111*1**11*1**111*11	0.9600165421 - 0.9824252510
		G:1111*11**1*1*1**111	0.9824252510 - 0.9833842249
		J:1111*1*11**1***1111	0.9833842249 - 1
13	6	A:1111**11*11*1*1*111	0 - 0.0250132023
		B:1111**11*1*1*11*111	0.0250132023 - 0.5568886832
		C:1111*1*1**11*11*111	0.5568886832 - 0.9726822943
		D:111*111*11*1**1*111	0.9726822943 - 0.9845644894
		E:111*11**1*1111*1*11	0.9845644894 - 0.9991885007
		F:111*1**111*111*1*11	0.9991885007 - 0.9999389894
		G:111111*1*1*11***11	0.9999389894 - 1
13	7	A:111*1*11*1*11**11*11	0 - 0.0128994408
		B:111*11**1*11*1*1*111	0.0128994408 - 0.1449709023
		C:1111*1**11**11*1*111	0.1449709023 - 0.7775640305
		D:111*111**1*11**1*111	0.7775640305 - 0.9771389517
		E:11111**11*1**1*1*111	0.9771389517 - 1
14	6	A:111*11*11*1*111**111	0 - 0.0255990753
		B:1111**11*11*1*11*111	0.0255990753 - 0.0269689880
		C:1111*11**11*11*1*111	0.0269689880 - 0.1397167664
		D:1111*11**11*1*1*1111	0.1397167664 - 0.8585536713
		E:11111**11*1*11*1*111	0.8585536713 - 0.9733966204
		F:11111*11*1*1*11**111	0.9733966204 - 0.9774612165
		G:1111*1**111*111*1*11	0.9774612165 - 1
14	7	A:111*1*11*11**11*1*111	0 - 0.0263720408
		B:111*11**11*1*11*1*111	0.0263720408 - 0.4156480746
		C:1111*1**11*1**111*111	0.4156480746 - 1
15	6	A:111*11*1*11*111*1*111	0 - 0.1140558268
		B:1111*1*11*11**111*111	0.1140558268 - 0.5238349212
		C:11111**1*11*1*111*111	0.5238349212 - 0.8916876889
		D:1111*11**1*1111*1*111	0.8916876889 - 1
15	7	A:1111*1*11**11*11*1*111	0 - 0.6176146226
		B:111*111*1**11*1*11*111	0.6176146226 - 0.6698669044
		C:11111*1**11*1**111*111	0.6698669044 - 0.9307720060
		D:1111**1*111**1*111*111	0.9307720060 - 0.9762530821
		E:111*1*1111*11**1*11*11	0.9762530821 - 1

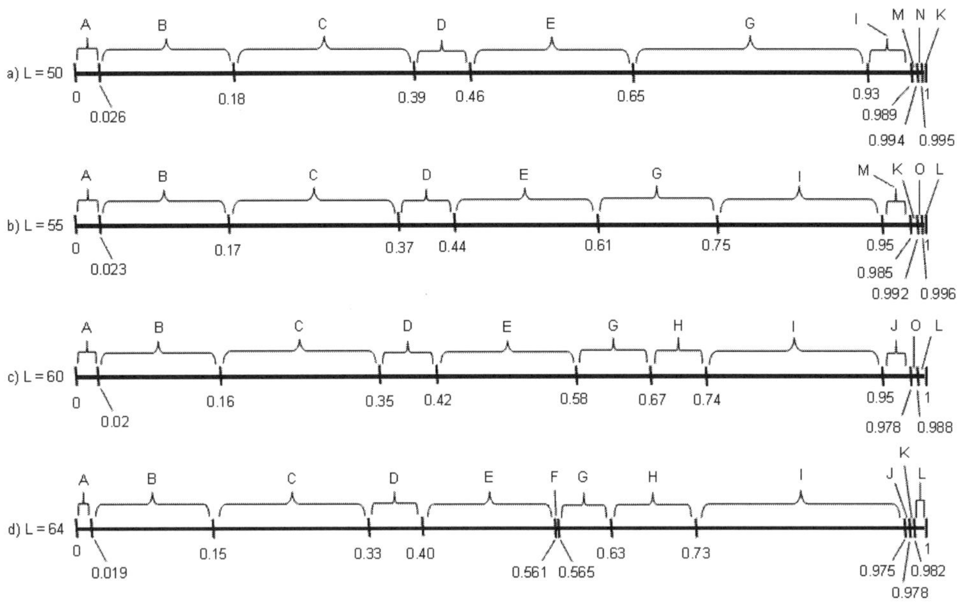

Figure 5. Optimal range of winning seeds of weight 11 at varying region lengths. a) region length 50, b) region length 55, c) region length 60, d) region length 64. A: 11111111111, B: 111111*11111, C: 1111*111*1111, D: 11111*1*11*111, E: 111*11*11*1*111, F: 1111*1*1**11*111, G: 111*111*1**1*111, H: (PH seed) 111*11**1*1**1*111, I: 111*1*1**11*1**111, J: 1111**1*1*111*11, K: 111*1111*1**11*1, L: 11111*1*1*11**11, M: 1111**1*111*111, N: 111*1**11*1*1**111, O: 111*1**11*1111*1

E, G, I, etc. If this property holds generally, it can be used to make the time to search for dominant seeds nearly linear in the number of seeds in the class (after preprocessing) rather than quadratic in that number. To search for dominant seeds at length L we would first choose the optimal seeds at length $L-1$ and then test the remaining seeds against only these to eliminate the non-dominant ones.

Optimality within probability equivalence classes. To say a seed is optimal for some homology model can be somewhat misleading. For example, the seed G (111*111*1**1*111) from figure 5 is optimal for the model with probability of matching = 60% at region length 64, but this might be construed to mean that it does best at

Table 3. Seeds with the most hits in equivalence classes (number of 1's) for weight 11 seeds at region length 64. Seed G from figure 5 is optimal for the model with probability of match equal to 0.60, but seed E does best at hitting alignments with 60% matches.

seed	most hits in equivalence classes	percentage of region length
E: 111*11*11*1*111	33-39	51%-61%
F: 1111*1*1**11*111	40-42	62%-66%
H: 111*11**1*1**1*111	43-47	67%-74%
I: 111*1*1**11*1**111	48-54	75%-85%

hitting alignments which contain 60% matches, which is not the case. The seed E is better for these alignments as shown in table 3. In fact, G is not best at hitting any individual equivalence class at length 64. This leads us to propose an alternative concept of optimality. We might be interested in seeds that do best in a particular probability equivalence class, rather than seeds that do best across all the classes. Our method for computing PECC vectors allows us to identify these winners, which by definition are dominant seeds.

7. Multiple Seed Sets

Here we explore the characterization of seed pair sensitivity using our PECC vector method. Seed pairs (and triplets, etc.) are more sensitive than single seeds and can be used efficiently in homology detection programs [12, 14, 2]. But, the calculation of dominant seed pairs is time intensive because the number of pairs is roughly quadratic in the number of seeds in the seed class. As an illustration of our method, we calculated the dominant and optimal seed *pairs*, for the class of seeds with 9 ones and 6 wildcards. Results are summarized for homologous region length 64 in table 4. Note that as with single seeds, we excluded mirrors from our calculation. For the seed pair A and B, the mirror is the pair: mirror of A and mirror of B. We also excluded redundant pairs, i.e., A with A. As observed with single seeds, only a miniscule number of seed pairs are dominant and/or optimal.

Because the long computation time prohibits easily calculating pairs for larger seed classes, we have explored the possibility of calculating dominant and optimal seed pairs from among only the set of dominant single seeds. We paired single dominant seeds from homologous regions of length 50 to 64 (since dominant seeds vary at different region lengths) and results are shown in table 5 for homology region length 64. We next compared the sensitivity of an optimal seed pair taken from the entire seed class to the

Table 4. Results for seed pairs in seed class for homologous regions of length 64. All computation was done on seed pairs after removal of mirrored pairs. Time is for computing all PECC vectors in a single run for regions of length 50 to 64. Lengths under 50 were also computed in the same run, but PECC vectors were not saved. Calculations were made on a dual 1GHz PIII processor with 2GB RAM.

class		number of seed pairs				Time
1's	*'s	all	minus mirrors	dominant	optimal	(hr:min:sec)
9	6	2,944,656	736,254	23	7	100:30:05

Table 5. Results for pairing single dominant seeds in seed class for homologous regions of length 64. All computation was done on seed pairs after removal of mirrored pairs. Time is for computing all PECC vectors in a single run for regions of length 50 to 64. Lengths under 50 were also computed in the same run, but PECC vectors were not saved. Calculations were made on a dual 1GHz PIII processor with 2GB RAM.

class		number of seed pairs				Time
1's	*'s	all	minus mirrors	dominant	optimal	(hr:min:sec)
9	6	528	256	8	7	00:02:18

Table 6. Comparison of optimal seed pair sensitivities at region length 64 at 70% matching. The optimal seed pairs of the entire seed class with 9 ones and 6 stars and from dominant seeds from the same seed class.

Homology Model		Optimal seed pair from entire seed class	Sensitivity optimal pair from dominant seeds
% match	% mismatch	111*11*1*1***11 + 1*111**1*11*11	111*1***11*1*11 + 111**1*1**11*11
70	30	0.85162	0.84953

sensitivity of the optimal pair calculated from only the dominant single seeds (table 6). At probability of match = 70%, the difference in sensitivity is less than 1% which is a fair tradeoff given the roughly 3000 fold decrease in computation time. Our results suggest that using only dominant single seeds to determine sensitive multiple seed sets can be a good heuristic.

8. Conclusion

We have presented a new, efficient preprocessing method to determine optimal seed sensitivity without first specifying the probability parameters for the homology model. This allows finding the optimal seed quickly once the parameters are specified. It reveals that the overwhelming majority of seeds have no chance of being optimal and allows us to describe the partition of probability space by optimal seeds. Finally, this method permits us to consider an alternative definition for optimality, and gives a good heuristic for designing near optimal sets of multiple seeds.

References

1. B. Brejová, D. Brown, and T. Vinar. Optimal spaced seeds for homologous coding regions. *Journal of Bioinformatics and Computational Biology*, 1:595–610, 2004.
2. D.G. Brown. Optimizing multiple seeds for protein homology search. *IEEE/ACM Trans. Comput. Biology Bioinform.*, 2(1):29–38, 2005.
3. J. Buhler, U. Keich, and Y. Sun. Designing seeds for similarity search in genomic DNA. *Journal of Computing and System Sciences*, 70:342–363, 2005.
4. K.P. Choi, F. Zeng, and L. Zhang. Good spaced seeds for homology search. *Bioinformatics*, 20:1053–1059, 2004.
5. K.P. Choi and L. Zhang. Sensitivity analysis and efficient method for identifying optimal spaced seeds. *Journal of Computer and System Sciences*, 68:22–40, 2004.
6. U. Keich, M. Li, B. Ma, and J. Tromp. On spaced seeds for similarity search. *Discrete Applied Mathematics*, 138(3):253–263, 2004.
7. M. Li, B. Ma, and L. Zhang. Superiority and complexity of the spaced seeds. In *SODA*, pages 444–453. ACM Press, 2006.
8. B. Ma, J. Tromp, and M. Li. Patternhunter: faster and more sensitive homology search. *Bioinformatics*, 18:440–445, 2002.
9. D. Mak, Y. Gelfand, and G. Benson. Indel seeds for homology search. *Bioinformatics*, 22(14):e341–349, 2006.
10. L. Noé and G. Kucherov. Improved hit criteria for DNA local alignment. *BMC Bioinformatics*, 5:149–158, 2004.

11. L. Pachter and B. Sturmfels. Parametric inference for biological sequence analysis. *PNAS*, 101(46):16138–16143, 2004.
12. Y. Sun and J. Buhler. Designing multiple simultaneous seeds for DNA similarity search. In *Proceedings of the Eighth Annual International Conference on Computational Molecular Biology, RECOMB*, pages 76–84, 2004.
13. Inc. Wolfram Research. *Mathematica: Solve function*. Version 5.2. Wolfram Research, Inc., 2005.
14. J. Xu, D. Brown, M. Li, and B. Ma. Optimizing multiple spaced seeds for homology search. In *Combinatorial Pattern Matching, 15th Annual Symposium, CPM*, pages 47–58, 2004.

FAST STRUCTURAL SIMILARITY SEARCH BASED ON TOPOLOGY STRING MATCHING*

SUNG-HEE PARK AND DAVID GILBERT

Bioinformatics Research Centre, Department of Computing Science, University of Glasgow, Glasgow, Scotland, G12 8QQ, UK

KEUN HO RYU

Chungbuk National University, Cheongju, 361-763, R..O.Korea

We describe an abstract data model of protein structures by representing the geometry of proteins using spatial data types and present a framework for fast structural similarity search based on the matching of topology strings using bipartite graph matching. The system has been implemented on top of the Oracle 9i spatial database management system. The performance evaluation was conducted on 36 proteins from the Chew and Kedem data set and also on a subset of the PDB40. Our method performs well in terms of the quality of matching whilst having the advantage of fast execution and being able to compute similarity search in polynomial time. Thus, this work shows that the pre-computed string representation of topological properties between secondary structure elements using spatial relationships of spatial database management system is practical for fast structural similarity search.

1 Introduction

The complexity of structure similarity search for 3D structures is NP hard [1]. Due to the complexity of similarity search and the exponential growth of the size of the 3D structure databases, the issue of speed becomes ever more important, and tools for fast structure similarity search have been required.

GRATH [2] and SSM [3] are examples of recent work dealing with the speed of structure similarity search which make full database searches possible. Some similarity search methods use fast pre-filtering before performing accurate alignment. Even in this case, due to the lack of pre-computed and stored data, each structure in the database has to be scanned at least once. Therefore, such an exhaustive search is expensive to perform.

The degree of abstraction influences the accuracy and performance of structure comparison programs. For example, DALI [4] and SSAP [5] which are based on atomic coordinate level data can be accurate but execute more slowly than less accurate approaches that use abstractions such as vectors and SSEs (e.g. VAST [6], LOCK [7], TOPS [8]).

We determined that the computation cost for geometrical features is known to be more expensive than that of I/O access to them whilst the opposite is true for general data. The computational cost of a topological match is less expensive than that of atomic-

* This work was supported by the Korean Research Foundation Grant funded by the Korean Government (MOEHRD). (KRF-2005-214-E00050).

coordinate based geometry matching, and the comparison of topological properties can identify global similarities. Our method is based on the spatial theory of GIS (Geographic Information Systems): topological properties of spatial objects are invariant even if geometric features such as length and angle are easily changed [9]. Topological properties are preserved in conserved structures and can help to identify the fold family in similarity searching. Thus, we consider that topological properties of proteins are suitable for fast structure comparison.

In this paper we describe in detail a method to derive abstract models of protein structures at the tertiary (fold) level, where secondary structure elements (SSEs) are represented using spatial data types. Our models comprise sequences of topology strings, each string describing the topological relationship between a pair of such SSEs using the 9IM (Intersection Matrix) [10]. As an application of our abstract model to protein structure comparison, we describe a framework for fast similarity search based on finding maximum matching pairs of topology strings using a bipartite graph matching algorithm implemented by join operations. We report experimental results on two significant data sets, comparing our approach with the TOPS system.

2 An Abstract Model of Protein Structures

2.1. Geometry Representation of Proteins

The geometry of SSEs is modelled as a polyline of 4–10 connected points (the normal range of the numbers of amino acids in SSEs). These polylines are implemented using the geometric types of SDO_GEOMETRY in ORACLE 9i Spatial by LINE STRING where points are connected by straight line segments. Figure 1 (a) shows an example of Helix 4 that starts from amino acid residue 348 and ends at residue 335; Figure 1 (b) is for a strand 2 of protein domain 2bopA0. The tertiary structure of a protein can be defined as a finite set of SSEs and the spatial topological relationships between them, and can thus be described by a list of polylines.

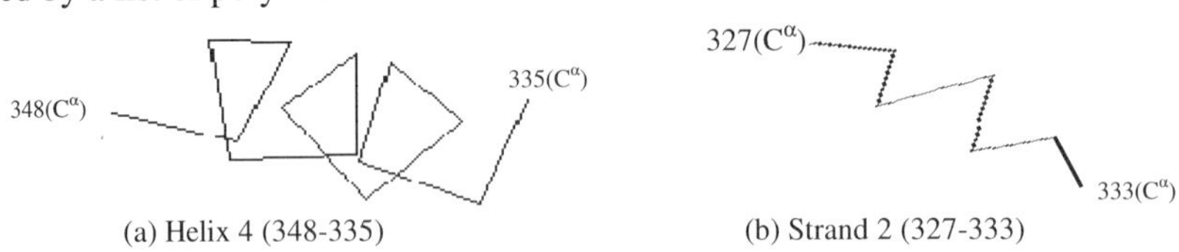

(a) Helix 4 (348-335) (b) Strand 2 (327-333)

Figure 1. Polylines for an Helix and a Strand of 2bopA0

2.2. Inference of Topological Relationships

The biological meaning of protein topology refers to the spatial relations of 3D folds, i.e. the spatial arrangements of SSEs. This can be described using the topological relationships of spatial geometry and spatial types. Thus, the problem of inferring the topological relationships between SSEs in 3D space can be transformed into that of inferring the spatial relationships between binary polylines in a spatial database. We employ spatial topological relationships [10] that have been previously used in GIS applications to infer relationships in 3D space. Spatial relationships between pairs of SSEs are represented by eight topological relations, defined by the 9IM (Intersection Matrix) [10].

Two lines can induce eight different types of topological relationships: *disjoint*, *contains*, *inside*, *equal*, *meet*, *cover*, *coveredBy*, and *overlaps*. In our approach we consider the three topological relations *touch*, *equal*, *overlaps* due to the fact that *overlaps* subsumes *inside, contains*, *cover, coveredBy*. These three topological relationships are enhanced by the addition of the types of SSEs, e.g. Helix and Strand. We thus generate the major nine topological relationships (e.g. Helix$\odot_{\text{overlaps}}$Helix, Helix$\odot_{\text{overlaps}}$Strand, Strand$\odot_{\text{overlaps}}$Strand, and etc.) as shown in Table 1. The three topological relationships (*overlaps, equal, touch*) are computed by a method that we have previously described [11], which uses topological operators in the ORACLE 9*i* Spatial. This system provides the spatial operator SDO_RELATE that implements a 9IM model between points, lines and polygons. The spatial operator is accomplished by the join operation of two R-tree indexes in a spatial query.

In the 9IM Model a binary topological relationship R(A, B) between two lines A and B is calculated by the comparison of A's interior (A°), boundary (∂A), and exterior (A−) with B's interior (B°), boundary (∂B), and exterior (B−). These six objects can be combined to form nine fundamental descriptions of the three topological relationships between two lines and can be concisely represented by the 3×3 matrix in Figure 2(a), called the 9-Intersection Matrix (9IM). The three topological relationships provide a mutually exclusive coverage with corresponding 9-intersection matrices in Figure 2(b) where 0 and 1 correspond to the empty set and non-empty set respectively.

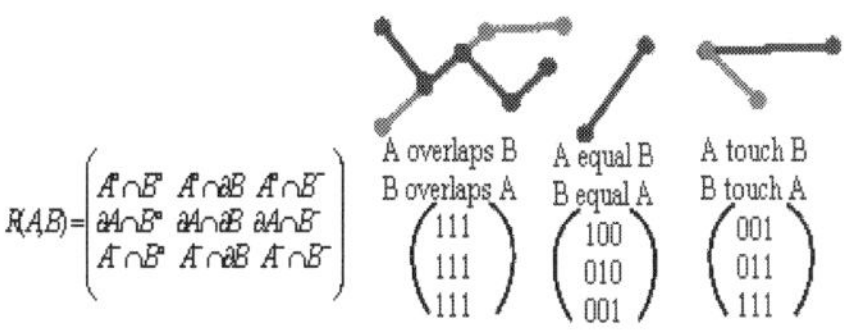

(a) 9IM (b) three relationships

Figure 2. Topological Relationships

Table 1. Enhanced Topological Relationships

Topological Relation (⊙)	SSE types of Secondary Structure		
	Helix ⊙ Helix	Helix ⊙ Strand	Strand ⊙ Strand
overlaps	A	F	K
equal	B	E	H
touch	D	I	N

The topological relationship *overlaps* identifies a pair of SSE in a protein, where the interiors and boundaries of two SSEs intersect. The relationship *touch* recognizes that the boundaries intersect but interiors do not intersect between two SSEs whereas *equal* finds two SSEs having the same boundary and interior.

We formalize the binary topological relations corresponding to SSE types using these three topological relationships. Given a protein P={ $S_1,\ldots, S_i, \ldots S_p$}, where $i \in 1, 2,\ldots, p$, $0 < i < j < p$, p is the number of SSEs in a protein P, S_i and S_j are SSEs, the type

(a) 2hbg00 (b) 1cdb00 (c) 1aa900

(a) 2hbg00:
TOPS: NhHhhHhHC 1:4R4:6R5:7R
TStringSet:{(Anf,2,12,10.04)(Anf,2,14,8.1)(Asf,4,6,11.90)(Anf,4,8,9.117)(Anf,4,12,8.6)(Anf,8,12,14.3)(Anf,8,14,15.2)(Anf,10,14,6.299)(Asf12.14.10.7)}

(b)1cdb00:
TOPS: NEEeeEC 1:5P2:3A2:4A4:5A
TStringSet: (Ksf,4,6,5.204)(Knf,4,8,3.601)(Ksf,8,10,5.202)}

(c)1aa900:
TOPS: NEheEhEhEhEhC 1:4P1:6P3:4A4:6R6:8Z8:10Z
TStringSet:{(Knf,2,8,3.92)(Knf,2,12,4.6)(Fsf,4,6,8.49)(Fnf,4,8,)(Anf,4,22,7.9)(Ksf,6,8,3.7)(Knf,12,16,3.92)(Anf,14,18.5)(Knf,16,20,5.58)}

Figure 3. 3D Structures and Textural Representation

of an SSE is $T(S_i) \in \{\alpha, \beta\}$ standing for helix (α) and strand (β), and a set of topological relationships R={*overlaps*, *equal*, *touch*}, then:

[Definition1] (a binary topological relation) Let $R^2 = \{(S_i \odot S_j)\}$ be a binary topological relation, where $S_i, S_j \in P$, $i \neq j$, $\odot \in R$ and $T(S_i), T(S_j) \in \{\alpha, \beta\}$. R^2_{ij} denotes a binary topological relationship ($S_i \odot S_j$) between SSEs S_i and S_j. The database terminology R^2 without subscripts is a binary topological relation over protein P, which is a set of binary topological relationships.

Most binary relationships are symmetric, but some are asymmetric. An asymmetric binary topological relationship, $R^2_{ij} = \{S_i \odot S_j\}$ is *forward* if $i < j$ and *backward* if $i > j$.

2.3. Notation of A Topology String and TStringSet

We represent our topological relationships by strings, called *topology strings* in order to reduce the complexity of the comparison problem, which is to find similar topological relationships between protein structures. A *topology string* is represented by a string of 3 characters, where:

- The alphabet of the first character denotes the type of topology relationship described in Table 1 (e.g. A, B, D, F, E, I, K, H, N).
- The second character encodes the order of the SSEs participating in a topology relationship. Sequential order, denoted by 's', means that SSE are adjacent along the backbone, whilst 'n' refers to neighboring order, i.e. one SSE is far from another in sequence (e.g. Asf or Anf).
- The third character denotes the direction of a topological relationship i.e. forward ('f') and backward ('b').

For instance, the string 'Anf' describes the Helix$\odot_{overlaps}$Helix topological relationship between two SSEs which are far from each other in the order of sequence, there being other SSEs in between the starting and ending helices in the relationship.

A protein structure is described by a (non-empty) set of topology strings, which we term a *TStringSet*. Each *topology string* together with its attributes are stored in a tuple. Our topology description can denote strand-strand and helix-helix relationships as well as helix-strand relationships that the current version of TOPS cannot represent.

2.4. Examples

Figure 3 illustrates examples of our representation of proteins and that of TOPS. SSE numbers and the distance between them are added to our topology strings. We observe that our string representation provides richer information about helix-helix interactions in protein 2hbg00 compared with TOPS.

3 Similarity Search

3.1. Topology String Match

A topology string match, i.e. matching a pair of SSE pairs, is the basic unit operation in this structure comparison. It evaluates the similarity between two topology strings.

[Definition 2] (a topology string match) The approximate string match function Match(a_i, b_j) returns string match score S_{ij} between given topology strings a_i and b_j..

Given a pair of topology strings (a_i, b_j), where $\{S_0, S_p\} \in a_i$, $\{S_q, S_r\} \in b_j$, and the topological relationships $R_{op} = (S_o \odot S_p)$ for the topology string a_i and $R_{qr} = (S_q \odot S_r)$ for b_j

Match(a_i , b_j) iff:

(i) $a_i = b_j$ if $T(S_o) = T(S_q)$, $T(S_p) = T(S_r)$ and $R_{op} = R_{qr}$

(ii) $0 \leq |D(a_i) - D(b_j)| \leq \sigma$

(iii) $0 \leq |L(a_i) - L(b_j)| \leq \varepsilon$

(iv) $Dir(a_i) = Dir(b_j)$ (1)

where,

σ: the tolerance for distance difference between the matched strings a_i, b_j ($0 < \sigma < 4$)

$D(a_i)$: distance between SSE S_0 and SSE S_p in a topology string a_i

$D(b_j)$: distance between SSE S_q and SSE S_r in a topology string b_j

ε: difference between the SSE length ratios for a pair of topology string a_i and b_j ($0 < \varepsilon < 0.5$)

$L(a_i)$: the length ratio of SSEs joined in a topology string a_i

$L(b_j)$: the length ratio of SSEs joined in a topology string b_j

$$L(a_i) = MinLen(S_0, S_p)/MaxLen(S_0, S_p) \text{ and } L(b_j) = MinLen(S_q, S_r)/MaxLen(S_q, S_r)$$

$MinLen(S_0, S_p)$: The smaller length of SSEs S_0 and S_p in a topology string a_i

$MaxLen(S_0, S_p)$: The greater length of SSEs S_0 and S_p in a topology string a_i

$MinLen(S_q, S_r)$: The smaller length of SSEs S_q and S_r in a topology string b_j

$MaxLen(S_q, S_r)$: The greater length of SSEs S_q and S_r in a topology string b_j

$Dir(a_i)$: a relative direction {P, A} of SSE S_0 and S_p in a topology string a_i

$Dir(b_j)$: a relative direction {P, A} of SSE S_q and S_r in a topology string b_j

The distance decides the degree of overlap between SSEs in a pair of topology strings. A low distance difference indicates a high degree of similarity between two pairs of SSEs.

In our experiments we used values of 2 for σ and 0.3 for ε, which are the default for these parameters when measuring the similarity between two proteins.

3.2. Pairwise Comparison

The similarity of a pair of protein structures in our representation is measured by the similarity between their corresponding *TStringSets*. A pairwise operation is employed to find the maximum subsets of matched pairs of topology strings between two sets of topology strings.

[Definition 3] (pairwise comparison: matching of two *TStringSets*) Given two sets of topology strings (*TStringSets*) for a query protein Q, Q= $\{a_1, \ldots, a_n\}$ and a target protein T in database D, T =$\{b_1, \ldots, b_m\}$, pairwise comparison S(Q, T) is an operation to find the maximum number of weighted matching pairs of topology strings $S(Q, T) = \{(a_1^1, b_1^1), (a_2^2, b_2^2), \ldots, (a_i^k, b_j^k)\}$, $1 \le k < n, m$ and returns the associated similarity score S.

The pairwise comparison problem is transformed into a maximum matching problem in a bipartite graph. Given the topology string sets Q and T, a directed weighted bipartite graph G = (V, E) can be constructed as follows: $V = V_Q \cup V_T$, $E = \{e_{ij}\}$. Each edge e_{ij} (i.e. $i = 1,2,\ldots, n$; $j = 1,2, \ldots, m$) corresponds to a weighted link between a_i and b_j, whose weight $w(e_{ij})$ is equivalent to the similarity S_{ij} between topology string a_i , b_j. A graph is bipartite if it has two kinds of vertices and edges are only permitted between vertices of different kinds. Thus the graph created is a weighted bipartite graph.

Our similarity score function for pairwise comparison is based on a compression measure over topology strings. It is calculated by the sum of topology string matching scores as follows:

[Definition 4] (Similarity score) The similarity score *S* for two sets of the topology strings for a query protein Q=$\{a_1, a_2, \ldots, a_i, \ldots, a_n\}$ and a target protein T=$\{b_1, b_2, \ldots, b_j, \ldots, b_m\}$, is computed by Eq. (2).

$$S = \sum_{k=1}^{n} \sum_{l=1}^{m} \left(\frac{S_{kl}}{Max(|Q|, |T|)} \right) \cdot (\log_e |M^{QT}|) \tag{2}$$

iff Match(k, l) and $|M^{QT}| \geq 2$

where,

k: a topology string in the query protein Q, i.e. $k \in \{a_1, a_2, \ldots, a_i, \ldots, a_n\}$

l: a topology string in the target protein T, i.e. $l \in \{b_1, b_2, \ldots, b_j, \ldots, b_m\}$

$|Q|$: total number of topology strings in a query protein Q

$|T|$: total number of topology strings in a target protein T in a database

$|M^{QT}|$: number of the matched topology strings between proteins Q and T

S_{kl}: The topology string match score S_{kl} for topology strings $k \in$ Q and $l \in$ T

High structural similarity scores do not always imply high significance of an alignment. Thus, we use the log ratio of the number of the matched strings in order to assign less weight for low numbers of matched topology strings. The best known strongly polynomial time bound algorithm for weighted bipartite matching is the classical

Hungarian method due to Kuhn [12], which runs in time $O(|V|(|E| + |V|)\log|V|)$. Weighted bipartite matching algorithms can be implemented efficiently, and can be applied to graphs of reasonably large size. The pairwise comparison can run in time $O(n(m + n\log n))$, where a bipartite matching graph $G = (V_Q \cup V_T, E)$, $n = |V_Q \cup V_T|$, $m = |E|$, n is number of vertex in query protein Q and a protein T in a database, m is the number of edges and the maximum edge capacity is 1. Our pairwise comparison algorithm can be solved in polynomial time.

4 Experiments

4.1. Datasets

In order to permit a large scale analysis, a protein 3D structure database was constructed from a subset of domains in the SCOP PDB40 [13] release 1.61 representative dataset. This release of the PDB40 contains proteins no two of which share >40 % sequence similarity, a total of 4,420 domains in the all-α, all-β, α/β, and α+β SCOP classes. We selected the subset of entries for which we had sufficient data to construct our models, a total of 2,654 domains.

Table 2. Data distribution in PDB 40

SCOP Class	Fold Family	DOM	HDP	% of HDP	NHDP	% of NHDP
All α	171	415	1,742	2.12	82,238	97.88
All β	124	457	5,074	5.35	94,810	94.65
α/β	172	904	10,104	2.56	395,008	97.44
α+β	284	878	2,609	0.69	376,983	99.31
Total	751	2,654	19,529	2.02	949,039	97.98

DOM: Number of unique protein domain entries
HDP: Number of homologous domain pairs
NHDP: Number of non-homologous domain pairs

Table 3. Fold Superfamily in Chew and Kedem

Supferfamily	Fold Class	Num. of DOM
1.1.1	Globin Like	16
1.107.1	All a	1
1.34.1	All a	1
2.1.1	All b	7
3.1.11	TIM barrels	3
3.1.15	TIM barrels	1
3.25.1	a/b	5
3.26.2	a/b	1

Our data set selected from SCOP PDB40 contains 751 SCOP fold superfamilies (Table 2); the distribution of SCOP fold superfamilies in our data set is diverse. We generated a set of all against all domain pairs for our 2,654 unique domain entries and selected a subset for evaluation of ~1M pairs comprising those pairs for which both domains belong to the same SCOP class. We also tested our method against the data set of Chew and Kedem [14] for a smaller scale evaluation of its accuracy. Table 3 shows 35 domain entries in eight different SCOP superfamilies of the three SCOP classes. Experimental data sets are available at http://www.brc.dcs.gla.ac.uk/~shpark/pcom/.

4.2. Evaluation

We compared our results with the SCOP classification hierarchy. We have constructed ROC curves and calculated the corresponding AUC (Area Under ROC Curves) values over different SCOP classes (Figure 4) for both data sets. To construct the ROC curve, we define that two domains are defined as homologous if at least their first three SCOP numbers are identical; the domains are non-homologous if only their first SCOP numbers are identical. We did not include protein domain pairs for which only the first two levels of their SCOP numbers match since the SCOP classification does not differentiate between homologous and non-homologous pairs at this fold level.

Chew & Kedem data sets: We evaluated the accuracy of our method by performing a one against all comparison for each of the 35 domains; the average AUC value for this is 87.93%. Figure 4(a-c) gives ROC curves for three illustrative one against all domain comparisons. Table 4 shows the results for our method regarding the coverage of the correct fold superfamily at the top of our resulting lists sorted by comparison score for those superfamilies that had more than one domain member. We repeated the same experiment for TOPS (Table 5). Overall, our system consistently finds the correct fold superfamily within the top 3 positions in the list with 85.7 % of certainty. Regarding the test for major SCOP fold classes, our algorithm performs better than TOPS for all-α proteins and the α/β Rossman folds (family 3.25.1). TOPS performs better for the all-β class and for the 3.1.11 α/β fold family (TIM barrels).

Table 4. Coverage of top K in our result lists

fold family	AUC value[a] (%)	Top K[b]=3		Top K[b] =4		Top K[b] =5	
		TP[c]	FP[d]	TP[c]	FP[d]	TP[c]	FP[d]
1.1.1	88.1	2.9	0.1	3.82	0.18	4.8	0.2
2.1.1	84.6	2.4	0.57	3	1	3.6	1.4
3.1.11	85.4	1.6	1.3	2	1	2	1
3.25.1	84.7	2.8	0.2	3.8	0.2	3.8	0.2

Table 5. Coverage of top K in TOPS

Fold family	AUC value[a] (%)	Top K[b]=3		Top K[b] =4		Top K[b] =5	
		TP[c]	FP[d]	TP[c]	FP[d]	TP[c]	FP[d]
1.1.1	82.15	2.9	0.1	3.8	0.2	4.8	0.2
2.1.1	97.85	3	0	4	0	5	0
3.1.11	94.58	2	1	2.3	0.7	2.6	0.4
3.25.1	81.9	2.6	0.4	3.2	0.8	3.2	0.8

[a]: Average AUC value in a corresponding family for one against all search
[b]: Number of top K domain in result lists of similarity score.
[c]: Number of average times ranks true positives that is homologous pairs ahead of non-homolog pairs
[d]: Number of average times ranks false positives

Our explanation is that since our topology strings encode geometrical features such as the distance between SSEs, and helix-helix and helix-strand interactions, our comparison performed well for the all-α class in the experiments. However, TOPS strings mainly encode β-strand connectivity [8]. Therefore TOPS works well for superfamilies in the all-β class as well as for those in the α/β class whose structures are classified according to β sheet topology (e.g. TIM barrels).

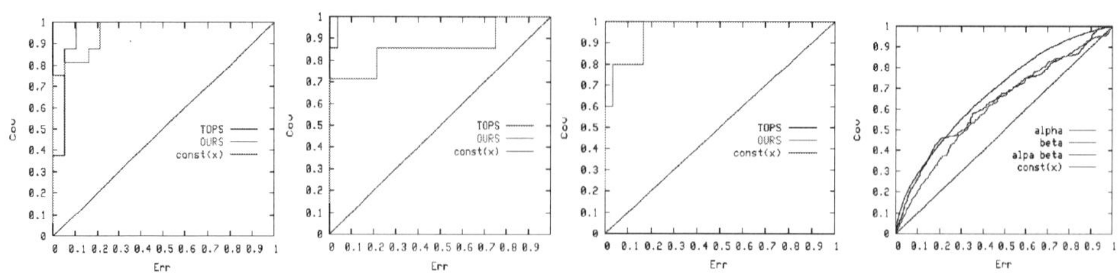

(a) 2hbg00 (1.1.1) (b) 1cdb00 (2.1.1) (c) 1aa900 (3.25.1) (d) SCOP PDB40

Figure 4 . ROC Curves

Subset of SCOP PDB40: The AUC values for α, β, and mixed α β classes are 68.5%, 62.6%, and 69.4% respectively, with an overall average of 66.83%. The corresponding ROC curves are shown in Figure 4(d). The reason that the accuracy is lower than for the Chew & Kedem data set is related to the distribution of the SCOP PDB40 data, which is extremely diverse and contains a very low number of homologous pairs.

4.3. Performance Analysis

Our similarity search algorithms ran on an IBM PC with 3GHz CPU and 2GBytes memory. We used the database mentioned above as our target database which was built on ORACLE 9*i* DBMS by using spatial types and topological operators running on a spatial index.

The ~1M pairwise comparisons took 10hr and 17 min and all against all for 35 domain entries took 8.1 sec. The average execution time for pairwise comparisons ranged from 12.17 ms in the small scale analysis to 38.45 ms in the large scale analysis. Although our method has a very fast comparison time, it took 1 hour and 20 minutes to represent the geometry of SSEs using spatial types and build an index over 2,645 protein domain having 34, 114 SSEs (Table 6). According to our experiments, it takes on average 1.08 sec to build the topology strings for one protein. Most of this time is in the discovery of binary topology strings.

Table 6. Execution time for computational task

Task name	Num. of Data	Total exe.time	AVG exe. Time
SSE representation	2,654 dom. ent.	1hr 20 m	1.8 sec
Generation of topology string	2,654 dom. ent.	1hr 05m	1.5 sec
All against all search	35 dom ent.	8.1 sec	12.17ms
pairwise comparison	968,568 dom. pairs	10hr17m	38.45ms

As shown in Table 6, the computation time for the identification of topological properties based on geometrical features is expensive to run. However, since the database of topology strings is pre-computed, the time required for the discovery of topology strings does not affect the execution time for structural similarity search. The matching of two topology string sets is performed in polynomial time.

5 Conclusions

We have developed a new fast similarity search based on topology string matching. We used a constrained string match algorithm for the comparison of linear topology strings encoding the non-linear topological relationships. Our method could be used as a filtering step prior to slower but more accurate similarity search and fold discovery methods based on all-atom approaches in a pipeline for automated structure classification.

Our results indicate that our algorithm performs reasonably well for structures with all α-helices and a mixture of α and β SSEs, but does not work so well for all-β proteins. The overall accuracy of our system is comparable to that of TOPS, which is a fast and accurate structure comparison method based on abstraction over SSE topologies. The weakness of TOPS is that it does not perform well on α class members; our method is superior in such cases.

Compared with existing methods, our method uses spatial characteristics of protein structures that are represented by using an existing spatial DBMS. The application of spatial databases in bioinformatics has many advantages; one can analyze and represent multi-dimensional information with spatial types, multidimensional index, and spatial operations that are implemented using algorithms from computational geometry.

We observe that the use of index structures will facilitate the retrieval of biological data, and extracted patterns from source databases. In future research, we will work on the development of index based similarity search over topology strings using spatial indexes.

Acknowledgements

We thank Mallika Veeramalai of the Bioinformatics Research Centre at the University of Glasgow and Juris Viksna of the University of Latvia for helpful discussions about use of bipartite graph matching.

References

1. Goldman,D. (2000) Algorithmic Aspects of Protein Folding and Protein Structure Similarity, PhD Thesis, Department of Computer Sciences, UC Berkeley.
2. Harrison, A., Pearl, F., Sillitoe, I., Slidel, T., Mott, R., Thornton J. and Orengo, C. (2003) Recognizing the fold of a protein structure, J. of Bioi., 19, 748.
3. Krissinel. E and Henrick. K. (2004) Secondary structure matching, a new tool for fast protein structure alignment in three dimentions, Crystallographica Section D Biological Crystallography, 60, 2258.
4. Holm, L. and Sander, C. (1993) Proein structure comparison by alignment of distance matrices, J. of Mol. Biol., 233, 123.
5. Orengo, C. A. and Taylor, W.R. (1996) SSAP: sequential structure alignment program for protein structure comparison, J. of Meth. in Enzy., 266, 617-635.
6. Gibrat, J-F., Madej, T. and Bryant, H. (1996) Surprising similarities in structure comparison, Curr.Opin.inStru.Bio., 6, 377.
7. Singh, A.P.and Brutlag, D.L. (1997) Hierarchical protein structure superposition using both secondary structure and atomic representations. Proc. Int. Conf. Intell. Syst. Mol. Biol. 5, 284.
8. D. Gilbert, D. R. Westhead, J. Viksna, and J. M. Thornton. A computer system to perform structure comparison using TOPS representations of protein structure. J.of Comput. Chem., 26:23-30, 2001.
9. Laurini, R.and Thompson, D. (1992) Fundamentals of Spatial Information Systems, Academic Press Ltd.
10. Egenhofer, M., Frank, A. and Jackson, J. (1989) A topological data model for spatial databases. In Proceeding of the Symposium on the Design and Implementation of Large Spatial Databases '89, Springer-Verlag, LNCS 409, 271.
11. Park, S.H., Ryu, K. H. and Gilbert, D. (2005) Fast Similarity Search for 3D Protein Structures using Topological Pattern Matching based on Spatial Relations IJNS, 15(4), 287.

12. Kuhn, W., (1955) The hungarian method for the assignment problem, Naval Research Logistics Quarterly, 2, 83.
13. Murzin, A.G., Brenner, S.E., Hubbard, T. and Chothia, C. (1995) SCOP: a structural classification of proteins database for the investigation of sequences and structures. J. Mol. Bio., 247, 536.
14. Chew,L.P. and Kedem, K. (2004) Finding consensus shape for a protein family. Algorithmica, 38, 115-129.

SIMPLE AND FAST ALIGNMENT OF METABOLIC PATHWAYS BY EXPLOITING LOCAL DIVERSITY

SEBASTIAN WERNICKE AND FLORIAN RASCHE

Institut für Informatik, Friedrich-Schiller-Universität Jena
Ernst-Abbe-Platz 2, D-07743 Jena, Germany
E-mail: {wernicke,m3rafl}@minet.uni-jena.de

An important tool for analyzing metabolic pathways is being able to do homology searches, that is, for a given pattern network one would like to find occurrences of similar (sub)networks within a set of host networks. In the context of metabolic pathways, Pinter et al. [Bioinformatics, 2005] recently proposed to solve this computationally hard problem by restricting it to the case where both the pattern and host network are trees. This restriction, however, severely limits the applicability of their algorithm. Here, we propose a novel algorithm that does not restrict the topology of the host or pattern network in any way; instead, we exploit a natural property of metabolic networks that we call "local diversity property," which allows us to obtain a very fast and simple algorithm for the alignment of metabolic pathways. Experiments on a testbed of metabolic pathways extracted from the BioCyc database indicate that our algorithm is much faster than the restricted algorithm of Pinter et al. and yet has a wider range of applicability and yields new biological insights.

1. Introduction

Motivation. Shifting attention from linear data to more complex functions and interactions, recent years have seen a surge in the availability of biological *network* data.[a] An important tool for analyzing these data is being able to search for homologous (sub)networks to a given pattern network: This promises to be useful, for example, for interaction predictions, functional annotation, data integration, knowledge transfers, and for developing a better understanding of biological network organization.[5,10] A recent survey by Sharan and Ideker[10] even advances the opinion that "network comparison techniques promise to take a leading role in bioinformatics [...]."

Unfortunately, the task of performing a network homology search turns out to be quite hard as it can be traced back to the NP-complete Subgraph Isomorphism problem.

> Subgraph Isomorphism
> **Input:** Two graphs G_P (the *pattern*) and G_H (the *host*).
> **Task:** Find whether G_H contains a subgraph that is isomorphic to G_P.

This problem is even NP-complete when restricted to graph classes that usually render NP-complete problems tractable—for instance, if the pattern is a forest and the host is a tree.[2] Polynomial-time algorithms are only known when the host graph has bounded treewidth ω

[a] We use the term "network" discussing biological aspects and the term "graph" for discussing algorithmic aspects.

and the pattern graph has either a high connectivity or bounded degree—in these cases, SUBGRAPH ISOMORPHISM can be solved in $O(n_P^{\omega+1} \cdot n_H)$ time for an n_P-vertex pattern and n_H-vertex host.[1,3,6]

To overcome the hardness of SUBGRAPH ISOMORPHISM when performing a network homology search on biological networks, various algorithms have been proposed:

- For protein interaction networks, Kelley et al.[5] presented an algorithm called PATHBLAST that, given a linear pathway as a query, randomly decomposes the host graph into linear pathways to find homologous pathways among these.
- For linear patterns, Shlomi et al.[11] proposed an algorithm that is based on random graph colorings.
- Pinter et al.[7] presented a homology search algorithm for metabolic pathways that is based on restricting the host and pattern graph to be trees, in which case polynomial-time algorithms are possible[8].

Notably, all of these algorithms basically take the same approach: They make network homology searches algorithmically feasible by restricting the topology of the network to be cycle-free. This, in turn, causes all of these algorithms to suffer from basically the same problems, namely at least one of the following three:

(1) *Limited Applicability.* Most biological networks of interest contain cycles and the proposed algorithms therefore cannot be directly applied to them.
(2) *Long Running Time.* To apply the existing algorithms to a network that contains cycles, the network must be decomposed in some way. Irrespective of whether these decompositions are randomized or deterministic, a great number of them is necessary in order to ensure that all good matches for the pattern are found. This usually leads to exponential running times that are only practical for very small pattern sizes (for example, PATHBLAST requires $O(\ell!)$ runs for a pattern of length ℓ).
(3) *Requirement of Manual Labor and Expert Knowledge.* One might choose to use expert knowledge and manually decompose networks into cycle-free subgraphs. Such an approach was chosen, for example, by Pinter et al.[7] to obtain some of the dataset for their pathway alignment tool (they also excluded some of the networks that have cycles).[9] It is clear that such a process is not always applicable (for example, when little information is known about a network beforehand or we have no idea what the result should roughly look like), tedious, and error prone.

Intriguingly, there is another thing that the existing network homology search algorithms have in common besides basically taking the same approach: They do not algorithmically expose the fact that vertices in biological networks are labeled; rather, the vertex labels are used only for scoring the similarity of the pattern graph to a given subgraph in the host. In the context of metabolic pathway alignment, this work proposes an approach that *does* expose vertex labels in order to obtain a network alignment algorithm that is simple, fast, and imposes no restrictions concerning the topology of the input networks.

Organization of this Work. After introducing some notation, Section 2 presents our new alignment algorithm for metabolic pathways in three steps: First, Subsection 2.1 formalizes our network alignment problem and presents a simple—yet impractical—algorithm for it called MATCH. Second, Subsection 2.2 introduces the local diversity property of metabolic networks which, third, is exploited in Subsection 2.3 by slightly modifying the MATCH algorithm so as to obtain our new metabolic pathway alignment algorithm FIT-MATCH. The FIT-MATCH algorithm has been implemented in C++; the source code is freely available at `http://theinf1.informatik.uni-jena.de/graphalignments/`. Section 3 reports experiments with our implementation on a testbed of metabolic pathways from the BioCyc database[4]. These indicate that our algorithm is much faster than the algorithm of Pinter et al.[7] and yet—because the topology of the input networks is not restricted—is simpler to use, yields new insights, and has a wider range of applicability.

2. A New Fast and Simple Pathway Alignment Algorithm

Before formalizing the problem of metabolic pathway alignment and discussing our new algorithm FIT-MATCH to solve this problem, it is useful to establish some notation:

Notation. We model metabolic pathways as connected directed graphs. Each vertex represents an enzyme and is labeled with the *Enzyme Commission number (EC number)* of that enzyme.[b] Two vertices u and v are connected by a directed edge (u, v) if a product of the pathway reaction catalyzed by u is a substrate of the reaction catalyzed by v. The *pattern* graph for which we seek a homolog is always denoted $G_P = (V_P, E_P)$, the *host* graph in which we seek an occurrence of the pattern graph is denoted $G_H = (V_H, E_H)$. A vertex with exactly one outgoing and one incoming edge (not counting self loops) is called a *path vertex*; all other vertices are called *branch vertices*.[c] A path in a graph that consists just of path vertices and where every vertex occurs at most once is called *simple*.

An *isomorphism* between two graphs $G = (V, E)$ and $G' = (V', E')$ is a one-to-one mapping $\Phi : V \to V'$ such that $(u, v) \in E \Leftrightarrow (\Phi(u), \Phi(v)) \in E'$ (note that this definition ignores the labels of the vertices). If there exists an isomorphism between two graphs, we call them *isomorphic*. Two graphs are called *homeomorphic* if we can subdivide their edges (that is, edges can be replaced by simple paths of arbitrary length in the same direction) in such a way that the resulting graphs are isomorphic. The corresponding *homeomorphism* is a function φ that bijectively maps the branch vertices of the two graphs onto each other.

2.1. *Formalization and a Simple Backtracking Algorithm*

In order to formalize the problem of metabolic pathway alignment, we must define two things, namely what we mean by "alignment" and what we view as a "high-scoring" align-

[b]The EC number is a four-level hierarchical scheme that classifies enzymes on a functional basis. Thus, each enzyme is classified by four numbers (as in "3.4.23.48"), the first number representing the top level classification and the three following numbers the subsequent refinements thereof.

[c]Although somewhat counterintuitive, to simplify the overall presentation we chose to use the term "branch vertex" also for vertices with degree one.

ment. Concerning a formalization of alignments, we follow Pinter et al.[7] and rely on a notion that is based on subgraph homeomorphism:

Definition 2.1. An *embedding* of a pattern graph G_P into a host graph G_H is a tuple (G'_H, φ) where G'_H is a subgraph of G_H that is homeomorphic to G_P and φ is a homeomorphism between G'_H and G_P.

We can use the notion of an embedding to phrase metabolic pathway alignment as a combinatorial problem called MAXIMUM-SCORE EMBEDDING.

MAXIMUM-SCORE EMBEDDING
Input: Two directed labeled graphs $G_P = (V_P, E_P)$ and $G_H = (V_H, E_H)$.
Task: Find the maximum-score embedding of G_P into G_H.

It remains to define the scoring scheme that we plug into this problem definition. Again, we follow Pinter et al.[7] and make use of a scoring scheme due to Tohsato et al.[12] that is based on mutual vertex–vertex similarities (observe that topological similarity is already ensured by relying on homeomorphisms). The similarity of two enzymes is calculated from their functional EC numbers—the more of this number two enzymes have in common, the more similar they are considered to be.[d] The scoring scheme also incorporates an information-theoretic consideration, namely that the similarity of two enzymes is more significant the less their common EC number prefix occurs among all enzymes.

Definition 2.2. Let the vertices u and v represent two enzymes e_1 and e_2, respectively. If the lowest common enzyme class of e_1 and e_2 as determined by their EC numbers contains h enzymes, then the *similarity* of u and v is defined as $sim(u, v) := -\log_2 h$.

Using the scoring for pairwise similarity, we can define a similarity score for two simple paths that is based on the notion of a sequence alignment.

Definition 2.3. Given two simple paths $p_1 = u_1 \ldots u_x$, $p_2 = v_1 \ldots v_y$ and a negative *gap penalty* g, their similarity $sim(p_1, p_2, g)$ is defined as the maximum possible score of a sequence alignment between p_1 and p_2 using g as the gap penalty and the scoring scheme of Definition 2.2 to evaluate pairwise similarities.

For the sake of simplicity in our presentation, let us assume from now on that there is at most one simple path between any two branch vertices and that neither the pattern nor the host is a simple cycle. Our implementation in Section 3 does not make any of these restrictions, but handling them explicitly in the remainder of this section obfuscates the main ideas. To render the scoring of an embedding precise, we use the following definition:

Definition 2.4. Given an embedding (G'_H, φ) of a pattern graph G_P in a host graph G_H, let $B(G_P)$ denote the branch vertices of G_P. For two branch vertices u and v let $p(u, v)$

---

[d]For some applications, a purely functional classification might be suspect and one might want to additionally include genetic similarity information for the enzymes; we do not consider this here, however.

be the simple path between u to v; if no such path exists, then $p(u,v)$ is the empty graph. Given a gap penalty $g < 0$, the *score* of (G'_H, φ) is defined as

$$\mathit{score}(G'_H, \varphi) := \sum_{v \in B(G_P)} \mathit{sim}(v, \varphi(v)) + \sum_{u,v \in B(G_P)} \mathit{sim}\big(\, p(u,v),\ p(\varphi(u),\ \varphi(v)\,)\,,\ g\,\big)\,.$$

Naïvely, MAXIMUM-SCORE EMBEDDING can be solved by a simple backtracking algorithm that exhaustively explores all possible embeddings of a given pattern graph G_P into a host graph G_H. Formally, this algorithm is best described by using the notions of a *partial embedding* and *extensions* thereof.

Definition 2.5. A *partial embedding* of a pattern graph G_P into a host graph G_H is an embedding of a connected subgraph G'_P of G_P into G_H. It is denoted by (G'_P, G'_H, φ) (where φ is the homeomorphism between G'_P and G'_H). Let p be a simple path in G_P that connects two branch vertices u and v such that at least one of these branch vertices is in G'_P but no path vertex of p. An *extension* of a partial embedding (G'_P, G'_H, φ) by p is a partial embedding of the subgraph induced in G_P by G'_P, u, v, and p that is identical to (G'_P, G'_H, φ) when restricted to the vertices of G'_P.

To illustrate the concept of a partial embedding and its extensions, consider the following example graphs G_P and G_H and a partial embedding of G_P into G_H:

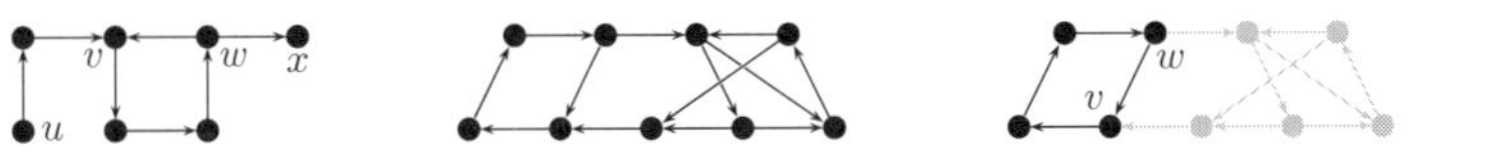

G_P $\qquad$ G_H $\qquad$ Partial embedding of G_P into G_H

The shown partial embedding has ten possible extensions by the path from u to v:

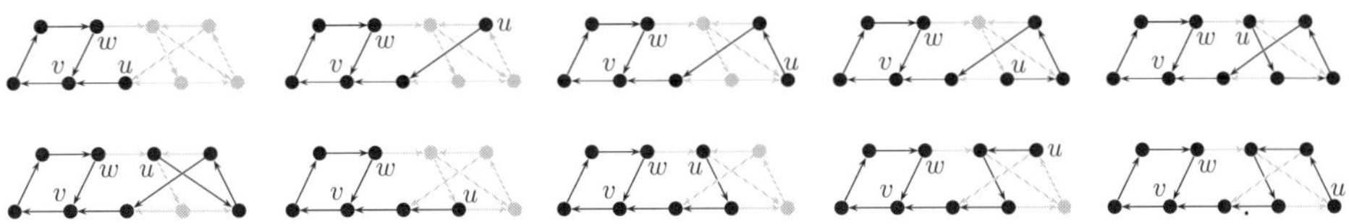

We can now describe our naïve backtracking algorithm for solving MAXIMUM-SCORE EMBEDDING. This algorithm, which we call MATCH, starts out by aligning a branch vertex of the pattern to a branch vertex in the host graph and then uses a recursive subprocedure EXTEND that takes as input a partial embedding and tries all possible extensions for it, thus enumerating all embeddings of the pattern graph into the host graph.

Algorithm: MATCH(G_P,G_H,g)
Input: Two labeled graphs $G_P = (V_G, E_G), G_H = (V_H, E_H)$ and a gap penalty g.
Output: A maximum-score embedding of G_P into G_H, if one exists.
Global variables: Graphs G_P and G_H, score *maxscore*, and embedding *best*.

```
best ← null ; maxscore ← −∞
u ← arbitrary vertex from V_G
for each v ∈ V_H do
    (G'_P, G'_H, φ) ← partial embedding by mapping u to v
    call EXTEND(G'_P, G'_H, φ)
return best
```

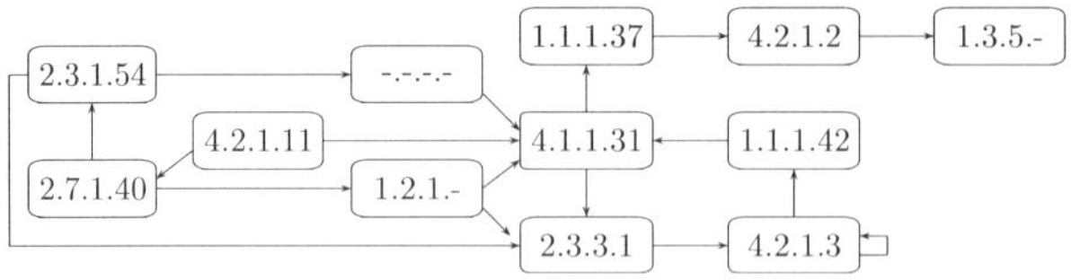

Figure 1. Anaerobic respiration pathway of *Escherichia coli* that illustrates the local diversity property. The label "-.-.-.-" denotes an unclassified enzyme.

```
EXTEND(G'_P, G'_H, φ)
 E1  if G'_P ≠ G_P then
 E2      p ← simple path in G_P not contained in G'_P such that at
             least one of the connected branch vertices is in G'
 E3      for each extension (G''_P, G''_H, φ') of (G'_P, G'_H, φ) by p do
 E4          call EXTEND(G''_P, G''_H, φ')
 E5  else if score(G'_P, G'_H, φ) > maxscore then
 E6      best ← (G'_P, G'_H, φ); maxscore ← score(G'_P, G'_H, φ)
 E7  return
```

Analysis of MATCH. The running time of MATCH is primarily determined by the number of recursive calls that are made in lines *05* and *E4* of the algorithm. While this number is upper-bounded by a constant—both the maximum path length and the maximum degree of a metabolic pathway are naturally bounded by some constant for biological reasons—it turns out to be rather large.[e] In our experiments, we have found that if the pattern graph consists of k simple paths, then the size of the search tree that is explored by MATCH is, on average, around 6^k. Considering that our dataset from the BIOCYC database contained a considerable amount of pathways with more than ten paths, this leads to a very long running time for MATCH.

2.2. *The Concept of Local Diversity*

As a typical example for a metabolic pathway, consider the anaerobic respiration pathway of *Escherichia coli* that is shown in Figure 1. The following observation can be made here which seems to hold for most metabolic pathways and is hence crucial to our approach:

Observation 2.1. Two paths that have the same starting vertex often carry out very different biological functions.

This observation describes what we refer to as the *local diversity property* of metabolic networks. There are plausible reasons why a metabolic network is expected to generally have this property: First, most metabolic products offer only very few possibilities where a certain reaction can chemically take place. Second, identical reactions for a certain substrate within a pathway are usually carried out by only one enzyme for reasons of efficiency.

[e] Note that *all* paths and not only simple paths in the host graph must be considered for an extension because a branch vertex in the host may become a path vertex in its subgraph.

Local diversity is an important property for the algorithmic alignment of metabolic pathways: Intuitively, SUBGRAPH ISOMORPHISM is hard because even very different graphs might appear similar based on local information. The local diversity property, however, means that metabolic pathways usually provide very rich and *diverse* local information that can be exploited to overcome this phenomenon.

2.3. *Exploiting Local Diversity*

When we compute all extensions of a partial embedding by a path p, some of these might not make sense from a biological perspective because the biological function of the pattern path p does not *fit* the biological function of the host path that it is aligned to. The key to making MATCH more efficient is to observe that the local diversity property implies that usually *a lot of* extensions of a partial embedding do not make sense from a biological perspective. Thus, to exploit local diversity and make MATCH more efficient, we need to devise a formal definition of "fitting biological function" for two given paths and then modify MATCH such that it only explores *fitting* embeddings.

Definition 2.6. Given a real number $0 \leq f \leq 1$, a gap score g, a simple x-vertex path p_1 and a simple y-vertex path p_2, we say that p_1 and p_2 *fit* if a maximum-score alignment between them aligns at most $\min\{\lceil(1-f)\cdot x\rceil, \lceil(1-f)\cdot y\rceil\}$ vertices to a gap. An extension of a partial embedding (G'_P, G'_H, φ) *fits* if every simple path between two branch vertices $u, v \in V'_G$ fits the corresponding simple path between $\varphi(u), \varphi(v) \in V'_H$.

As an illustration, if we have a fitting parameter of $f = 0.50$, then a four-vertex path fits no path that consists of seven or more vertices; a higher fitting parameter of $f = 0.75$ would cause it to fit no path that consists of six or more vertices.[f] To exploit local diversity, we now modify MATCH so that it only explores fitting embeddings. For this purpose, lines *05* and *E4* need to be modified so that the EXTEND-subprocedure is only called for fitting extensions. We name the resulting algorithm of this modification FIT-MATCH.

Analysis of FIT-MATCH Experiments show that, indeed, exploring only fitting extensions is a very effective pruning strategy due to the local diversity property of metabolic networks. More precisely, they show that whereas MATCH explored a search tree of size around 6^k to align a k-path pattern, even a conservative fitting parameter of $x = 0.5$ reduces this to around 2.5^k, "conservative" meaning that we found no meaningful alignment in our experiments that is missed by this setting.

3. Experiments on Metabolic Networks

We implemented FIT-MATCH in C++ to test its practical performance; the source is available at `http://theinf1.informatik.uni-jena.de/graphalignments/`.

[f]For some applications, Definition 2.6 might be considered too strict in its handling of very short paths: In particular, a one-vertex path never fits a length-3 path, regardless of the fitting parameter. While we have not found this property to be an issue in practice, one can easily circumvent it by introducing a minimum number of gaps that is always allowed regardless of the path lengths or fitting parameter.

Table 1. Runtimes of our FIT-MATCH implementation for all-against all alignments between the five datasets described in the text. For each combination of host and pattern, we show the total runtime including I/O overhead and excluding I/O overhead. All values are given in seconds.

Pattern	Run time of FIT-MATCH in seconds (including / excluding I/O overhead)				
	B. subtilis	*E. coli*	*H. sapiens*	*S. cerevisiae*	*T. thermophilus*
B. subtilis	82 / 0.41	120 / 2.25	102 / 2.25	95 / 0.29	147 / 2.28
E. coli	120 / 0.02	121 / 0.22	112 / 0.19	151 / 0.02	227 / 0.20
H. sapiens	107 / 0.02	120 / 0.19	89 / 0.20	130 / 0.02	190 / 0.29
S. cerevisiae	93 / 0.06	141 / 0.09	121 / 0.09	114 / 0.08	172 / 0.10
T. thermophilus	140 / 0.02	135 / 0.22	107 / 0.23	167 / 0.03	264 / 0.24

Method and Results. Our testing machine is an AMD Athlon64 3400+ with 2.4 GHz, 512 KB cache, and 1 GB main memory running under Debian GNU/Linux 3.1. Sources were compiled with the GNU g++ 4.2 compiler using the option "-O3".

To evaluate the performance of FIT-MATCH, metabolic pathways were extracted from the BioCyc database[4] for five different organisms, yielding 145 pathways of *B. subtilis*, 220 pathways of *E. coli*, 190 pathways of *H. sapiens*, 176 pathways of *S. cerevisiae*, and 267 pathways of *T. thermophilus*. If the full EC number of an enzyme was not specified, the unknown part of the code was treated as "don't care", meaning that the enzyme is scored as if it were identical to every enzyme for which the known part of the codes match. All 25 possible all-against-all inter- and intra-species alignments between the five datasets where performed, resulting in a total of 996 004 homology searches.

Following the suggestion of Pinter et al.[7] to set the gap score to about one third of the worst vertex–vertex similarity score, we set $g = -4.5$. The fitting parameter x was set to .50 as a conservative choice, meaning that we never encountered an interesting alignment that is only found with a lower fitting parameter in some preliminary experiments. The obtained runtimes are shown in Table 1; some sample alignments are shown in Figure 2.

Discussion. The experiments show that our FIT-MATCH implementation is capable of quickly aligning metabolic pathways; the complete dataset can be aligned in under an hour on our testing machine (including the I/O overhead, which turned out to consume far more time than the algorithm itself). This is much faster compared to the pathway alignment tool of Pinter et al.[7]: Their implementation (called MetaPathwayHunter) requires some hours alone to align the simplified trees of the *E. coli* and *S. cerevisiae* pathways whereas FIT-MATCH can align the corresponding unsimplified data in roughly seven minutes.

The alignments shown in Figure 2 exemplify some interesting application scenarios where FIT-MATCH can efficiently be used:

- *Pathway Comparison.* Figure 2a shows the highlighting of alternative metabolic pathways by comparing the classical TCA cycle with a more complex variant (note how the complex variant uses more pathways and the succinate dehydrogenases 1.3.99.1 instead of 1.3.5.1).
- *Enzyme Classification.* In Figure 2b, our results align all unclassified enzymes

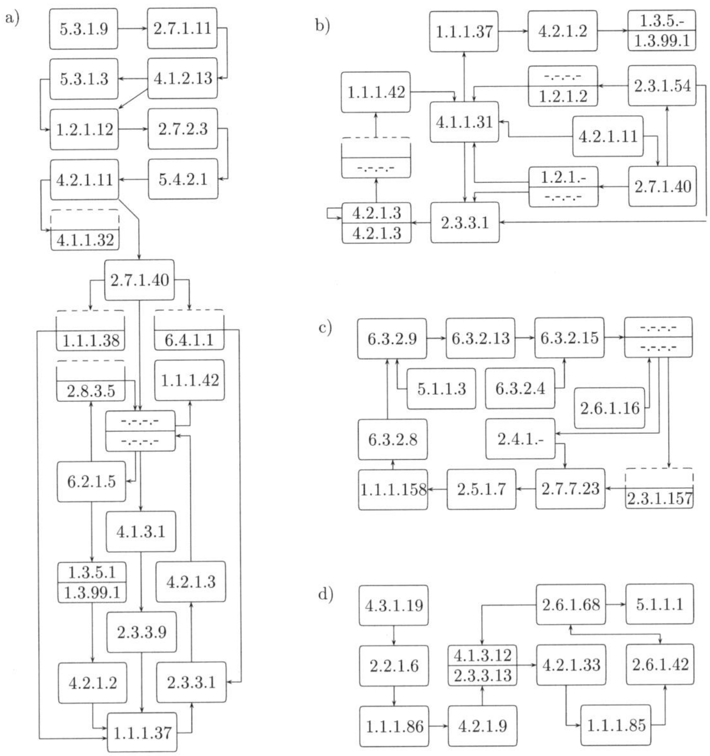

Figure 2. Four examples for the alignments that were found by the FIT-MATCH algorithm. In all graphs, the vertices are not split if they have the same label in the host and the pattern, otherwise, the pattern enzyme is shown at the top and the host enzyme at the bottom. A dashed top half indicates that a vertex is only present in the host graph. The four alignments (pattern/host) that are shown are **a)** superpathway of of glycolysis, pyruvate dehydrogenase, TCA, and glyoxylate bypass versus Embden-Meyerhof pathway in *B. subtilis* **b)** anaerobic respiration pathway of *E. coli* versus the same pathway in *B. subtilis* **c)** peptidoglycan and lipid A precursor biosynthesis in *B. subtilis* versus the same pathway in *T. thermophilus* **d)** superpathway of leucine, valine, and isoleucine biosynthesis in *E. coli* versus the same pathway in *T. thermophilus*.

(denoted "-.-.-.-") with already known enzymes, possibly hinting at their function.

- *Identifying Enzyme Complexes* The pathways shown in Figure 2c are almost identical, except that *B. subtilis* does not possess the enzyme 2.3.1.157 (an acyltransferase) but is rather aligned to a gap. The preceding enzyme is unclassified in both organisms. We can derive from the alignment that the unclassified enzyme in *B. subtilis* fulfills a task that requires two enzymes in T. thermophilus.
- *Data Integration.* Figure 2d shows an example where we can use FIT-MATCH to detect the consistency of a database: The two enzyme classification numbers that are seemingly totally different are the result of a change in nomenclature.

The results we found moreover demonstrate that the topological restrictions imposed by the algorithm of Pinter et al.[7] cause relevant alignments to be missed in several cases. For example, if the methylglyoxal pathway and the chorismate superpathway of *E. Coli* are aligned, MetaPathwayHunter does not produce any results whereas FIT-MATCH finds an alignment. Or, as a second example, MetaPathwayHunter misses the possible alignment between the cobalamin biosynthesis and the KDO_2 lipid biosynthesis superpathway of *E. coli* (which FIT-MATCH found).

4. Conclusion

We have presented the concept of local diversity for metabolic networks and shown how this property can be exploited to obtain a simple alignment algorithm FIT-MATCH for metabolic pathways that is both faster and more generally applicable than previous approaches. We are currently turning the FIT-MATCH implementation into a graphical tool for the discovery and analysis of metabolic pathway alignments.

All biological networks carry labels at their vertices. We think that the concept of local diversity is likely to occur in other types of biological networks than metabolic networks and could thus be exposed for alignment algorithms there, too. Given the nice properties of FIT-MATCH, this certainly seems worthwhile to investigate.

Acknowledgments. This work was supported by the Deutsche Telekom Stiftung (Sebastian Wernicke) and the Deutsche Forschungsgemeinschaft (DFG) project PEAL (parameterized complexity and exact algorithms, NI 369/1) (Florian Rasche). The authors are grateful to Rolf Niedermeier (Jena) for discussions and comments.

References

1. A. Dessmark, A. Lingas, and A. Proskurowski. Faster algorithms for subgraph isomorphism of k-connected partial k-trees. *Algorithmica*, 27(3):337–347, 2000.
2. D. S. Garey, M. R.and Johnson. *Computers and Intractability*. Freeman, 1979.
3. M. Hajiaghayi and N. Nishimura. Subgraph isomorphism, log-bounded fragmentation and graphs of (locally) bounded treewidth. In *Proceedings of 27th MFCS*, volume 2420 of *LNCS*, pages 305–318. Springer, 2002.
4. P. D. Karp, C. A. Ouzounis, C. Moore-Kochlacs, et al. Expansion of the BioCyc collection of pathway/genome databases to 160 genomes. *Nucleic Acids Research*, 19:6083–6089, 2005.
5. B. P. Kelley, B. Yuan, F. Lewitter, et al. PathBLAST: a tool for alignment of protein interaction networks. *Nucleic Acids Research*, 32(web server issue):83–88, 2004.
6. J. Matoušek and R. Thomas. On the complexity of finding iso- and other morphisms for partial k-trees. *Discrete Mathematics*, 108(1-3):343–364, 1992.
7. R. Y. Pinter, O. Rokhlenko, E. Y. Lotem, and M. Ziv-Ukelson. Alignment of metabolic pathways. *Bioinformatics*, 21(16):3401–3408, 2005.
8. R. Y. Pinter, O. Rokhlenko, D. Tsur, and M. Ziv-Ukelson. Approximate labelled subtree homeomorphism. In *Proceedings of 15th CPM*, volume 3109 of *LNCS*, pages 59–73. Springer, 2004.
9. O. Rokhlenko. Personal communication, 2006.
10. R. Sharan and T. Ideker. Modeling cellular machinery through biological network comparison. *Nature Biotechnology*, 24(4):427–433, 2006.
11. T. Shlomi, D. Segal, E. Ruppin, and R. Sharan. QPath: a method for querying pathways in a protein–protein interaction network. *BMC Bioinformatics*, 7:199, 2006.
12. Y. Tohsato, H. Matsuda, and A. Hashimoto. A multiple alignment algorithm for metabolic pathway analysis using enzyme hierarchy. In *Proceedings of 8th ISMB*, pages 376–383, 2000.

COMBINING N-GRAMS AND ALIGNMENT IN G-PROTEIN COUPLING SPECIFICITY PREDICTION

BETTY YEE MAN CHENG[†]

Language Technologies Institute, School of Computer Science, Carnegie Mellon University, 5000 Forbes Ave., Pittsburgh, PA 15213, USA

JAIME G. CARBONELL

Language Technologies Institute, School of Computer Science, Carnegie Mellon University, 5000 Forbes Ave., Pittsburgh, PA 15213, USA

G-protein coupled receptors (GPCR) interact with G-proteins to regulate much of the cell's response to external stimuli; abnormalities in which cause numerous diseases. We developed a new method to predict the families of G-proteins with which it interacts, given its residue sequence. We combine both alignment and n-gram features. The former captures long-range interactions but assumes the linear ordering of conserved segments is preserved. The latter makes no such assumption but cannot capture long-range interactions. By combining alignment and n-gram features, and using the entire GPCR sequence (instead of intracellular regions alone, as was done by others), our method outperformed the current state-of-the-art in precision, recall and F1, attaining 0.753 in F1 and 0.796 in accuracy on the PTbase 2004 dataset. Moreover, analysis of our results shows that the majority of coupling specificity information lies in the beginning of the 2nd intracellular loop and over the length of the 3rd.

1 Introduction

G-protein coupled receptors (GPCR) are a diverse superfamily of proteins characterized by their structure of 7 transmembrane alpha helices separated by alternating intracellular and extracellular loops. They are responsible for signal transduction across the cell membrane and are the targets of 60% of all drugs[1]. Their extracellular domains are capable of recognizing a wide range of ligands such as ions, hormones and neurotransmitters. The binding of these ligands causes the receptors to change their conformation, particularly in their intracellular domains exposing sites critical for the subsequent coupling with specific G-proteins.

G-proteins consist of α-subunits bound to βγ complexes and are classified into 4 families by their α-subunits: $G_{i/o}$, $G_{q/11}$, G_s and $G_{12/13}$. G_s and $G_{i/o}$ activates and inhibits adenylyl cyclase respectively, while $G_{q/11}$ activates phospholipase C. The function of the last family, $G_{12/13}$, remains unknown. Through their coupling with G-proteins, GPCRs regulate the cell's response to external stimuli. Abnormalities in this regulation lead to numerous diseases.

To date, most of the known G-protein coupling specificity information has been obtained through experimental approaches, a survey of which can be found in [2]. An

[†] To whom correspondence should be addressed: ymcheng@cs.cmu.edu

accurate method to predict the G-protein families a given GPCR sequence can couple with is of immense value to pharmaceutical research for three reasons. First, the development of such a method can elucidate the physiological mechanisms underlying the response mediated by a GPCR in diseases. Second, the coupling specificity of a GPCR is needed to identify its activating ligands because the appropriate G-protein needs to be present in the cell while one passes potential ligands (tissue extracts or libraries of chemical compounds) over the cells and watches for the suitable response. Finally, information learned in a study on coupling specificity prediction is likely to be applicable to the more general problem of protein-protein interaction prediction as well.

To the best of our knowledge, there have been 4 previous studies on predicting G-protein coupling specificity from the receptor sequence, although only one of them[3] considered and validated their approach on receptors coupling to multiple families of G-proteins, as we do in our study. Each of these studies focused on the intracellular domains of the receptor, using either alignment information[3, 4], n-grams[5] or physiochemical properties of the amino acids[6]. Alignment-oriented approaches have been very popular in computational biology. They utilize biological domain knowledge via the use of amino acid similarity matrices and account for some long-range interactions but they have inherent limitations due to their assumption that contiguity of homologous segments is conserved[7]. This assumption contradicts the genetic reshuffling and recombination that occurs through evolution[8, 9], and as a result, sequence alignments become unreliable when sequences have less than 40% similarity[10] and are unusable below 20% similarity[11, 12].

Moreover, since protein-protein interactions occur in 3-d space, only the orientation of the motifs for coupling specificity need to be conserved for the interaction to occur, and not the ordering of the motifs in the linear sequence. N-grams have the potential to capture the presence and absence of coupling specificity motifs (but not their 3-d orientation) without imposing a restriction on their ordering in the primary sequence. However, since the dimension of the n-gram feature space increases exponentially with the length of the n-gram, n-grams tend to be short and do not account for long-range interactions. Hence, with the complimentary pros and cons of sequence alignment and n-grams in mind, we developed a new G-protein coupling specificity prediction method that uses both sequence alignment and n-grams.

All previous studies have used only the intracellular domains, ignoring the extracellular and transmembrane domains as those do not make physical contact with G-proteins. While it is intuitive that the majority of coupling specificity information would lie in the intracellular regions, our preliminary results showed that including the extracellular and transmembrane regions can improve the prediction accuracy because the predicted transmembrane boundary positions are not entirely accurate and there may be relayed effects in these regions from the GPCR becoming activated by the ligand, as shown in our preliminary experiments (data not shown). Thus, our approach made use of whole sequence alignment and n-grams extracted from the whole sequence. In addition, we explored which areas within the intracellular domains contain the most discriminative coupling specificity information.

2 Methods

To address prediction of coupling to multiple families of G-proteins, we defined the coupling specificity problem as a set of 3 binary classification problems, one for each G-protein family to determine whether the given receptor couples to proteins from the family. We developed an n-gram based prediction module and an alignment based prediction module, each outputting a probability of the GPCR coupling to the particular G-protein family. Each module can be used independently to make a prediction by setting a probability threshold above which a coupling is predicted to occur. Alternatively, the modules can be combined, as we have in our hybrid prediction method, to utilize both n-gram and alignment information in making the prediction.

2.1 N-gram Based Module

The approach in our n-gram based module is analogous to the bag-of-words approach to document classification in the language technologies domain and has been successfully applied to GPCR family and subfamily classification[13]. Each GPCR sequence is represented as a vector of n-gram counts where unigrams, bigrams, trigrams and tetra-grams are extracted at each reading frame from the whole sequence. Preliminary results showed n-grams from the whole sequence yielded more accurate predictions than n-grams from the intracellular domains alone (data not shown). We used a 21-letter alphabet, 20 for the known amino acids and 1 for amino acid x, giving us a vector length of 204204.

A high dimensional feature space can confuse a classifier with irrelevant features. By using only features that are informative to the task, we can optimize our prediction accuracy while reducing the running time. Various feature selection methods have been developed for this purpose in the machine learning field, such as information gain, mutual information and chi-square. For each G-protein family C, we employed chi-square feature selection (section 2.1.1) to derive the p most discriminative binary features from n-gram counts in differentiating between receptors that can and those that cannot couple with proteins in C. The feature vectors were then converted to vectors containing only those p features and a k-nearest neighbors classifier (k-NN) was applied on them.

To predict whether a given sequence d_i couples to proteins from G-protein family C, the k-NN classifier finds the k "closest" sequences to d_i as defined by the normalized Euclidean distance between their feature vectors and computes their majority vote weighted by their inverse distance to d_i. The normalized Euclidean distance between 2 vectors $\vec{u}$ and $\vec{v}$ is defined as follows:

$$d(\bar{u},\bar{v}) = \left[\sum_i (\tilde{u}_i - \tilde{v}_i)^2\right]^{1/2} \text{ where } \tilde{u}_i = \frac{u_i - \min(i)}{\max(i) - \min(i)},\ \tilde{v}_i = \frac{v_i - \min(i)}{\max(i) - \min(i)} \quad (1)$$

and min(i) and max(i) are the minimum and maximum observed values of the ith attribute. We attempted 3 weighting functions, uniform weighting, 1–distance and inverse distance, and found inverse distance to be the best (data not shown). The score of the vote is then normalized to lie between 0 and 1 to yield the probability of d_i coupling to proteins in C.

2.1.1 Chi-square feature selection

The chi-square statistic measures the dependence between a given binary feature x and a classification category c. We chose to use chi-square in our study because it is one of the most effective feature selection methods in text classification[14] and because it has been successfully applied in GPCR subfamily classification[13].

In our task, the classification category c is the group of GPCRs which can couple to proteins in G-protein family C. Twenty binary features x are derived from each n-gram count by considering whether the n-gram has occurred at least *i* times in the sequence, where i = 5, 10 ... 100 for unigrams and i = 1, 2 ... 20 for all other n-grams. We computed the chi-square statistic for each feature x as the normalized square of the difference between the "expected" e(c,x) and observed o(c,x) number of objects in c with feature x. The "expected" number is the number of instances in c with feature x if x had a uniform distribution over all categories. Thus, the formula for chi-square statistic is

$$\chi^2(x,c) = \frac{[e(c,x) - o(c,x)]^2}{e(c,x)} \quad \text{where} \quad e(c,x) = \#\text{GPCRs in } c \times \frac{\#\text{GPCRs having } x}{\#\text{GPCRs in dataset}} \tag{2}$$

Next, for each n-gram j, we found the value i_{max} such that the binary feature x_j^* of having at least i_{max} occurrences of j has the highest chi-square statistic out of the 20 derived binary features associated with j. The n-grams were then sorted in decreasing order by the chi-square value of their respective x_j^*. The top p n-grams were selected where p is tuned from data, and each feature vector was transformed into one of length p where the components were the derived binary features x_j^*.

2.2 Alignment Based Module

Like our n-gram based module, our alignment based module utilized the k-NN classifier, with the bit score from Basic Local Alignment Search Tool (BLAST)[15] as its similarity metric. Given a test sequence d_i, we retrieved the top k training set sequences with the highest BLAST bit score. Unlike previous studies which used only the intracellular domains, we used the alignment of the whole receptor sequence, since other parts of the sequence prove to be informative as well. The probability of the GPCR sequence d_i coupling to proteins from G-protein family C was computed as the fraction of retrieved sequences which couple to proteins in C.

2.3 Hybrid Prediction Method

Our hybrid prediction method combines n-gram and alignment information in making the coupling specificity prediction by utilizing the probabilities from both the n-gram based module and the alignment-based module. Given a GPCR sequence and a G-protein family C, the method predicts the receptor to couple to proteins in C if either the probability of the interaction occurring computed by the alignment-based module is 1 or if the probability computed by the n-gram based module is above the trained threshold.

3 Data

We compared the performance of our approach to the current state-of-the-art[3] on their own dataset derived from the *2001 Trends in Pharmaceutical Sciences (TiPS) Nomenclature Supplement*[16] with added sequences from the authors of the study. We were able to replicate the entire test set but only 81% of the training set. The test set sequences had 49.4% sequence identity on average with the most similar training set sequence. We used WEKA[17] implementation of k-NN in our n-gram based module.

In addition, we assessed our hybrid method in a ten-fold cross validation and performed feature analysis to determine the location of coupling specificity information in the GPCR sequence on a more recent dataset, the Pharmacological Targets Database (PTbase)[18]. We used only the human sequences from PTbase which includes all the unique GPCRs in the database and yields test sets having 54.7% sequence identity on average with the most similar training set sequence. While the PTbase dataset contains significantly more sequences than the dataset from [3], it is not a superset of the latter. To test robustness, we also evaluated our method on 2 subsets of PTbase dataset, removing either the 8 sequences having higher than 75% sequence identity or the 4 sequences having higher than 80% with any training set sequence. In working with PTbase, we used an implementation of k-NN designed for sparse datasets developed by Paul Bennett.

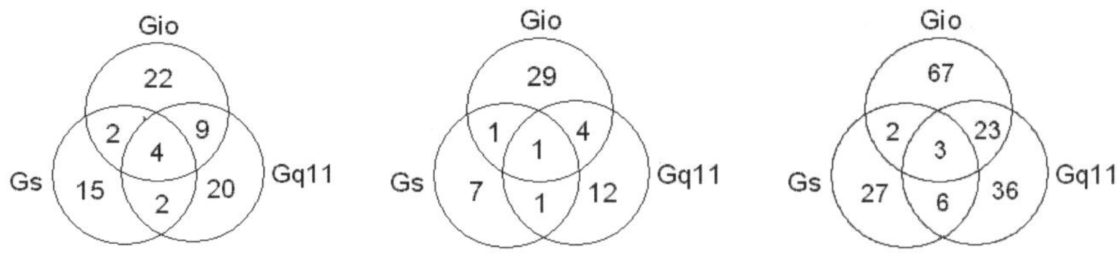

Figure 1. Distribution of training (left) and test (middle) sets from [3] and PTbase 2004 dataset (right).

4 Results and Discussion

4.1 Comparison to Current State-of-the-Art

Using the dataset from the current state-of-the-art[3] study, we assessed our n-gram and alignment based modules independently and as part of our hybrid prediction method to explore whether any advantage was gained by combining the two types of information. Since the training and test sets were not representative of each other and Cao *et al.* had optimized their parameters using the test set, we examined the performance of our method on the test set at various parameter settings (Table 1). The number of neighbors K used by each k-NN in our n-gram module was tuned on the training set constraining $K \leq 5$.

Compared to the state-of-the-art[3], our n-gram based module outperformed it in F1 and matched either its precision or recall but not both at the same time. Our alignment based module surpassed it in precision, recall and F1 all at once. Moreover, our hybrid method outperformed both modules in precision, recall and F1, demonstrating an advantage in combining n-gram and alignment information in the coupling specificity prediction task. Optimizing our parameters on the training set to avoid overfitting the prediction model, our hybrid method attained 82.4% in accuracy compared to the current state-of-the-art's reported accuracy of 72% by optimizing on the test set.

Table 1. Comparison of our n-gram based module, alignment based module and hybrid prediction method against current state-of-the-art[3]. Prob. Thres.: probability threshold in decision criterion.

Method	Prob. Thres.	Precision	Recall	F1
N-gram Module	0.26	0.514	0.889	0.651
	0.34	0.658	0.794	0.719
Alignment Module	0.50	0.630	0.921	0.748
Hybrid Method	0.66	0.698	0.952	0.805
Cao *et al.* [3]		0.577	0.889	0.700

4.2 Evaluation on Current Dataset

Having shown our hybrid method outperformed the current state-of-the-art in G-protein coupling specificity prediction, we evaluated our method on the more recent PTbase dataset[18]. Our evaluation protocol is a ten-fold cross validation, where in each trial, 8 folds were used as the training set, 1 fold as the validation set to optimize parameters for maximum F1 and 1 fold as the test set. In order to determine the number of features , we performed ten-fold cross validations with the n-gram based module at varying number of features. Figure 2 shows the average validation set F1 against the number of features. The n-gram based module attained optimal validation set F1 at 1375 features. The hybrid method using the top 1375 chi-selected features scored 0.749 precision, 0.763 recall, 0.753 F1 and 0.796 accuracy on the test set.

To test robustness, we evaluated the hybrid method on two subsets of the PTbase dataset. It attained 0.793 accuracy and 0.752 F1 on sequences less than 70% identical with any training sequence and 0.792 accuracy and 0.748 F1 on sequences less than 80%.

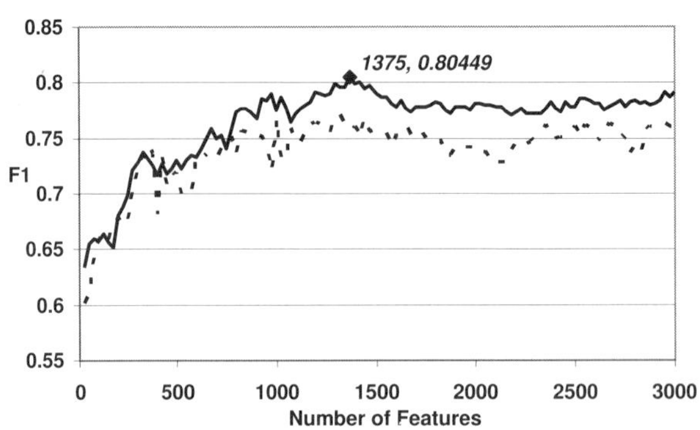

Figure 2. N-gram based module's validation set (solid) and test set (dotted) F1 versus the number of features.

5 Biological Analysis

One advantage of our prediction method is its simplicity and modularity which allows us to make biological interpretations of the data and our predictions. Of particular interest is the location of coupling specificity information in the GPCR sequence. Since the GPCR makes physical contact with the G-protein via its intracellular domains, it is likely the intracellular domains would contain the majority of the coupling specificity information. Our goal is to determine which intracellular domains and which areas within them contain the most information.

5.1 Domain Combination Analysis

To determine which intracellular domains together provide the most information in predicting coupling specificity, we compared the performance of our alignment based module in a ten-fold cross validation on each of the 15 possible combinations of the 4 intracellular domains. In each trial, we reserved 1 fold as test set and 1 fold as validation set to optimize parameters for maximum F1 while training on the other 8 folds. From Table 2, the 2nd and 3rd intracellular domains together yielded the best F1. Using a single intracellular domain, the 2nd domain generated the highest F1, followed by the 1st, 3rd and 4th domains in order. However, the 1st and 2nd domains together yielded a lower F1 than the 2nd domain with either the 3rd or 4th. This suggested that the coupling specificity information in the 1st intracellular domain overlaps largely with the information in the 2nd.

Table 2. Performance of alignment based module on different intracellular domain combinations in ten-fold cross validation on PTbase dataset. IC: intracellular domains

IC	Precision	Recall	F1	Accuracy	IC	Precision	Recall	F1	Accuracy
1	0.782	0.703	0.739	0.796	2, 3	0.837	0.825	0.828	0.861
2	0.820	0.799	0.808	0.845	2, 4	0.828	0.816	0.821	0.853
3	0.661	0.721	0.682	0.730	3, 4	0.773	0.807	0.788	0.821
4	0.632	0.755	0.670	0.694	1, 2, 3	0.822	0.814	0.816	0.850
1, 2	0.820	0.805	0.811	0.847	1, 2, 4	0.807	0.809	0.807	0.843
1, 3	0.799	0.765	0.780	0.825	1, 3, 4	0.792	0.807	0.797	0.832
1, 4	0.780	0.755	0.765	0.807	2, 3, 4	0.839	0.820	0.828	0.861
					1, 2, 3, 4	0.824	0.813	0.817	0.853

5.2 Motif Location Analysis

We examined the portions of intracellular domains that contain the majority of coupling specificity information by finding the maximally discriminative (maximally predictive) n-grams. For each G-protein family, we applied chi-square feature selection to identify discriminative binary features derived from n-gram counts which can differentiate between receptors that couple to the G-protein family and those that do not.

A feature is considered highly discriminative if its presence in receptors that couple to the G-protein family is much more prevalent than its presence in receptors that do not, or vice versa. For the majority of the top 100 selected features, we observed that the presence of the selected features for $G_{i/o}$ is indicative of not coupling to $G_{i/o}$ while the presence of the selected features for $G_{q/11}$ and G_s is indicative of coupling to $G_{q/11}$ and G_s

respectively (data not shown). Thus, we examined the location of the selected features for $G_{i/o}$ in receptors that do not couple to $G_{i/o}$ and the location of the selected features for $G_{q/11}$ and G_s in receptors that couple to $G_{q/11}$ and G_s.

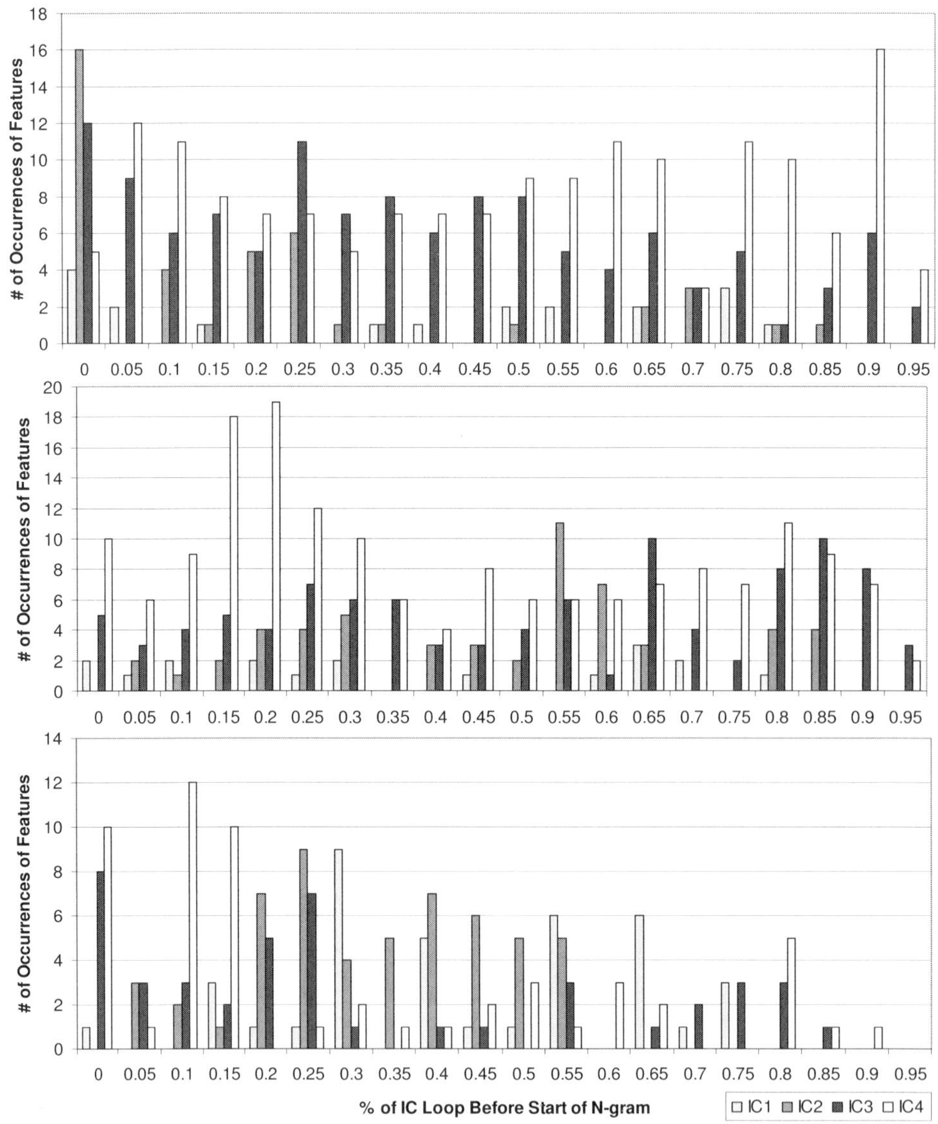

Figure 3. Histograms of the locations of the top 50 features selected by chi-square to distinguish GPCRs that couple to $G_{i/o}$ (top), $G_{q/11}$ (middle) and G_s (bottom) proteins, in receptors not coupling to $G_{i/o}$ (top) and receptors coupling to $G_{q/11}$ (middle) and G_s (bottom), respectively.

Figures 3 shows histograms of where in the intracellular domains the top 50 n-grams start for $G_{i/o}$, $G_{q/11}$ and G_s respectively. For instance, if the intracellular domain is 25 amino acids long and the n-gram on which the selected feature is based starts at the 6th

amino acid, then 5% of the domain lies before the n-gram and this contributes a count towards the '0.05' column. The selected n-grams for all 3 G-protein families were concentrated in the 2nd, 3rd and 4th intracellular domains. In the 2nd intracellular domain, the selected n-grams for $G_{q/11}$ were spread across the entire length of the domain, while those for $G_{i/o}$ and G_s were concentrated in the first 30% and 60% of the domain respectively. The selected n-grams in the 3rd intracellular domain were spread across the entire length of the domain for all 3 families. The concentration of discriminative n-grams in the beginning portion of the 2nd intracellular loop and over the entire length of the 3rd intracellular loop correlates with the physical positioning of the G-protein with the GPCR in 3-dimensional space when they couple.

6 Conclusions

G-protein coupled receptors are involved in numerous diseases due to their role with G-proteins in regulating the cell's response to external stimuli. Studies on the interaction between GPCR and G-proteins can lead to insights and potential drug targets for these diseases. In this paper, we developed a new method to predict the families of G-proteins a GPCR can interact with given the receptor sequence, that outperformed the current state-of-the-art[3]. Analyzing the features used by this method, we found the coupling specificity information to be concentrated in the beginning of the 2nd intracellular loop in the GPCR and over the entire length of the 3rd loop.

Our method differs from previous G-protein coupling specificity prediction methods in two major ways. First, previous studies focused only on the intracellular domains of the receptor as those are the regions having direct contact with the interacting G-protein. We found evidence of coupling specificity information in the non-intracellular domains of the receptor sequence in our preliminary studies and developed our method to utilize information from the whole receptor sequence instead of the intracellular regions alone.

Second, features derived from n-grams and sequence alignments are commonly used in many prediction problems in bioinformatics. Previous coupling specificity studies have all used either n-grams or alignment information but not both. Yet, the two types of features have complimentary strength: alignment can capture long-range interaction information but is unreliable below 40% sequence similarity due to its assumption that the linear ordering of conserved segments is preserved, while n-grams makes no such assumption but cannot capture long-range interactions. By combining the information in both types of features, our prediction method outperformed the current state-of-the-art[3] with only 81.3% of the training data and attained 0.753 F1 and 0.796 accuracy on the PTbase 2004 dataset[18]. Moreover, our method suffered less than 0.005 drop in accuracy and F1 when sequences sharing more than 75% sequence identity were removed. This demonstrates the potential in combining multiple representations of sequence or other information in prediction problems.

Acknowledgments

The authors would like to thank Dr. Paul N. Bennett for the use of his efficient k nearest neighbors program for sparse feature vectors, and Dr. Judith Klein-Seetharaman for sharing her GPCR expertise. This material is based upon work supported by the National Science Foundation under grant no. 0225656.

References

1. G. Muller, Towards 3D structures of G protein-coupled receptors: a multidisciplinary approach. Curr Med Chem, 2000. **7**(9): p. 861-88.
2. J. Wess, Molecular basis of receptor/G-protein-coupling selectivity. Pharmacol Ther, 1998. **80**(3): p. 231-64.
3. J. Cao, et al., A naive Bayes model to predict coupling between seven transmembrane domain receptors and G-proteins. Bioinformatics, 2003. **19**(2): p. 234-40.
4. B. Qian, et al., Depicting a protein's two faces: GPCR classification by phylogenetic tree-based HMMs. FEBS Lett, 2003. **554**(1-2): p. 95-9.
5. S. Moller, J. Vilo, and M.D. Croning, Prediction of the coupling specificity of G protein coupled receptors to their G proteins. Bioinformatics, 2001. **17 Suppl 1**: p. S174-81.
6. A. Henriksson, Prediction of G-protein Coupling of GPCRs - A Chemometric Approach, in Engineering Biology. 2003, Linkoping University: Linkoping. p. 79.
7. S. Vinga and J. Almeida, *Alignment-free sequence comparison-a review.* Bioinformatics, 2003. **19**(4): p. 513-23.
8. M. Lynch, *Intron evolution as a population-genetic process.* Proc Natl Acad Sci U S A, 2002. **99**(9): p. 6118-23.
9. Y.X. Zhang, et al., Genome shuffling leads to rapid phenotypic improvement in bacteria. Nature, 2002. **415**(6872): p. 644-6.
10. C.H. Wu, et al., *Protein family classification and functional annotation.* Comput Biol Chem, 2003. **27**(1): p. 37-47.
11. W.R. Pearson, *Effective protein sequence comparison.* Methods Enzymol, 1996. **266**: p. 227-58.
12. W.R. Pearson, Empirical statistical estimates for sequence similarity searches. J Mol Biol, 1998. **276**(1): p. 71-84.
13. B.Y. Cheng, J.G. Carbonell, and J. Klein-Seetharaman, *Protein Classification based on Text Document Classification Techniques.* Proteins: Structure, Function and Bioinformatics, 2005. **58**(4): p. 955-70.
14. Y. Yang and J. Pedersen. A comparative study on feature selection in text categorization. in 14th International Conference on Machine Learning. 1997. Nashville, US: Morgan Kaufmann Publishers, San Francisco, US.
15. S.F. Altschul, et al., *Basic local alignment search tool.* J Mol Biol, 1990. **215**(3): p. 403-10.
16. S. Alexander, et al., *TiPS Receptor and Ion Channel Nomenclature Supplement.* Trends in Pharmacological Sciences, 2001.
17. I.H. Witten and E. Frank, Data Mining: Practical Machine Learning Tools with Java Implementations. 2000, San Francisco: Morgan Kaufmann.
18. PTbase, *PTbase.* 2004, BioMedNet.

AUTHOR INDEX